A Practical Guide to
Understanding, Managing,
and Reviewing

ENVIRONMENTAL RISK ASSESSMENT REPORTS

A Practical Guide to Understanding, Managing, and Reviewing

ENVIRONMENTAL RISK ASSESSMENT REPORTS

EDITED BY
SALLY L. BENJAMIN
DAVID A. BELLUCK

CRC Press
Taylor & Francis Group
Boca Raton London New York

CRC Press is an imprint of the
Taylor & Francis Group, an **informa** business

First published 2001 by Lewis Publishers

Published 2019 by CRC Press
Taylor & Francis Group
6000 Broken Sound Parkway NW, Suite 300
Boca Raton, FL 33487-2742

First issued in paperback 2020

© 2001 by Taylor & Francis Group, LLC
CRC Press is an imprint of Taylor & Francis Group, an Informa business

No claim to original U.S. Government works

ISBN 13: 978-0-367-57880-0 (pbk)
ISBN 13: 978-1-56670-448-9 (hbk)

This book contains information obtained from authentic and highly regarded sources. Reasonable efforts have been made to publish reliable data and information, but the author and publisher cannot assume responsibility for the validity of all materials or the consequences of their use. The authors and publishers have attempted to trace the copyright holders of all material reproduced in this publication and apologize to copyright holders if permission to publish in this form has not been obtained. If any copyright material has not been acknowledged please write and let us know so we may rectify in any future reprint.

Except as permitted under U.S. Copyright Law, no part of this book may be reprinted, reproduced, transmitted, or utilized in any form by any electronic, mechanical, or other means, now known or hereafter invented, including photocopying, microfilming, and recording, or in any information storage or retrieval system, without written permission from the publishers.

For permission to photocopy or use material electronically from this work, please access www.copyright.com (http://www.copyright.com/) or contact the Copyright Clearance Center, Inc. (CCC), 222 Rosewood Drive, Danvers, MA 01923, 978-750-8400. CCC is a not-for-profit organization that provides licenses and registration for a variety of users. For organizations that have been granted a photocopy license by the CCC, a separate system of payment has been arranged.

Trademark Notice:
Product or corporate names may be trademarks or registered trademarks, and are used only for identification and explanation without intent to infringe.

Visit the Taylor & Francis Web site at
http://www.taylorandfrancis.com

and the CRC Press Web site at
http://www.crcpress.com

Library of Congress Card Number 00-048667

Library of Congress Cataloging-in-Publication Data

A practical guide to understanding, managing, and reviewing evironmental risk assessment reports / Sally L. Benjamin, David A. Belluck, editors.
   p. cm.
   Includes bibliographical references and index.
   ISBN 1-56670-448-0 (alk. paper)
   1. Environmental risk assissment. I. Benjamin, Sally. II. Belluck, David.

GE145 .P73 2000
333.7'14—dc21   00-048667

Dedication

To our fathers, Louis Belluck and Norton James Benjamin, for their love of books.

Acknowledgments

We wish to thank the 31 other professionals, who shared our vision of a comprehensive, general guide to environmental risk assessment, for their dedication to the idea and for their patience as our manuscript went through several iterations. We owe special thanks to Ruth Hull, who gave freely of her ideas, professional contacts, and support. Without her, this book would not have been possible. We thank our peer-reviewers and readers, most especially Dr. Hiai Rothmann. We also acknowledge the contribution of Steven David, who has successfully implemented many of the ideas in this book.

Disclaimer

Extreme care has been taken in preparation of this work. However, neither the publisher, editors, nor authors shall be held responsible or liable for any damage resulting in connection with or arising from the use of any of the information in this book.

Contributors

George Anderson, B.A., M.A., C.H.M.M., is the Director of Environmental Compliance & Safety at US Filter Recovery Services Inc., 2430 Rose Place, Roseville, Minnesota 54701, one of Minnesota's largest hazardous waste recycling, treatment, and storage facilities. Mr. Anderson has a Bachelors degree in Biology and Chemistry and a Master's degree in Biology from St. Cloud State University. Mr. Anderson has 29 years of professional experience in industry, consulting, government, and utilities. He currently serves as the President of the Minnesota Chemical Technology Alliance, the Chemical Manufacturers Association, State of Minnesota Affiliate. Mr. Anderson has testified on hazardous waste management issues before the Wisconsin State Legislature, the Minnesota Waste Management Board, and has represented the waste management industry before the U.S. EPA and the U.S. Congress.

Carol Baker, M.S., M.A. is Senior Consultant at ENTRIX, Inc., 5252 Westchester, Suite 250, Houston, Texas 77005. She is Manager of Environmental Sciences and, for the past 10 years, she has worked in human health risk assessments under CERCLA, RCRA, UST and Voluntary Cleanup Programs. Ms. Baker earned her Bachelor of Arts degree in Wildlife Management from North Carolina State University, a Master of Science in Fisheries Science/Aquaculture from Louisiana State University, and a Masters of Arts in Science Education from North Carolina State University.

David A. Belluck, B.S., Ph.D., is a toxicologist and risk assessor with more than 30 years experience in public health and environmental science. Dr. Belluck is principal toxicologist for Risk Writers, Ltd., 3108 46th Avenue South, Minneapolis, Minnesota 55406. Dr. Belluck provides expert services for litigation in the areas of toxicology, environmental risk, and the history of science. Dr. Belluck publishes extensively on risk assessment, toxicology, and groundwater protection. His current research interests include historical toxicology of manufactured gas plants and improved project management techniques for risk assessment and public decision-making. He is a member of the National Advisory Committee to the U.S. EPA on Ambient Exposure Guideline Levels (AEGLs). Dr. Belluck earned his Bachelor of Science degree from Cornell University and his Ph.D. from the University of Illinois.

Sally L. Benjamin, M.S., J.D., is an environmental scientist and attorney with more than 25 years of professional experience in environmental policy, natural resources management, and public dispute resolution. Ms. Benjamin's firm, Risk Writers, Ltd., 3108 46th Avenue South, Minneapolis, Minnesota 55406, conducts historical and technical research for environmental litigation and provides expert services in toxicology, environmental science, and the history of science. Ms. Benjamin earned her Bachelor of Arts in Biology, Mansfield State College, Pennsylvania, her Master of Science from the Institute for Environmental Studies, University of Wisconsin-Madison, and her Juris Doctorate from the University of Minnesota. She is a member of the Hennepin County Bar and of the Minnesota and the Wisconsin State Bar Associations. Her research interests include environmental impacts of local land use planning, interrelationships of transit, traffic and urban air quality, manufactured gas plant history, and the role of environmental science in litigation, public policy, legislation, and regulation.

Bruce Braaten, P.E., J.D., works for the Minnesota Pollution Control Agency in Winona, Minnesota, and he teaches environmental law at the University of Minnesota. Mr. Braaten earned his Juris Doctorate from the William Mitchell School of Law and his Bachelor of Science and Master of Science degrees in engineering from the University of Minnesota.

Jan W. Briede, Ph.D. (New Mexico State University) is a biologist with Dames & Moore, 644 Linn Street, Suite 501, Cincinnati, Ohio 45203. He has more than 20 years of experience in ecology and ecosystem modeling world-wide (Africa, the Middle East, Asia, Europe, and the U.S.). Presently, Dr. Briede supports clients in fields such as: project management, ecological risk assessments, decision support systems, ecological inventories, wetland issues, ecosystem restoration, permitting, and environmental site assessments.

Gary J. Burin, Ph.D., D.A.B.T., Associate Director of the Toxicology Division, Technology Sciences Group, Inc., 1101 17th Street, N.W., Suite 500, Washington, D.C. 20038, earned his Bachelor of Science in Biology and M.P.H. in Toxicology from the University of Michigan and his Ph.D. in Biology, with a Pharmacology minor, from George Washington University. He coordinated much of the international work surrounding OECD, WHO and the EC, particularly the harmonization of data requirements, study interpretation, and risk assessment. Dr. Burin authored the WHO document, "Environmental Health Criteria Document 104 — Principles for the Toxicological Assessment of Pesticide Residues in Food," and the U.S. EPA's "Standard Evaluation Procedures" for the evaluation of chronic reproductive and developmental toxicity studies. He has served on national committees under the National Toxicology Program and the Office of Science and Toxicology and has lectured at the Universities of Sienna, Italy, and Surrey, England, as a Diplomate of the American Board of Toxicology.

Rick D. Cardwell, Ph. D., Parametrix, Inc., 5808 Lake Washington Blvd. N.E., Kirkland, Washington 98052, is an ecotoxicologist with 30 years experience studying the fate and effects of chemicals and wastes in the environment. He has authored dozens of ecological risk assessments, including two primers on ecorisk assessment methodologies.

Robert Craggs, M.S., J.D., is Senior Director of Environmental Services for R.W. Beck, Minneapolis Office, 1380 Corporate Center Curve, Suite 305, St. Paul, Minnesota 55121. Mr. Craggs earned his Juris Doctorate and Master of Science degrees from the University of Iowa.

John P. Cummings, Ph.D., C.H.M.S., R.E.A., R.E.P., J.D., P.O. Box 2847, Fremont, California 94536, is a chemist, environmental engineer, teacher, and attorney with over 30 years of experience in environmental project management, remedial design, and implementation of environmental programs and policies. He has a strong background in hazardous/toxic waste management, UST work, including pollution abatement implementation, solid waste management, resources recovery and recycling, asbestos assessment, lead audits and clean-up, and legal aspects of OSHA and U.S. EPA litigation and product liability. He also has an extensive technical background in chemistry, ceramic, paper, and plastic materials. He is a patentee and author of more than 40 papers.

Maxine Dakins, Ph.D., is an Assistant Professor of Environmental Science at the University of Idaho, Idaho Falls, Idaho, 83402. Dr. Dakins teaches courses in uncertainty analysis, the sampling and analysis of environmental contaminants, and natural resources policy. Her research interests include various aspects of uncertainty analysis including Bayesian Monte Carlo Analysis and uncertainties related to measuring contaminants at trace levels.

Clifford S. Duke, M.A., Ph.D., a Senior Environmental Analyst at The Environmental Company, Inc., 1611 North Kent Street, Suite 900, Arlington, Virginia 22209, earned his Bachelor of Science degree in Biology at the University of Vermont and graduate degrees in public policy analysis (M.A.) and botany (Ph.D.) at Duke University. He has prepared ecological risk assessments and managed National Environmental Policy Act documents for Department of Energy and Department of Defense facilities nationwide. Dr. Duke is a past-president of the Ohio Valley Chapter of the Society of Environmental Toxicology and Chemistry and is an active participant in risk assessment standardization efforts of the American Society for Testing and Materials.

Nava C. Garisto, Ph.D., is Senior Scientist at SENES Consultants Limited, 121 Granton Drive, Unit 12, Richmond Hill, Ontario, Canada L4B 3N4. Dr. Garisto has 20 years of scientific and consulting experience and has published more than 60 journal publications and reports relating to environmental model development, mass transport of radionuclides and toxic contaminants and environmental risk assessment.

Guy L. Gilron, M.Sc., R.P.Bio., is a Senior Project Manager (Ecotoxicology) with ESG International Inc., 361 Southgate Drive, Guelph, Ontario, Canada, N1G 3M5. Mr. Gilron has conducted environmental effects assessment of natural resources in North America, the Caribbean, South America, and the Middle East. Moreover, he has conducted ecotoxicological valuations in large- and small-scale environmental programs. He is an experienced environmental biologist, with a focus on aquatic toxicology and ecology, and with expertise in ecological risk assessments, aquatic community structure analyses, toxicological research, ecological inventories, and water quality assessments for aquaculture.

Michael E. Ginevan, Ph.D., is president of M.E. Ginevan & Associates, 307 Hamilton Avenue, Silver Spring, Maryland 20901. Dr. Ginevan's firm provides interdisciplinary statistical consultation for the health and environmental sciences. 301-585-4951; Fax: 301-585-1350; e-mail: mginevan@cais.com or MGINEVAN@worldnet.att.net.

Laura C. Green, Ph.D., D.A.B.T., is a Senior Scientist and President of Cambridge Environmental, Inc., 58 Charles Street, Cambridge, Massachusetts 02141, and Lecturer in the Division of Toxicology at the Massachusetts Institute of Technology. Dr. Green has performed original research, published, and consulted in the areas of chemical carcinogenesis, toxicology and pharmacology, food chemistry, analytical chemistry, risk assessment, and regulatory policy. Prior to founding Cambridge Environmental, Dr. Green was Senior Vice President at Meta Systems Inc. and the founder and director of Meta's Environmental Health and Toxicology group.

She also served as Research Director of the Scientific Conflict Mapping Project at the Harvard University School of Public Health, during which time she coauthored the text, *In Search of Safety: Chemicals and Cancer Risk*. Dr. Green currently specializes in: performing qualitative and quantitative assessments of health and environmental risks; providing toxicologic and other technical expertise designed to aid in regulatory compliance and in decision-making; providing and directing scientific support for litigation and other matters; and teaching toxicology. Dr. Green holds a B.A. from the Department of Chemistry at Wellesley College (1975) and a Ph.D. from the former Department of Nutrition and Food Science (currently the Division of Bioengineering and Environmental Health) at the Massachusetts Institute of Technology (1981). She is a diplomate of the American Board of Toxicology (D.A.B.T.).

Carol "Griff" Griffin, M.S., Ph.D., is an Assistant Professor in Natural Resources Management at Grand Valley State University, 218 Padnos Hall, Allendale, Michigan 49401-9403. Professor Griffin teaches courses in natural resource policy, water resources, environmental policy, environmental science, and environmental ethics. Her research interests include public participation in natural resource management, nonpoint source pollution modeling, and the role of error in GIS modeling. Dr. Griffin earned her M.S. and Ph.D. in Environmental Science from the State University of New York — College of Environmental Science and Foresty.

Donald R. Hart, M.S., Ph.D., Senior Ecologist at Beak International Incorporated, 14 Abacus Road, Brampton, Ontario, Canada, L6T 5B7, has 15 years of post doctoral research and consulting experience and over 30 journal publications and reports in aquatic ecology, ecotoxicology, and ecological risk estimation. Dr. Hart earned his Ph.D. in Environmental Biology from Tulane University and both his Master of Science and Bachelor of Science degrees in Zoology from the University of Manitoba.

Ruth N. Hull, M.Sc., is a Risk Assessment Specialist at CANTOX ENVIRONMENTAL INC., 2233 Argentia Road Suite 308, Mississauga, Ontario, Canada, L5N 2X7. Ms. Hull earned a Masters of Science in Ecotoxicology from Concordia University and a Bachelor of Science in Biology from the University of Waterloo, Waterloo, Ontario. Ms. Hull oversaw contractor-produced risk assessment reports while working for the State of Minnesota's Pollution Control Agency and Oak Ridge National Laboratory (ORNL). She has been the lead ecological risk assessor on several risk assessments for contaminated sites across North America.

Colleen J. Dragula Johnson, M.S., D.A.B.T., 5815 Redford Drive, #E, Springfield, Virginia 22152, provides general toxicology services relating to the Food and Drug Administration and the U.S. EPA, including summarizing toxicity data for investigation of new drug applications.

Wendy Reuhl Jacobson, B.S., M.S., of Colorado Springs, Colorado, earned her degrees in Natural Resources from the University of Wisconsin in poultry genetics and in Natural Resources Management from the University of Alaska, Fairbanks, where she investigated the readability and graphic content of federal environmental impact statements. Ms. Jacobson's research article, "The Typography of Environmental Impact Statements: Criteria, Evaluation, and Public Participation" was published in *Environmental Management* in January 1993.

Robert A. Kreiger, M.S., Ph.D., 9414 North 84th Street, Stillwater, Minnesota 55082, earned his Master of Science Degree in Epidemiology from the University of Michigan and a Doctor of Philosophy in Environmental Health from the University of Minnesota. Dr. Kreiger has over a decade of experience in risk assessment and has conducted epidemiological studies of worker cohorts and residential populations, and has researched the use of chromosomal damage biomarkers to characterize occupational and public exposures and the use of immunoassay methods for screening worker exposure to pesticides.

Jeanette H. Leete, Ph.D., is the Supervisor of the Technical Analysis Group in the Ground Water Unit of the Minnesota Department of Natural Resources in St. Paul Minnesota 55155. Dr. Leete is a Licensed Professional Geologist in the State of Minnesota, a certified Professional Hydrogeologist (American Institute of Hydrology) and a Certified Professional Geologist (American Institute of Professional Geologists). Dr. Leete received her doctoral degree from the University of Minnesota.

Kathy Malec, M.S., is an environmental librarian with more than a decade of experience in technical library research. Ms. Malec earned her Master of Science degree in Library Science from the University of Minnesota. She is a Librarian/Information Specialist with the Minnesota Pollution Control Agency and has worked as a librarian for private consulting firms and in academia, prior to entering state service.

Wayne Mattsfield, B.S., has over a decade of professional experience in state government, as a Minnesota Department of Health environmental laboratory analyst, laboratory certification officer and quality assurance officer, and Minnesota Pollution Control Agency quality assurance coordinator for federal and state Superfund sites. Mr. Mattsfield earned his Bachelor of Science degree in Biology, with emphasis in Microbiology, from St. Cloud State University. Mr. Mattsfield is currently in private consulting and can be reached at 16123 Harvard Lane, Lakeville, Minnesota 55044.

William Phillips, B.A., M.S., is Senior Project Director & General Manager, Environmental Strategies Corporation, 123 North 3rd Street, Suite 706, Minneapolis, Minnesota 55401. Mr. Phillips is widely experienced in environmental and regulatory toxicology, risk assessment, environmental claims, investigation and remediation of hazardous waste sites, environmental fate and transport, and evaluation of environmental impairment. He earned his Master of Science in Environmental Toxicology from the University of Minnesota and his Bachelor of Arts degree in Biology and History from Macalester College. He publishes in environmental remediation and provides expert services for environmental litigation.

Mark W. Rattan, J.D., is an attorney with Litchfield Cavo, 303 West Madison Street, Suite 200, Chicago, Illinois 60606. Mr. Rattan practices insurance coverage and insurance defense litigation. He earned his Juris Doctorate from Loyola University. He is licensed to practice in Wisconsin and Illinois. He is a member of the State Bar of Wisconsin, the Chicago Bar Association, and the Milwaukee Bar Association.

Bruce T. Rodgers, M.Sc., P.Eng., Senior Environmental Engineer, Beak Consultants Limited, 14 Abacus Road, Brampton, Ontario, Canada, L6T 5B7. Mr. Rodgers specializes in the analysis of natural receiving water systems. He has been involved extensively in environmental impact assessments, numerical model studies, and field monitoring planning for marine and freshwater environments. In particular,

Mr. Rodgers has been involved in the application and interpretation of multi-dimensional numerical models for predicting physical, hydraulic, thermal, and water quality characteristics of receiving waters. He earned his Bachelor of Science in Geotechnical Engineering from the University of Toronto and his Master of Science in Coastal Engineering from Queen's University.

Richard A. Rothstein, C.C.M., Q.E.P., is Senior Air Quality Consultant for RAR Associates, 46 Liberty Street, North Andover, Massachusetts 01845. Mr. Rothstein earned his Bachelor of Science in Meteorology from Rutgers University, and his Master of Science in Meteorology and Air Resources Engineering from New York University. He has more than 25 years of diversified project management and technical expertise in air quality impact assessment; environmental pollution control and permitting; facility siting and design; and regulatory review, compliance planning and agency negotiation for waste processing, power generation, and industrial projects.

Bradley E. Sample, Ph.D., is a Senior Wildlife Toxicologist and Ecological Risk Assessor with CH2M HILL (2485 Natomas Park Dr., Suite 600, Sacramento, California 95833). Dr. Sample holds a Bachelor of Science in Wildlife Biology from West Virginia University, a Master of Science degree in Entomology from the University of Delaware, and a Doctorate of Philosophy in Wildlife Toxicology from West Virginia University. Dr. Sample has more than a decade of experience in wildlife toxicology and ecological risk assessment emphasizing development of models and methods for estimation of bioaccumulation, exposure, and effects in birds and mammals, as well as other ecological receptors.

David Weitz is a Public Affairs Manager with the Wisconsin Department of Natural Resources, N902 910th St., Mondovi, Wisconsin 54755. Mr. Weitz worked for thirteen years as a newspaper reporter. During more than two decades as a public information officer and public affairs manager he has managed communications on numerous emergency response efforts, including large train derailments, massive tire fires, and chemical releases. He has served as information officer for the Wisconsin Law Enforcement Task Force on tribal fishing and for the U.S. Forest Service on major forest fires and Texas Forest Service in emergency fire prevention programs.

Jeanne C. Willson, Ph.D., D.A.B.T., M.B.A, is a toxicologist with her own firm, Global Environmental Strategies, P.O. Box 3492, Englewood, Colorado 80155, http://www.mindspring.com/~jwillson. Dr. Willson earned her Ph.D. at Cornell University and her M.B.A. (International Business) at the University of Colorado, Denver. She has over 15 years of experience in toxicology, chemical safety, risk assessment, and risk management. Her work at mining, smelting, and metals disposal sites has included metals risk assessment, blood lead and urinary arsenic studies, metals bioavailability, IEUBK (lead) model evaluation and application, plant uptake of metals, quantitative uncertainty analysis, and evaluation of ecological benchmarks and measurements. She is also experienced in radiological risk assessment (especially radon), air emissions and deposition modeling, groundwater quality, and carcinogen risk assessment. Dr. Willson has managed or reviewed hundreds of human health and ecological risk assessments of hazardous waste and occupational chemical exposure at Superfund and RCRA sites and operating facilities of all types, including landfills and incinerators. Her interest in international environmental policy has taken her to Russia many times, including a visit for a NATO conference on air pollution. jwillson@mindspring.com

Stephen G. Zemba, Senior Engineer, Cambridge Environmental, is an Adjunct Professor at Tufts University and University of Massachusetts–Lowell, and Lecturer at Harvard University School of Public Health. Dr. Zemba has performed original research, published, and consulted in the areas of air pollution phenomenology, fate and transport modeling, and risk assessment. He has investigated such topics as acid rain, dense-gas plume dispersion, indoor air dispersion modeling, ocean disposal of carbon dioxide, evaluation of methods to estimate exposure point concentrations, and vapor transport of contaminants in soils. Dr. Zemba currently specializes in performing qualitative and quantitative assessments of health and environmental risks, with emphasis on modeling of pollutant fate and transport. His recent work includes the design and implementation of multi-pathway exposure assessments for air pollution sources and the assessment of contaminated waste disposal sites. Dr. Zemba teaches courses on air quality management and practical applications of air dispersion modeling. Dr. Zemba holds a B.S. from Carnegie-Mellon University, and an M.S. (1985) and Ph.D. (1989) from the Massachusetts Institute of Technology, all in the field of Mechanical Engineering.

Contents

Part I
The Risk Assessment Process

Chapter 1
Introduction 3
David A. Belluck and Sally L. Benjamin

Chapter 2
Human Health Risk Assessment 29
David A. Belluck and Sally L. Benjamin

Chapter 3
Ecological Risk Assessment 79
Ruth N. Hull and Bradley E. Sample

Chapter 4
Risk Assessment Project Planning (Phase I) 99
David A. Belluck and Sally L. Benjamin

Chapter 5
Managing Risk Assessment Report Development (Phase II) 173
David A. Belluck and Sally L. Benjamin

Chapter 6
Concluding a Risk Assessment Contract (Phase III and IV) 221
David A. Belluck and Sally L. Benjamin

Part II
Primers

Chapter 7
Legal Context of Environmental Risk Assessment 233
Bruce Braaten

Chapter 8
Risk Assessment Contract Formation 245
Robert Craggs and Sally L. Benjamin

Chapter 9
Ecological Risk Assessment Review 257
Clifford S. Duke and Jan W. Briede

Chapter 10
Environmental Chemistry 265
John P. Cummings and Sally L. Benjamin

Chapter 11
Analytical Quality Assurance/Quality Control for Environmental Samples
Used in Risk Assessment 277
Wayne Mattsfield and David A. Belluck

Chapter 12
Environmental Sampling Design 301
Rick D. Cardwell

Chapter 13
Sampling for Ecological Risk Assessments 319
Jan W. Briede and Clifford S. Duke

Chapter 14
Ecotoxicity Testing in Risk Assessment 325
Guy L. Gilron and Ruth N. Hull

Chapter 15
Epidemiology and Health Risk Assessment 339
Robert A. Kreiger

Chapter 16
Surface Water Modeling 351
Bruce T. Rodgers

Chapter 17
Groundwater Modeling in Health Risk Assessment 357
Jeanette H. Leete

Chapter 18
Air Toxics Dispersion and Deposition Modeling 369
Richard A. Rothstein

Chapter 19
Using Statistics in Health and Environmental Risk Assessments 389
Michael E. Ginevan

Chapter 20
Uncertainty Analysis 413
Maxine Dakins and Carol Griffin

Chapter 21
Risk Communication 425
David Weitz and Sally L. Benjamin

Chapter 22
Clear Communication in Risk Assessment Writing 439
Wendy Reuhl Jacobson

Chapter 23
Scientific Library Risk Research for Risk Assessment 447
Kathy Malec and David A. Belluck

Chapter 24
Risk Assessment of Airborne Chemicals 465
Jeanne C. Willson

Chapter 25
Radiation Risk Assessment 479
Nava C. Garisto and Donald R. Hart

Chapter 26
Remediation Risk Assessment 497
William Phillips

Chapter 27
Facility Risk Assessment 505
George Anderson

Chapter 28
CERCLA and RCRA Risk Assessments 515
Carol Baker

Chapter 29
International Health Risk Assessment Approaches for Pesticides 527
Colleen J. Dragula Johnson and Gary J. Burin

Chapter 30
Historical Toxicology and Risk Assessment 537
David A. Belluck, Mark W. Rattan, and Sally L. Benjamin

Chapter 31
Special Topics in Risk Assessment: Models and Uncertainties 551
Stephen G. Zemba and Laura C. Green

Appendix: Risk Assessment Resources Guide 563

Index 637

PART I

The Risk Assessment Process

CHAPTER 1

Introduction

David A. Belluck and Sally L. Benjamin

CONTENTS

I. Introductory Remarks 4
II. You Need This Book 4
III. Introduction to Environmental Risk Assessment 6
   A. Common Terms 6
   B. Risk Assessment Controversy 7
IV. Who is Technically Qualified to Produce a Risk Assessment? 12
   A. Different Risk Assessments Need Different Experts 12
   B. Technical Credentials Needed to Perform Expert Tasks 12
V. Risk Assessment as a Multidisciplinary Endeavor 13
   A. Mandated Science 13
   B. Team Work in Risk Assessment 13
   C. Roles in Risk Assessment Teams 15
   D. Teams Establish Performance Standards 17
VI. An Overview of the Risk Assessment Process 20
   A. Phase One — Planning a Risk Assessment 20
   B. Phase Two — Managing a Risk Assessment
      (Including Iterative Review) 23
   C. Phase Three — Accepting a Risk Assessment
      (Including Iterative Review) 24
   D. Phase Four — After a Risk Assessment 25
   E. Risk Assessment Planning Form 25
VII. Conclusion 25
   References 26

I. INTRODUCTORY REMARKS

This is a very different risk assessment book. Many risk assessment books target risk assessment practitioners exclusively, providing them with greater technical insights and complex methodologies to aid in professional practice. Other risk assessment books provide brief overviews of the risk assessment process and technical inputs for a lay audience.

In contrast, this book is intended to introduce environmental risk assessment and to also provide sufficient technical, procedural, and methodological knowledge to empower every reader with tools and information to participate in a risk assessment team, communicate effectively with colleagues, manage a risk assessment report, direct work of expert consultants, and critically review a completed risk assessment report. How is this done?

This book is essentially divided into two functional parts. Part One begins by introducing risk assessment as a process. Next, it discusses team building to plan a risk assessment report and hire a consultant to perform risk assessment work. Then, it discusses managing a consultant to prepare a risk assessment report. Finally, Part One concludes by discussing how to formally complete a risk assessment project. Part Two, presents a series of primers, succinct treatments of key risk assessment topics, to assist readers in conversing knowledgeably with risk assessment team members. Reviewing the risk assessment, in its parts and as a whole, is discussed throughout this book.

II. YOU NEED THIS BOOK

You need this book if you are not an expert in every facet of risk assessment generation and review. While you may be expert in certain fields, you are likely to still need to understand, communicate, and work with other disciplines to complete a successful risk assessment. One of the great weaknesses of risk assessment is the lack of interdisciplinary linkage among its components.

It is common when preparing risk assessment reports for one expert to hand off a work product to another expert in a different field. Since each part of a risk assessment hinges on earlier parts, this is logical. Unfortunately, one great weakness of risk assessment originates when work products of one discipline are used by another, without the technical result of the exchange being checked. For example, an emissions expert produces a table listing those chemicals the emissions expert believes to be important, based solely on emission rates. However, a toxicologist might add or delete chemicals from the list, based solely on toxicity. The end-product of each discipline's independent view of important chemicals for the risk assessment is insufficient. A better approach, is for these experts to collaborate and arrive at a joint, shared vision of the important chemicals list.

It is, therefore, critical for all experts involved in a risk assessment to understand each other's decision logic, so where work intersects, they can collaborate successfully. When collaboration does not occur at the borders of disciplines involved in a risk assessment, erroneous results can propagate throughout a report, producing false risk findings. This book is intended for persons who want to better collaborate on a risk assessment process to reduce preventable errors.

It is also intended for persons who want an introduction to risk assessment. Risk assessment literature is extensive. Excellent technical papers, guidance documents, and treatises exist for each scientific discipline involved in environmental risk assessment. Nevertheless, a gap exists. No single book presents a comprehensive treatment of practical issues routinely encountered by people who develop, review, or use environmental risk assessment reports.

Why was this book written? It is intended as a plain English discussion of what it takes to prepare a risk assessment report on time, within budget, and with sufficient technical credibility to be defensible. It provides step-by-step instructions on how to push through technical "smoke-and-mirrors" to determine whether risk assessors make a technically defensible case for their risk findings.

We intend this book to fill a gap in environmental risk assessment literature by presenting a comprehensive discussion of this important process and offering strategies for developing credible risk assessment reports on-time and within budget. Toward this end, we attempt to explain the risk assessment process in simple terms, introduce basic tools of project management, and offer concepts and techniques for managing many problems routinely encountered on risk assessment projects. This book is no substitute for technical risk assessment publications. It provides guidance on how to integrate documents on technical guidance, management and review, in order to develop a high quality risk assessment report.

This book is written by risk assessment practitioners for anyone who wants to understand, manage, or review a human health or ecological risk assessment report. While certain information in this book might be found in other documents, no book brings it all together as a single publication aimed at making every reader conversant in risk assessment.

As noted earlier, literature on the risk assessment process, and its component technical disciplines, is voluminous. Scattered across government publications (including websites, formal and informal guidance documents, library catalogues, and microfiche collections), academic writing (journals, books, theses, and conference publications), practical handbooks and field references, and trade publications, all this information cannot possibly be collated into a single source. However, we have compiled one of the most extensive collections of reference materials to be found in one book. Specifically, practitioners and general readers alike should refer to the Appendix (additional resources include Chapter 23, Scientific Library Risk Research for Risk Assessment, and the end of each chapter for a collection covering both recent materials and seminal works in risk assessment-related disciplines). Use of these book sections should save a reader enormous amounts of time, may lead to resources rarely listed by other finding tools, and will provide some indication of the vast reach of the risk assessment field, with all its multifaceted parts.

A novice risk assessor and risk assessment reviewer may encounter certain technical areas that they are uncertain how to even start researching. This book eases the learning curve by providing the process, discipline, and data categories necessary to consider when performing, understanding, managing, or reviewing a risk assessment report and indicating where essential information can be found.

As you will see repeated again and again throughout our book, it is our intention to help our readers understand how to start from zero and build and manage development of an acceptable risk assessment report or review a completed report. We do not hope to supplant or compete with the numerous technical risk assessment volumes currently in print. First, we will introduce the concepts of environmental risk assessment.

III. INTRODUCTION TO ENVIRONMENTAL RISK ASSESSMENT

A. Common Terms

The term "risk assessment" refers to both the risk assessment process and documents that result from that process. Procedurally, risk assessment is "an organized process used to describe and estimate the likelihood of adverse health outcomes from environmental exposures to chemicals. The four steps of risk assessment are hazard identification, dose-response assessment, exposure assessment, and risk characterization."* In risk assessment, risk assessors use data of known quality in a standardized analytical framework to estimate type and degree of risks posed by environmental contaminants. These estimates are referred to as "risk estimates" or "risk findings." The result of the risk assessment process is a document, also termed a risk assessment, which presents risk findings and describes how they were generated (see Chapters 2 and 3).

"Risk assessors," usually experts in toxicology or a related scientific discipline, are responsible for technical aspects of producing risk assessments. Risk assessors work closely with a project manager to ensure that data, assumptions, methods, and analytical framework used to generate environmental risk estimates meet current technical and regulatory standards. "Project managers" are responsible for managing a risk assessment project. They may have a science background, but need not be technical specialists. Instead, good project managers understand leadership, politics, and negotiation. They can work with a diverse set of technical and scientific experts, as well as with parties with opposing interests.

The primary purpose of environmental risk assessment is to provide risk managers with all available information in a form that facilitates scientifically informed decisions. "Risk managers" are those persons responsible for making a decision regarding environmental risk. "Risk management is the process of identifying, evaluating, selecting, and implementing actions to reduce risk to human health and to ecosystems. The goal of risk management is scientifically sound, cost-effective, integrated actions that reduce or prevent risks, while taking into account social, cultural, ethical, political, and legal considerations."** Risk managers use risk estimates, derived through risk assessment, to determine whether a process, activity, or site poses significant risks to human health or the environment. Risk managers may

---

* From the Presidential/Congressional Commission on Risk Assessment and Risk Management, 1997, Framework for Environmental Health Risk Management, Final Report, vol. 1, p. 61.
** From the Presidential/Congressional Commission on Risk Assessment and Risk Management, 1997, Framework for Environmental Health Risk Management, Final Report, vol. 1, p. 61.

decide, for example, that estimated risks are acceptable, and no action is required, or that risks are too high and require remediation, mitigation, regulation, reduction, or prohibition. Risk managers tend to be non-scientists and may view risk estimates as indicators of "real risks," rather than mere estimates of risk. Risk managers should understand that risk estimates are one component in a multi-faceted decision making process.

Ideally, risk managers use "risk communication" as part of environmental risk decision-making. Risk communication is a means of establishing meaningful two-way communication with people concerned about risk estimates and risk management decisions that use these estimates. Two-way communication provides a risk manager with information about important social factors (such as economics, law, ethics, cultural norms, and politics) and better informs the risk management decision. It also provides information about a risk assessment process, risk estimates, risk decisions, and reasons for the decision to people concerned about risk management decisions (see Chapters 21 and 22).

Environmental risk assessment can come into play at every level of environmental decisionmaking. It has been used by lawmakers to develop statutes and by regulators to write rules, to formulate regulatory guidance, and to grant or deny permit applications (see Chapter 7). Private companies, as well as government agencies and other public entities, may use risk assessment to evaluate environmental effects of projects, both to assess potential liability and to demonstrate project safety to regulators.

Risk assessments can become controversial because of concerns for health, financial, legal, or other impacts. These concerns can create high degrees of controversy, the subject of the next section.

B. Risk Assessment Controversy

Environmental risk assessment reports often generate controversy. Controversy stems from three sources:

- Important issues at stake
- Conflicting expectations for risk assessment reports
- Pressure to perform

1. Important Issues at Stake

Risk assessment deals with a contentious subject: how society balances potential dangers posed by environmental contaminants (some with potential to cause cancer, birth defects, neurological damage, or species extinction) against our appetite for raw materials and saleable products, and inexpensive waste disposal. Risk assessment reports play a central role in risk management decisions on whether to require risk reduction activities to reduce human or ecological risks or to allow a site, activity, or facility to remain unchanged. Thus, environmental risk assessment occurs within a highly political realm with potential for serious outcomes affecting human health and environmental quality, on one hand, and affecting financial well-being of a corporation or community and imposing legal liability or regulatory enforcement, on the other.

2. Conflicting Expectations for Risk Assessment Reports

Controversy is heightened by certain characteristics of risk assessment. In addition to being highly technical, and, thus, difficult to discuss, risk assessments often fail to meet commonly-held, but erroneous, expectations. Some citizen activists, for example, hope a risk assessment process will present an opportunity to kill a project. In contrast, project proponents may expect the report to provide irrefutable proof of the safety of a proposed project. The next sections will attempt to disabuse readers of some common misconceptions that result in conflicting expectations for risk assessment reports.

a. Risk Assessment Provides True Risk Levels

Many persons expect the results of a risk assessment to provide true estimates of risk. This is a false expectation. Risk assessment can provide an estimate of risks within the framework and limitations of the risk assessment process, no more. Risk assessment is not a crystal ball. It cannot be used to predict exact risks. It cannot say that you will or will not be the person to have their health effected by a chemical, process, activity, or site. It can give risk estimates with associated limitations and uncertainties.

b. Risk Decisions are Based Solely on Scientific Facts and Risk Certainties

Many persons, including some risk managers, believe that risk management decisions are dictated solely by risk findings. While many regulators choose to make risk management decisions strictly in line with risk findings, because of political considerations, this is not necessarily how risk assessment findings are supposed to be used. Risk findings are intended to be combined with nonrisk considerations, including economics and political factors, to determine whether a risk estimate will lead to some type of risk reduction action or prevent some type of action from occurring (e.g., issuance of a facility permit to emit air pollutants).

c. Risk Assessment Is a Research Activity

Neither pure science nor pure policy, risk assessment does not entirely conform to either world. Environmental risk assessors bring science to bear in the world of environmental regulation, a world governed by both scientific principles and social values, as expressed in laws, rules, policies, and personal ideals. The result is an irksome alloy, guaranteed to leave everyone involved less than fully satisfied with the outcome.

d. Risk Assessment Findings are Unimpeachable, as Pure Science

Although technical in nature, risk assessment is not pure science. This simple fact is often overlooked by risk managers and scientists alike.

On one hand, risk managers prefer an unassailable basis for their decisions and, therefore, they press for "scientifically defensible" risk assessments, reports that are sure to withstand all technical, political, and legal challenges because they have undergone the highest level of peer review and employ testable hypotheses. This is natural because they rely on risk assessment reports to make decisions with highly political and emotional consequences, as well as significant legal and regulatory ramifications.

On the other hand, environmental scientists also forget that risk assessment is not pure research science, especially when defending their professional work. Early in the education of environmental scientists, they learn to value technical rigor and the formal scientific process (hypothesis testing, peer review, and control of variables). When challenged, an honest scientist must agree that risk assessments fail to achieve the rigor of pure science. Many scientists face criticisms of risk assessment rigor by redoubling their efforts to perform a scientifically defensible assessment, but such efforts are doomed.

The problem does not stem from inherent flaws in risk assessment, but from a failure to recognize the difference between environmental risk assessment and research science. Whereas a research scientist articulates a hypothesis and then conducts tests under controlled conditions to learn about the natural world, risk assessment functions within a totally different process with a different purpose. The environmental risk assessment process does not control variables or test (or even articulate) a null hypothesis. Risk assessment acquires specific types of data for use in a standardized analysis in order to generate a risk estimate and discuss the uncertainties surrounding that estimate. Once this distinction is made, risk professionals can view challenges to risk assessment rigor in a new way. Specifically, they will see that, while it is appropriate to improve environmental risk assessment, if possible, it is inappropriate to hamstring the environmental decision-making process in a quixotic quest for scientific rigor equal to that demanded of research science. However, where science is employed, it must be current, applicable, and technically correct.

e. Risk Assessment is Junk Science

Risk assessment is not junk science. It is not intended to meet academic levels of research and analysis because a risk assessment cannot be evaluated using common scientific hypothesis testing techniques. It is simply a regulatory and governmental analysis scheme to evaluate potential risks in a systematic and reviewable manner. Thus, although components within a risk assessment may achieve research levels of rigor, the whole report cannot. Expectations that risk assessments should meet hypothesis testing levels of performance are at best disingenuous and at worst junk logic.

f. Risk Management Decisions can Ignore Risk Assessment Findings

Risk management decisions cannot ignore risk assessment findings in order to achieve a predetermined decision based on hidden agendas or political expediency. Court cases have shown that risk management decisions by administrative agencies not based in risk assessment findings cannot withstand judicial scrutiny.

g. Risk Assessment Guidance and Methods can be Ignored and Still Produce a Credible Risk Assessment

International, national, and local risk assessment guidance, methods, data, techniques, and court decisions cannot be ignored. To do so jeopardizes institutional and risk assessment credibility as well as professional reputations. Risk assessment reports must meet generally accepted standards of risk assessment or fail critical review, with all its consequences.

h. Citizens Cannot Understand, Review, or Contribute to a Risk Assessment Report

Given the chance and the information provided in this book, anybody can participate in a risk assessment in a meaningful capacity. The input-output analysis presented in this report allows the reader to critically evaluate all data put into a report to determine if it is properly generated, used, and interpreted.

i. All Data Used in a Risk Assessment are Equal

All data are not created equal. Some are better than others. Data from a peer reviewed report can be of much better quality and, therefore, more reliable, than data generated by a party directly affected by a risk assessment report, especially since such data sets are unlikely to have been peer reviewed. Thus, reviewers must check that data of the highest available quality have been used in a risk assessment report. Where lesser quality data are used, the reviewer must ensure that their limitations for use in the risk assessment, and all uncertainties associated with their use, are fully articulated.

j. All Models to be Used in a Risk Assessment are Equal

All models are not created equal. Some are useful for some situations and may not be suitable for others. Many models have never been fully evaluated to ensure that their outputs reasonably reflect reality. Any model used in a risk assessment should have a proven technical track record before it is accepted for a specific use. Reviewers must determine that this evaluation process has occurred for every model used in a risk assessment report.

k. Much of the Information and Data Presented in a Risk Assessment is too Complicated to Explain

All information and data should be presented in a risk assessment in such a way that an educated lay person can understand the technical process, determine the source and validity of data inputs, and check the math. If this cannot be done, with a few notable exceptions (e.g., all the calculations done by a computer modeling program — however, the validity of the model, its inputs and outputs can be reviewed), then the risk assessment is not complete. Good science does not excuse bad writing or weak logic. All information, data, inputs, and outputs in a risk assessment should be presented in such a manner that it can be readily reviewed.

3. Pressure to Perform

Risk assessment functions under tight timelines, with limited budgets, and under constant pressure to produce results that are relevant to nonscientists. Pressure to be timely and cost-effective, and to still create a high quality report, invariably causes friction.

In the recent past there has been persistent pressure to make risk assessment less expensive and time-consuming. This consistent pressure occurs, despite the fact that risk assessments often represent a fairly small part of the total time spent in reaching a risk management solution.

a. Conflicting Demands

Conflicting demands to reduce costs, shorten production time, and improve technical rigor, place those who produce risk assessments in a thankless situation. The result has been greater use of generic data, models, canned "risk assessment" software, or default assumptions. This can result in criticism that risk findings are unrealistic. Selecting the proper level of technical rigor in a risk assessment (and commitment to the resulting time lines, costs, and confidence in risk findings), often turns on the need for stringent analysis against the need for cost savings and efficient use of time. In practical terms, this balance of rigor against cost is usually based on a sense of the project's likely political or legal consequences, not on a scientist's need to prepare a technically defensible report capable of withstanding peer review, litigation, or public scrutiny.

b. Why Bother?

So, why bother with risk assessment? For one thing, risk assessment is a process embraced by regulatory agencies, legislative bodies, and courts. For another, although environmental risk assessment will never achieve the rigor of pure science, it is a valuable and essential tool to lead to informed risk management decisions as society seeks to balance environmental safety against industrial growth and economic development. Risk assessment forms the technical underpinnings for risk management, a decision-making process by which society decides whether to accept or reject risks posed by a site, activity, or facility. It is a key component of environmental decision-making and regulation in technologically advanced nations, including the U.S. When those involved in risk assessment recognize that a legitimate purpose of risk assessment is to bring science into public policy-making, they will be prepared to meet its challenges and may take pride in their ability to work with limited data, limited time, and limited budgets to create reasonable, clear, and honest appraisals of environmental risk.

IV. WHO IS TECHNICALLY QUALIFIED TO PRODUCE A RISK ASSESSMENT?

A. Different Risk Assessments Need Different Experts

Environmental risk assessments address risk to either human health (Human Health Risk Assessments, termed HHRAs) or ecological systems (Ecological Risk Assessments, termed ERAs). HHRAs characterize the nature and magnitude of risks to human health from exposure to hazardous substances, pollutants, or contaminants. Risk characterization can be quantitative (describing risk as a number) or qualitative (describing risk in relative terms, such as high or low). ERAs estimate impacts or potential risks to living things other than humans. An ERA may consider stress from habitat alterations and ecosystem disruption, as well as exposure to potentially toxic substances. Since ERAs might deal with potential risk to entire populations or ecosystems, as well as to individual organisms, they may require far more complex analysis than HHRAs, which typically deal with risks to individuals. Each risk assessment type requires different experts who are trained and experienced to perform the specialized and different tasks in an HHRA or ERA.

B. Technical Credentials Needed to Perform Expert Tasks

Technical training and experience required to conduct HHRAs and ERAs differ. HHRAs require expertise in human health-related disciplines. ERAs require expertise in wildlife biology, ecology, botany, or other disciplines focused on health and interrelationships of nonhuman organisms. Although professionals probably exist with adequate cross-training to handle both HHRAs and ERAs, most risk assessment professionals specialize in one area. In fact, demand for sophisticated analysis in risk assessment may limit a professional's expertise to certain narrow aspects of a human health or ecological risk assessment.

An essential step in obtaining a quality analysis is to match professional credentials and experience to the type of risk assessment to be performed. Significant problems occur when unqualified individuals conduct risk analyses. There is an unfortunate trend for professionals without biological training, such as engineers and hydrologists, to treat health risk assessment as a type of physical science where a correct answer can be generated simply by plugging data into equations and calculating a result. Unfortunately, such simplistic analyses disregard the complexity and subtlety of the biological world and result in questionable risk estimates.

V. RISK ASSESSMENT AS A MULTIDISCIPLINARY ENDEAVOR

The following discussion emphasizes HHRA, an emphasis that reflects the history of environmental risk assessment. HHRA has enjoyed a longer and more in-depth technical treatment, although an ERA paradigm was recently developed by the U.S. Environmental Protection Agency (U.S. EPA). Compared to HHRAs, a generally accepted technical guidance on ERAs is recent, and somewhat limited.

A risk assessment project is a multidisciplinary endeavor. A project manager leads a project, coordinating a team of experts from technical disciplines and non-technical professions. The precise mix reflects project needs. The core of a risk assessment project is typically analysis of environmental movement of chemicals and of their toxic effects on human or ecological health. This analysis requires environmental modeling, sampling, and data quality assurance and quality control (QA/QC), and involves toxicologists, ecologists, environmental chemists, modelers, statisticians, and experts in chemical procedures and analytics. A project may also benefit from involvement of a variety of other professionals. Attorneys, for example, may contribute to a project by drafting contracts that define and enforce project performance standards. Technical writers and editors help a team write a report that is both accurate and understandable. Risk communicators help a team explain risk estimates in meaningful ways to risk managers, political leaders, and concerned citizens. Planning, accounting, team-facilitation, and dispute resolution skills may also be required to produce a quality risk assessment report, on-time, within-budget, and in a useable form.

A. Mandated Science

Risk assessment is a mandated science (see Figure 1). Neither pure science nor pure public policy, risk assessment reports are a hybrid of both. A risk assessor usually works on a multidisciplinary team of regulatory scientists under direction of a project manager. The goal is to generate a risk assessment report that provides credible risk estimates (see Figure 2).

B. Team Work in Risk Assessment

A project manager must appreciate the importance of teams to successfully manage a complex environmental risk assessment project. This is true because risk assessments pose particular challenges to teamwork.

First, success of the project hinges on full participation by experts from a variety of disciplines. Each discipline brings its own paradigm, language, assumptions, and skills to the project, as does each individual. Such diverse views can lead to confusion and friction in a team setting. If a team is to generate a truly acceptable* final risk assessment report, a project manager must send a clear message that, although credentials and disciplines differ on a team, all team members have an equal duty

---

* An "acceptable" risk assessment report is more than "merely acceptable" in the common sense of the term. Here, "acceptable" requires a risk assessment report to meet or exceed all performance standards (e.g., all math and science is correct and can be verified by critical reviewers).

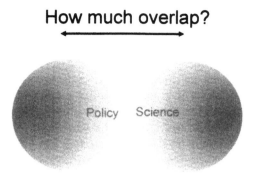

Figure 1 Mandated science at the intersection of policy and science. (Adapted from Mandated Science, 1988.)

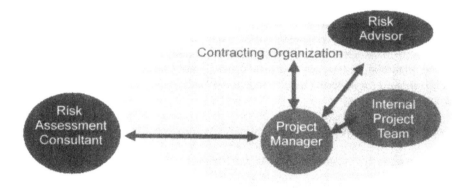

Figure 2 Risk assessment teams.

to voice concerns, and to respond to concerns with respect. All team members must employ methods that allow all technical work to be verified and reviewed.

Some experts may resist teamwork, believing that there is one right answer and that their only task as a scientific expert is to determine that answer, not to explain how they perform tasks, and why, nor to debate ideas or consider alternate views. No matter what their credentials, such people will make poor team members. Arrogance will prevent them from helping a team to integrate their expertise into a project. This attitude can destroy teamwork and must be curtailed by a project manager. Otherwise, the power of teamwork will be lost.

Second, mixed loyalties arise when people involved serve two masters — an organization that pays them and a risk assessment team. Environmental risk assessment participants usually have differing goals. For example, an environmental risk assessment normally draws experts from several divisions of an organization, especially in large organizations, each division with a slightly different view of the project. Also, outsiders are sometimes involved, such as regulators or other government officials, citizen activists, or community leaders, or even industrial competitors.

Organizations may hire environmental consultants to provide specialized technical expertise. When team goals conflict with goals of their principal employer, team members will feel a degree of stress. A project manager, who typically lacks direct authority over team members, must acknowledge the stress, attempt to reconcile conflicting goals and, thus, win team member cooperation and support for the risk assessment process.

A third challenge to teamwork on a risk assessment project results from prior relationships among participants. People involved in an environmental risk assessment project — as project sponsors, affected parties, or reviewing authorities of a final product — are likely to know one another from involvement on other projects. Naturally, prior relationships affect expectations about roles, tactics, and agendas. If previous interactions were productive, a project manager is lucky. However, more often, prior interactions occurred in a win-lose setting. If so, a project manager must establish a new way for people to interact with each other. This requires a project manager to address assumptions and make explicit every aspect of how a report will be developed — including the basis of team work: team roles, project priorities, and working rules.

Although most professionals have experience with meetings, it takes more than meeting etiquette to create a team environment that allows members to contribute fully to the process. A project manager must help team members agree upon a legitimate purpose for a team. Then, based on its purpose, a team can identify roles team members should fill. Rules for working together must be developed, agreed upon, and enforced. Finally, a team should consider potential project outcomes and establish realistic project expectations that achieve a team's purpose.

Although much of how a team works is negotiable, there are issues not open to negotiation. Laws, rules, guidance documents, and generally accepted technical and scientific principles are clear examples of items not open to a group consensus-building process. Negotiating items that a professional and general populace accept as "given," wastes time and resources. It also endangers success of a project and undermines morale and professional credibility of those associated with the risk assessment. Negotiation of nonissues is a signal that certain players controlling a project are either not technically qualified or hope to kill the project.

Consensus-building in a team setting must never be used as a means to squelch expert input and determinations. Teams must recognize and respect expert opinions. Teamwork is a process to smooth the development of complex tasks, such as preparation of a risk assessment report. Consensus-building must not be used as a bludgeon to silence or marginalize an expert working within their field of expertise. For example, the opinions of four hydrologists do not outweigh the views of one toxicologist if the issue is toxicology.

C. Roles in Risk Assessment Teams

Although team members may be equals within a team, a project manager must recognize that different team members play different roles in a risk assessment process. Certain roles will be assigned with specific responsibilities. For example, a project manager and risk advisor play unique vital roles on a project. These roles

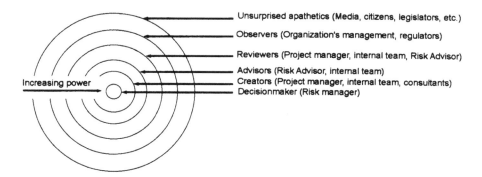

Figure 3 Roles in risk assessment project development. (Adapted from Synergy, 1986.)

are discussed below. A project manager might work differently with internal team members versus outside experts. Staff, project proposers, and other paid participants will typically fill different roles than volunteers. Team members who are on loan may be less involved than team members who work for a project manager.

Certain generic roles can be identified for any project. It is useful to identify which role each participant may occupy on a risk assessment project (see Figure 3). As this figure indicates, most active participants occupy roles close to the center. Roles introduced below are discussed in more detail in Chapters 4 through 6.

1. Project Manager

Project managers manage a risk assessment project. They oversee project communications, administer a work schedule, and budget for contractors and a project team, and ensure that resulting work meets performance standards.

2. Internal Experts

In-house expertise is a tremendous asset to a risk assessment project. Depending on the nature and degree of internal expertise, an internal team may either perform risk assessment work, or oversee work performed by a contractor with specialized risk assessment expertise.

Even when a consultant is employed, internal experts play a vital technical role on a risk assessment project. As members of an internal project team, they help formulate a scope of work, review work plan adequacy, and set project performance standards. An internal project team can help a project manager anticipate and solve problems. A team can also provide oversight by reviewing interim and final deliverables to assure that consultant work meets process and product standards, as required under a project contract.

Support of internal experts can greatly enhance project credibility and speed internal acceptance of a risk assessment report; opposition can defeat a project. Internal experts bring technical expertise and organizational savvy to a project team.

They serve as both trustworthy sources of technical knowledge and as internal reality checks on outside consultants' views of a project. Therefore, a risk assessment project manager must make every effort to recruit and earn support from internal technical experts.

3. Risk Advisor

A risk advisor is a person who has mastered the risk assessment process through experience on several successful projects. The exact role of a risk advisor is defined by an organization's needs. A risk advisor serves as mentor to a novice project manager, as a sounding board to an experienced project manager, and as a watchdog over outside consultants in areas where internal expertise is lacking. A risk advisor can also function as a technical liaison between internal-project staff, who may lack in-depth understanding of risk assessment techniques, and technical consultants. A risk advisor may be found within an organization, but often is hired from an environmental consulting firm. A risk advisor's first duty is to advance the contracting organization's interests. Due to an adversarial relationship between a Risk Advisor and external consultants, a Risk Advisor should not be an employee of a consulting firm hired to conduct a project (see Chapters 4, 5, and 6).

4. Consultants

Since few organizations possess internal technical capacity required to conduct a credible risk assessment project, organizations in need of an environmental risk assessment hire consultants to perform technical risk assessment services. Consultants typically work under the guidance of a contracting organization's project manager with review by an internal-project team and risk advisor, discussed above.

The precise role of a consultant will vary somewhat depending on performance standards established for a project. However, in order to fulfill the basic role, a firm and individuals assigned to a project must be technically and ethically credible. Specifically, a consulting firm must either have technical experts on staff who are capable of performing required work or it must demonstrate professional affiliations sufficient to cover any gaps in expertise through subcontracting. A credible consultant will be prepared to prove technical expertise through statements of staff credentials and prior project descriptions. A reputation for honest dealing should be required of any consultant. An experienced firm will be able to provide names of satisfied clients. Individuals assigned to a project must also be trustworthy. Although this is more difficult to determine, it is important. Any ethical or legal breach will reflect badly on a project and on an organization represented by the consultant and its staff.

D. Teams Establish Performance Standards

The purpose of an environmental risk assessment project is to define and generate an acceptable risk assessment report. An "acceptable" risk assessment report is defined as a report that meets all performance standards for a project, discussed in the following section. A team will define a complete set of performance standards that articulates needs of the organization. A team will also ensure that the project adheres to these standards, as it proceeds.

1. Performance Standards

A team's first, most important, task is to establish "performance standards." Performance standards articulate a process a risk assessment project will follow, termed "process standards," and attributes of interim and final work products, termed "product standards." Every project has a timeline and a budget, for example. A precise project schedule and details of the budget should reflect specific project demands. A project schedule and budget are two basic performance standards. A team's analysis must typically go far beyond basic performance standards of schedule and budget. This is accomplished by articulating the purpose of environmental risk assessment and then, keeping that purpose firmly in mind, identifying all decisions necessary to accomplish that purpose.

For example, what degree of technical accuracy is required? An appropriate degree of accuracy depends on the expected use of a risk assessment. Is it for litigation and, thus, must it be highly defensible? Or, is it for planning, and will estimates and qualitative analyses be acceptable? Most risk assessment reports fall somewhere between these extremes. If litigation is a purpose of a risk assessment, it is realistic to expect aggressive scrutiny in court. A risk assessment report will need to be scientifically accurate and technically defensible to survive: models must be current and must be generally accepted, default values and assumptions must be realistic (or their use must be minimized), and data must be of the best quality. On the other hand, a high level of technical rigor may not be required, or appropriate, in a risk assessment report intended merely to aid internal planning. High levels of technical rigor, where it is not needed, may be a waste of resources (see Chapters 2 through 6).

2. Process Standards

Process standards address "how" questions. They define how a risk assessment will be conducted and managed and they define acceptable behaviors of project participants.

One fundamental process standard establishes how a contractor will be managed, by a proactive or reactive management approach. If a "proactive" contract management strategy is used, project work will undergo iterative review, comment, and approval throughout a project. "Iterative review" requires a consultant to submit each interim work product for team review as soon as a deliverable is completed. Each interim work product must meet all relevant standards before a product is accepted and a consultant is allowed to begin work on the next deliverable. If project management is reactive, product review starts only after delivery of a draft final report (see Figure 4).

A second important set of process standards will govern how communication will occur on a project. Specifically, how will communication occur within a project team,* between a consultant and project manager, and with outsiders (such as

---

* Throughout this book, use of the term "project team" always refers to staff of an organization that hires a risk assessment contractor. Contractor staff may, in actuality, also constitute a separate project team, but we refer to contractor staff collectively as "contractors" to avoid confusion.

INTRODUCTION

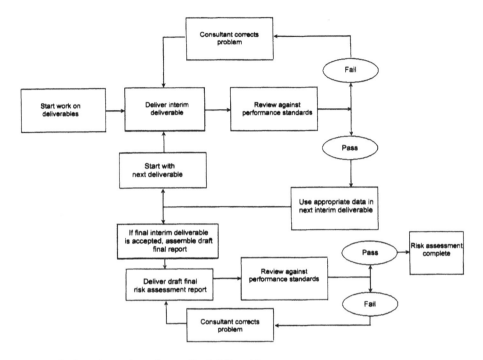

Figure 4 Iterative review of consultant deliverables.

interested staff and managers within the organization, political leaders, citizens, and the media). In order to develop process standards for communication, a team first articulates internal and external communication needs, then selects appropriate techniques and, finally, assigns responsibility for maintaining communications channels (see Chapter 21).

Project review and communications are just two examples of many procedural matters a risk assessment team will address through process standards. Each decision on process standards affects how a project will proceed and how it will be judged.

3. Product Standards

Product standards address "what" questions and, thus, articulate characteristics required from an acceptable work product. Product standards define the quality of a final product. They may also define quality of interim work products. Product standards establish the scope of a risk assessment — human health, ecological risk, or both? They also address the type of assessment to be performed — a quantitative or qualitative assessment — and a level of scientific rigor. They mandate rigor of technical review; they set the clarity and style of writing and editing; and they may specify a style and consistency of document layout, as well as myriad other non-procedural aspects of a risk assessment.

4. Teams Apply Performance Standards

After performance standards are established, the main work of a project manager and project team will be to ensure that a project meets these standards (see Part I). During the course of a project, however, certain performance standards may require modification. A consultant might identify unmet standards, for example. If so, a project manager should require a consultant to document reasons for failing to meet each standard and, based on justification, determine whether to drop, amend, or enforce a requirement. Unmet standards will also be discovered when a project manager and team review work products. Again, the issue is why a failure occurred and whether it matters.

VI. AN OVERVIEW OF THE RISK ASSESSMENT PROCESS

Now that you understand the basics of environmental risk assessment and the role of teams and experts, we will integrate this information into practical methods to produce a risk assessment report.

There are four phases in risk assessment report development: planning, managing, accepting, and dealing with results. Chapters 4 through 6 discuss major steps in developing a risk assessment report. The process is capsulized in Table 1. This table can be used as it is presented, but it will function best if it is expanded or simplified to reflect specific project needs. Whether an expanded or simplified version of this form is used, a project manager and internal project team will need to perform, or oversee, all outlined steps.

A. Phase One — Planning a Risk Assessment

Planning is the first phase of a risk assessment project. Planning deserves careful attention because it reduces "preventable problems." Preventable problems are those obstacles that could have been easily avoided or removed, if someone had anticipated them. After deciding to perform a risk assessment, an organization selects a project manager. The project manager then recruits a project team. A project team works with a project manager to develop a scope of work. A scope of work describes each important facet of a risk assessment project and serves as the basis for a Request for Qualifications (RFQ) or a Request for Proposals (RFP), and for project performance standards. An organization distributes or publishes an RFQ/RFP to notify contractors that it seeks services they may offer. Contractors respond by submitting bids, which a project manager reviews with an internal project team. A project manager selects a contractor, based on qualifications, project needs and cost, and then negotiates with a prospective contractor on specific contract terms and a project work plan. Parties sign a contract when they agree on a contract and work plan. If negotiations break down, a project manager may decide to negotiate with another qualified contractor.

INTRODUCTION

Table 1 Generic Risk Assessment Planning Form

Step	Actions
Phase One — Planning a Risk Assessment	
Is risk assessment needed?	Consider why the risk assessment is being done. Is it required, requested, or voluntary? Identify the site, activity, or facility to be assessed.
Staff the risk assessment	Build a project team. Assign staff to serve as project manager and project team members. Determine your role in the process. Assess skills and technical specialties needed to generate a risk assessment report and determine which skills are available in-house. Consider using a risk advisor to supplement team and project manager skills. Consider need for consultants to perform part/all of the risk assessment.
Fund risk assessment	Estimate required funding needed for the project. Determine actual/likely funding available. Encumber the financial resources (or develop alternate strategies for obtaining support, personnel, resources).
Determine report end-user needs	Set appropriate project goals and expectations. Establish clear performance standards to evaluate and demonstrate project success and failure.
Scope the risk assessment	Develop a risk assessment scope of work that includes project performance standards, including timelines and budget.
Distribute RFQ/RFP	Write, issue, publish, and distribute the Request for Qualifications (RFQ)/Request for Proposals (RFP) (if contractors are needed).
Hold a project kick-off meeting	Invite interested contractors and other interested parties to attend a project overview and ask questions.
Evaluate proposals	Evaluate submissions based on criteria outlined in the scope of work, especially project performance standards.
Select contractor	Select contractor(s) with skills to produce an HHRA or ERA and notify the firm of their opportunity to negotiate a contract.
Negotiate contract and contractor work plan	Negotiate a contract that includes a contractor work plan. Base acceptability of both documents on project performance standards.
Phase Two — Managing a Risk Assessment (Including Iterative Review)	
Mobilization	Initiate work. This assumes use of proactive development process illustrated in Figure 4 above to generate five deliverables.
Hazard evaluation	Collect and evaluate data. Produce a draft Chemicals of Potential Concern (COPC) and a final Chemicals of Concern (COC) list. For each COC, produce a source concentration or emission rate for use in the exposure assessment. Iteractive review requires submission of a draft hazard evaluation for review by the internal risk assessment review team. Failures to meet performance standards are identified and the contractor is notified of insufficiencies requiring correction. A deliverable that meets all performance standards is accepted and the contractor receives approval to initiate work on the next step.

Table 1 continued

Exposure assessment	Chemical-specific source concentrations or emission rates are used in fate and transport models, or environmental monitoring data are used, to calculate the concentration of each chemical in a given environmental medium at a location where organisms will be exposed. Exposure equations are used to calculate chemical specific uptakes or intakes. The draft Exposure Assessment is submitted an interim deliverable for iterative review and approval, as described above.
Toxicity assessment	Chemical-specific and chemical-mixture toxicology information is gathered. Chemical-specific toxicity values are obtained or derived from data found in the open literature. This information is used with exposure levels from the exposure assessment to characterize risks. The draft toxicity assessment is submitted as an interim deliverable for iterative review and approval, as described above.
Risk characterization	Exposure levels and toxicity values are coupled to calculate risks and impacts. The draft risk characterization is submitted as an interim deliverable for iterative review and approval, as described above.
Review draft report	Review of the report should be minimal if iterative review by the internal risk assessment team was thorough.
Phase Three — Accepting a Risk Assessment (Including Iterative Review)	
Accept final draft	Final review should focus on report clarity, completeness of explanatory materials, and integration of the interim deliverables into a coherent report. The conclusions, uncertainty analysis, and executive summary bear special scrutiny because they will not yet have been reviewed and they synthesize the reports various pieces. When using reactive risk assessment development process, all aspects of report must be evaluated. Any problems identified by reviews must be corrected prior to acceptance of report. This may require several iterations and considerable time.
Close contract	Bring closure to the contract and the professional relationships developed on the project by hosting a formal meeting where report findings are presented to the group that generated the report, to those who will accept the report and those who will use the results. Conduct a series of private exit interviews with both internal team members and contractors to learn how the process can be improved. Final copies of the report are deliv-ered to the contracting organization. The contractor is paid.
Phase Four — After a Risk Assessment	
Risk communication	Use formal acceptance of the report as a transition into the risk management and risk communication phase. Emphasize rigorous process of review and clear performance standards used to generate the report to highlight its technical credibility. For most projects, it is best to conduct risk communication throughout the risk assessment project, as well, using citizen input to provide information on the type of land use, exposure routes, and other aspects of the project. Use of such information can improve report assumptions and credibility, as well as public acceptance.

INTRODUCTION

Table 1 continued

Risk management	Use a formal evaluation methodology to generate and support risk management options. Generate a risk management decision document that provides all risk management decisions with their associated data and logic, including uncertainties and limitations. Coordinate this activity with participants in the production of the risk assessment and other appropriate interested parties.
Defending the risk assessment report	Present and defend risk estimates at public meetings, public hearings, administrative actions, and court proceedings, as required.

Note: An actual risk assessment project can have greater or fewer steps, depending on project needs.

B. Phase Two — Managing a Risk Assessment (Including Iterative Review)

A second phase of a risk assessment project involves technical work; a project manager must oversee work of a contractor, facilitate review by a project team, and manage communication and disputes on a project. Work planning and scoping processes that occurred in Phase One will have delineated process and product standards that come into play in Phase Two. Therefore, a project manager will have developed a grasp of major aspects of a project, such as what work products are to be produced (interim and final products); how they will be produced (who will do the work, what resources will be used, when each work product will be delivered); how progress will be tracked, and how work will be reviewed and evaluated for sufficiency. We recommend using a proactive approach. This calls for a series of discrete interim deliverables. Each deliverable must pass review before work begins on subsequent deliverables.

After a contract is signed, a contractor starts work, guided by performance standards set forth in the project contract and work plan. A formal risk assessment process begins with data collection and evaluation (also known as hazard assessment). Contractors accumulate all existing data relevant to a site, activity, or facility and then determine whether sufficient information exists to develop a risk assessment report. If time or funding is limited, risk assessors may evaluate quality and quantity of available data to determine what level of risk evaluation can be done. Data quality must be properly matched to the level of risk analysis rigor (e.g., qualitative, semi-quantitative, and quantitative). If available data is of suitable quality for required risk analysis, no additional data are gathered. If not, additional data must be collected and analyzed. Project managers decide how to collect and analyze additional data in consultation with other team professionals.

After a contractor gathers all relevant and acceptable data, data are statistically evaluated to generate source concentrations (e.g., for each water or soil contaminant, and emission rates for each air contaminant). Environmental contaminants pose no risk unless they move to a point where an organism will be exposed. If there is no exposure, there is no risk. While it is possible to measure environmental contaminant concentrations at an exposure point some distance from its source, risk assessments generally rely on mathematical environmental fate and transport models and calculate exposure point concentrations in environmental media (e.g., soil, air, water, food), rather than collecting data. This makes sense when using "potential to emit" estimations for proposed facilities.

Next, movement from environmental media at a given location into an exposed organism is considered. All relevant exposure pathways are evaluated. Standardized exposure equations are used to calculate exposure levels, i.e., intake and uptake (see Chapter 2 IV. C). Chemical intakes and uptakes are compared to toxicological values to calculate chemical-specific risks. Risks are then considered by grouping chemicals with similar toxic effects. For example, all risks are summed for all carcinogen exposures; this value is compared to an acceptable cancer-risk yardstick. For noncarcinogens, all risks are summed for all pathways for chemicals with similar toxic effects and exposure duration; this value is compared to acceptable noncancer risk yardsticks.

After completing these steps, a contractor organizes numerical findings into a series of summary tables. A quantitative or qualitative uncertainty analysis is also provided in narrative form. If the risk assessment was financed by the interested party, or their contractor, they might wish to include a chapter that presents their editorial comments on their mandated risk assessment.

Summary tables provide a better understanding of the basis of a report's risk estimates, and uncertainty analysis clarifies a risk assessment project's rigor and points out limitations of its findings.

C. Phase Three — Accepting a Risk Assessment (Including Iterative Review)

In the third phase of a risk assessment report development process, a final report is critically reviewed by the project manager and risk assessment project team. It is corrected as necessary. When it meets all performance standards, work is accepted.

If a proactive contract management strategy was used, Phase Three is relatively simple. As discussed above, previous project work will have already undergone iterative review and final review requires detailed examination of only the last set of interim deliverables, and of integration of all interim deliverables into a consistent, cogent final report.

If project review was reactive, review is delayed until all work is completed and delivered as a draft final report. This will undoubtedly make Phase Three more difficult.

Reactive review is a favorable situation for consultants. It allows them to maximize use of consulting staff because there is no predetermined order in which work is done. As consultant staff finds time, work is performed on a risk assessment. Eventually, all pieces are integrated into a draft report for review. A project manager and project team are, however, disadvantaged by a consultant's use of reactive management. First, problems with interim work are not remedied before they are integrated into other work. Second, serious problems can lead to serious delays toward the end of a project, when time is running out. Third, a project manager is at a disadvantage when negotiating with a consultant to fix problems near the end of a project. A contractor will have scheduled other projects to begin as a risk assessment concludes. New project demands will make a contractor far less likely to cooperate at the end of a risk assessment project than at the beginning.

In most cases, passing final review concludes a contract, unless public comment requirements are required, precipitating additional changes to a report. Contract provisions should delineate this work and make clear that contractual obligations are not concluded until public comments have been incorporated into a final risk assessment report.

D. Phase Four — After a Risk Assessment

In the fourth phase of the process, risk managers receive risk report findings and use them, along with nonrisk factors (e.g., technical feasibility of risk reduction measures, economics, politics, and cost/benefit analyses) to arrive at a risk management decision. Risk management options are evaluated and risk communication strategies are determined. Risk management decisions are explained to interested parties through risk communication.

E. Risk Assessment Planning Form

A Risk Assessment Planning Form, presented in Table 1, provides a detailed treatment of the risk assessment process. A project manager may use this form to quickly establish time lines, interim and final deliverables, and other routine scheduling and budgeting items. This table combines elements of a risk assessment performed using resources within an organization and one where consultants are hired to perform a risk assessment. Depending on the specific situation, sections of this table may be omitted or supplemented. This abbreviated approach cannot replace in-depth risk assessment report planning. If there is absolutely no other way to meet a mandate to initiate a risk assessment, however, abbreviated planning is better than no plan.

VII. CONCLUSION

Risk assessment is a standardized method for evaluating and presenting potential health risks and environmental impacts from potentially toxic substances released to the environment. It serves as a framework to force science into constraints of societal needs, and of political and legal mandates. Risk assessments follow procedural rules established by regulatory and scientific organizations. An extensive body of federal and state guidance outlines risk assessment requirements and standard methods. Guidance documents are also being produced by international organizations. In practice, however, implementation of this generally accepted risk assessment paradigm varies greatly.

Unfortunately, although detailed guidance exists on technical aspects of assessing environmental risk, little heed has been paid to improving day-to-day development of risk assessment reports and how environmental risk estimates are communicated. Reports are often confusing, logic is muddled, math and modeling can not be checked, and terms are obtuse and undefined. As a result, even people well-versed in environmental risk assessment find it difficult to understand the basis for risk estimates, to review adequacy of their supportive reports, or to judge the validity of science and assumptions used in an environmental risk assessment. Thus, an important aspect of the scientific method, the ability to check and verify technical work, becomes impossible. This has resulted in a perception that risk assessment is "smoke and mirrors" and, thus, unreliable. This is, arguably, the fault of risk assessment practitioners, not an inherent flaw in the discipline.

A risk assessment cannot be quick, comprehensive, and cheap. Every risk assessment project manager is probably asked, at some time, to produce a high-quality, low-budget, scientifically-rigorous risk assessment using a contractor. In such circumstances, at least one of three ideal attributes — speed, thoroughness, or cost effectiveness — will be sacrificed. If an organization requires a risk assessment that is both fast and cheap, it must recognize that thoroughness will suffer.

While limitations inherent in risk assessment will probably not be completely eliminated, they can be minimized through use of procedures presented in this book. Our following chapters provide methods to control quality of risk assessment reports, to manage the process, and to critically evaluate risk assessment work products. Understanding gained from this book will prepare a reader to make better use of information from a wealth of technical documents relating to environmental risk assessment and to build a common understanding of risk assessment. Techniques offered in this book can help a project manager keep report development on track, manage and control consultants, and create a report that people can understand, review, use, and trust. Finally, methods discussed in this book can allow effective critical review of risk assessment reports.

REFERENCES

Belluck, D.A., et al., Defining scientific procedural standards for ecological risk assessment, in *Environmental Toxicology and Risk Assessment*, 2nd Volume, STP 1216, Gorsuch, J.W., et al., Eds., American Society for Testing and Materials, Philadelphia, 1993, 440.

McVey, M., et al., *Wildlife Exposure Factors Handbook*, Vols. I and II, Office of Health and Environmental Assessment, Washington, 1993.

The Presidential/Congressional Commission on Risk Assessment and Risk Management, *Framework for Environmental Health Risk Management, Final Report, Vol. 1*, Washington, 1997.

Salter, L., *Mandated Science*, Kluwer Academic Publishers, Dordrecht, Holland, 1988.

Syngery, Training materials presented at the Wisconsin Department of Natural Resources, Madison, WI, 1986.

U.S. Environmental Protection Agency, *A Descriptive Guide to Risk Assessment Methodologies for Toxic Air Pollutants*, Office of Air Quality Planning and Standards, Washington, 1993.

U.S. Environmental Protection Agency, *Health Effects Summary Tables*, Office of Research and Development, Washington, 1994.

U.S. Environmental Protection Agency, *Human Health Evaluation Manual, Supplemental Guidance: Standard Default Exposure Factors*, memo from Timothy Fields and Bruce Diamond to various EPA directors, Washington, 1991.

U.S. Environmental Protection Agency, *Proposed Guidelines for Carcinogen Risk Assessment*, Office of Research and Development, 1996.

U.S. Environmental Protection Agency, *A Review of Ecological Assessment Case Studies from a Risk Assessment Perspective*, Office of Emergency and Remedial Response, Washington, 1993.

U.S. Environmental Protection Agency, *Risk Assessment Guidance for Superfund, Vol. I, Human Health Evaluation Manual (Part A), Interim Final*, Office of Emergency and Remedial Response, Washington, 1989.

CHAPTER 2

Human Health Risk Assessment

David A. Belluck and Sally L. Benjamin

CONTENTS

I.	Introduction to Human Health Risk Assessment	30
II.	Hazard Assessment	32
	A. Defining Acceptable Data Quality	32
	B. Defining Data Needs	33
	C. Defining Chemical Background Concentrations	33
	D. Defining Acceptable Sampling and Analytical Plan	34
	E. Defining Quality Assurance/Quality Control (QA/QC) Methods	36
	F. Defining Methods for Pooling Sampling Data	36
	G. Defining Data Sources	37
III.	Hazard Assessment Conservatism	39
	A. Problems Associated with Developing a COPC and COC List	39
IV.	Exposure Assessment	41
	A. Fate and Transport Analysis	42
	B. Exposure Equations	49
	C. Chemical Intake and Uptake	50
V.	Exposure Assessment Conservatism	51
VI.	Toxicity Assessment	53
VII.	Regulatory Toxicology and the Science of Toxicology	53
	A. Types of Tests	53
	B. Physical and Chemical Properties	54
	C. Pharmacokinetic Properties	54
	D. Use of Regulatory Toxicology in Toxicity Assessment	55
	E. Exposure Routes	56
	F. Exposure Duration	59
	G. Absorption, Distribution, Metabolism, and Excretion	59

	H.	Target Organs 59
	I.	Using Toxicological Understandings in Toxicity Assessment 63
VIII.	Risk Characterization 67
IX.	Concluding the HHRA 70
X.	Presenting HHRA Data 70
	A.	U.S. EPA's Standard Tables for Superfund Risk Assessments 71
	B.	Variable Selection Tables 75
	C.	Decision Logic or Criteria Tables 75
XI.	Conclusion 75
	References 76

I. INTRODUCTION TO HUMAN HEALTH RISK ASSESSMENT

HHRA reports provide risk findings, estimates of human health risks associated with a site, activity or facility. Risk managers use HHRA risk findings for many purposes. Risk findings guide risk reduction measures. For example, they help determine a need for site cleanup, define cleanup levels, and aid in establishing facility permit conditions to limit environmental releases and, thus, limit risks.

HHRA risk findings are often numerical* and are compared to numerical regulatory criteria (e.g., bright lines), official or informal yardsticks of acceptable and unacceptable risk. If HHRA numerical risk findings do not exceed numerical criteria, risks are typically deemed "acceptable" or "insignificant." Risk findings that exceed applicable risk criteria are typically considered "unacceptable" or "significant." Exceeding risk criteria may pose serious legal and economic results for a regulated entity because these numbers serve as triggers for regulatory action. Exceeding them may trigger remediation, denial of a permit, or enforcement action.

Government agency use of terms discussed in previous paragraphs are often confusing and inconsistently applied. For example, some regulatory and health protection programs may use different bright line values (e.g., cancer risks from one-in-ten thousand to one-in-one million) to determine when risks are too high. When using these bright line values for carcinogens, it is reasonable to expect that exceedance of a bright line will result in cancer health risk concerns, whereas risks at, or below, a bright line value will not result in cancer health risk concerns. In practice, however, application of bright lines is highly variable; there is no uniform black or white, unsafe, or safe application of a bright line concept. Determining when a risk estimate moves from acceptable to unacceptable is merely a value judgment made by risk managers (e.g., government regulatory agency senior- or middle-management), not by risk assessors. Risk managers use risk findings as a single input into a complex decision-making process that balances calculated risks with broader considerations, including economics, social impacts, and politics. Thus, a purely technical finding of unacceptable risks from a risk assessment report (e.g., risk estimate exceeds a bright line) can still be negated, resulting in a risk management

---

* Quantitative risk assessment reports yield numerical risk estimates, whereas qualitative risk assessment reports characterize risk in relative terms, such as "high," "medium," and "low."

finding of acceptable risks. Risk findings and risk management decisions of health concerns make legal implications of a risk assessment difficult to predict.

Risk assessment involves four formal steps: Hazard Assessment (also referred to as Data Collection and Evaluation, Hazard Evaluation, or Hazard Identification), Exposure Assessment, Toxicity Assessment (e.g., quantitative dose-response relationships) and, ultimately, Risk Characterization. The following discussion will provide a thumbnail sketch of a generic HHRA development process and is not designed to duplicate or replace the voluminous library of government guidance documents and technical reports on risk assessment. This information provides readers with context for the remainder of our book.

The first step in HHRA process is hazard assessment. Hazard assessment begins with collecting existing data on a site, activity, or facility of concern. This analysis may reveal a need for additional data collection prior to initiating risk assessment calculations. When sufficient data of known quality have been collected, a list is produced of all potentially toxic chemical substances that may result from a site, facility, or activity, termed COPCs.* A list is narrowed to a final list of COCs, those chemicals slated for quantitative evaluation in the next three steps of an HHRA (some authors use COPC and COC interchangeably).** A concentration term (or emission rate***) is calculated (or obtained) for each COC at its source. Source concentrations (or emission rates) are used in fate and transport mathematical models in the next step, exposure assessment.

Exposure assessment, the second step in an HHRA process, determines chemical concentration in soil, air, or water at locations where humans may be exposed, termed receptor points. In some cases, actual chemical residue data can be collected at a receptor point. Since it may be difficult or impossible to obtain field collected media-specific (e.g., soil, water, air, food) chemical contaminant concentrations, especially for proposed facilities, mathematical models are used to calculate chemical-specific exposure levels. Chemical source concentration terms (or emission rates) are used in environmental fate and transport equations or computer models to calculate chemical concentrations at receptor points by calculating decrease in a chemical's concentration from its source to potential human receptors at a given location. This step in HHRA is very complex and typically relies heavily on data derived from literature or generated using models. This step in the process produces numerical exposure levels.

Toxicity assessment is the third step in HHRA. It may be conducted concurrently with exposure assessment. Toxicity data are collected on each COC in this step. Chemicals are classified as either carcinogens or noncarcinogens and their toxic properties and numerical toxicity values are determined.

Risk characterization, the fourth and final step of HHRA, generates risk levels based on exposure levels and toxicity data. Although methods of calculating carcinogenic and noncarcinogenic risk differ, numerical expressions of both types of risk

---

* A chemical of potential concern (COPC) is a chemical known or suspected to be associated with a site, activity, or facility under review. A chemical of concern (COC) is a chemical that will be evaluated in the next three steps of a risk assessment.
** Chemicals not evaluated quantitatively, for example because they lack a toxicity value, still should undergo qualitative evaluation in the uncertainty analysis.
*** "Emission rate" refers to an air concentration of a COPC or COC.

are compared to appropriate risk criteria to determine whether calculated risks exceed an acceptable risk threshold.

The next four sections discuss each of the four HHRA steps in detail. Information presented in these sections is a broad overview of each subject, intended to familiarize readers with the HHRA process, and assist in day-to-day work with other members of a risk assessment team and in reviewing a risk assessment report. It does not replace a need to rely on qualified risk assessment professionals or source materials that risk assessment practitioners use to conduct and review a risk assessment.*

In order to avoid later confusion, readers should note that risk assessment guidance documents and books differ in where they place a given activity. Thus, for a given risk assessment process, scoping document, or report, an exact location of a specific risk assessment task may vary. In final analysis, it is inclusion of all required parts of a risk assessment that is crucial, not necessarily their precise order.

II. HAZARD ASSESSMENT

Hazard assessment is the first step in a formal evaluation of potential risks posed by environmental releases of chemicals. To conduct an HHRA, the names and concentrations of chemicals known, or expected to be released to the environment, must be determined. Data used to generate chemical release levels must either meet minimal data-quality requirements, or be of known quality (e.g., acceptable, marginal, unacceptable). All existing data relating to identity of COPCs and their source concentrations is collected for a site, activity, or facility that is subject to risk assessment. Existing data sets are then evaluated or grouped as to their adequacy for determining identities of COPCs. During evaluation, data quality is checked and data sets may be combined, analyzed, and statistically manipulated to yield chemical concentration terms (or emission rates) at a source of each COPC.

If existing data are inadequate, data collection is required. A sampling and analysis plan assures statistical relevance of data collection. New data sets can be used alone or combined with existing data sets. Sufficient data must be amassed to evaluate each COPC and determine whether to list it as a COC to undergo quantitative risk assessment. Various methods can be used to develop a COC list from a COPC list. These are discussed later in this chapter.

For each COC, concentrations are calculated for water, soil, or other media; emission rates are calculated for air contaminants. These environmental concentrations serve as inputs to environmental fate and transport models in Exposure assessment. Risk assessment findings are only as reliable as chemical-specific data inputs. Our following sections describe issues influencing data reliability.

A. Defining Acceptable Data Quality

Data quality and usefulness varies. Some data points can be unusable because of sampling or laboratory analysis problems or errors. Data usefulness relates directly to its anticipated use. Data Quality Objectives (DQOs) ensure that only data of

---

* Many of the technical aspects discussed in this chapter are portable for use in ERAs.

quality required for HHRA purposes are used in an HHRA. The DQO process identifies risk assessment data needs, objectives, and uses. Sampling approaches and analytical options are established and a data collection program and methods are designed to obtain data acceptable for its intended use.

B. Defining Data Needs

Several generic data types are used in an HHRA. Existing information is gathered on chemical identities and their concentrations in environmental media (e.g., soil, air, water, food, organisms). Data are gathered on environmental characteristics that could influence fate, transport, and persistence of released chemicals, probable or known exposed individuals or populations, and properties and degradation pathways of chemicals of potential concern. Comprehensive data collection, and analysis of these data sets, requires time and resources.

C. Defining Chemical Background Concentrations

Background concentrations (sometimes also referred to as ambient concentrations), by definition, cannot be attributed to a site, activity, or facility under review. There are two different types of chemical background concentrations. Naturally occurring levels are ambient concentrations of chemicals in the environment that are not caused by human activity. In contrast, anthropogenic levels are chemical concentrations that are a result of human activities. A given background level of a chemical can have a localized spatial distribution or it can be ubiquitous. Appropriate background sampling is conducted to establish naturally occurring levels of chemicals and anthropogenic levels, to distinguish these levels from those associated with a site, activity, or facility of concern. Some professionals use "ambient concentrations" to describe actual conditions measured in the field (e.g., city air chemical concentration levels).

Background samples are collected at or near a site, activity, or facility in areas that are not contaminated from such operations or activities. Sampling areas and sample size are specific to each case. Background chemical levels cannot be defined by measuring so-called "clean areas" within a zone of impact or contamination. For example, soil concentrations at a suspected hazardous waste site may not be deemed of regulatory concern, until it is shown to exceed both background or regulatory concentrations. In other cases (e.g., air pollutant levels in cities), background levels are considered to be those that typically exist. These levels could be of regulatory concern. Unless background concentrations are exceeded, there may be no scientifically valid basis for performing a risk assessment.

A valid sample size is required, both to establish background concentration of a particular chemical and to properly differentiate it from greater concentrations. Statistics are used to set a valid sample size. An appropriate degree of statistical certainty (e.g., $\alpha = 0.01, 0.05, 0.10$) is selected on a case-specific basis. Statistical analyses of background samples may be necessary to differentiate them from non-background sites.

After background concentrations are calculated, they are compared to a "contaminated medium" to determine whether that medium is truly contaminated. If a medium is found to have chemical concentrations significantly higher than background or regulatory concentrations, a risk assessment can be performed. In some cases, background concentrations of a chemical (such as natural arsenic levels in some midwestern aquifers) are already above levels of health concern. In such cases, a risk assessment may be used to estimate total risks from exposure to all contaminants found in the groundwater.

1. Regulatory Concentrations

State, federal, and international organizations often establish different regulatory concentrations, i.e., concentration at which a chemical or substance may be of health concern. Regulatory concentrations are numerical expressions relating to risk posed by exposure to chemical- or mixture-specific concentrations. Exceeding a regulatory concentration may pose unacceptable risks to exposed organisms. Regulatory concentrations, however, are not necessarily based solely on toxicological or risk assessment factors (e.g., U.S. Environmental Protection Agency Drinking Water Standards). Social values or environmental policies, for example, may influence risk management decisions that are reflected in regulatory concentrations.

"Regulatory standards" are legally enforceable regulatory concentrations. These numbers define maximal permissible levels of single chemicals or mixtures in a given medium. Government agencies also generate guidance concentrations. Unlike standards, guidance concentrations are not legally enforceable, but are often used as if they have legal force. There are innumerable names given by government agencies for guidance concentrations (e.g., action levels, action limits, etc.).

Precisely which regulatory concentrations apply in a particular situation depends on the experience of a regulator, applicable laws, and nature of a risk assessment project. In Superfund, for example, regulatory concentrations that are considered for a site cleanup are termed "Applicable or Relevant and Appropriate Requirements" (ARARs). Three types of ARARs are recognized: chemical-specific, location-specific, and action-specific. ARARs can be selected from among many possibly applicable state and federal standards and guidance concentrations (see Table 1).

D. Defining Acceptable Sampling and Analytical Plan

Sampling and analytical plans should be prepared before new data are collected. These plans address all relevant human exposure routes and points (see Table 2), exposure pathways, transport media mechanisms and chemical-specific factors (see Table 3), media of concern, areas of concern, contaminant types, routes of contaminant transport, environmental media characteristics, analytical chemistry requirements, and organisms of concern.

Goals of a project govern details of sampling plans. Sampling locations, for example, can be chosen with a purpose (such as to identify all contaminants), or they may be random (for unbiased sampling) or systematic. Project goals also influence choice of sample types (grab samples or composite samples*), use of field screening analytical methods, and time and resources allocated to sampling.

---

* Composite samples combine subsamples from different locations or times.

Table 1 Examples of Common Regulatory Standards and Guidelines

Standard / Guideline	Purpose
U.S. EPA Drinking Water Health Advisory Concentrations	Maximally recommended concentrations of individual drinking water contaminants for 1-day, 10-day, longer-term (~7 years) and lifetime exposures
U.S. EPA Maximum Contaminant Level (MCL)	Maximum permissible level of a contaminant in water that is delivered to public water systems
U.S. EPA Water Quality Criteria	Recommended maximum concentrations in surface water of a pollutant consistent with protection of aquatic organisms, human health, recreational activities, and other specified uses
OSHA Permissible Exposure Limits (PELs)	Establish safe concentrations of air contaminants in work places.
National Institute for Occupational Safety and Health Recommended Exposure Limits (RELs)	Exposure to potentially hazardous airborne substances in work places
National Ambient Air Quality Standards (NAAQS)	Protect public health or welfare. Not directly enforceable
National Emission Standards for Hazardous Air Pollutants (NESHAPs)	Chemicals not covered by NAAQS
Food and Drug Administration Action Levels	Maximum allowable levels of poisonous and deleterious substances in food
U.S. EPA Tolerance Levels	Control levels of pesticide residues in raw or processed agricultural products and processed food
RCRA Appendix VIII and IX, Superfund Target Substances	Enforceable point source discharge limits
Clean Water Act Priority Pollutants	Enforceable point source discharge limits
State Groundwater Standards	May be enforceable concentrations
State Surface Water Standards	May be enforceable concentrations
State Air Standards	May be enforceable concentrations
State Medium-Specific Cleanup Standards and Guidance Concentration	May be enforceable concentrations
State Drinking Water Standards	May be enforceable concentrations
State Fish Flesh Contaminant Advisories	Designed to minimize risk from eating fish but allow sport fishing to occur

Sampling plans also address physical factors, such as meteorology of a project area, and physical/chemical characteristics of environmental media to be sampled. Some environmental sample matrices are difficult to sample and require specialized collection. Others are easy to sample, but yield samples that are difficult to analyze in the laboratory and require special analytical chemistry procedures. Sampling plans are applied through sampling protocols which define objectives of a sampling study

Table 2 Examples of Exposure Routes and Points by Environmental Medium

Environmental Medium	Exposure Points	Exposure Routes
Groundwater	Municipal and private water wells, swimming pools, discharge zones to surface water, irrigation, springs, sinkholes	If used as a drinking water source: direct ingestion, dermal and ocular contact, inhalation of chemicals volatilized from water
Surface Water	Locations where water bodies used for recreational purposes	Direct ingestion, dermal and ocular contact, inhalation of chemicals volatilized from water
Soil	Hazardous waste sites, residential soil surfaces, excavations, dust	Direct ingestion, dermal and ocular contact, inhalation of volatilized chemicals and dust
Air	Indoor or outdoor exposure to dusts, aerosols, gases, and particulates in respirable air	Inhalation of volatilized chemicals, dermal contact with aerosolized chemical droplets
Food	Chemical contaminants on food as a residue or in food via food chain uptake and distribution	Ingestion of food products containing chemical contaminants in their tissues or on their surfaces, dermal contact with contaminated food products

and, in combination with QA/QC methods, govern each step in sample collection, preservation, transportation, and analysis.

E. Defining Quality Assurance/Quality Control (QA/QC) Methods

QA/QC methods ensure data quality through proper sampling, handling, storage, and preservation. Sampling protocols define objectives of a sampling study and articulate procedures for sample collection, preservation, handling and transport, and analysis. Data collected under sampling and analysis plans should be reviewed as they become available to ensure that data meet project needs. This helps eliminate data gaps and limits problems to be addressed in the data evaluation phase.

F. Defining Methods for Pooling Sampling Data

Available data are evaluated to determine whether they can be combined for use in an HHRA. It is important to define quality of available data sets. Analytical chemists review available data, determine its reliability, and can apply a letter data qualifier to each reported data point. Each "data indicator" indicates a chemist's degree of certainty about a chemical's reported identity and concentration. Data qualifiers can also note data problems. Risk assessors rely on data qualifiers to judge whether a data point can be used in a quantitative risk assessment and, if so, how much reliance on data is appropriate. Rigor, reliability, and credibility of numerical risk assessment findings relate directly to quality of data sets used in a risk assessment.

Table 3 Examples of Transport Media, Transport Mechanisms, and Chemical Specific Factors that Could Affect Environment Transport of Chemical Contaminants

Environmental Medium	Transport Mechanisms	Chemical-Specific Factors Affecting Transport
Groundwater	Groundwater movement	Density, water solubility, organic carbon partition coefficient (K_{oc})
	Volatilization	Water solubility, vapor pressure, Henry's Law Constant
	Adsorption to soil particles	Water solubility, octanol/water partition coefficient (K_{ow}), K_{oc}
	Precipitation out of solution	Water solubility K_{ow}, K_{oc}
	Biological uptake	K_{ow}, bioconcentration factor
Surface Water	Overland flow	Water solubility, K_{oc}
	Volatilization	Water solubility, vapor pressure, Henry's Law Constant
	Move to groundwater	Density
	Adsorption to soil particles	Water solubility, K_{ow}, K_{oc}
	Sedimentation of particles	Density, water solubility
	Biological uptake	K_{ow}, bioconcentration factor
Soil	Runoff by soil erosion	Water solubility, K_{oc}
	Leaching	Water solubility, K_{oc}
	Volatilization	Vapor pressure, Henry's Law Constant
	Suspension	Density, particle size
	Biological uptake	Bioconcentration factor
Air	Aerosolization	Water solubility
	Atmospheric deposition	Particle size
	Volatilization	Henry's Law Constant
Biota	Bioaccumulation	Bioconcentration factor

Adapted from ATSDR, 1990.

G. Defining Data Sources

Chemical identity, concentration, or emission rates can be obtained from various sources. Actual data can be collected and pooled for an existing site, activity, or facility. When this is not possible, however, surrogate data sets must be obtained from models or existing sources of environmental releases. For example, surrogate data may be used when an HHRA involves risks associated with a facility that has not yet been built; surrogate data sets will probably be comprised of data gathered at existing facilities that are identical or similar to a proposed facility. Chemical identities and release information can be derived from Material Safety Data Sheets, published literature, monitoring data, or mathematical models, using projections for proposed facility operations. As a source of chemical identity and release information becomes less specific to a site, activity, or facility of concern, uncertainties increase in an HHRA.

When a risk assessor has collected sufficient data of acceptable quality, a list of all COPCs is developed. A concentration*, or emission term, is statistically generated for each chemical at its source using location-specific data or surrogate data sets.

In the past, qualitative or quantitative methods have been used to reduce an exhaustive list of COPCs to a shorter list of COCs. RAGs 1989, pages 5-23 to 5-24, provides a detailed discussion of this topic. One way to generate a COC list is to use a chemical concentration-toxicity screen. EPA provides the following equation for calculating Individual Chemical Scores:

$$Rij = (Cij)(Tij)$$

where Rij = Risk factor for chemical i in medium j, Cij = Concentration of chemical i in medium j and Tij = Toxicity value for chemical i in medium j (i.e., either a slope factor or 1/RfD).

Risk factors are generated for individual COPCs by multiplying a chemical's concentration in a particular medium by its toxicity value (noncarcinogenic or carcinogenic). Risk factors are summed for all COPCs to generate a total score for each medium. A percentage of total risk attributable to each chemical is then determined by dividing each chemical-specific risk factor by a total score for each medium evaluated.

Chemicals posing an insignificant percentage of a total risk may, in some cases, be eliminated from further consideration. Those representing a significant percentage undergo full analysis. Chemicals representing the lowest 1% of a risk might be eliminated from a list of chemicals of concern, for example, while those representing 99% of risk undergo complete risk analysis. Chemicals included in a COC list represent a majority of risks from a site, activity, or facility and they have readily available emission, concentrations, and numerical toxicity values. COPCs screened out of quantitative analysis, because of inadequate data, no numerical toxicity value, or because they seem to pose insignificant risk, are not included in a final COC list. These chemicals still deserve qualitative analysis and should be discussed in an uncertainty analysis section of a risk characterization.

In other cases, all identified chemicals with toxicity values are addressed throughout an entire report. No chemicals are eliminated from evaluation.

---

* Concentration terms can be generated using an arithmetic average concentration for a contaminant, based on a set of sampling results, and the 95% upper confidence limit (UCL) of an arithmetic mean. This approach compensates for uncertainties associated with ascertaining a true average concentration at a sampling area. Averages are used because carcinogenic and noncarcinogenic toxicity criteria are based on lifetime average exposures. An average concentration is considered most representative of a concentration that would be expected at a location over a lifetime. When chemicals are expected to be present, but are not detected, they may be assigned a numerical value other than zero, such as a percentage of a detection limit. However, defining a concentration term is often a function of which methods are preferred by those producing or reviewing a report.

III. HAZARD ASSESSMENT CONSERVATISM

Chemical screening to reduce risk assessment production time and costs is no longer considered a routine practice and is disfavored by many regulatory agencies. Risk assessors can rapidly generate credible risk estimates as a result of significant productivity improvements in risk assessment methods, techniques, and tools during the past decade. Risk assessors, who used pencils and hand calculators in years past, now use powerful computers able to run sophisticated risk assessment and fate and transport modeling programs. They are also able to obtain environmental and toxicological data from on-line databases. Although technical means to generate risk estimates have improved, many cost- and labor-saving methods adopted in early days of risk assessment still linger. Concentration-toxicity screening, described above, is one such holdover.

Risk assessment software, commercial spreadsheets, and toxicological values readily available from U.S. EPA's internet or hard copy accessible Integrated Risk Information System (IRIS) and Health Effects Summary Table (HEAST) databases (for most common contaminants) negate a need to limit quantitative analysis to an abbreviated list of COCs. Risk assessors no longer must perform laborious calculations by hand. Instead, they use computers to perform calculations required to generate risk estimates. Thus, there is little justification to eliminate chemicals, unless a COPC lacks a concentration/release term or a toxicity value, or it is shown not to be relevant to a specific risk assessment. If data exists for all COPCs, a complete quantitative evaluation is possible. In cases where a COPC with known human health effects lacks an approved toxicity value, a risk assessor can either generate a toxicity value or evaluate a chemical qualitatively in uncertainty analysis of a risk characterization section.

A. Problems Associated with Developing a COPC and COC List

Certain problems commonly occur during preparation of a hazard assessment section of a risk assessment report. If these problems are not addressed, a result could be a COPC or COC list that can mischaracterize environmental releases and, consequently, underestimate exposures and risks. Common problems include:

- Failure to adequately describe chemical processes occurring at a facility. When inadequate analysis of an activity, facility, or site occurs, chemical identification can suffer (e.g., large numbers of chemicals known or expected to be released from a facility are missed and not included on a COPC or COC list). Adequate description of all chemical processes helps to formulate a comprehensive list of COPCs and COCs.
- Failure to adequately review available literature. All too often an incomplete review of site records, industry literature, government literature, or peer-reviewed literature results in a hazard assessment that fails to list all chemicals known or expected to be produced at a given type of facility. A robust COPC and COC list can only be produced when a comprehensive review of relevant literature is done.
- Failure to use engineers and chemists. Chemists and engineers working at a site, facility, or activity have special knowledge about the chemicals that go into and

out of their work location. For example, at facilities involving high-temperature processes or combustion, combustion chemists and engineers can help predict identities and estimate amounts of chemicals that may be released. Such specialists provide a valuable means for identifying chemicals that might be released directly from facility activities or that may materialize as a result of physical or chemical reactions in a waste stream (e.g., gas condensation from smoke stacks).

- Failure to review analytical chemistry methods to ensure that releases have been adequately evaluated. If erroneous methods are used (e.g., sampling, extraction, digestion, and analytical methods) or selected analytical techniques are unable to detect chemicals at levels of health concern, chemicals moving off-site could go undetected or underreported. Standard methods exist that should be followed to ensure generation of reliable data.
- Failure to evaluate all relevant operating units on a site. Some sites contain many different operating units with different chemical processes and environmental releases. If each unit is not fully evaluated, many chemicals being released to the environment could be missed in a risk assessment. All operating units should be evaluated for chemical releases by trained and experienced personnel.
- Failure to obtain certifications of work from hazard assessment preparation contractors or permittees. One common way to ensure that quality work has been performed by a contractor or permittee is to have them sign a certification statement that all work was conducted and performed to standards of relevant disciplines. Lacking such signed statements, hazard assessment reviewers may not fully understand who prepared documents and how they were prepared, bringing their credibility into question.
- Failure to adequately evaluate literature used in development of a COPC or COC list. When data on a particular site, facility, or activity are limited, a risk assessor may be forced to rely on literature of limited quality and reliability. For example, some literature does not list chemicals if they are less than a certain percentage of total mass, regardless of their presence or their toxicity. As a result, highly toxic chemicals in very small amounts may not be included in a given type of literature, whereas low toxicity, high concentration materials may be listed.
- Failure to establish environmental release criteria that are relevant to establishment of a COPC and COC list. Inclusion of chemicals in a risk assessment is sometimes linked to estimated emission rates or concentrations, on-site or off-site. Specifically, chemicals are not included in a COPC or COC list if their concentrations do not exceed some set value. If a calculation of this value is not strictly defined and related to health effects (e.g., average versus peak air concentrations), chemicals could be excluded from a COPC and COC lists for wrong reasons.
- Failure to establish performance standards for development of a COPC and COC list. Without performance standards, COPC and COC lists of various levels of quality and reliability are generated.
- Failure of toxicologists and risk assessors to design and implement rigorous chemical selection processes. In some organizations, toxicologists and risk assessors are not responsible for designing how COPC and COC lists will be generated. Results of this management decision can drastically alter risk findings.
- Failure to review hazard assessment documents provided by regulated parties for technical accuracy. Many times hazard assessments are provided to government by parties with vested interests in an outcome of a risk assessment. These hazard assessments must be rigorously reviewed before they are accepted to ensure risk assessment integrity.

- Failure to combine site-specific and generic information sources to generate a COPC and COC list. By conducting a comprehensive review of literature and conducting interviews with relevant experts, a robust COPC and COC list can be produced. Without such an effort, a COPC and COC list may be of little value in development of a credible risk assessment.
- Failure to gather extensive lists of toxicity values from state, national, and international sources. Often, chemicals are not quantitatively evaluated in a risk assessment because there is no numerical carcinogen or noncarcinogen toxicity value listed for them among a limited number of sources. Obtaining a comprehensive library of toxicity value sources ensures that all relevant chemicals with appropriate toxicity values can be evaluated quantitatively in a risk assessment.
- Failure to evaluate secondary effects. While there is no standard method to quantitatively evaluate secondary toxic effects of a chemical (i.e., primary or critical toxic effects are used to establish numerical toxicity values), cumulative secondary effects of several chemicals may pose significant, if unrecognized, health risks when their release rates and exposure levels are combined. Unfortunately, the authors are aware of no practical solution to this problem at this time.
- Failure to establish COPC and COC list criteria for use in multipathway risk assessment. In an effort to reduce risk assessment complexity, costs or eliminate generation of unacceptable risk findings, some organizations use "exclusionary" risk assessment tools. Rather than develop a robust list of COPCs and COCs, based on actual case conditions, managers mandate use of methods and techniques that reduce risk assessment scope and limit COPC and COC lists to consider only a single approach (e.g., inhalation exposure only). As a result, chemicals that might pose risks via ingestion or dermal exposure may not be evaluated at all, unless they happen to pose an inhalation risk as well. Many times exclusionary risk assessments rely on emission, concentration, or toxicity tables linked to acceptable risk levels established by a regulatory agency or other government office. Non-risk assessors compare these emission or concentration values from these tables to values provided by permittees or engineering staff. Not fully aware of complexities of risk assessment, untrained staff cannot evaluate toxic chemical interactions, environmental chemistry, or validity of values they are provided (e.g., values in such tables may be out-of-date or based on calculation methods or regulatory values for one medium that cannot legitimately be used for another medium). Thus, rejecting, by fiat, use of hazard assessment techniques to produce COPC and COC lists for a multipathway risk assessment can routinely underestimate total incremental risks from an activity, facility, or site, placing receptors at unknown risk.

IV. EXPOSURE ASSESSMENT

Exposure assessment, the second step in HHRA, follows hazard assessment and may be performed concurrently with a toxicity assessment. Exposure assessment produces numerical exposure levels.

Exposure occurs when a chemical of concern contacts an outer boundary of a receptor organism, either at a chemical's source or some distance from a source. Exposure assessment evaluates movement of a chemical from its source to a potential human receptor by identifying potential exposure pathways. In moving from its source to a receptor organism, a chemical concentration generally decreases by processes of dilution, dispersion, and degradation and, as a result, a receptor typically receives less than a concentration of a chemical in an environmental medium. Degradation may increase risks, however, if breakdown product toxicity is greater. Exposure assessment quantitatively evaluates this process. This step in HHRA typically relies on data found in technical literature or generated by using models.

First, exposure setting is characterized. This requires an examination of physical setting of a site, activity, or facility: its climate, meteorology, geological setting, vegetation, soil types, groundwater hydrology, and surface water features. Potentially exposed populations are identified, including populations of special concern such as children, elderly people, pregnant women, people with chronic illnesses, and other potentially sensitive subpopulations. Current and future land uses are characterized, in part to locate and identify potentially exposed populations and to project characteristics and location of populations that may move into an area at some future time.

Next, exposure pathways are identified. Exposure pathways describe movement of a COC from its source to human receptors. As much as possible, every step is identified in potential exposure pathways. These include:

- Sources of chemical contaminants: such as a waste pile, smokestack, automobile, and leaking drum
- Mechanism of environmental release: such as volatilization, fugitive dust generation, surface runoff, overland flow, leaching, and groundwater seepage
- Environmental medium to hold or transport chemicals: such as air, surface water, soil, groundwater, sediment, and biota
- Human exposure point: such as on- or off-site, backyard, and shower
- Exposure routes: ingestion, inhalation, or dermal exposure — "direct exposure" or "indirect exposure." (Direct exposure might occur by ingestion of contaminated water, whereas, indirect exposure might occur through consumption of contaminated fish)

After identifying potential exposure pathways, a risk assessor evaluates likelihood that a pathway will be completed. Usually, only those exposure pathways likely to be completed undergo further analysis; others are eliminated from consideration. In special circumstances, risk assessment may go farther and address potential future pathways.

A. Fate and Transport Analysis*

Environmental fate and transport models** simulate environmental behavior of a chemical when monitoring is not possible or practical. A concentration of a COC at its source, termed chemical source concentration, is a starting point. A modeler uses a series of equations to project change in concentration for each COC as it moves from its source along likely exposure pathways. This analysis yields a plausible estimate of each COC concentration, termed an exposure level, likely to reach a location where human exposure is expected, termed a receptor point (see Figure 1).

---

* Risk assessment treatises vary in their treatment of chemical fate and transport. It may be discussed either in hazard evaluation or exposure assessment. We deal with it as part of exposure assessment.
** "Model" signifies both mathematical equations and computer models, unless otherwise noted.

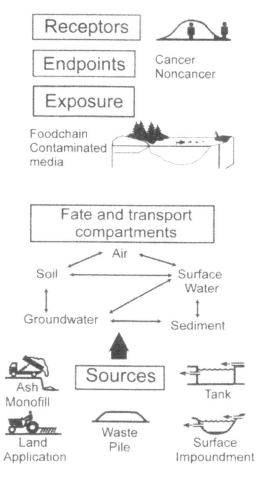

Figure 1 Human health risk assessment multipathway analysis. (Adapted from U.S. EPA, 1995, Development of Human Health Based and Ecologically Based Exit Criteria for the Hazardous Waste Identification Project, Figure 1-1, pages 1–6.)

1. Chemical Movement Depends on Physical and Chemical Properties

Chemicals released move within and between environmental compartments (such as water to air and back, water to soil/sediment and back, and soil to air and back) and from the physical environment into living organisms and back into the environment.

Chemicals can exist in three physical states, as solids, liquids, and gases. Chemicals can shift physical state by undergoing a "phase change." For example, water is solid at 32°F; it is liquid between 32°F–212°F, and at 212°F it starts to boil and enters a gaseous phase. Some chemicals, such as carbon dioxide, can move directly from solid (dry ice) to gas phase without going through a liquid phase. This is called "sublimation."

Chemical movement in the environment is also related to a chemical's affinity to a media in which it is found. For example, chemicals that bind strongly to a medium tend to stay in that medium (such as dioxin in soils). Chemicals weakly bound to a medium tend to move out of that medium into other media (such as volatile chemicals moving from soil particles or water to air). Chemicals that are released to air can disperse in air or they can enter other environmental media where they can concentrate.

Chemicals in the environment can be altered through "abiotic" (no organisms involved) or "biotic" (living organisms involved) processes. These processes include chemical hydrolysis; oxidation, reduction, and conjugation; photolysis or photooxidation; and biological degradation reactions. These general principles apply to movement of environmental contaminants.

A study of distribution of chemicals in the environment based on their chemical properties is called "chemodynamics." Knowledge about environmental fate chemistry of a contaminant is important, since environmental fate can change as chemical structure is altered. Thus, a chemical of moderate potential to bioaccumulate/biomagnify can be altered by biotic or abiotic processes into a chemical with very high potential to bioaccumulate/biomagnify. Toxicity can also change through even seemingly minor alterations in chemical structure. Environmental contaminants have numerous chemical and physical properties that dictate their environmental fate and how they are transported in the environment (see Table 4).

Knowledge of how a chemical moves in the environment is acquired through "fate and transport" analysis. Physical and chemical data for environmental contaminants directly affects their fate and transport in the environment and such data are used in fate and transport models. Models are a mathematical abstraction of a physical system used to predict concentration of specific chemicals, as a function of space and time subject to transport, inter-media transfer, storage, and degradation in the environment. Computer simulations, such as a Fugacity Model, are used to predict how a chemical will move in the environment, to which compartment or medium it will move, and what percent of released chemicals will enter and be found in each environmental compartment or medium.

2. Steps in Fate and Transport Analysis

At each step in the analysis, a fate and transport model must account for environmental factors capable of influencing COC movement. Environmental interactions may transform a COC physically, chemically or biologically, affecting how and where it travels. If a COC changes physical state, it will exhibit different characteristics. As a result, it may move through an entirely different series of environmental compartments. Transformations due to chemical reactions or biological interactions can convert COCs into new substances with distinct physical, chemical, and toxicological properties.

Chemical transformations may also occur as a COC interacts with the environment. For example, as a chemical is discharged to air from a stack, do chemical reactions occur? If so, what new substances are created? What are their chemical properties? How much of a COC transforms by chemical reaction? Does any remain?

HUMAN HEALTH RISK ASSESSMENT

Table 4   Examples of Physical Properties Affecting Chemical Environmental Fate and Transport.

Boiling point	Definition: Temperature in degrees Celsius at which vapor pressure of a constituent in aqueous form is equal to atmospheric pressure.
	Effect: Some chemicals have boiling points far below ambient temperatures. Boiling points provide information on how a chemical will behave in the environment at a given temperature. Inhalation exposure is most common route of exposure for low-boiling liquid, in contrast to high-boiling liquids which enter a body via direct contact.
Chemical structure	Definition: Chemical formula drawn to show relative arrangement of molecules.
	Effect: Chemical structures provide important clues to toxicity and environmental fate characteristics of a chemical.
Cosolvency	Definition: Ability of one chemical to enhance solubility of another in water.
	Effect: Change fate and transport of chemicals in soils, sediment, and ground water.
Degradation rates	Definition: Expressed in terms of half-lives, time required for a chemical, under defined conditions, to reach half of its initial concentration.
Density	Definition: Weight of a substance divided by its volume.
	Effect: Density measurements provide clues to a chemical's environmental behavior. Very dense liquids (DNAPLs or Dense Nonaqueous Phase Liquids) move to deepest confining layer of an aquifer. Materials of lesser density dissolve in water (LNAPLs) or form layers on top of an aquifer (Light Nonaqueous Phase Liquids).
Empirical formula	Definition: States number of each type of atom in a molecule.
Henry's Law Constant	Definition: Ratio of equilibrium concentration (in atmospheres) of a constituent in air relative to its concentration (in moles/cubic meter) in water at referenced temperature.
	Effect: Often termed "air-water partition coefficient," it describes relative volatility of chemicals. Henry's Law Constant less than 10^{-7} atm-m^3/mol indicates a chemical of low volatility, greater than 10^{-7} atm-m^3/mol, but less than 10^{-5} atm-m^3/mol, indicates slow volatilization into air, values greater than 10^{-5} atm-m^3/mol but less than 10^{-3} atm-m^3/mol indicate volatilization is an important mechanism of loss to air. Values exceeding 10^{-3} atm-m^3/mol indicate rapid volatilization.
Log K_{oc}	Definition: Ratio of absorbed chemical in soil/sediment to an aqueous solution concentration.
	Effect: Also called "soil/sediment partition coefficient," it provides information on relative attraction of a chemical for soil/sediment in comparison to water. Chemicals with high values typically have low water solubilities while chemicals with low values have high water solubilities.
Log K_{ow}	Definition: Log of ratio of equilibrium concentration of constituent in octanol relative to its concentration in water.
	Effect: This metric is also known as "n-octanol/water partition coefficient." Chemicals with higher Log K_{ow} values tend to partition into fatty tissue, compared to those with lower values and also have a higher tendency to bioaccumulate/biomagnify than those with lower values. This is a key parameter to predict environmental fate of organic chemicals.

Table 4 continued

Melting point	Definition: Temperature in degrees celsius at which a chemical in solid phase is in equilibrium with liquid phase at atmospheric pressure.
	Effect: Melting point data provides information on physical state of a chemical under local conditions.
Molecular weight	Definition: Molecular or formula weight of constituent in grams/mole.
Partition coefficient	Definition: Ratio of any two chemical species in two phases (e.g., water and oil) that are in equilibrium with each other.
	Effect: Partitioning occurs between two immiscible solvents. For example, in case of water/n-octanol partition coefficients, chemicals that partition more into water phase than oil phase (n-octanol), are not expected to bioaccumulate or biomagnify, whereas those that partition more into oil can be expected to readily bioaccumulate or biomagnify.
Solubility	Definition: Concentration of chemical (in mg/L) that is required to form a saturated solution in water at referenced temperature. Solubility is tendency of a chemical to move from solid form into solution.
	Effect: Solubility relates to chemical and physical properties of solute (chemical contaminant) and solvent (water, benzene).
Specific gravity	Definition: Ratio of density of a chemical to density of water. An alternate method of expressing weight density. It is density of a liquid or solid divided by density of water. Water has a specific gravity of one. Substances with specific gravities greater than one will sink (such as steel, at 7.8), those less than one will float (such as oil at 0.8). Knowledge of a chemical's specific gravity provides information on how a chemical will behave in presence of water.
Vapor pressure	Definition: Pressure (in mm Hg) of vapor phase of a chemical that is in equilibrium with its liquid or solid phase at referenced temperature.
	Effect: Vapor pressure describes tendency of a chemical to escape from a solid or liquid matrix. A variable input, used to calculate Henry's Law Constant.
Water solubility	Definition: Solids, liquids, and gases can be dissolved, to a degree, in water. Degree to which a chemical can be dissolved in water is its water solubility.
	Effect: A chemical's solubility can range from miscible (soluble at all proportions) to partially soluble to immiscible (insoluble). Water solubility data provides information on how a chemical will behave in the environment at a given temperature.

Adapted from U.S. EPA, 1992, *Risk and Decision Making, A Workshop in Risk Assessment, Risk Management, and Risk Communication*, Office of the Senior Science Advisor, Region IX.

Does a COC transform biologically, as well? If it settles back to earth, does it stay on the surface where children or animals may ingest it? Does it wash into nearby lakes and streams where it enters tissues of aquatic animals or plants? If so, what are its biological effects? How much is taken into each level of the food chain? Do organisms metabolize it, further altering its physical, chemical, and biological properties?

Environmental fate and transport modeling generates estimates of COC concentrations at receptor points. This concentration is extrapolated from an initial COC concentration using a series of estimates of amounts of chemical that enter each environmental compartment (soil, water, air, and food chain) and amounts lost and gained along exposure pathways due to physical, chemical, and biological transformation.

Fate and transport models are essentially a complex inventory of chemical gains and losses in each environmental compartment. Gains in a compartment result from new sources of a chemical. Losses in a compartment reflect chemical transformation, decay, and transport to other compartments. Thus, exposure point concentration in each compartment is extrapolated from an initial concentration, using a series of estimates of chemical allocation, transformation, and movement.

Concentration, or emission terms, calculated for use in COC screening processes are used in environmental fate and transport models to calculate an exposure concentration for each chemical, in each media of concern, at a receptor point, which is at some given distance and location from a chemical's source. Chemical concentrations in a medium at a receptor point are termed exposure point concentrations. They can be estimated through use of exposure point monitoring data (e.g., monitoring stations, sampling sites, samples); mathematical model concentration outputs (e.g., output of a fate and transport model); surrogate data (e.g., data from literature or gathered at an analogous site); or default values (e.g., values assigned by government policy or practice).

3. Limitations of Fate and Transport Models

All fate and transport models have limitations. They are limited by uncertainties related to input data, model assumptions (such as model complexity, simplifying assumptions, and model sensitivity to changes in input variables), and model validation for an exposure scenario under consideration.

Fate and transport modeling results are no better than their inputs and default assumptions. Specifically, quality of model results relates to how well they mimic actual environmental conditions and processes. Model results are confirmed or validated when model predictions match environmental measurements. (There is a shift from model "validation" to "confirmation" in recognition of the idea that no model will be truly "valid" for all locations and times.)

Minimum data-quality standards are delineated to meet DQOs, which in turn reflect risk assessment purpose. Minimum data-quality standards are used to establish sampling approaches and analytical options, and to design a data collection program, and methods capable of producing data of acceptable quality. Selection and appropriate use of models, and receptor points, requires input of trained risk assessment professionals. It is essential that data of highest quality be used as numerical inputs into these models and that models are only used if they have been accepted by the scientific community as valid representations of reality. Furthermore, data quality should match model complexity. Many risk assessors believe that complex models using crude data produce unreliable results.

Assumptions can influence model reliability in several ways. First, as model complexity increases, aggregate uncertainty increases. Due to this phenomenon, simple models may be more reliable than complex models. Second, models must be confirmed to assess their reliability in a given setting. Often risk assessors assume a model is reliable, simply because it is in common use. Unfortunately, even models well-known to risk assessors may not have been checked for reliability, in general, much less in a specific setting. Finally, no model integrates every factor at play in actual field conditions. There will always be some simplifying assumptions which reduce any model to a mere estimate of reality.

Thus, fate and transport models provide only best estimates of chemical behavior in the environment. Several general types of fate and transport models are used in risk assessment reports.

4. Examples of Fate and Transport Models

a. Groundwater Modeling

Groundwater models can generate estimates of future exposure concentrations for application in future use scenarios based on current groundwater monitoring data or expected groundwater contaminant concentrations. However, groundwater models are complex. Considerable professional judgment is required to select an appropriate model for a particular application and to interpret sampling data and model outputs. Special considerations in use of groundwater monitoring data and models include sample pH; use of filtered versus unfiltered samples; well location, depth, and construction materials; sampling devices and protocols; transport and handling of samples; analytical methods; and laboratory QA/QC procedures and results.

b. Soil Contaminant Modeling

Soil monitoring data can be used as a direct estimate of current exposure concentrations. However, such data may not be suitable for future exposure scenarios because concentrations change over time due to leaching, volatilization, photolysis, biodegradation, wind erosion, and surface runoff. Modeling can be used to predict future exposure concentrations.

c. Surface Water Exposure Modeling

Surface water exposure models treat each step in movement of a chemical in the environment as a compartment. Mathematical equations quantitate movement of a chemical from one compartment to other compartments. Four types of models are used to quantitatively define contaminant source and water-quality relationships. These are direct spatial definitions, simple empirical models, simple deterministic models, and complex deterministic models.

d. Sediment Contaminant Modeling

Monitoring data are an excellent source of information for estimating exposure concentrations for sediment contaminant modeling. Models are available for calculating exposure concentrations.

e. Food Exposure Modeling

Contaminant physical and chemical data can be used to estimate chemical concentrations in human food sources, such as fish and shellfish, plants, and terrestrial animals.

f. Contaminant Air Modeling

There are three standard methods of estimating air exposure concentrations. They are: ambient air monitoring; contaminant source emission measurements and dispersion modeling; and emission and dispersion modeling. Atmospheric transport of contaminants as particulates, gas, vapor, and aerosols can occur from both active and inactive industrial facilities. Direct measurement of ambient atmospheric contaminant levels (e.g., sampling devices with appropriate trapping media) is the preferred sampling method. When this is not possible, however, it is appropriate to model air contaminant concentrations. Dispersion models use known, or modeled, air emissions to calculate contaminant concentrations at a potential receptor point. Emission models predict contaminant release rate to air from a source. Models calculate downwind concentrations of emitted chemicals as a function of several factors including emission rates, distance of receptor from contaminant source, atmospheric dispersivity, stack height and diameter, and terrain features. Gaussian plume models are used to estimate airborne concentrations of a chemical near its source. Long-range atmospheric transport models calculate chemical concentrations over geographical regions. Gaussian plume dispersion models require calculation of an emission rate (mass of substance/unit of time). In contrast to these two types of continuous release models, puff models are used to calculate emission concentrations after episodic or short-duration releases.

B. Exposure Equations

Exposure point concentrations, concentration of a substance in a medium at a receptor point, are used in exposure assessment intake and uptake equations to calculate human exposure to a potentially toxic chemical. Exposure point concentrations can be derived from actual biological monitoring data (e.g., fish fillets); from biomonitoring that involves collecting and analyzing human samples (e.g., breath, blood, fat, nails, hair, and urine) to determine total internal doses; from biomarkers that use biochemical and cytochemical markers to measure an organism's biological and physiological responses to a stressor, such as toxic chemicals; from ambient monitoring involving sampling a site's environmental media (i.e., soils, water, or air); or from modeling results, surrogate data, and default values.

Exposure is quantified through uptake and intake equations. Standard intake/uptake equations and supplies suggestions for variable values that risk assessors can use with these equations are provided by the U.S. EPA. Site-specific factors, and a measure of professional judgement, influence which variable value is selected from a range of values that could apply. All exposure equations are variations on the following theme:

$$I = C \times \frac{CR \times EFD}{BW} \times \frac{1}{AT} \times AF$$

Where I = intake (or uptake) of a chemical, C = chemical concentration (average concentration contacted over an exposure period — outputs of actual monitoring data or fate and transport models), CR = contact rate (amount of contaminated medium contacted per unit of time), EFD = exposure frequency and duration (how long and how often exposure occurs, often calculated as EF [exposure frequency] and ED [exposure duration]), BW = body weight (average body weight over an exposure period), AT = averaging time (period over which exposure is averaged), AF = absorption factor ≤ 1.

This generic exposure equation uses chemical related variables (i.e., exposure concentration), exposed population variables (i.e., chemical contact rate, exposure frequency and duration, and body weight), and assessment determined variables (i.e., time over which exposure is averaged). Other commonly used exposure equations include ingestion of chemicals in drinking water or surface water while swimming, dermal contact with chemicals in water, ingestion of chemicals in soil, dermal contact with chemicals in soil, inhalation of air (vapor phase) chemicals, and ingestion of contaminated food. These are all variations on the general equation, discussed above.

C. Chemical Intake and Uptake

"Intake" occurs when chemicals cross an external boundary (through the mouth by eating, drinking, or breathing; through the nose by breathing; and through the skin by direct contact), but have not passed an absorption barrier (e.g., gut wall or lung tissue) to enter the bloodstream for distribution to organs and tissues. "Uptake" involves absorption of a chemical through skin or other exposed tissue (such as the eye). Uptake also occurs within a body when a chemical passes through an absorption barrier (e.g., gut wall or lung tissue) to enter the bloodstream for distribution to organs and tissues. Thus, uptake can occur following intake.

During exposure assessment, a risk assessor calculates an amount of a COC available at a point of exposure. An amount of a COC in the environment at a receptor point probably far exceeds the amount of chemical actually available to cause toxicity within an exposed organism because chemical and physical factors tend to allow only a fraction of a COC present in the environment to enter the body. First, the amount taken into a body (intake) is only a fraction of the amount present in the environment. Then, only a fraction of that taken in is absorbed (referred to as an "internal dose"). Internal dose equals intake multiplied by an absorption factor. A body may deal with an internal dose in a variety of ways, each with potential to alter a chemical's toxic effect. A chemical may be metabolized, stored, excreted, or transported to other parts of a body. Thus, a fraction of internal dose, termed "delivered dose," is transported to a particular organ, tissue, or fluid. Finally, a fraction of delivered dose, termed "biologically effective dose," reaches a cell, membrane, or other site where adverse effects actually occur. Thus, at each step — from the environment through a body, to a target site — a COC's concentration is attenuated. It is important to acknowledge that two factors may still provide for a toxic effect, despite attenuation. First, although a biologically effective dose may seem attenuated in comparison to a source concentration, a biologically effective dose may represent a sufficient dose to have a significant health effect. Second, metabolic processes may either decrease or increase chemical toxicity.

Unfortunately, the complex process by which chemicals enter and move through a human body is poorly quantified. Lacking suitable data and models to describe the process for most COCs, risk assessors generally do not quantify exposure beyond calculation of intakes or absorbed doses. Exposure assessment rarely considers internal dose, delivered dose, and biologically effective dose.

V. EXPOSURE ASSESSMENT CONSERVATISM

HHRAs differ in their level of risk conservatism, depending on exposure cases employed to calculate exposures and risks. Exposure scenario and exposure case are conceptually different, but often confused.

"Exposure scenarios" are site-specific representations of real or hypothetical situations that define a source, individual(s), pathway(s) of exposure, and variables that affect exposure pathways. Exposure scenarios are a collection of facts, assumptions, and inferences about how exposure occurs. An exposure scenario aids a risk assessor in evaluating, estimating, or quantifying exposures. Innumerable exposure scenarios have been created by risk assessors to match statutory requirements, rules, or specific cases. The type of exposure scenario selected for a risk analysis affects conservatism of the analysis. For example, exposure scenarios can be residential or commercial. Residential scenarios are more conservative than commercial scenarios since the former looks at sensitive human receptors (e.g., children, pregnant women) at work and play in areas with unrestricted access, while the latter looks at exposures in persons who have access to restricted areas (see Table 5).

In contrast, an "Exposure Case" defines a level of risk conservatism that risk assessors will strive to achieve by selection of appropriate methods and numerical variables (such as worst case, maximum exposed individual, Reasonable Maximum Exposure [RME], and upper bound) that are used in risk equations or computer models. Each exposure case uses a different set of facts, assumptions, and inferences about how exposure occurs. These assumptions influence variable selection and use in a risk assessment report. For a given set of chemicals and environmental releases, the more conservative an exposure case, the higher the calculated risks and more likely that calculated exposures will generate unacceptable risks. Will an HHRA use a model of a highly conservative situation, such as worst case, maximum exposed individual, high-end, or RME or a less conservative central tendency exposure case?

Risk assessors achieve a given exposure case, or percentile distribution, by selecting an appropriate mix of variables — such as body weight, exposure time, ingestion rate — for use in an exposure equation. Each variable has been studied to determine what percentage of a population possesses each attribute. For example, information has been developed on how long a person is expected to live at one residence, how much fish they will eat daily, and how fast they breathe. These studies generate distribution curves that tell percentage of a population that exhibits a certain

Table 5 Comparison of Typical Values used in Residential and Commercial/Industrial Exposure Scenarios

Exposure Scenario	Variable		Central Tendency	High End
Residential	Soil ingestion (mg/day)	Child	200	800
		Adult	60	100
	Air inhalation (m3/day)		20	30
	Drinking water ingestion (l/day)	Child	1	1
		Adult	1.4	2
	Exposure frequency (days/year)		350	350
	Exposure duration (years)		9	30
	Body weight (kg)	Child	15	15
		Adult	70	70
Industrial/ Commercial	Soil/dust ingestion (mg/day)		60	480
	Air inhalation (m3/workday)		15	20
	Drinking water ingestion (l/workday)		1	2
	Exposure frequency (days/year)		250	250
	Exposure duration (years)		25	40

Adapted from U.S. Environmental Protection Agency, 1992, *Risk and Decision Making. A Workshop in Risk Assessment, Risk Management, and Risk Communication*, Office of the Senior Science Advisor, Region IX.

behavior or characteristic. For each variable in an exposure equation, risk assessors can select a variable value at some point along a distribution curve. The higher a value on a distribution curve, the more conservative the value and the greater percent population represented by the value. By carefully selecting each variable value, risk assessors generate a level of exposure conservatism that matches an exposure case. Federal and state agency documents and peer reviewed literature are sources of exposure variable values. Each exposure case provides a numerical exposure level directly related to how high, or low, on a range of exposures, an exposure case is designed to emulate (see Table 5).

Exposure case may be mandated by statute, regulation, or agency guidance, or it may be left to a risk assessor's judgment. Defining an exposure case is a difficult task. Exposure cases define a level of risk conservatism to employ in order to calculate exposures and risks. Choice of exposure case controls level of overall exposure to be calculated by mathematical models. A risk assessor must quantify exposure case definitions that are expressed in qualitative terms and must understand the mathematical meaning of each exposure case.

VI. TOXICITY ASSESSMENT

Toxicity assessment is the third step in HHRA. It is performed after hazard assessment and may be performed before, during, or after exposure assessment. During toxicity assessment, a risk assessor gathers qualitative and quantitative toxicity data for COCs, identifies exposure periods for which toxicity values are necessary, determines toxicity values for carcinogens and noncarcinogens, and then classifies COCs by toxic effect. Risk assessors weigh available toxicological data to evaluate potential human health effects of COC alone and in combination. Two steps of toxicity assessment (Hazard Identification and Dose-Response) are discussed below. To understand toxicity assessment, however, it is necessary to understand the science of toxicology, discussed in our next section.

VII. REGULATORY TOXICOLOGY AND THE SCIENCE OF TOXICOLOGY

"Toxicology" is the science of evaluating toxic effects of substances on living organisms and assessing relationships of dose and observed effects. It is the study of poisons. Regulatory toxicology is a branch of toxicology in which government officials evaluate toxicological properties and risks of chemicals, and regulate their use or environmental presence.

A. Types of Tests

1. Toxicity Testing

Toxicity tests and evaluations are performed to obtain dose-response relationships for toxic effects. Toxicity tests require a toxicological agent (e.g., a chemical) in some sort of vehicle (e.g., oil or water) be delivered to a test organism under controlled conditions. Test organisms exposed to a toxicant (via ingestion, inhalation, or dermal exposure) are observed for signs of a toxic reaction for some period of time and, after observation, are examined for any physical signs of toxic effects. A variety of test organisms exist; choice of organism depends on regulatory needs or site-specific concerns. Chemicals entering an organism may be distributed, biotransformed, stored, or excreted. Each of these processes can make a chemical more or less toxic to test organisms. Types and rates of chemical distribution, transformation, storage, and excretion are usually specific to a chemical and to an exposed organism. Several types of studies are used in toxicity testing to define potential carcinogenic and noncarcinogenic effects.

2. Epidemiological Studies

Epidemiological studies define distribution and occurrence of disease in a human population. In the case of chemical risk assessment, epidemiological studies are used to describe a relationship between human exposure to chemical substances and subsequent illness or death in persons in an exposed population. Positive results from well-planned and properly-conducted epidemiological studies are strong evidence for linking a chemical exposure to a specific health effect. Negative results from epidemiological studies do not necessarily mean that a substance under investigation does not cause measurable health effects. Epidemiological studies provide statistical evidence, a correlation, between chemical exposure and disease.

Epidemiologists use both descriptive and analytical studies to evaluate human health impacts from chemical exposures. "Descriptive studies" characterize distribution and occurrence of disease in an entire population, while "analytical studies" (i.e., case-control and cohort studies) are used to define cause and effect relationships.

3. In Vivo *Toxicological Studies*

In vivo toxicological studies are conducted to determine effects of a chemical exposure on living organisms. *In vivo* studies done on nonhuman species are based on an assumption that any effects observed in test animals are relevant to human exposures. Many animal studies have been experimentally validated to ensure that it is appropriate to extrapolate from animals to humans in developing human health risk estimates. Exposure durations can vary from acute, short-term tests to long-term cancer bioassays.

4. In Vitro *Toxicological Studies*

In vitro toxicological studies are conducted to determine effects of a chemical exposure on cell cultures rather than whole living animals. *In vitro* studies done on cell cultures are based on an assumption that any effects observed in test animals may be relevant to human exposures. *In vitro* studies provide supportive data on potential human health effects of single chemicals or chemical mixtures.

B. Physical and Chemical Properties

Studies of physical and chemical properties of a substance provide clues to its toxic potential. They may include studies of structure/activity relationships (SARs) exhibited by a substance. These studies assume that a known toxic potential of a chemical can provide clues to toxic potential of a chemical of similar structure whose toxic properties are not known.

C. Pharmacokinetic Properties

Pharmacokinetic properties studies investigate absorption, distribution, metabolism, and excretion of substances in living organisms. These properties influence how a chemical will enter an organism, distribute within an organism, be biotransformed, exert a toxic effect, and be eliminated.

There are numerous ways to express human and animal toxicity. Regardless of how it is expressed, toxicity is a function of dose and effect (see Table 6). Toxicity tests can generally be discussed as either acute, subchronic, or chronic studies. Studies may start with a ranging experiment to determine dosage levels for full

Table 6   Example of Relationship of Qualitative to Quantitative Expressions of Toxicity

Lethality	Extreme	High	Moderate	Low
Oral LD50	<50 mg/kg	50–500 mg/kg	500–5000 mg/kg	>5000 mg/kg
Dermal LD50	<200 mg/kg	200–2000 mg/kg	2000–20,000 mg/kg	>20,000 mg/kg
Inhalation LC50	<200 mg/m3	200–2000 mg/m3	2000–20,000 mg/m3	>20,000 mg/m3

Adapted from U.S. EPA, Course materials for personal protection and safety 165.2 and hazardous material incident response 165.5. Office of Emergency Response, Environmental Response Program; Klaassen, C.D., and Doull. Evaluation of Safety: Toxicologic evaluation. In *Cassarette and Doull's toxicology, the basic science of poisons*, second edition, Doull, J., Klassen, C.D., and Amdur, M.O., Eds., 1980, pages 11-27; U.S. Environmental Protection Agency, *Workshop on risk assessment and communication*, Lake Geneva, Wisconsin, May 30 - June 1, 1989, Air Risk Information Support Center, Research Triangle Park, NC, 1989.

experiment. Animals are selected and treated with several dosage levels (usually a series of three, elevated by multiples of three) of a substance by one route of exposure and then observed for a period of time. During a toxicity experiment, some animals may be sacrificed to determine their health status. During exposure, animals may be observed for behavioral effects, as well as other frank effects. At the end of an experiment, all remaining animals are usually sacrificed and examined for biochemical, physiological, functional, and morphological effects. Types of toxicity studies include acute studies, such as oral and dermal LD_{50} studies and inhalation LC_{50} studies. Subchronic and chronic toxicity studies include animal feeding and inhalation studies to evaluate carcinogenic and noncarcinogenic effects. Reproductive studies may address chemical effects on human reproductive and developmental cycles, whereas mutagenicity studies employ microbial and animal cell bioassays.

D. Use of Regulatory Toxicology in Toxicity Assessment

Our world is a chemical soup to which people are constantly exposed. Substances in this soup are of both natural and anthropogenic origin. Exposure to many of these substances at typical concentrations poses no significant human health threat. However, some chemicals, alone or in combination, do pose significant human health risks due to certain factors. Toxic responses may be a function of physical, chemical, and biological properties of a substance; concentration of a substance; exposure duration, route, or presence of other chemicals; heredity, age (e.g., child vs. adult), sex; or hormonal, nutritional, or medical status of exposed individuals.

Toxic responses can be categorized by dose needed to elicit an adverse response. There is a wide spectrum of toxic response to chemicals (see Table 7). Adverse effects include mortality (i.e., death), morbidity (e.g., observable illness), pathophysiology (e.g., tissue damage, changes in structure or function, irritation), physiological changes of uncertain health significance, and exposure and dose of uncertain health significance. Adverse effects can occur immediately after exposure (i.e., acute effect) or after a long period of exposure (i.e., chronic effects). Some effects can be delayed over long periods of time such as an allergic response or cancer.

Responses to toxicant exposures can vary in persons and populations. Persons exposed to potentially toxic substances can absorb them, distribute them to organs and tissues in a body, metabolize substances (to create more, less, or equally toxic substances), or excrete them. A body can show toxic effects from exposure at a site of contact (termed a "local effect") or some where else (termed a "systemic effect"). For many substances, severity of injury tends to increase with increasing dosage. For example, a chemical, that kills cells and destroys an organ at high doses, will show lesser or different toxic responses at lower doses, until a dose is reached where no observable effect is seen. Increased severity of response is caused by increased damage at higher doses. For other substances, severity of toxic response may not change, but number of organisms exhibiting toxic effects can increase with increasing dosage. Increased incidence of toxic effects can be attributed to differences in sensitivity of individuals in a population to a toxicant. Toxic responses can range from those that disappear when exposure stops to those that are permanent and irreversible (e.g., birth defect).

Most environmental exposures involve more than one chemical, although current toxicology databases have little data on chemical mixture health effects. Chemical mixtures can result in less toxicity (termed "antagonism"), equal toxicity (termed "additivity") or greater toxicity than expected from each chemical alone (termed "synergism" or "potentiation").

E. Exposure Routes

Potentially toxic substances in solid, liquid, gas, or vapor form, can enter a human body via four primary routes: inhalation, ingestion, dermal, and injection.

By inhalation, substances enter via lungs as a result of respiration. Toxic substances can be inhaled as gas, vapor, dust, fumes, mist, smoke, aerosols, and particles. Deposition of particles within a body is size dependent. Particle behavior in the respiratory tract is discussed below. After inhalation, potentially toxic substances can cause direct tissue damage and can be absorbed into blood and travel via the blood stream to tissue and organs. Lungs can be cleared of these substances by coughing, by mucocilliary action, or by cleansing by macrophages and the lymphatic system.

By ingestion, substances enter a body when contaminated materials are consumed. Sources include contaminated food, water, or soil, and substances cleared from the respiratory tract and swallowed. Ingested substances can pass through a body unabsorbed and be excreted or can be absorbed across the gastrointestinal-tract lining and moved via the blood stream.

In dermal exposure, substances may cause direct contact injury and they may enter a body through skin contact, eye contact, puncture wounds, or other breaks in skin. Dermal contact can cause tissue destruction or can lead to absorption and distribution through the blood stream. Sources of direct contact include gases,

Table 7 Examples of Types of Toxic Effects

Type	Effect
Allergens and allergic sensitizers	Allergens are substances that induce an allergic response characterized by bronchoconstriction and pulmonary disease. Allergic sensitizers do not result in a toxic effect on initial exposure. However, subsequent exposures to a substance can result in significant toxic effects at much lower levels. Allergic responses can be from the same chemical or a chemical with a similar structure.
Anesthetics and narcotics	Anesthetics and narcotics depress Central Nervous System and can cause dizziness, drowsiness, weakness, fatigue, incoordination, unconsciousness, respiratory system paralysis, and death. Many substances induce this effect, including many hydrocarbons and organic substances.
Asphyxiants	Asphyxiants are gases that deprive body tissues of oxygen. Simple asphyxiants are physiologically inert gases that can cause suffocation, unconsciousness, and death by displacing oxygen in air. Chemical asphyxiants are gases that prevent body from using oxygen in air.
Behavioral toxicants	Behavioral toxicants are substances that cause changes in normal behavior patterns.
Carcinogens	Carcinogens can cause development of cancer in an exposed individual. Carcinogenesis is a multistep process that is thought to include: an initiation step where DNA damage occurs; a promotion stage where physical changes or damage occurs that can cause cellular and genetic damage to adjacent cells; a progression stage where a neoplastic (an abnormal tissue mass or tumor that is benign or malignant) cell line proliferates; and a transformation stage where a visible tumor appears. Substances can be divided into genotoxic carcinogens that interact directly with genetic materials that cause changes in DNA and epigenetic carcinogens that do not directly interact with genetic material but cause carcinogenesis by some other mechanism (e.g., immunosuppression, hormonal imbalance, cytotoxicity). For risk assessment purposes, it is assumed that there is no threshold for carcinogenic risk and that every exposure to a carcinogen has some level of associated risk.
Developmental toxicants and genotoxicants	Developmental toxicants cause adverse effects in a developing organism resulting from exposure to either parent prior to conception, during prenatal development, or postnatally, to time of sexual maturation. These effects can be expressed or seen at any time during life span of an organism. Developmental toxicity manifestations include death of a developing organism (e.g., embryotoxicity and fetotoxicity), structural abnormalities (e.g., a malformation that is a permanent structural change), altered growth (e.g., change in offspring organ or body weight or size), functional deficiency (e.g., changes in ability of an organism or organ system), and variations (e.g., structural changes greater than normal range that may or may not result in adverse effects). Genotoxicants cause changes in cellular DNA that can be expressed upon cell replication as mutagenicity or carcinogenicity.
Fibrosis producers	Substances, such as silicates and asbestos, cause tissue to become fibrotic. High levels of fibrosis may block air passages and decrease lung capacity.

Table 7 continued

Type	Effect
Idiosyncratic toxicants	Certain substances produce an adverse effect in individuals genetically disposed to react abnormally to material. Individuals with this genetic predisposition can be either highly sensitive at low doses or very insensitive at high doses.
Immunotoxicants	Immunotoxicants can cause immune-mediated responses to toxicant exposures or can impair immune system function.
Irritants	Certain substances that dissolve natural oils in skin can cause dermatitis. Repeated contact with these substances can cause skin to dry, become cracked, inflamed, and possibly infected. Irritant responses can range from mild reddening of skin or eyes, to tissue corrosion and second or third degree chemical burns. Tissue responses are a function of tissue type and concentration of irritant substances. Other substances can irritate air passages and can cause constriction of air passages leading to edema (i.e., lungs fill with fluid) and infection.
Necrosis producers	Necrosis producers cause cell death and edema.
Mutagens	Mutagens cause inheritable changes in DNA that are not due to normal recombination processes. A mutation is an altered gene that may be nonfunctional, dysfunctional, or functionally unchanged. Mutagens are considered a subset of genotoxins. Heritable mutagenic changes are of great concern and include point mutations (i.e., changes in base sequence of DNA), structural aberrations (e.g., deficiencies, duplications, insertions, and translocations), and numerical aberrations (e.g., gains or losses of whole chromosomes or sets of chromosomes). Mutagenic effects can occur through direct action of chemicals on DNA or through interference with normal DNA synthesis.
Neurotoxicants	Neurotoxicants cause adverse effects on the nervous system. Effects include acute neurotoxicity (e.g., inhibition of chemical acetylcholinesterase, which breaks down chemical acetylcholine, which conducts nerve impulses across gaps between nerve cells), chronic neurotoxicity (e.g., changes in electroencephalographic patterns), and delayed neurotoxicity (e.g., toxicity exhibited some time after exposure).
Photosensitizers	Photosensitizers increase sensitivity to light, so that less exposure can cause same or greater cell damage.
Teratogens	Teratogens adversely affect sperm, ova, or fetal tissue; alter development in ways to produce defects in developing embryo or fetus; cause death during development; or produce offspring with physical or behavioral defects.
Tolerance	Tolerance occurs when substances decrease responsiveness to subsequent exposures to that substance.

Adapted from Doull, J., Klassen, C.D., and Amdur, M.O., Eds., *Cassarette and Doull's Toxicology, the Basic Science of Poisons*, 2nd ed.

liquids, and solids that are purposefully applied or accidentally introduced to skin, including through eyes or skin lesions.

Injection poses a less commonly considered exposure route for environmental risk assessment, although it is of great concern to medical professionals. Contaminated objects may penetrate or puncture skin and introduce toxic contaminants into blood for distribution to tissues and organs.

F. Exposure Duration

Exposure duration influences toxic effects for both single chemicals and chemical mixtures. Exposure duration is commonly described as acute, subchronic, or chronic. However, many different exposure periods have been associated with each term. Acute has been defined as less than 24 hours, as well as 14 days or less. Subchronic has been defined as exposure for 3–6 months or for 15–364 days. Chronic exposure has been defined as lasting more than 6 months, or as lasting 365 days or more. Care should be taken to understand the precise meaning of these terms as they are used by various authors and researchers.

G. Absorption, Distribution, Metabolism, and Excretion

Potentially toxic substances must pass through an absorption barrier, usually skin, lungs, or gastrointestinal tract. This occurs by active or passive diffusion, filtration, facilitated diffusion, or by cellular engulfment. Assessing toxic effects of a substance requires an understanding of "pharmacokinetics/toxicokinetics," a process by which a body absorbs potentially toxic substances and distributes them (e.g., in blood, ability to cross membranes, tissue affinity), chemically alters them in tissues and organs (i.e., two processes affect chemical toxicity in organisms: detoxification and metabolic activation), and excretes them (e.g., substances are excreted in their original form or as metabolites through urine, feces [including liver produced bile], milk, exhalation, sweat, and saliva).

After a substance is absorbed, it moves through a body via the blood stream to tissues and organs. There toxicants are metabolized, stored, or excreted. Exactly how a substance moves via blood to target sites depends on its affinity for a target site, ability to pass through membrane barriers, its physical and chemical properties, and on blood flow rates to target sites. Metabolism may increase or decrease a substance's toxicity. Potentially toxic substances may also interact with other biological molecules or become localized in certain tissues. Time required for half of a substance to clear from a body is termed its "half-life," expressed as "$t_{1/2}$." Amount of a substance in a body is termed "body burden."

H. Target Organs

Our next section, discussing several important toxicant targets in a human body, is offered merely as an introduction to this area of study. It is not comprehensive.

1. Respiratory Tract

Lungs transfer oxygen and carbon dioxide between blood and air. Divided into nasopharyngeal (i.e., nose to larynx), tracheobronchial (i.e., trachea, bronchi, and bronchioles), and pulmonary acinus (i.e., basic functional unit of the lung, composed of respiratory bronchioles, alveolar ducts, and alveoli), a human respiratory tract is in constant and direct environmental contact. Lungs have 70–100 square meters of exposed surface area, in contrast to skin, which has 2 square meters, and to a human's digestive system which has 10 square meters. Respiratory tract deposition of potential toxicants relates to respired particle size. Particles 5 μm, or larger, deposit primarily on nasal or oropharyngeal mucosa, then are expelled or swallowed. Particles 2–5 μm deposit in tracheobronchioli and are cleared by mucocilliary escalation and swallowed. Particles 1 μm, or less, in diameter can penetrate to alveoli, the deepest part of lungs. If a particle dissolves, its constituent chemicals will readily pass over the pulmonary capillary bed, to the blood stream. Types of respiratory system damage include irritation, constriction, allergic reactions, hypersensitivity, cell death, edema, fibrosis, emphysema, and cancer.

The respiratory tract can absorb inhaled substances or expel them. Expelled material may leave a body entirely or it may be swallowed and enter the gastrointestinal tract, where absorption may occur. Inhaled substances can be absorbed through the respiratory tract, enter blood and be transported to tissues and organs where they can cause systemic effects. Acute effects are mostly localized, causing injury to airways or lung tissue. Acute effects include airway irritation and obstruction due to swelling or constriction of bronchi or accumulation of fluid in alveolar air spaces, termed "pulmonary edema." Acute systemic effects can also occur if chemicals are absorbed into the blood stream. Inhalation of some substances (e.g., hydrocarbon solvents and fuels) can cause acute pneumonic reaction. Chronic exposures to particulate matter can cause pneumoconioses that are characterized by inflammation, scarring, and fibrosis of lung tissue.

2. Skin

Skin separates our inner body from the outside world. It is comprised of an outer nonvascularized layer, termed "epidermis," and an inner vascularized layer, "dermis." Intact epidermis comprises our "stratum corneum," a cohesive membrane made up of dead epidermal cells. This is a major barrier to infectious agents and absorption of potentially toxic substances, although all parts of dermis, including soles and palms, absorb pesticides. Beneath the stratum corneum is living tissue, or epidermis, where cells rapidly proliferate and totally replace cells in the stratum corneum every 2–3 weeks. Next, is an area of skin, termed dermis, containing fat tissues, nerve endings, capillaries, sweat glands, sebaceous glands, hair erector muscles, hair shafts, and papillae of growing hair. In general, potentially toxic substances must pass through epidermis to reach dermis and blood vessels. Some direct movement to dermis can occur, however, through sweat glands, sebaceous glands, and hair follicles. Dermal toxicants can cause irritation, rashes, itching, damage to hair follicles, sensitization, phototoxicity, photoallergy, changes in pigmentation, chloracne, skin hardening or scaling, ulcerations, and cancer. These adverse effects on skin can occur by either direct contact or systemic exposure.

3. Eyes

Eyes are complex structures that provide visual input to the brain. Eye tissue is a sensitive tissue. Following a toxic exposure, eye tissue can exhibit instant tearing (i.e., lacrimation), chemical burns, optic nerve damage, retinal damage, corneal burns, iris irritation, ulceration, cataracts, optic nerve damage, perforation, and cornea (or lens) clouding. Typical substances that cause eye tissue damage include acids, alkalies, and organic solvents. Some substances are inhaled, ingested, or absorbed through skin, but move from blood to eye tissue where they can cause damage.

4. Nervous System

A human nervous system has two main components: the Central Nervous System (CNS), including brain and spinal cord, and the Peripheral Nervous System (PNS), including nerves connecting to the spinal cord, sense organs, glands, blood vessels, and muscles. Our nervous system controls and coordinates movement, vision, thought, hearing, speech, heart function, respiration, and other physiological processes. Physical control and coordination are accomplished through a network of nerve processes, neurotransmitters, hormones, receptors, and channels. "Neurons," or nerve cells, are the most fundamental, functional nervous system structures. Neurons conduct electrical nerve impulses along long cell processes, termed "axons." An insulating "myelin sheath" covers each nerve cell and assists in nerve impulse transmission. Gaps, termed "synapses," exist between one nerve cell and "dendrites," the beginning of the axon of adjacent nerve cells. Electrical impulses in an axon stimulate release of a "neurotransmitter" into a synapse. This chemical substance transmits electrical impulses across the synapse to dendrites of the next nerve cell in the series. This stimulation also occurs where nerves and muscles meet.

Nerve cell exposure to potentially toxic substances can cause structural changes in cellular and subcellular morphology; cell destruction or swelling; damage to neuronal bodies (termed "neuropathy"), axons (termed "axonopathy") or myelin sheaths (termed "myelinopathy"). It can also cause slow deterioration of a nerve cell body or axon degrading motor and sensory activities, altering emotional state or behavior (such as anxiety, nervousness, depression, sleep difficulties, memory loss, loss of appetite, speech impairment, bizarre behavior, hallucinations, and convulsions), impairing integrative functions (such as learning and memory), or causing death. Some toxicants interfere with nerve impulse conduction or synaptic transmission.

The nervous system has only limited ability to replace damaged cells and, thus, is especially susceptible to injury. Blood-brain and blood-nerve barriers can offer some nervous system protection.

5. Liver

The liver is the primary site where our bodies biotransform chemicals. Through a process termed "metabolism," the liver alters materials for use as nutrition or storage, or for detoxification or excretion. "Hepatocytes" are primary functional liver cells and are involved in most liver metabolic functions. Toxicants in blood can reach the liver and be metabolized. Hepatotoxins damage liver cells and can impair or destroy metabolic function, since liver cells do not readily regenerate. Hepatotoxic substances can cause lipid accumulation. Thus, a fatty liver indicates organ injury. Hepatotoxins can lead to liver dysfunction, resulting in "jaundice" (where yellow bile pigments are not excreted), cancer, or necrosis of liver cells, termed "cirrhosis." In cirrhosis, chronic cell destruction results in replacement of normal liver cells with altered cells and connective tissue, such as collagen. Enzyme production can increase in livers exposed to foreign substances. Increased enzyme levels may either result in faster detoxification or in production of more toxic metabolites, depending on substances involved.

6. Kidney

Kidneys produce urine, arguably the main route of toxicant excretion from a human body. Kidneys filter blood, eliminate waste, and retain important nutrients. "Nephrons" are functional units of kidneys. They receive large amounts of blood flow and, thus, toxicants in the bloodstream tend to reach kidneys quickly. At a kidney, toxicants are either concentrated or metabolized, to form more or less toxic substances. Harmful substances to a kidney are termed "renal toxicants." They can change a kidney's ability to produce chemicals necessary for homeostasis, alter fluid flow through a kidney, form kidney stones, or dilate or constrict passages. Necrosis and cell death can also occur.

7. Circulatory System

The human circulatory system transports oxygen, carbon dioxide, and other substances. It is comprised of the hematopoietic system,* platelets that help form blood clots, white blood cells, ** red blood cells,*** and plasma. Exposure to potentially toxic chemicals can change blood cell production, damage existing or developing blood cells, and change oxygen carrying capacity of red blood cells.

8. Reproductive System

Our reproductive system produces gametes† and, as in all mammals, delivers reproductive cells to a female's vagina and uterus for conception, implantation, gestation, and birth. Lactation provides offspring with milk, a source of nutrients and immunological protection. Although reproductive systems differ physiologically and biochemically in male and female mammals, exposure to toxic substances can interfere with reproductive capabilities in both sexes and can cause sterility, infertility, abnormal eggs or sperm, low sperm count or motility, hormonal changes, impaired ability to conceive, conceptus death, behavioral changes, and abnormal offspring. Younger animals are generally believed to be more sensitive to toxic substances than older animals.

---

* The hematopoietic system is composed of bone marrow, the source of most blood components, the heart, and the spleen, which filters bacteria and particulate matter from the blood.
** White blood cells, also termed "leukocytes," defend against foreign organisms and substances.
*** Red blood cells are also termed "erythrocytes" and contain hemoglobin, which is used to transport oxygen.
† "Gametes" refers to reproductive cells, i.e., sperm in males and eggs in females.

9. Immune System

The immune system recognizes foreign substances and protects our bodies by reacting to organisms, cells, and chemicals. Numerous types of immune system cells are produced in bone marrow. These cells travel to other sites and differentiate into specific classes of immune system cells. "Immunotoxicology" studies interactions between toxicants, and how the immune system interacts with substances, including heightened or lessened protection against foreign substances.

10. Cardiovascular System

The cardiovascular system is comprised of a heart, which pumps blood through a network of vessels, termed the vascular system. "Myocardial cells," or heart muscle, are the heart's functional units. Exposure to toxic substances can alter the heart's depolarization potential and induce irregular heart rhythms. Toxins can also dilate vessels, leading to hypotension, interstitial edema, fibrosis, or necrosis. Blood vessels exposed to toxicants may also exhibit increased capillary permeability, vasoconstriction, degenerative changes, fibrosis, hypersensitivity reactions, or tumors.

I. Using Toxicological Understandings in Toxicity Assessment

In hazard identification a risk assessor determines whether exposure to a COC has potential to cause an increased incidence of a particular adverse health effect and whether an effect is likely to occur in humans. A risk assessor gathers information on a COC's potential to cause adverse health effects in humans, considering two broad types of toxic effects: carcinogenic and noncarcinogenic effects. This distinction is important, because cancer-causing COCs are assessed differently from those that cause other effects. Carcinogens are generally considered "nonthreshold toxicants" and, consequently, risk assessors assume that every exposure to a carcinogen results in an associated level of increased risk. In contrast, most noncarcinogenic substances are believed to have a "threshold," a concentration below which there is no measurable toxic effect.

Hazard identification characterizes evidence that COCs cause particular health effects, considering the type of evidence and its strength. Evidence may come from many sources. Toxicology data sources include primary toxicological literature (dose/response studies in peer reviewed journals or government reports); secondary government literature (review documents such as those produced by ATSDR); U.S. EPA's IRIS, and HEAST reports that provide a summary of IRIS verified and non-verified toxicity values from other U.S. EPA programs. Credible toxicity data may also be generated by state agencies, international governments or organizations, industry, or interest groups.

Perhaps best known as an example of classifying toxicity evidence is a system developed by U.S. EPA and in common use by state regulatory agencies to rank carcinogens. It assigns carcinogens into one of five letter designations, using a weight of evidence approach based on quantity and quality of scientific evidence that a chemical causes cancer in humans. In this system, Group A, termed "known human carcinogens," are chemicals with sufficient evidence of human carcinogenicity. Group B, "probable human carcinogens," are subdivided into Group B_1 (those substances with limited human data of carcinogenicity) and Group B_2 (substances with sufficient evidence of carcinogenicity in animals, but inadequate or no evidence of carcinogenicity in humans). Group C is comprised of "possible human carcinogens," Group D are chemicals not classifiable as to human carcinogenicity, and Group E are chemicals with evidence of noncarcinogenicity for humans.

According to U.S. EPA, their 1996 Proposed Guidelines for carcinogen risk assessment propose a new weight-of-evidence approach intended to better inform risk managers. This approach summarizes key evidence, describes toxicological mode of action and conditions of hazard expression, and recommends dose-response approaches. A narrative highlights significant strengths, weaknesses, and uncertainties of contributing evidence and presents an overall conclusion regarding likelihood of human carcinogenicity, by route of exposure. Instead of six alphanumeric categories (A, B_1, B_2, C, D, E) previously used, and described above, three new descriptors classify human carcinogenic potential: "known/likely," "cannot be determined," and "not likely." Subdescriptors within these categories further differentiate carcinogenic potential.

According to U.S. EPA, its 1986 cancer guidelines did not take conditions of hazard into account. If an agent was carcinogenic by inhalation, for example, it was assumed to pose a cancer risk by any route of exposure. Under 1996 Proposed Guidelines, hazard characterization is added to integrate data analysis of all relevant studies into a weight-of-evidence conclusion of hazard, to develop a working conclusion regarding a chemical agent's mode of action in leading to tumor development, and to describe conditions under which a hazard may be expressed (e.g., route, pattern, duration, and magnitude of exposure).

For carcinogenic risk assessments, a dose-response relationship must be established between toxicant exposure and cancer induction. This regulatory toxicology dose-response value, termed a slope factor (SF), is generated to perform a human health risk assessment for carcinogens. SFs are numerical values. They represent a calculated dose of a carcinogen and its biological response (cancer). When linearized multistage mathematical modeling is used, SF equals increased risk per unit dose, or risk per mg/kg-day. Toxicity values for carcinogens can also be expressed in other ways. The upper 95th percent confidence limit of slope of the dose response curve is one such way, expressed as $(mg/kg\text{-}day)^{-1}$.

Dose-response data are rarely available for humans or animals at exposure levels of regulatory concern. Animal and human dose-response data available within literature for a given toxicant generally is derived by dosing experimental organisms at far higher exposure levels than those set by regulatory agencies to protect human health. Therefore, regulatory agencies typically extrapolate from available high-dose data to calculate lower exposure levels that provide a margin of safety for potentially exposed individuals. Without sufficient human health effects databases for potential carcinogens, regulatory scientists rely on laboratory animal toxicology experiments. If animal data are used to estimate human health risks, human equivalent doses must be calculated to account for differences between humans and animals, and a conversion factor must be applied. Animal test data are fit into one of several mathematical models that extrapolate from high dose data sets generated by animal bioassays to much lower dose levels expected for probable human exposure scenarios. Extrapolation models may use very different biological assumptions to generate a numerical SF and, thus, different models yield different slopes and significantly different risk estimates, even using identical data sets.

Carcinogenicity models differ mostly in how they estimate carcinogenic response as dose approaches zero, an area where there is no measured dose-response data. Five common model types for calculating a cancer SF are linear (e.g., Linear Model), mechanistic (e.g., 1-Hit, Multihit, Multistage, and Linearized Multistage Model), tolerance distribution (e.g., Log-Probit, Logit, Weibull, and Gamma-Multihit models), time-to-tumor (e.g., Lognormal Distribution, Weibull Distribution, Armitage-Doll, and Hartley-Seilkin models), and biologically-motivated (M-V-K Model). Risk assessors generally rely on linearized multistage modeling, unless there are compelling reasons not to do so. When linearized multistage modeling is used to calculate a cancer SF, SF is also referred to as q_1^*, pronounced "Q one star."

U.S. EPA's 1986 cancer guidelines are also limited in their approach to dose-response assessment; they allowed for only one default approach (i.e., linearized multistage model for extrapolating risk from upper-bound confidence intervals). Under U.S. EPA's 1996 Proposed Guidelines, mode of action is emphasized both to reduce uncertainty in describing likelihood of harm and in determining a proper approach to dose-response. Biologically based extrapolation model is a preferred approach for quantifying risk. Since U.S. EPA expects necessary data for parameters used in such models to be unavailable for most chemicals, its 1996 Proposed Guidelines allow for alternative quantitative methods, including several default approaches.

Dose-response assessment is a two step process. In step one, response data are modeled in a range of observation and, in step two, a determination of point of departure (or range of extrapolation below the range of observation) is made. In addition to modeling tumor data, U.S. EPA's 1996 Proposed Guidelines call for use and modeling of other kinds of responses, if they are considered measures of carcinogenic risk. Three default approaches are used: linear, nonlinear, or both. Curve fitting in the observed range should be used to determine effective dose corresponding to the lower 95% limit on a dose associated with 10% response (LED_{10}). This LED_{10} then serves as a point of departure for extrapolation to origin as the linear default or for a margin of exposure (MOE) discussion as the nonlinear default. The LED_{10} is a standard point of departure, but others may be used, if deemed more reasonable, given the data set (e.g., a No-Observed-Adverse-Effect-Level [NOAEL]).

In support of discussion of anticipated decrease in risk associated with various MOEs, biological information concerning human variation and species differences, dose response slope at point of departure, background human exposure (if known), and other pertinent factors are taken into consideration. U.S. EPA recommends describing major default assumptions and criteria for departing from them, and claims this provides an incentive for generating information needed to reduce default assumptions used in risk assessment.

Slope factors are calculated so there is only a 5% likelihood of carcinogenic response from exposure to a substance greater than estimated by experimental data and modeling. Toxicology data are fit to an appropriate model to calculate the upper 95th percent confidence limit of resulting dose-response curve slope. Thus, SFs are considered conservative estimates of dose and carcinogenic response, generated for chemicals in Group A and B and, sometimes, Group C.

Noncarcinogens are deemed threshold toxicants for risk assessment purposes. It is believed that noncarcinogens must overcome a body's protective mechanisms before they can cause an adverse effect. First, noncarcinogenic health effects are characterized as follows. Human and animal data sets are reviewed and a critical study and toxic effect are selected. Based on this critical study and effect, a risk assessor selects a NOAEL or, lacking a NOAEL, a Lowest-Observed-Adverse-Effect-Level (LOAEL), and then uses it to generate a noncarcinogenic toxicity factor. For ingestion exposures, it is termed a "Reference Dose" (RfD) and, for inhalation exposures, a "Reference Concentration" (RfC).

A chronic RfD is defined as an estimate (with uncertainty spanning perhaps one order of magnitude or greater) of a daily exposure level for a human population, including sensitive populations, that is likely to be without an appreciable risk of deleterious effects during a lifetime.

A RfC is an estimate (with uncertainty spanning one order of magnitude) of continuous exposure to a human population (including sensitive subgroups) through inhalation that is likely to be without appreciable risk of deleterious effect during a lifetime.

RfDs and RfCs are calculated by dividing a NOAEL, or LOAEL, by a series of "uncertainty factors" (UFs). Each UF represents a specific area of uncertainty inherent in extrapolation of available data. UFs of ten are commonly employed. UFs account for general population variability; protect sensitive subpopulations, such as children or elderly or immunocompromised people; account for extrapolation from animal data to humans; and address interspecies variability. UFs are also used to address shortcomings in available data, such as when a NOAEL of 10 is used from a subchronic study instead of a chronic study, when a LOAEL is used instead of a NOAEL, and to account for uncertainty associated with extrapolating from LOAELs to NOAELs. "Modifying factors," up to ten are also used. Modifying factors reflect qualitative professional assessment of additional uncertainties in critical studies and in a chemical's entire database. A default value for a modifying factor is 1.

Recent state, national, and international regulatory concentration development efforts have begun to use lesser UFs in their calculations. For example, U.S. EPA's Acute Exposure Guideline Levels (AEGLs) for exposure periods ranging from ten minutes to eight hours, are using UFs that are significantly lower than those typically used by U.S. EPA for chronic studies. This reduction in UFs occurs both in individual and cumulative UFs. Similar reductions in UF magnitude are also occurring at state level generation of ambient air inhalation risk values.

Credibility of a risk assessment report depends on use of toxicity values that are acceptable to scientists and regulators. Usefulness of a risk assessment report is reduced (or even destroyed) by use of outdated, miscalculated, poorly researched, or otherwise dubious toxicity values. Use of inadequate toxicity values can wildly overestimate or underestimate risk. It is essential, therefore, that a risk assessor use only current toxicity values of acceptable quality.

As discussed above, chemical mixtures can result in antagonism, additive effects, or synergy. Risk characterization can use a regulatory convention of summing risks of known or probable human carcinogens and, for noncarcinogens, summing risks of chemicals possessing similar toxic effects with appropriate and compatible exposure pathways and routes of exposure. Where synergism exists, however, this convention can lead to an underestimate of chemical mixture risks.

VIII. RISK CHARACTERIZATION

Risk characterization, the fourth (and final) step in risk assessment, involves calculation of carcinogenic and noncarcinogenic risks by combining exposure intake and uptake levels with toxicity values. Numerical results of a risk characterization must be accompanied by text to fully explain risk assessment findings.

Whereas earlier U.S. EPA risk assessment guidance documents gave little guidance for risk characterization (that component of a risk assessment report that describes potential human risk, strengths and weaknesses of data, size of risk, and confidence of conclusions for a risk manager), newer publications provide direction on how to present overall conclusion and confidence of risk for a risk manager, and call for clear explanations of all assumptions and uncertainties.

An effective risk characterization is essential. Risk characterization must be clear in order to preserve a risk assessment report's credibility. Writing an effective risk characterization section requires risk assessors to interact with document end-users during all phases of planning and report generation. Complete risk characterization, generation of legitimate numerical risk estimates, and a clear overall description of the situation are important goals. In some cases, it might be helpful to use several different risk cases to explore how different models and exposure assumptions provide an array of possible risk outcomes. Risk characterization may also need to address risk perceptions and social values, important factors in most risk management decisions. Several key characteristics of a well written risk characterization include: a discussion of relevance of exposure scenarios used in the assessment to real world experience; clear writing and consistent presentation; balanced presentation of scientific judgements made during the assessment; and a level of detail appropriate to information needs and understanding of primary end-users of a report.

Risks can be expressed in a variety of ways, depending on the nature of risk, purpose of a report, and needs of risk analysts and risk managers. Several techniques may be useful. One measure in common use is "Individual Lifetime Risk."* Individual lifetime risk can be defined as an increase in probability that an individual will experience a given adverse effect resulting from exposure to a toxic substance. This expression of risk is presented as a probability of an adverse effect (e.g., one-in-one million lifetime cancer risk) for carcinogens, and a linear dose-response relationship is assumed. As threshold toxicants, by definition, noncarcinogens pose no risk to health if exposure is below a threshold concentration. A second expression of risk is "population risk" ** or "societal risk." Population risk connotes morbidity

---

* Individual Lifetime Risk = Dose × Potency
** Population Risk = Individual Lifetime Risk × Exposed Population

or mortality that occurs after a year of exposure, or number of cases occurring in one year. Population risk assumes a linear dose-response relationship, which may be a faulty assumption. "Relative risk"* offers a third representation of risk. It compares risks in exposed populations to nonexposed or differently exposed populations. "Standardized Mortality (or Morbidity) Ratio" ** is a fourth measure of risk. It represents numbers of deaths or disease cases observed in an exposed population divided by numbers of deaths or illnesses expected in a general population. Standardized means that factors such as age and exposure period have been taken into account. "Loss of Life Expectancy Days,"*** or years of life lost due to a given activity or exposure, is a fifth descriptor of risk.

Risk characterization provides an opportunity for risk assessors to ensure that a risk assessment report is scientifically and procedurally consistent with current risk assessment standards. To do so, risk assessors gather and organize exposure and toxicity data. This data will include calculated intakes for various exposure durations for each chemical, mathematical modeling assumptions (such as chemical concentration at an exposure point, frequency and duration of exposure, absorption assumptions, and characterization of exposure uncertainties), current carcinogen SF, weight of evidence classification, type of toxic effect and site of toxicity, exposure duration toxicity factors (e.g., RfD, RfC), uncertainty and modifying factors used to derive toxicity values, expression of toxicity values (as absorbed or administered doses), and uncertainties associated with toxicity assessment. After all necessary data has been gathered, it is evaluated for accuracy and consistency. Risk assessors then quantify carcinogenic and noncarcinogenic risks, separately.

Carcinogenic risks are quantified as:

$$Risk = Chronic\ Daily\ Intake \times SF$$

where Risk = a unitless probability (e.g., 2E-5) of an individual developing cancer., Chronic Daily Intake (CDI) = chemical intake averaged over 70 years and expressed as (mg/kg-day),$^{-1}$ Slope Factor (SF) = numerical expression of upper-bound probability of an individual developing cancer as a result of a lifetime of exposure to a particular level of a carcinogen.

Using a SF, an estimated daily intake averaged over a lifetime of exposure is converted into incremental risk (i.e., cancer risk over background) of an individual developing cancer. This assumes a linear dose-response relationship between carcinogen exposure and cancer induction (when using linear multistage modeling). Based on this assumption, SF is a constant and risk is directly related to intake. It is reasonable to assume that "true risk" will not exceed risk estimated by this model. When risks exceed 0.01, however, use of linear multistage modeling may not be appropriate.

---

* Relative Risk = $\dfrac{\text{Incidence Rate in Exposed Group}}{\text{Incidence Rate In General Population}}$

** Standardized Mortality or Morbidity Ratio = $\dfrac{\text{Incidence Rate in Exposed Group}}{\text{Incidence Rate in General Population}}$

*** Loss of Life Expectancy = Individual Lifetime Risk × 36 years (average remaining lifetime)

HUMAN HEALTH RISK ASSESSMENT

Noncarcinogenic risks are quantified by comparing daily intakes to toxicity values for a specified exposure period and similar toxicity endpoints. Noncarcinogenic effects are not expressed as probability of an individual suffering an adverse effect. Rather, potential for significant effects is expressed when noncancer hazard quotient (HQ) is greater than one, as illustrated below:

$$\text{Noncancer HQ} = \frac{\text{Exposure Level or Intake}}{\text{RfD or RfC for a given period of time}}$$

Unlike cancer SFs, RfDs and RfCs are not probabilistic values and they provide no information on the slope of a dose-response curve. Steep dose-response curves indicate that a small amount of chemical causes a relatively large toxic response, above a threshold concentration, whereas a shallow dose-response curve means a relatively small toxic response occurs at concentrations above threshold. Thus, for a given dose of a chemical above its threshold concentration, resulting toxic response would be much greater for a chemical with a steep dose-response curve than for a chemical with a shallow dose-response curve. Unfortunately, dose-response data are not provided with RfCs and RfDs. As a result, risk assessors cannot state relative probability of morbidity or mortality occurring when HQs exceed one. However, benchmark doses under development by regulatory agencies take into account relative probability of morbidity or mortality. Benchmark dose development is often hampered by insufficient toxicological data.

A receptor can be exposed to one, or more, chemicals at a time. For single chemicals, HQs (measured or estimated single chemical concentrations divided by its regulatory concentration) or calculated cancer risk levels are compared to an appropriate risk yardstick. For noncarcinogens, this is often an HQ of one. For carcinogens, it usually ranges from a one-in-ten thousand to one-in-one-million excess lifetime cancer risk level. When a single chemical HQ or cancer risk level exceeds its appropriate risk yardstick, risk assessors can become concerned that a significant risk level has been exceeded and risk reduction measures are required.

Slightly different methods are used to evaluate multiple chemical exposures. When a receptor is exposed to more than one chemical at a time, risk assessors generally assume additivity and sum individual chemical HQs. For example, noncarcinogen individual HQs are summed by duration of exposure, similar toxic endpoint (e.g., liver toxicants), and across exposure pathways and exposure routes. Summing HQs generates a Hazard Index (HI). HIs greater than one typically represent a potential for significant noncarcinogenic risks, however, there are organizations that consider significant risks to occur at HIs of less than one. For carcinogens, carcinogenic risks of all carcinogens are added to yield a total carcinogenic risk. This risk is then compared to an appropriate risk yardstick, as described above.

Risk characterization sections are usually accompanied by an uncertainty analysis. Uncertainty analysis may be qualitative, semi-quantitative, or quantitative. Uncertainties discussed in uncertainty analysis include site-specific UFs (such as likelihood of exposure pathways and land uses actually occurring); ramifications of eliminating chemicals from quantitative risk analysis; model applicability, assumptions, and weaknesses; significant gaps in site data and significant data uncertainties; potential uncertainty magnification through assessment; quantitative uncertainty analysis that involves statistical manipulation of data in exposure model (e.g., Monte Carlo simulation); uncertainties associated with fate and transport exposure; multiple chemical exposures (e.g., synergism or antagonism); use of surrogate data sets; and mathematical manipulation of codependent variables in exposure equations.

IX. CONCLUDING THE HHRA

Risk assessors complete a risk assessment by summarizing risk characterization, and explaining risk findings in terms of significant risk yardsticks. These findings are typically restated in a report's executive summary section. Such a summary discusses uncertainties and weaknesses of risk assessment. Stating level of confidence (e.g., low, medium, high) in the report findings is advisable, as well. Such summaries and evaluations aid reviewers, readers, and risk managers in determining a risk assessment's reliability and credibility of its findings.

X. PRESENTING HHRA DATA

Having presented a basic review of the HHRA process, we next turn to how best to present risk assessment report information. While our next sections focus on HHRA, many of our suggestions (such as generic table types) apply equally well to ERA, with little or no modification. Certain tables geared specifically to ERAs will be required, however, to address community effects, population effects, bioassay results, etc.

Each risk assessor seems to have individualized ways of presenting data in a table. While interesting, creative presentations can confuse a reviewer or reader. Risk assessment report authors (and reviewers) should answer three fundamental questions:

- Is all necessary data presented?
- Is information presented in a form easily followed from first table to last?
- Is all science and math verified and verifiable using only information provided in tables (i.e., tables can stand alone)?

Often, one or more answers to these three key questions is negative. Consequently, standardized tables are gaining favor. In addition to providing a familiar format across risk assessments, use of standardized tables focuses efforts of risk assessors on their technical work, rather than their artistic and creative talents.

We do not present a sample risk assessment report, within this book, because of space constraints and its limited value to most readers. We do, however, cite several risk assessments available from U.S. EPA in the Appendix, Additional Resources. Instead, we present types of tables, figures, formats, and data that we feel are obligatory in most, if not all, risk assessment reports, and ways to evaluate their

HUMAN HEALTH RISK ASSESSMENT

contents and technical validity. Our next section describes U.S. EPA's standardized tables for Superfund risk assessments, followed by a discussion of other generic table types that should make risk assessment reports more understandable and easier to review.

A. U.S. EPA's Standard Tables for Superfund Risk Assessments

Risk assessment contractors and organizations that hire them often spend enormous amounts of time negotiating or arguing over the form of tables to be used in a risk assessment report. This can be a very expensive and time consuming process. To remedy this problem, many government agencies, including U.S. EPA, have developed standardized risk assessment table formats. For example, in 1998, U.S. EPA published the fourth part (Part D) in the series *Risk Assessment Guidance for Superfund: Volume I — Human Health Evaluation Manual (RAGS/HHEM)*. Part D complements guidance provided in Parts A, B, and C of this series and presents a standardized approach to risk assessment planning, reporting, and review, including a series of standardized risk assessment tables. These standardized tables can serve as templates for many types of risk assessments. They can also be used by risk assessment scoping teams to better understand types of data and formats that will be required to complete their contracted report (see Chapter 4). Scoping teams can use these tables to assist in scoping and work planning activities. Without uniform tabular presentations, each risk assessment can feel like a freshman writing seminar where each page is a voyage of discovery.

1. Standard Tables

A great deal of data will be amassed and a wide range of technical decisions will be made during a risk assessment project. Information collection and manipulation, begun during project scoping, continues throughout each step of a project. Managing this information can be particularly challenging at first. Proper management can save valuable time and enhance project efficiency. Fortunately, in 1998, U.S. EPA developed a set of ten standardized tables, for use in Superfund risk assessments, that can help:

- Standard Table 1 — Selection of Exposure Pathways
- Standard Table 2 — Occurrence, Distribution, and Selection of Chemicals of Potential Concern
- Standard Table 3 — Medium-Specific Exposure Point Concentration Summary
- Standard Table 4 — Values Used for Daily Intake Calculations
- Standard Table 5 — Noncancer Toxicity Data (Oral/Dermal, Inhalation, Special Case Chemicals)
- Standard Table 6 — Cancer Toxicity Data (Oral/Dermal, Inhalation, Special Case Chemicals)
- Standard Table 7 — Calculation of Noncancer Hazards
- Standard Table 8 — Calculation of Cancer Risks
- Standard Table 9 — Selection of Exposure Pathways
- Standard Table 10 — Summary of Receptor Risks and Hazards for COPCs

Although intended for Superfund risk assessment work, this set of tables can serve as a useful organizing tool for any project team planning a human health risk assessment. They can also serve as examples for developing similar tables for ecological risk assessment. Standard tables, electronic software, and instructions for completing them are available from U.S. EPA through the Internet or by mail. Purpose and contents of each Standard Table are discussed next. Please note that some government agencies use COPC and COC terms interchangeably. We use COPC as an all-inclusive list of chemicals and COC to connote chemicals to undergo quantitative evaluation.

Standard Table 1, Selection of Exposure Pathways, complements use of a site conceptual model. On this table, a project manager and team present all possible receptors, exposure routes, and exposure pathways and state their reasons for selecting or excluding each exposure pathway. This table is also useful to show which exposure pathways will be qualitatively and quantitatively evaluated. Such a transparent presentation helps to communicate risk information to interested parties outside of a project as well as serving as a helpful organizing system for a project team.

Standard Table 2, Occurrence, Distribution, and Selection of Chemicals of Potential Concern (COPCs), provides information adequate to give a sense of what chemicals have been detected at a site and potential magnitude of site problems. It also provides chemical screening data and states a rationale for selection of COPCs and COCs. Specifically, it presents statistical information about chemicals detected in each medium, detection limits of chemicals analyzed, toxicity screening values for COPC selection, and identifies whether a chemical is selected as a COC or deleted. In other words, it identifies whether a chemical will be a COC. One iteration of this table is completed for each unique combination of scenario timeframe, medium, exposure medium, and exposure point and given a unique table number. Even though some versions may present identical data, U.S. EPA recommends preparation of separate tables to ensure transparency in data presentation and appropriate information transfer for each exposure pathway.

Standard Table 3, Medium-Specific Exposure Point Concentration Summary, summarizes information about exposure point concentrations by environmental medium. Specifically, it provides reasonable maximum and central tendency medium-specific exposure point concentrations (Medium EPCs) for measured and modeled values. It also presents statistical information used to calculate Medium EPCs for chemicals detected in each medium and states reasons for selecting statistics for each chemical (i.e., discuss statistical derivation of measured data or approach for modeled data). Whereas Medium EPC does not change for a particular medium, regardless of exposure route, Route EPC considers transfer of contaminants from one medium to another. One copy of Standard Table 3 is completed for each unique combination of scenario timeframe, medium, exposure medium, and exposure point that will be quantitatively evaluated and will be identified by unique numbering.

Standard Table 4, Values Used for Daily Intake Calculations, sets forth exposure parameters used for RME and Central Tendency (CT) intake calculations for each exposure pathway (scenario timeframe, medium, exposure medium, exposure point,

HUMAN HEALTH RISK ASSESSMENT

receptor population, receptor age, and exposure route). It also provides intake equations or models used for each exposure route/pathway. It documents values used for each intake equation for each exposure pathway and provides references and rationale for each, as well as the intake equation, or model, used to calculate intake for each exposure pathway. One copy of this table is completed for each unique combination of six fields to be quantitatively evaluated:

- scenario timeframe
- medium
- exposure medium
- exposure point
- receptor population
- receptor age

Each table is identified by unique numbering.

Standard Table 5, Noncancer Toxicity Data (Oral/Dermal, Inhalation, Special Case Chemicals), provides information on RfDs, target organs, adjustment factors, and references for noncancer toxicity data. Specifically, this is a set of three standard tables. Standard Table 5.1, Noncancer Toxicity Data — Oral/Dermal, presents RfDs for each chemical of potential concern, organ effects of each COPC, as well as modifying factors and oral to dermal adjustments. Standard Table 5.2, Noncancer Toxicity Data — Inhalation, provides information on RfCs, RfDs, target organs, and RfC to RfD adjustment factors. It also verifies references for noncancer toxicity data, presents organ effects of each COPC, and provides references for RfCs and organ effects cited. Similarly, Standard Table 5.3, Noncancer Toxicity Data — Special Case Chemicals, provides information for unusual chemicals or circumstances that are not covered by other tables.

Standard Tables 6.1, 6.2, and 6.3 deal with cancer toxicity data. Each of these tables presents similar information — toxicity values, accompanied by references or sources of information, to provide weight of evidence/cancer guideline descriptions for each COPC. In addition, Standard Table 6.1, Cancer Toxicity Data — Oral/ Dermal, provides methodology and adjustment factors used to convert oral cancer toxicity values to dermal toxicity values. Standard Table 6.2, Cancer Toxicity Data - Inhalation, provides methodology and adjustment factors used to convert inhalation unit risks to inhalation cancer SFs. Standard Table 6.3, Cancer Toxicity — Special Case Chemicals, deals with "special case" chemicals. For example, a toxicity factor derived specifically for an individual risk assessment would be documented in Table 6.3.

Standard Table 7, Calculation of Noncancer Hazards, summarizes values chosen for variables used to calculate noncancer hazards: exposure point concentration, noncancer intake, reference doses, and reference concentrations. It states a noncancer hazard quotient for each COPC for each exposure route/pathway. It also presents EPC (medium-specific or route-specific) and intake used in noncancer hazard calculations, i.e., output from calculating each exposure route/pathway for each COPC, and total hazard index for all exposure routes/pathways for each scenario timeframe, exposure medium, and receptor presented in the table. One table is completed for each unique combination of fields to be quantitatively evaluated:

- scenario timeframe
- medium
- exposure medium
- exposure point
- receptor population
- receptor age

Each table is identified by unique numbering.

Standard Table 8, Calculation of Cancer Risks, provides a summary of variables used to calculate cancer risks. It shows EPC (medium-specific or route-specific) and intake used in cancer risk calculations, cancer risk value for each COPC for each exposure route/pathway, and total cancer risks for all exposure routes/pathways for scenario timeframe, exposure medium, and receptor. One table is completed for each unique combination of six fields to be quantitatively evaluated:

- scenario timeframe
- medium
- exposure medium
- exposure point
- receptor population
- receptor age

Tables are identified by unique numbering.

Standard Table 9, Selection of Exposure Pathways, presents cancer risk and noncancer hazard information, including primary target organs, for all COPCs and media/exposure points quantitatively evaluated in risk assessment. One version of Table 9 is completed for each unique combination of scenario timeframe, receptor population, and receptor age that will be quantitatively evaluated.

Standard Table 10, Summary of Receptor Risks and Hazards for COPCs, provides a summary of cancer risks and noncancer hazards for "risk drivers," those COCs that trigger cleanup. If all risks are below actionable levels, i.e., there are no risk drivers, this table simply summarizes information that demonstrates reasonableness of a "No Action" decision. Table 10 presents cancer risk and noncancer hazard information for those chemicals and media/exposure points that trigger a cleanup. It documents information on cancer risk and noncancer hazard to each receptor for each COC by exposure route and exposure point; total cancer risk and noncancer hazard for each exposure pathway for risk drivers; cancer risk and noncancer hazard for each medium across all exposure routes for risk drivers, and primary target organs for noncarcinogenic hazard effects. One version of Table 10 is completed for each unique combination of scenario timeframe, receptor population, and receptor age that will be quantitatively evaluated.

Although standard tables serve primarily as a framework for actually conducting risk assessment work, they also serve as excellent organizing tools and reminders of information that will be required during the project.

B. Variable Selection Tables

A variable selection table is another important generic table. While a variable selection table takes on several forms, it essentially provides a risk assessment writer, reviewer, and reader with a systematic presentation of reasons why a given numerical value was selected for use. For example, in the heading of a variable selection table an equation under consideration is presented, along with a citation showing the source document with page number for this equation. Columns present each important factor, such as:

- Each variable symbol
- Variable name
- Range of values that might be selected for each variable for use in the equation, along with a source citation for each value
- The value actually selected for each variable
- Where each selected value falls on a distribution curve of available values for each variable
- A brief explanation of why a specific value was properly selected, and a source citation for the data of the value selected

Using such tables provides a transparent view of logic and data used by a risk assessor for a particular equation.

C. Decision Logic or Criteria Tables

A variation of a variable selection table is a decision logic, or criteria table. Here, all discretionary and nondiscretionary decisions made in a risk assessment are arrayed to provide a clear understanding of what decisions were made and the logic behind each decision. In this way, all decisions made in a risk assessment are transparent. Decision logic, or criteria tables are as simple as listing a decision to be made, stating the actual decision, and outlining rationale, or criteria, used for the decision. This is a simple and effective aid to risk assessment transparency.

XI. CONCLUSION

HHRAs conducted in the U.S. follow a systematic four-step process. Starting with data collection and evaluation, hazard assessment provides data of mandated, or known, quality to generate a COC list, source concentration, or emission terms. This information is used in exposure assessment fate and transport models to generate a concentration for each chemical in media where human exposure is known or expected to occur. Chemical–specific medium concentration data is used in exposure equations to generate exposure levels for each chemical. Carcinogenic and noncarcinogenic toxicology data for individual chemicals and chemical mixtures, along with numerical toxicity values for individual chemicals is gathered during toxicity assessment. Numerical and qualitative toxicity data is coupled to exposure levels in risk characterization to generate risk estimates that can be compared to appropriate acceptable/unacceptable risk criteria. Risk managers use numerical risk findings and comparisons with acceptable/unacceptable risk criteria and their understanding of risk assessment uncertainties and limitations to make risk management determinations. Risk management determinations are usually not made by risk assessors and are decisions that result in risk reduction activities, initiation of legal actions, and other regulatory actions.

HHRAs can be of varying scientific rigor but, within a given level of scientific rigor, they must be scientifically and mathematically correct. Using risk evaluation methods and levels of risk conservatism required by regulatory agencies, or selected using professional judgement, risk assessors use data of known quality to produce a report containing numerical carcinogen and non-carcinogen risk estimates with clearly defined levels of credibility.

HHRA reports must meet performance standards to ensure report quality. Factors to include in performance standards include a requirement that a report be understandable to its intended audience (e.g., any educated person, as well as technical experts), that it provide all assumptions, logic, and mathematics used to generate numerical risk findings, and that reviewers be able to rapidly check all mathematics and science, i.e., that it is a seamless and transparent document. HHRAs meeting minimum report performance standards can withstand peer review and, as a result, present credible and defensible risk findings that can be compared to appropriate risk criteria for risk management determinations.

REFERENCES

ATSDR, *Health Assessment Guidance Manual Draft*, U.S. Dept. Health and Human Services, Agency for Toxic Substances and Disease Registry, Atlanta, 1990.

Brown, J., *Handbook of RCRA Ground-Water Monitoring Constituents, Chemical and Physical Properties (40 CFR Part 264, appendix 9)*, Office of Solid Waste, U.S. Environmental Protection Agency, Washington, 1992.

Cohrssen, J.J. and Covello, V.T., *Risk Analysis: A Guide to Principles and Methods for Analyzing Health and Environmental Risks*, Executive Office of the President of the United States, Washington, 1989.

Federal Emergency Management Agency, *Handbook of Chemical Hazard Analysis Procedures*, U.S. Department of Transportation; U.S. Environmental Protection Agency, Washington, 1990.

Helfand, J.S. and Mancy, K.H., Carcinogenesis risk assessment model for environmental chemicals, *Wat. Sci. Tech.*, 27, 7-8, 279, 1993.

Klaassen, C.D. and Doull, J., Chapter 2. Evaluation of Safety: Toxicological Evaluation, in Klaassen, C.D., et al., Eds., *Toxicology: The Basic Science of Poisons*, 2nd ed., Macmillan Publishing Co., Inc., New York, 1980.

Montgomery, J.H., *Groundwater Chemicals Desk Reference*, Lewis Publishers, Boca Raton, FL, 1996.

Rand, G.M. and Petrocelli, S.R., Eds., *Fundamentals of Aquatic Toxicology, Methods, and Application*, Hemisphere Publishing Corporation, Washington, 1985.

U.S. Environmental Protection Agency, *Guidance for Data Usability in Risk Assessment, Interim Final*, Office of Emergency and Remedial Response, Washington, 1992.

U.S. Environmental Protection Agency, *Human Health Evaluation Manual, Supplemental Guidance: Standard Default Exposure Factors*, Washington, 1991.

U.S. Environmental Protection Agency, *Methodology for Assessing Health Risks Associated with Indirect Exposure to Combustor Emissions, Interim Final*, Environmental Criteria and Assessment Office, Office of Health and Environmental Assessment, Office of Research and Development, Cincinnati, 1990.

U.S. Environmental Protection Agency, *Proposed Guidelines for Carcinogen Risk Assessment*, National Center for Environmental Assessment, Washington, 1996.

U.S. Environmental Protection Agency, *Risk Assessment Guidance for Superfund, Volume I: Human Health Evaluation Manual (Part A), Interim Final*, Office of Emergency and Remedial Response, Washington, 1989.

U.S. Environmental Protection Agency, *Risk Assessment Guidance for Superfund, Volume I: Human Health Evaluation Manual (Part B, Development of Risk-based Preliminary Remediation Goals), Interim Final*, Office of Emergency and Remedial Response, Washington, 1991.

U.S. Environmental Protection Agency, *Risk Assessment Guidance for Superfund, Volume I: Human Health Evaluation manual (Part C, Risk Evaluation of Remedial Alternatives), Interim Final*, Office of Emergency and Remedial Response, Washington, 1992.

U.S. Environmental Protection Agency, *Risk Assessment Guidance for Superfund: Volume I. Human Health Evaluation Manual (Part D, Standardized Planning, Reporting and Review of Superfund Risk Assessments), interim*, Office of Emergency and Remedial Response, Washington, 1998.

U.S. Environmental Protection Agency, *The Risk Assessment Guidelines of 1986*, Office of Health and Environmental Assessment, Washington, 1987.

World Health Organization, *Chemical Risk Assessment: Human Risk Assessment, Environmental Risk Assessment and Ecological Risk Assessment: Training Module No. 3*, Inter-Organization Programme for the Sound Management of Chemicals, Geneva, 1999.

World Health Organization, *Assessing Human Health Risks of Chemicals: Derivation of Guidance Values for Health-based Exposure Limits*, International Programme on Chemical Safety, Geneva, 1994.

World Health Organization, *Risk Assessment Guidance for Superfund: Volume I: Human Health Evaluation Manual , Part A*, Office of Emergency and Remedial Response, U.S. Environmental Protection Agency, Washington, 1990.

CHAPTER 3

Ecological Risk Assessment

Ruth N. Hull and Bradley E. Sample

CONTENTS

I. Introduction 80
II. Technical Aspects of Ecological Problem Formulation 80
III. Ecological Exposure Assessment 84
   A. Fish Community 86
   B. Benthic Macroinvertebrate Community 86
   C. Soil Invertebrate Species 86
   D. Terrestrial Plants 86
   E. Terrestrial Wildlife 86
IV. Ecological Effects Assessment 88
   A. Fish Community 89
   B. Benthic Community 89
   C. Soil Invertebrate and Plant Communities 89
   D. Wildlife 90
   E. Sampling 90
   F. Sources of Other Effects Information 94
V. Ecological Risk Characterization 94
   A. Uncertainties 95
VI. Comparisons with Other Studies 96
VII. Concluding the ERA 96
VIII. Conclusion 96
   References 96

I. INTRODUCTION

The four major components of the ERA paradigm are problem formulation, exposure assessment, effects assessment, and risk characterization (U.S. EPA 1997; 1998; 1992; Suter et al. 2000). An ERA begins with problem formulation. Activities occurring during this phase include: defining the goals and spatial and temporal scale of the ERA; development of a site conceptual model; endpoint and nonhuman receptor species selection; and preliminary identification of contaminants of potential concern. Exposure assessment and effects assessment follow and can be performed simultaneously. Exposure assessment evaluates the fate, transport, and transformation of chemicals in the environment, and quantitative uptake and intake of these substances in receptor organisms. Effects assessment establishes the relationship between exposure levels and toxic effects in receptors. Risk characterization is the last step in the ERA and is where exposure and toxic effect information are combined to describe the likelihood of adverse effects in receptors.

Many of the evaluation criteria needed to evaluate an ERA are identical to those presented for HHRA in Chapter 2. This chapter focuses primarily on the unique aspects of ERAs and will not repeat material covered under HHRA that applies to both subjects.

II. TECHNICAL ASPECTS OF ECOLOGICAL PROBLEM FORMULATION

Determining how many data are needed to address the ERA goals is termed the DQO process. All risk assessment stakeholders (e.g., the U.S. EPA, the State, the Fish and Wildlife Service, etc.) should be involved in this process. The DQO process is conducted at the beginning of an assessment, to define both the amount and quality of data required to complete the assessment. Scheduling time to complete DQOs at the beginning of the ERA may save the project time and money in the end. Once the goals and DQOs have been determined, the remainder of the problem formulation may be conducted. The ultimate goal of problem formulation is the site conceptual model.

A wide range of ecosystem characteristics may be considered during problem formulation. These include abiotic factors (e.g., climate, geology, soil/sediment properties) and ecosystem structure (e.g., abundance of species at different trophic levels, habitat size, and fragmentation). The environmental description may be documented using recent photographs and maps. Plant and animal species lists should be compiled.

The scale of the assessment is especially important if a large, complex site has been subdivided into several smaller sites. It also is not uncommon for Superfund sites to be located adjacent to each other. Hence the areal extent of the assessment must be defined. For example, is an off-site area included in the assessment, and to what distance off-site? The development of the site conceptual model and the selection of assessment endpoints will be directly related to the spatial scale. For example, due to their large home ranges, effects of soil contamination on deer would not be assessed if the site encompasses only two acres; assessment of endpoint species with smaller home ranges, such as small mammals, would be more appropriate.

It is necessary to decide if the assessment must consider temporal changes. All historical information should be evaluated. Then, it may be determined how much new information is needed to adequately evaluate impacts and risks. Certain parts of the year may need to be included in the sampling season for the assessment. For example, environmental exposures may change over the course of a year, or over several years, due to various seasonal influences in either chemical form or organism behavior (e.g., salmon returning to a contaminated river to spawn; migrating birds making temporary use of a site).

The site conceptual model (SCM) describes a series of working hypotheses regarding how contaminants or other stressors may affect ecological receptors (ASTM, E1689). An SCM clearly illustrates the contaminated media, exposure routes, and receptors for the risk assessment. In addition to a written description, a diagrammatic SCM is easy to understand and is useful for ensuring that no relevant component is omitted from the assessment.

During SCM development, all contaminant sources are identified (e.g., landfills, burial grounds, lagoons, air stacks, effluent pipes), and all contaminated media are represented (e.g., soil, water, sediment, air, biota). Groundwater usually is not considered an exposure medium, until it becomes surface water, but is a medium that allows migration of contaminants from soil to surface water and biota. An exception is shallow groundwater or seeps where plants may be exposed via their roots. All exposure pathways are represented, unless adequate rationale can be provided to exclude a pathway from the assessment. For example, an effluent pipe releasing metals into a stream would not need an air exposure pathway, and the only soils that would need to be considered are those of the floodplain. Thus, terrestrial receptors would be exposed by direct contact with or drinking from the stream, living in floodplain soils, or obtaining contaminated food from the stream and floodplain. An appropriate food web must be presented. A food web going from contaminated soil to earthworm to shrew may be appropriate for a 1 acre site, but a significantly larger site may require the food web to continue up to larger predators which have larger home ranges (see Figure 1).

For nonchemical stressors such as water level or temperature changes, or habitat disturbances, the SCM describes which ecological receptors are exposed to the physical disturbance, and the temporal and spatial scales of the alterations.

The idea behind the SCM is that although many hypotheses may be developed during problem formulation, only those that are expected to contribute significantly to risks at the site are carried through the remainder of the ERA process. The SCM does ensure that all exposure scenarios have been considered, and allows for full documentation of the rationale behind selection and omission of pathways and receptors.

ERAs may have more than one SCM. In predictive ERAs, impacts on different components of the ecosystem from various activities may require several SCMs. In retrospective ERAs, a hypothetical future scenario often requires assessment. For example, an area which is currently industrial and which provides little habitat for wildlife (and hence little exposure and little risk) may in future become covered in

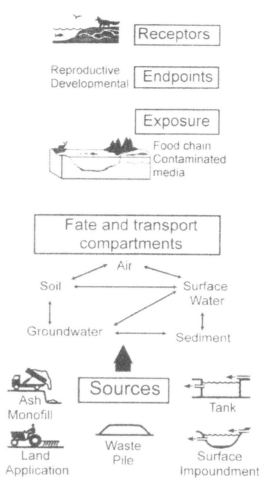

Figure 1 Environmental risk assessment multipathway analysis. (Adapted from U.S. EPA, 1995, Development of Human Health Based and Ecologically Based Exit Criteria for the Hazardous Waste Identification Project, Vol. 1, Figure 1-1, pgs. 1–6.)

vegetation. It is then more attractive as wildlife habitat, and hence the risk of exposure to contaminants becomes greater. Similarly, a plume of contaminated groundwater which has not yet reached a pond, may do so in several years. This future risk must be evaluated.

Before the SCM can be completed, the assessment endpoints of the ERA must be defined and rationale given for their selection. An assessment endpoint is the actual environmental value that is to be protected (Suter, 1989; Suter, et al. 2000). An example of an assessment endpoint would be "no less than a 20% decrease in the survival, growth, or reproduction in the largemouth bass population in the creek." Desirable characteristics for assessment endpoint species include (Suter, 1989; Suter et al., 2000):

- An assessment endpoint must be relevant to decision-making.
- The structure and function of components of the ecosystem must be understood in order to determine the ecological relevance or importance of the endpoint. Species that control the abundance and distribution of other species, and those that are involved in nutrient cycling and energy flow, are generally considered to be ecologically relevant.
- Selection of endpoints may be influenced by societal involvement and concern.
- Only species that are present, or likely to be present at the site, should be used to evaluate risks, regardless of the value or importance of the species.
- Since only some species at a site can be evaluated, endpoint species must be selected which are sensitive to the contaminants at the site, and are likely to receive high exposures. In this way, other species that may be less sensitive or receive lower exposures will also be protected. Other information necessary for each receptor species includes: diet composition; habitat preference/needs; home range size; intake rates of food, water, sediment, air, and soil; and body weight.
- Finally, an assessment endpoint must be able to be measured or modeled. If there is no method available to measure or model effects on an endpoint, evaluation of risk cannot be completed.

Because there are so many species and other ecosystem characteristics from which to choose assessment endpoints, all stakeholders (e.g., risk assessors, managers, regulators, the public) must agree on the appropriate assessment endpoints early in the ERA process. The remainder of the assessment cannot be completed until these have been chosen. After assessment endpoints have been selected, ecological risk assessors can select appropriate measurement endpoints for each assessment endpoint. "Measures of exposure and effect" are measurable environmental characteristics related to the valued characteristic chosen as an assessment endpoint (Suter, 1989; Suter et al., 2000). There are three categories of measures (U.S. EPA, 1989). "Measures of effect" are measurable changes in an attribute of an assessment endpoint in response to a stressor to which it has been exposed (formerly referred to as "measurement endpoints"). "Measures of exposure" are measures of stressor existence and movement in the environment and theis contact or co-occurrence with the assessment endpoint. "Measures of ecosystem and receptor characteristics" are measures of ecosystem characteristics that influence the behavior and location of assessment endpoints, the distribution of a stressor, and life history characteristics of the assessment endpoint that may affect exposure or response to the stressor. These three difference measures are especially important when completing a complex ERA.

ERAs that involve Superfund remedial actions must meet federal and state standards, requirements, criteria or limitations that are ARARs (U.S. EPA, 1989). ARARs which may need to be considered at a site include: Clean Water Act; Clean Air Act; Endangered Species Act; Fish and Wildlife Conservation Act; Wild and Scenic Rivers Act; Migratory Bird Treaty Act; and many others. If numerical ARARs exist, modeled or measured chemical concentrations in site media cannot exceed these values.

During problem formulation, historical data and/or site investigation data are used to prepare a preliminary list of Contaminants of Potential Ecological Concern (COPEC). In order to obtain a meaningful ERA, selection of COPECs must ensure that all contaminants that may contribute significantly to risk are included. Reasoning must be provided for exclusion of chemicals from the COPEC list. In this initial screening of contaminants, valid reasons may include (but not be limited to): contaminant concentrations at or below background levels; concentrations below ARARs, other regulatory concentrations, or toxicity benchmarks; or chemicals infrequently detected. Exclusion of COPECs because the HHRA excluded them is not a valid reason. This is because protection of human health does not guarantee protection of nonhuman biota. Several reasons for this are described in Table 1.

III. ECOLOGICAL EXPOSURE ASSESSMENT

ERA has several considerations that HHRA lacks. One of the most important factors affecting the exposure assessment is the spatial and temporal scale of the assessment. Spatially, exposure estimates must take into account the home range of, and the availability of, suitable habitat for the receptor species, relative to the areal extent of contamination. Temporal considerations include whether the receptor species is a resident or migrant species, and whether contaminant concentrations vary over the course of the year due to seasonal changes.

Another concept that is not often addressed in HHRA is the different level of protection afforded to different species. HHRAs are designed to protect individuals. In ERA, only threatened and endangered species, or other species of special legal (e.g., migratory birds) or public concern are evaluated for impacts at the individual level. For other species, protection is primarily afforded at the population level. For example, it is important to protect a population of deer at a site; individual deer will not be protected. Practically, this means that impacts on measures relevant to the population as a whole, such as survival and reproduction, are evaluated. Individual quality of life is not considered.

As in HHRA, for an exposure pathway to be complete, there must be a contaminated medium, a transport medium, receptor species, and an exposure route which enables the contaminant to enter the organism (e.g., ingestion, inhalation, root uptake, etc.). However ERA has unique exposure routes, such as fish respiration of water.

In the exposure assessment, contaminant concentrations at an exposure point are determined, or intake rates calculated. In the risk characterization, these concentrations are related to toxicological benchmarks; which are contaminant concentrations that are assumed not to be hazardous to the receptor species.

The exposure scenario in an ERA may not be the same scenario as the HHRA. ERA does not have a default "residential scenario," or "industrial scenario." However, hazardous waste sites often are industrial in nature. Scenarios are developed which are appropriate to the current land use. Like the human health assessment, the ERA may make assumptions regarding future land use. This future scenario may assume the site is abandoned and undergoes natural succession. Therefore, it is unreasonable to assume that the same wildlife species will be present in the current and future scenarios, especially if the habitat changes. All assumptions regarding exposure scenarios must be documented early in the ERA process.

Table 1   Differences Between Human Health and Ecological Risk Assessments

Component	Human Health Risk Assessment	Ecological Risk Assessment
Institutional controls	Institutional controls may be considered when selecting exposure parameters	Nonhuman organisms are not excluded from waste sites by controls, such as fences or signs.
Standard exposure factors	The U.S. EPA provides standard exposure parameters and toxicological benchmarks for humans	Risk assessors must generate their own exposure parameters and toxicity data.
Receptor species	Humans only	Nonhuman organisms (flora and fauna) and ecosystem properties (e.g., nutrient flow)
Exposure routes	Ingestion of food and water, incidental ingestion of soil, inhalation of contaminants from air, dermal contact, ingestion of fish fillets	As well as the exposure routes common to HHRA, other routes exist, such as fish respiring water, benthic organisms consuming sediments, small mammals burrowing in soil leading to enhanced exposure, fish-eating wildlife consume the entire fish and chemicals accumulate to a different degree in different organs.
Chemical form	Total metals in water are assumed to be available to humans.	Dissolved metals are available to aquatic biota for gill uptake.
Spatial scale	Often assumes a residential scenario at the site, regardless of appropriateness.	Scale is important, since a small site (e.g., a few acres) cannot support a population of larger organisms (e.g., deer, hawks), but could support small animal populations (e.g., shrews).
Temporal scale	Often only considered when seasonality may change chemical concentrations.	Seasonality is more important in ERA, often because of habitat changes or changes in organism behavior.

During characterization of the exposure environment, the relationship between the receptor species and the environment is detailed. Ecosystem characteristics can modify the nature and extent of contaminants. Chemicals may be transformed by microbial communities or through physical processes such as hydrolysis and photolysis. The environment also may affect bioavailability of contaminants. Physical stressors such as stream siltation and water temperature fluctuations may have considerable impact on ecological risks, and, therefore, must be described.

As part of the characterization of the exposure environment, it is also important to consider both the habitat requirements of receptor species and the amount of suitable habitat available at the site. Availability of habitat will determine the amount of use that a site receives. Because exposure cannot occur if receptor species are not present and receptor species will not be present if suitable habitat is not available, it is important to identify habitat requirements and availability early in the exposure assessment.

Selecting exposure routes depends on the endpoints to be evaluated. Several examples of endpoints and exposure routes are discussed below.

A. Fish Community

Fish are exposed to contaminants in surface water through respiration and dermal absorption. They also may be exposed through the consumption of contaminated sediment or food. There are two important considerations for the fish community. The first is that for inorganic contaminants, it is the dissolved fraction of the contaminant in the surface water that the fish are exposed to by inhalation (i.e., gill uptake). Practically speaking, this involves filtering the water sample through a 0.45 µm filter prior to analysis. HHRA calculates exposures using the total inorganic concentration in water. However, the particulate-bound fraction is not available to fish at the gill. Secondly, dermal absorption as a separate exposure route is not evaluated, because existing toxicity data for fish were generated either by feeding contaminated food to fish or exposing fish to contaminants in the water, without attempting separate evaluations of the various uptake routes.

B. Benthic Macroinvertebrate Community

Benthic macroinvertebrates live in or on contaminated sediments. They may be exposed through ingestion of the sediment or contaminated food. Also, benthic organisms may respire overlaying water or the sediment pore-water. Special considerations for this endpoint include the need for bulk sediment contaminant concentrations and pore water analyses, in order to compare these concentrations to benchmark concentrations (see below). For nonionic/nonpolar organic contaminants, bulk sediment concentrations are used. The organic carbon content of the sediment is also required. For ionic/polar organic contaminants, the sediment pore water must be analyzed. For inorganic contaminants, either analysis is adequate.

C. Soil Invertebrate Species

Soil invertebrates, such as earthworms, are in direct contact with contaminated soil. Also, the earthworm ingests large amounts of soil during feeding. Contaminants are in contact with and may be absorbed by the gut of the worm.

D. Terrestrial Plants

Plants are in direct contact with soil. Contaminants may be taken up from the soil at the root. Also, contaminants in shallow groundwater may be taken up by the plant roots. Airborne contaminants also may enter the plant through the leaf stomata.

E. Terrestrial Wildlife

As terrestrial wildlife move through the environment, they may be exposed to contamination via three pathways: oral, dermal, or inhalation. Oral exposure occurs through the consumption of contaminated food, water, or soil. Dermal exposure occurs when contaminants are absorbed directly through the skin. Inhalation exposure occurs when volatile compounds or fine particulates are respired into the lungs. While methods are available to assess dermal and inhalation exposure to humans, data necessary to estimate dermal and inhalation exposure are generally not available for wildlife However, these routes are generally considered to be negligible relative to other routes. Because contaminant exposure experienced by wildlife through both the dermal and inhalation pathways may be negligible, the majority of exposure is attributed to the oral exposure pathway. It should be noted that for some contaminants, dermal, and inhalation exposure may be significant. If these compounds are present, special attention should be paid to these pathways.

All sites should have more than one measurement of contaminants in each medium. Ideally, seasonal data would provide the most complete evaluation of contaminants present in the environment. Wherever possible, site-specific data should be used, rather than modeled data. Where EPCs must be modeled, the same methods and considerations are applicable to ERA as in HHRA.

EPCs are developed differently according to endpoint. For the fish community, the concentration of contaminant in water or sediment is used as the EPC. No exposure models are required. The upper 95% confidence limit on the mean water concentration may be used instead of the mean or maximum detected concentration. This is because chronic exposures of the maximally exposed aquatic organisms would be to spatially and temporally varying contaminant concentrations.

For the benthic, soil invertebrate and plant communities, the concentration in the sediment or soil at each sample location is used as the EPC. Again, no exposure models are required. However, in each of these cases, the maximum concentration in the sediment or soil should be used as the EPC because these organisms are not particularly mobile. The entire community could be exposed to the maximum concentration present in the medium.

For wildlife species, contaminant concentrations in food, water and soil are used in exposure models to estimate dose. Because wildlife are mobile, use various portions of a site, and are exposed through multiple media, the upper 95% confidence limit on the mean best represents the spatial and temporal integration of contaminant exposure wildlife will experience.

Exposure estimates for wildlife are usually expressed in terms of a body weight-normalized daily dose or mg contaminant per kg body weight per day (mg/kg/d). Exposure estimates expressed in this manner may then be compared to toxicological benchmarks for wildlife, or to doses reported in the toxicological literature.

Very few wildlife consume diets that consist exclusively of one food type. To meet nutrient needs for growth, maintenance, and reproduction, most wildlife consume varying amounts of multiple food types. Because it is unlikely that all food types consumed will contain the same contaminant concentrations, dietary diversity is of one of the most important exposure modifying factors.

To account for varying contaminant concentrations in different food types, exposure estimates should be weighted by the relative proportion of daily food consumption attributable to each food type, and the contaminant concentration in each food type. Each parameter in a wildlife contaminant intake equation must be obtained from the literature because few site-specific values are likely to be available. U.S. EPA's *Wildlife Exposure Factors Handbook* (U.S. EPA, 1993) contains a compilation of values for parameters such as diet composition, food intake rate, body weight, and home range for 15 birds, 11 mammals, and 8 reptiles and amphibians. The primary and secondary literature must be consulted for any parameter values not contained in this document or if the values provided are not appropriate for the site or become outdated.

One advantage that ERA has over HHRA is the ability to sample the receptor species itself. Rather than introducing modeling uncertainties, fish, benthic macroinvertebrates, soil invertebrates, plants, and some wildlife species (e.g., small mammals) can be sampled directly to give an indication of the bioavailability of environmental contaminants. Of course, it is not acceptable to destructively sample many species, such as rare, threatened, and endangered species, or those with high societal value or low abundance. However, when possible the additional sampling and analytical costs will be worth the added certainty in the exposure assessment and risk characterization.

Ideally, contaminant analysis of whole fish are used when conducting an exposure assessment on piscivorous species. However, fish body burdens may be estimated using bioaccumulation factors.

Professional judgement is required when selecting a parameter value for the exposure model. Full rationale for the selection of any parameter value must be provided in the exposure assessment. Exposure assessments will use a variety of data with varying degrees of uncertainty associated with them. Each assumption made will be a result of professional judgement but will still have some uncertainty. It is important that the exposure assessment document and characterize each source of uncertainty, including those associated with analytical data, exposure model variables, contaminant distribution and bioavailability, receptor species presence and sensitivity, and other incomplete exposure information.

IV. ECOLOGICAL EFFECTS ASSESSMENT

An ecological effects assessment includes a description of ecotoxicological benchmarks used in the assessment, toxicity profiles for contaminants of concern, and results of the field sampling efforts. The field data may include field survey information and toxicity test results.

Ecotoxicological benchmarks represent concentrations of chemicals in environmental media (i.e., water, soil, sediment, biota) that are presumed not to be hazardous to biota. There may be several benchmarks for each medium and each endpoint species, which allows for estimation of the magnitude of effects that may be expected based on the contaminant concentrations at the site. For example, there may be a benchmark for a "no-effect level," a "low-effect level," "chronic-effect level," a "population-effect level," and an "acute-effect level." Using all of these benchmarks will provide more information for decision makers than any one of the above.

There are few federal or state benchmarks currently available in the U.S. or elsewhere. Criteria that are used as benchmarks are the National Ambient Water

Quality Criteria for the Protection of Aquatic Life (NAWQC) (U.S. EPA, 1986). These are ARARs, and are used as benchmarks for the fish community and other water-column species (e.g., invertebrates such as daphnids). However, not all contaminants have these criteria. Therefore, other benchmarks are needed. Benchmarks for the fish, benthic, soil invertebrate, and plant communities, and wildlife are described briefly below. The primary source of toxicity information used in the development of these benchmarks is the open literature.

A. Fish Community

The acute and chronic NAWQC or state water quality criteria are ARARs and must be used as benchmarks. However, these were developed as broadly-applicable values, and thus it may be more appropriate to determine benchmarks for the geographical location and species present at the site. The literature should be reviewed for chronic values in systems similar to that at the site, whether it be a freshwater, estuarine, marine, hard-water, or soft-water system. Laboratory toxicity tests have been conducted on many different aquatic species for many contaminants. In fact, the aquatic system currently has the largest readily-available data base of contaminant concentration/effects data.

B. Benthic Community

There are several methods that may be used for calculating sediment benchmarks for the benthic community. For nonionic/nonpolar organic contaminants, the equilibrium partitioning approach is often employed. For inorganic contaminants, existing bulk sediment toxicity values from the literature may be used, or pore water concentrations of contaminant may be compared to existing NAWQC. Unfortunately, the database of single-contaminant exposure/ effects data for sediments is limited. The majority of the data come from contaminated sites and, therefore, multiple contaminants were present. However, sediment contamination is receiving more attention, and risk assessors and managers must stay current with respect to advances in the areas of sediment toxicology and policy.

C. Soil Invertebrate and Plant Communities

The plant community plays a dominant role in energy flow and nutrient cycling in ecosystems. Soil invertebrates and plants form the bases of many food webs. There is an extensive database for soil contaminants. However, the majority of endpoints used by researchers have been food crop species. While this information is crucial to human health risk assessors, it is not directly applicable to ecological risk issues.

The primary literature will be the major source of toxicity information that must be used in the development of toxicity benchmarks. Soil contamination impacts on plant, invertebrate, and even microbial communities are recent important issues. Again, this is an area within ERA in which it is imperative to remain current.

D. Wildlife

Wildlife benchmarks are particularly complicated because wildlife may be exposed to contaminants in their drinking water, the soil around them, and in their diet (both from plant and animal sources). Therefore, wildlife benchmarks must account for these multiple exposure routes. Benchmarks may be derived for each exposure route separately (for cases where exposure is through only one route) and also for total exposure. In the case of exposures from multiple routes, a benchmark (e.g., NOAEL, LOAEL) expressed as a dosage (e.g., mg contaminant/kg body weight/day) is used. The dosage is used rather than a concentration (e.g., mg contaminant/kg soil). Benchmarks for wildlife are species specific, in order to account for different species sensitivities, body weights, foraging habits, and diets. In the selection of appropriate benchmark values, the toxicological literature must be consulted, with emphasis on reproduction endpoints.

Contaminant toxicity profiles assist risk assessment readers to clearly understand the toxic effects of contaminants in the environment. Toxicity profiles in a risk assessment can provide a concise summary of relevant toxicity information. It is worth repeating the fact that the information must be relevant to the waste site and endpoints of concern. That is, the profile should not simply be a list of LD_{50}s for rats and mice. Dose/response information should be compiled for the contaminants that are found at the site, and for the receptor species of interest there.

Toxicity profiles also are useful for helping risk assessors and risk managers evaluate the extent and magnitude of risk. Because there are so many receptor species requiring evaluation in ERA, biological effects data for the species of interest must be presented if it is available, and data on surrogate species only when necessary, or if it will add to the reliability of the receptor species data. Contaminant concentrations at which lethal and sublethal effects (including behavioral modifications) are observed should be presented (i.e., dose/response information). Information such as the mobility of the chemical (e.g., water solubility, soil sorption, octanol/water partition coefficient), persistence in the environment (e.g., degradation half-life, bioconcentration factor), and its interactions with other contaminants will help risk managers make an informed decision and educate the public so that they may better understand, and hopefully feel more comfortable with, the decisions made about the site.

E. Sampling

Although general sampling issues will have necessarily been addressed before the ERA reached the effects assessment stage, it is worthwhile to note a few of them here. This will ensure that the risk assessor has mentioned and considered the potential impacts of these issues. Field surveys, toxicity tests, and ambient media chemical analyses are also addressed.

Before determining sample locations, sampling "reaches" must be defined. These are areas that may be impacted by specific contaminant sources. For example, one stream may have several contaminant sources along its length; a reach may be defined as that area between two sources. Sampling in reaches allows for the determination of the relative contribution of various sources to observed toxicity.

It is important not to forget to sample an appropriate background (or reference) site. In fact, it is better to have a few reference sites, to account for natural variability in the environment. In the past, there was a distinction between background (meaning pristine) and reference (meaning not impacted by this particular site). However, this distinction is losing popularity. It is necessary to know which definition is being used.

One facet of field sampling that is often forgotten when schedules are set is the problem of seasonality in field parameters. For a large portion of the country, winter hinders sampling efforts. For example, it is difficult to sample worms or fish when the ground and creeks are frozen. Also, bats hibernate during the winter, birds migrate, and rare plants are more difficult to identify when they are not in bloom. It is better to delay completion of a risk assessment than to collect data at an inappropriate time.

A waste site investigation will necessarily involve the coordination of a variety of investigators covering the various sampling tasks. The coordination is important in order to obtain results useful for the ERA. Some examples of necessary coordination include water, sediment, or soil toxicity tests being taken at the same time and from the same location as that taken for chemical analysis. It is less critical to coordinate other activities, such as collection of sediment samples, because, whereas water concentrations may change dramatically over a short period of time, sediment concentrations integrate contamination over a longer period of time.

1. Field Surveys

Field surveys have the advantage of giving a real-world indication of effects. However, the cause of any observed effects is likely to be unknown. For example, a decrease in young of the year fish may be due to contaminants that impact fish eggs or larvae, or may be due to natural causes, such as a storm event which caused increased water flow that eroded the spawning beds. Another disadvantage is that small changes are unlikely to be detected. Usually a greater than 20% decrease in a field parameter (e.g., population size, number of species) is necessary for it to be detected. Field surveys may be further complicated because without appropriate and comparable reference sites, interpretation of effects observed at the site is extremely difficult.

In the case of predictive ERAs, field surveys provide information on the environment that may receive contaminants in the future. It is important to have this information in order to document any future adverse impacts. Surveys may include wetland surveys, threatened and endangered species surveys, and aquatic and terrestrial community surveys. Each of these is discussed briefly below.

2. Wetland Survey

In the U.S., a wetland survey must be done for the site to identify and, if necessary, delineate wetlands. Note, it is easier (and less expensive) to identify than to delineate wetlands. It would only be necessary to delineate a wetland if remediation or other activities necessitated the destruction of all or part of the wetland.

3. Threatened and Endangered Species and Habitat Surveys

In the U.S., a survey must be done for threatened and endangered species and their habitat. The Endangered Species Act requires that the ERA assess threats to these species, sensitive habitats, and critical habitats of species protected under this legislation.

4. Aquatic Species and Habitats

Aquatic habitats may be sampled to determine the impacts on the fish community. Please note, the public often has concerns about fish sampling techniques such as electroshocking, because it sounds like a destructive technique. In fact, only a very few fish are killed using this technique. A few fish may be taken to the laboratory for chemical analysis if bioaccumulation of contaminants is considered a potential problem at the site. In addition to fish community structure, specific population parameters may be studied as well, such as age/class structure. This is important because a particular life stage of the organism (e.g., egg or larvae) may be more sensitive to the contaminants which may result in an absence of younger fish in the population. The benthic macroinvertebrate community, which is composed of organisms that live in or on the bottom sediment such as crayfish, aquatic worms, leeches, snails, shell fish, and insect larvae, also may be sampled. This is important because these organisms are an important source of fish food, and because these organisms are in contact with potentially-contaminated sediments. Benthic macroinvertebrates are not as mobile as fish, and hence are a good indication of contamination conditions at a particular reach of the water body. These organisms may be sampled destructively (e.g., preserved, taken back to the laboratory, identified, and counted) without public pressures to the contrary, and without concern for the invertebrate community which will quickly recolonize the sampled area.

5. Terrestrial Habitats

Terrestrial habitats often prove more difficult to sample than aquatic habitats. This is because most wildlife species are widely dispersed and generally secretive. This is not so, however, for plants and soil invertebrates. These receptors have little or no mobility and they represent the foundation of most terrestrial food webs. Sampling of plants and soil invertebrates, therefore, is critical for defining foodweb transport of contaminants at many affected sites. Because of the diversity of the terrestrial species that may be sampled or surveyed, many different sampling techniques are needed for these habitats.

6. Predictive and Retrospective Assessments

Toxicity tests are relied upon heavily for predictive assessments, and are valuable for retrospective assessments. In the latter case, toxicity tests give an indication of the toxicity of ambient media. Most often they are conducted in the laboratory, but they also may be done *in situ* in the field. Toxicity tests have an advantage over literature-derived toxicity information because most toxicity literature was derived using single chemicals. Waste sites typically have more than one chemical, and it is largely unknown how mixtures of chemicals affect various organisms. Therefore, a toxicity test may be used to determine if the mixture of chemicals at a site are toxic to biota. If impacts are recorded in the field surveys, toxicity tests may be used to confirm that contaminants in the medium are the cause of the observed effects. In predictive assessments, toxicity tests provide dose-response information for major COPECs.

Toxicity tests do have limitations. Typical exposure durations in a toxicity test are several days to a few weeks, which is unrealistic in terms of the exposures of organisms in the environment. However, it usually is not feasible to conduct a toxicity test throughout the life cycle of the organism. Also, there are very few standard toxicity tests using few species, and hence results must be extrapolated to the species of interest at the site.

Federal regulatory agencies as well as the American Society for Testing and Materials (ASTM) are continuing to develop guidance for conducting toxicity tests. Tests may be acute (short-term, usually with lethality as the endpoint) or chronic (longer-term, usually with growth, reproduction, or some other endpoint) (see Chapter 22).

7. Chemical Concentrations in Ambient Media

Samples of ambient media do not refer exclusively to ground water, surface water, sediment, soil, and air. This also includes the biota. Human health risk assessors cannot sample people, but ecological risk assessors can sample the biota in order to evaluate contaminant exposure and effects. This is an important source of information available to ecological risk assessors which may allow greater certainty in the ERA results.

Information on the speciation of the chemical in various media may be useful for contaminants, such as arsenic or chromium that have species with very different relative toxicities. Before sending the samples for analysis, ensure that the analytical method used will have detection limits below the regulatory concentrations of interest (e.g., ARARs) and the concentration that would produce an unacceptable risk, unless this is not technically or economically feasible. If these detection limits cannot be met, there will be added uncertainty in the risk assessment, because it will not be known whether these contaminants are present or not, and hence whether they constitute a risk. Chemical concentrations in media at a site, along with the abundant single chemical toxicity data available in the literature, may be used to determine the specific causes of the impacts observed in the field surveys or toxicity tests, and define the sources of the contamination. These data are used in predictive ERAs to model effects of contaminant exposures. However, the measured concentrations may not be indicative of the bioavailable fraction (e.g., chemicals may be bound to soil particles and hence not be available for uptake by organisms). As mentioned before, there is little toxicity information for chemical mixtures, and toxicity studies reported in the literature often used common laboratory organisms. This information, used in conjunction with toxicity test data and/or field surveys can allow the risk characterization to be completed using a weight-of-evidence approach.

F. Sources of Other Effects Information

Supplementary information that may be useful in the interpretation of ecological data includes an analysis of biomarkers. Biomarkers serve as sensitive indicators in individual organisms of exposure to contaminants or other sublethal stressors. They are typically physiological or biochemical responses, such as enzyme concentrations, genetic abnormalities, histopathological abnormalities or body burdens of contaminants. While biomarkers give an indication of exposure to stressors, they rarely yield information on the impacts of this exposure on the population. That is, if a fish has an elevated level of liver enzymes, what does this mean to the fish? Ecological risk assessment is concerned primarily with the viability of organism populations, not physiological effects in a single individual. However, some biomarkers are chemical-specific, and hence may provide valuable information on the potential cause of observed toxic effects. For example, increased blood levels of the enzyme delta-aminolevulinic acid dehydratase (ALAD) indicates exposure to lead.

V. ECOLOGICAL RISK CHARACTERIZATION

Historically, the most common approach to risk characterization was the calculation of hazard quotients. This was adopted from the HHRA field, where this approach is still used. Simply, it compares chemical concentrations in ambient media to some toxicity benchmark. If the quotient exceeds 1, there is a potentially unacceptable risk. While this approach is simple, it is relatively meaningless in ERA. It has found use in predictive assessments, and screening level (otherwise known as preliminary or tier I) retrospective ERAs. In the screening level assessments, the quotient method is used to refine the contaminant of concern list and focus a subsequent, more detailed assessment. However, for a baseline ERA, this approach should be used with caution. It is especially important to realize that the magnitude of the exceedance in the hazard quotient has no quantitative relation to the magnitude of potential toxic effects. Calculating several hazard quotients using different benchmarks (e.g., derived from different toxicity data, such as acute, chronic, or population level effects) has more direct applicability than using a single benchmark.

Because ecological effects can be measured in a retrospective ERA, an epidemiological, weight-of-evidence approach can be used. This approach depends upon weighing multiple lines of evidence, such as those provided by the field surveys, toxicity tests, and ambient media chemical analyses and literature toxicity data. Risk assessors, risk managers, and the public will have more confidence in a risk assessment that uses the weight-of-evidence approach, because it integrates all sources of information, attempts to reconcile conflicting data, and can account for the bioavailable fraction of chemicals in the environment, and the effects of multiple contaminants.

The primary line of evidence in the weight-of-evidence approach is the field survey data. Field surveys monitor actual ecological impacts, and therefore are the most credible line of evidence. However, as discussed in the Ecological Effects Assessment section, field surveys have their limitations. Also, many ERAs will not have the budget necessary to conduct field surveys, and some species are not easily surveyed (e.g., nocturnal, migratory, secretive, or wide-ranging species). Also, small impacts are not readily apparent in field surveys. Therefore, other lines of evidence are used as support.

Toxicity tests give an indication of whether ambient media are toxic. When several contaminants exceed benchmarks and there is an impact in the toxicity tests or field surveys, it is important and necessary to evaluate the magnitude of the effect caused by the contaminants which exceeded benchmarks. Using media contaminant analysis and the information provided in the toxicity profile (See Ecological Effects Assessment section), an evaluation is conducted of which contaminants could be responsible for the observed toxicity. Combining all of these lines of evidence will present a picture of actual or potential impacts at the site, and contaminants responsible for the impacts. In some cases, benchmarks may indicate unacceptable risk while field observations show no measurable impacts. Therefore, the weight of evidence suggests no unacceptable risks to a community, even though contaminant concentrations exceeded benchmarks. Reconciling multiple lines of evidence is difficult, and requires experience and understanding of the ecosystem being evaluated.

A. Uncertainties

Uncertainties are inherent in all risk assessments. The nature and magnitude of uncertainties depend on the amount and quality of data available, the degree of knowledge concerning site conditions, and the assumptions made to perform the assessment.

For example, there is uncertainty associated with the toxicity values selected as benchmarks. Because there is no one single benchmark for each contaminant, medium, and receptor, it is necessary to document any limitations in the use of a particular benchmark value.

Incomplete or absent toxicity information must be acknowledged. Several contaminants may not have any toxicity information. Toxicological benchmarks and profiles will not be available for these contaminants and, therefore, risks cannot be assessed.

Uncertainties associated with the bioavailability of contaminants must be discussed, especially if toxicity and field survey data are lacking for the assessment. These latter types of data do provide an indication of contaminant bioavailability. Field survey techniques may have specific uncertainties associated with them that must be documented.

Uncertainty in the risk characterization often comes from the lack of multiple lines of evidence in many assessments. The fewer the lines of evidence, the less confidence in the risk characterization. Uncertainties associated with the extrapolation of toxicity test results to effects on endpoint species must be addressed. Toxicity tests typically use only a few common species that are easy to rear and maintain in the laboratory. Often, these are not the assessment endpoint species in the ERA. Species may vary widely in their sensitivity to contaminants. For example, rainbow trout, brown trout, and brook trout have very different sensitivities, although they are all trout species.

Quantitative uncertainty analysis may not be necessary if risk calculations indicate that the risk is clearly below a level of concern. However, if quantitative analysis is warranted, simple models or computer-assisted numerical approaches may be used. One common numerical approach is the Monte Carlo method (see Risk Assessment Forum, 1996, 1997, 1999).

VI. COMPARISONS WITH OTHER STUDIES

Results of the risk assessment may be compared with results obtained from other sites in a similar environment and with similar contamination, or previous investigations at the same site. While not a mandatory component of the ERA, this exercise may help in the interpretation of results, and aid in the evaluation of remedial alternatives, or in the analysis of potential environmental impacts. This is especially true if a similar site has already undergone remediation, because the efficacy of the chosen alternative may be evaluated.

VII. CONCLUDING THE ERA

At the end of an ERA, conclusions and recommendations are often requested by managers and, therefore, are provided. In this section, it is determined if all DQOs have been met. Preliminary remedial action objectives may be calculated, which are concentrations of contaminants identified as the key contributors to risk, in order to protect the environment. The risk managers then use this information, in combination with other considerations (e.g., public, legal, regulatory issues, cost), in order to identify remedial options or pollution prevention/control strategies.

VIII. CONCLUSION

A quality ERA must be completed by a qualified ERA team. Good planning at the beginning of the ERA, including the development of DQO, will help ensure an acceptable product. Documentation of exposure assumptions is essential. Collection of field survey and toxicity test data, along with ambient chemical concentration data, will allow the use of the weight-of-evidence approach to risk characterization. Risk estimates using all available data and a documentation of uncertainties will provide the risk managers with enough information to make credible, supportable decisions.

REFERENCES

American Society for Testing and Materials, *Guide for Developing Conceptual Site Models for Contaminated Sites,* Philadelphia, E1689.

Risk Assessment Forum, *Summary Report for the Workshop on Monte Carlo Analysis*, U.S. Environmental Protection Agency, Washington, 1996.

Risk Assessment Forum, *Guiding Principles for Monte Carlo Analysis*, Washington, 1997.

Risk Assessment Forum, *Report of the Workshop on Selecting Input Distributions for Probabilistic Assessments*, Washington, 1999.

Suter, G.W., II et al., *Ecological Risk Assessment for Contaminated Sites*, Lewis Publishers, Boca Raton, FL, 2000.

U.S. Environmental Protection Agency, *Quality Criteria for Water*, Office of Water, Washington, 1986.

U.S. Environmental Protection Agency, *Risk Assessment Guidelines for Superfund, Vol. II, Environmental Evaluation Manual*, Washington, 1989.

U.S. Environmental Protection Agency, *Framework for Ecological Risk Assessment*, Risk Assessment Forum, Washington, 1992.

U.S. Environmental Protection Agency, *Wildlife Exposure Factors Handbook, Vol. I*, Office of Research and Development, Washington, 1993a.

U.S. Environmental Protection Agency, *Ecological Risk Assessment Guidance for Superfund: Process for Designing and Conducting Ecological Risk Assessments, Interim Final*, Emergency Response Team, Washington, 1997.

U.S. Environmental Protection Agency, *Guidelines for Ecological Risk Assessment*, Risk Assessment Forum, Washington, 1998.

CHAPTER 4

Risk Assessment Project Planning (Phase I)

David A. Belluck and Sally L. Benjamin

CONTENTS

I. Introduction 100
    A. Building a Foundation for Contracting a Risk Assessment 100
    B. Documents Generated Prior to Beginning
       a Risk Assessment Report 100
II. Phase I: Project Planning 101
    A. Determine Need for a Risk Assessment 101
    B. Select a Project Manager 103
    C. Build a Risk Assessment Project Team 105
    D. Organize a Project Management Team 107
    E. Document Project Expectations 108
    F. Identify Project Limitations 109
    G. Scoping a Project 110
    H. Fund the Project 126
    I. Solicit Contractor Qualifications or Proposals 127
    J. Host a Kick-off Meeting for Potential Contractors 129
    K. Evaluate Bids 132
    L. Select a Contractor 132
    M. Negotiate a Contract 152
    N. Negotiate a Work Plan 153
    O. Hire the Contractor 166
III. Conclusion 166
    References 166

I. INTRODUCTION

A. Building a Foundation for Contracting a Risk Assessment

As this book will show, developing a useful and enforceable risk assessment contract can be complex. While no single approach is ideal for all situations, it is possible to take an organized approach to developing a Scope of Work* and request for proposals, and to contracting with a risk assessment firm. Experience indicates an organized approach helps a risk assessment project succeed. Detailed planning for a risk assessment, with its concomitant generation of planning reports and memos, increases likelihood of all parties involved fully understanding responsibilities and sharing performance expectations. As planning proceeds into contracting, effort and detail expended on document production can increase dramatically.

It is absolutely essential to understand roles of players in a planning process. Central to a project, by definition, is a project manager, who functions on behalf of an organization that needs a risk assessment, shepherding the entire process. This "contractee" project manager manages work by the "contractor" risk assessor, and organizes others into an effective team, including recruiting a project team and hiring a risk advisor. This chapter discusses planning a risk assessment, including:

- Determining whether a risk assessment is necessary, selecting a project manager, and building and organizing a risk assessment project team
- Understanding project expectations and limitations
- Scoping and funding a project, and soliciting and evaluating contractor proposals
- Negotiating a contract and work plan
- Hiring a contractor

It presents all of the steps between recognizing a need for a risk assessment and actually conducting a risk assessment, as discussed in Chapter 5.

B. Documents Generated Prior to Beginning a Risk Assessment Report

Examples provided in this book illustrate methods, and associated documentation, that a project team can produce in preparation for hiring a risk assessment contractor. These include, in order of generation:

- Team briefing document (optional)
- Project planning tables (optional)
- Risk assessment project planning document (optional)
- Scope of Work
- RFP/RFQ
- Work plan
- Contract

---

* A Scope of Work may also be termed a Work Scope or a scoping document. In this book we use Scope of Work and scoping document interchangeably.

Following contract signature by all relevant parties, the contractor mobilizes their staff and begins the work of writing the risk assessment report.

The next three chapters will address how to plan, manage, and conclude a risk assessment project, using iterative review. This chapter will present planning for a risk assessment project. Chapter 5 will describe management and Chapter 6 will address how to draw a risk assessment project to a close.

II. PHASE I: PROJECT PLANNING

Phase I of the risk assessment deals with the first fifteen steps; these steps comprise project planning. From the decision to undertake a risk assessment project, project planning proceeds through defining the project purpose, organizing the process to be followed, and determining the work products essential to achieve that purpose. Note that Phase I, Project Planning, represents slightly more than half of all steps in a risk assessment project (see Table 1). Although planning may seem like a luxury when time is short and resources are scarce, think of it as the foundation that will support all other project work. Remember, each part of a risk assessment contributes to subsequent report sections. A faulty report section will weaken the entire report and may even render it technically (or politically) inadequate.

A. Determine Need for a Risk Assessment

A risk assessment project should only proceed for very good reasons. Determining whether an HHRA or ERA is needed can be simple or very complex. Simple decision-making happens when a government agency, court, or law requires generation of a risk assessment. Complex decision-making happens when generating a risk assessment is a discretionary process. Complexity arises when one has to determine appropriate risk assessment type (human health, ecological, or both), and establish costs, timelines, levels of effort, and technical rigor. In discretionary situations, political concerns constitute a key aspect of decision-making. Of course, if sufficient information is available showing no or very limited environmental releases of chemicals, or no or minimal habitat alteration, a risk assessment may not be needed.

A decision-maker might feel overwhelmed when facing so many factors impinging on a decision to undertake a discretionary risk assessment. Decision matrices are one simplifying tool. Such matrices can help organize a complex situation by focusing on critical factors and, thus, leading a decision-maker through a series of logical choices, to a conclusion to either proceed with a project, or not. This creates a decision framework and helps clarify and organize key information about a site, activity, or process of concern, including chemical fate and transport and human health or ecological toxicology. Decision matrices may be case-specific or generic, i.e., applicable to any site, facility, or activity. A decision maker often begins with a generic decision matrix and proceeds to consider case-specific factors, before finally deciding (see Table 2).

In all cases where a risk assessment is required, or determined to be needed, parties to the risk assessment process must determine whether to conduct an HHRA,

Table 1 Iterative Review Comment and Approval Process for Human Health Risk Assessment

Phase I: Planning a Risk Assessment Project	
Step 1.	Determine need for a risk assessment
Step 2.	Select a project manager
Step 3.	Build a risk assessment project team
Step 4.	Organize a project management team
Step 5.	Document project expectations
Step 6.	Identify project limitations
Step 7.	Scope the project
Step 8.	Fund the project
Step 9.	Solicit contractor qualifications or proposals
Step 10.	Host a kick-off meeting for potential contractors (optional)
Step 11.	Evaluate bids
Step 12.	Select a contractor
Step 13.	Negotiate a contract
Step 14.	Negotiate a work plan
Step 15.	Hire the contractor
Phase II: Managing Risk Assessment Report Development	
Step 16.	Begin contractor work
Step 17.	Implement iterative review, comment, and approval of interim deliverables
Step 18.	Hazard Assessment
Step 19.	Exposure Assessment
Step 20.	Toxicity Assessment
Step 21.	Risk Characterization
Step 22.	Conduct a final review of the draft risk assessment report
Step 23.	Accept the final draft
Phase III: Concluding a Risk Assessment Contract	
Step 24.	Close Contract
Phase IV: Follow-up Studies and Activities	
Step 25.	Risk Management and Communication
Step 26.	Post-Risk Assessment Report Activities and Studies

ERA, or both types of risk assessments. After determining the types of risk assessment to perform, persons scoping the risk assessment must decide how much effort must go into the risk assessment for it to withstand the expected level of review. This is a pure judgement call and is not scientific at all. It is usually wise to conduct a very rigorous risk assessment when a project has a high political profile; very toxic chemicals; or high quantities of one or more chemicals on, in, above, or moving off the site to points where people or animals are exposed; or a valued resource may

Table 2 Simple Key to Decide Whether to Undertake a Risk Assessment

A	Does the site, activity, or facility actually/ probably:	Result in human exposure to potentially toxic substances?	Yes	Perform an HHRA
			No	Go to B
		Result in nonhuman exposure to potentially toxic substances?	Yes	Perform an ERA
			No	Go to B
B	Is a HHRA or ERA required by a government agency?		Yes	Perform required risk assessment
			No	Go to C
C	Will performing an HHRA or ERA constitute acceptable "due diligence" for legal liability purposes?		Yes	Perform required risk assessment
			No	Go to D
D	Will HHRA or ERA assuage community health concerns?		Yes	Perform a risk assessment
			No	No risk assessment needed

be altered. When these factors are not important, a less rigorous risk assessment might suffice.

Although human health and environmental protection are the primary reasons for risk assessment, the bottom line in determining the need for a given type of risk assessment is often meeting the letter or intent of applicable laws, attempting to minimize bad publicity, or taking a defensive posture toward possible litigation or regulatory intervention. Those who pay for a risk assessment or request a risk assessment be done should ask two questions:

- Can the projected risk assessment survive expected peer review, media, academic, neighborhood, and government scrutiny?
- If not, what are the ramifications associated with developing an unacceptable risk assessment?

All these are value-based or political judgements, rather than scientific judgements. Thus, it all comes down to reading the legal, political, and economic situation correctly for a given project. Matching the level of risk assessment technical rigor to a project is more of an art than a science.

Finally, who will do the risk assessment, how much it will cost, and when will it start and end, become important questions.

At some point, despite less than perfect knowledge, a decision must be made to either perform a risk assessment, or not. Otherwise, analysis of case-specific factors continues, until the essence of a risk assessment is completed on the need for a risk assessment. A decision matrix can help the decision-maker avoid this bind.

B. Select a Project Manager

If the situation justifies a risk assessment, the next step is to select a person to manage the risk assessment project, termed a "project manager." A project manager is at the center of all activities during every step of the project. A project manager's duties vary, depending on the project, but typically include:

- Building and managing a project team
- Obtaining project resources
- Defining the project purpose
- Selecting and managing an external contractor, if external expertise is required
- Ensuring delivery of a risk assessment of acceptable quality, on-time, and within-budget

Success in this high-pressure, high-profile position requires good political instincts, solid technical credentials, stellar ethics, and aptitude for organization and human relations.

Ideally, a project manager's duties, roles, and powers are clearly defined by top management. This is important, since a typical project manager must beg, borrow, or lure staff and other resources for the risk assessment project from on-going programs. Efforts to acquire staff and resources necessary for a risk assessment project can be curtailed if managers of long-standing programs, with established power bases, fail to sense unambiguous support for the project from top management.

A project manager must consider the nature and extent of a project and establish general project parameters, identifying:

- Site, activity, or facility of concern
- Type of risk assessment required (human health risk assessment, ecological risk assessment, or both)
- Potentially-exposed populations
- Key decision-makers (individuals and organizations) expected to use the risk assessment report or its risk estimates
- Performance standards required for the report and for individual tasks (see below)
- Depth and breath of the risk analysis
- Project budget
- Project time frame

Articulating these basic project parameters is the essence of initial risk assessment project scoping, discussed in detail below.

After defining the general nature of a risk assessment project, a project manager determines whether a project can be accomplished with in-house expertise or whether a risk assessment contractor will be needed. Typically, a contractor is hired. Human health and ecological risk assessment report production is a complex, highly specialized discipline with elaborate regulatory requirements. Aside from environmental consulting firms and government agencies, few organizations possess the in-house expertise to produce an acceptable risk assessment report. Due to the costs of training and supporting all the technical disciplines required to produce a risk assessment report, even government agencies hire consultants to supplement their internal risk assessment capabilities. Thus, public and private organizations of all sizes and all levels of sophistication tend to rely, to some degree, on contractors to produce risk assessment reports.

C. Build a Risk Assessment Project Team

The third step is to build an internal team to work on the risk assessment, a "project team." A project team will assist with project management and review of consultant deliverables. Project managers can usually find technical experts to serve on a project team within their organization.

If an organization has adequate staff to assist with each phase of the project, these persons should be tapped for a project team. Organizations rarely have sufficient internal expertise to provide all the skills and time needed for a risk assessment project. So, environmental risk assessment contractors are hired as project consultants to supplement internal resources. Contractors may be hired both as project consultants and as "risk advisors." Project consultants produce the risk assessment report. Risk advisors review the work of project consultants and, thus, help ensure that the project consultants do a good job for their client.

Internal experts, serving on a project team, must function as a team. This requires them to have a shared goal for the project and to agree on a process for achieving that goal. This is the essence of teamwork. If a project manager is lucky, the organization will have team-building specialists on staff who can help create a cohesive project team. If not, a project manager should undertake team-building as an important project management duty. Team-builders use many ingenious techniques. In essence, most of these techniques are structured discussions to define the team's goal, identify appropriate member roles, clarify group expectations, and establish working rules aimed at encouraging collaboration through a planned process. Most people find team work awkward, at first, but will adjust, if team-building occurs in a reasonable and respectful manner.

A project manager must build a project team with members who possess proper technical qualifications, work well in a team, and are willing to commit to the project. Few risk assessment project managers are empowered to hire staff. Instead, a project manager typically forms a project team by negotiating with management for permission to staff a team by drawing from existing personnel or, perhaps, from staff of sister organizations or agencies. Regardless of their origin, most internal experts on a risk assessment project team are "on loan" from somewhere and, therefore, must balance risk assessment project demands with those of their direct supervisor.

Risk assessment project managers should seek the best available technical and administrative personnel. Unfortunately, such capable staff are also people that managers consider indispensable. So, a project manager must negotiate skillfully for their services.

1. Identifying Required Skills

Staffing a risk assessment project team is a process of identifying ideal staff, negotiating to recruit these people, and compromising between required skills and available personnel. A project manager starts by considering what technical skills a project team requires in each team member and listing desired skill levels, experience, and educational backgrounds. Then, a project manager identifies persons with desired expertise. If some areas of expertise are not available in-house, a project manager considers whether experts are available from other offices or from a sister organization.

Attributes other than technical skills may make certain people more desirable team members than others. Does a certain person have special influence, as well as technical expertise? Is balanced representation important? Do some people always cause friction? A project manager must consider these human factors.

Finally, a project manager considers how to recruit the best people possible. This will be challenging and probably not entirely successful. Managers are rarely eager to share their best staff members. When negotiating for loan of staff, first, a project manager should recognize that these people are probably already committed to other important projects. Second, if project needs don't coincide with those of management, there is no incentive for other managers to loan staff members, even for a limited time period. Therefore, a project manager should emphasize how project needs align with other managers' needs. This requires some insight, investigation, and a strategy.

Project managers should be creative, considering what a risk assessment project can offer in exchange for use of staff and listening carefully during negotiations for clues of what another manager needs. Possible incentives include:

- An opportunity for junior staff who are loaned to a project to gain experience or training
- An opportunity for technical specialists to learn other skills, such as management, negotiations, or other technical skills
- Internal recognition of a manager for cooperation on an important project
- A chance to earn chits for use in future deals

Project managers should avoid merely accepting a grudging offer of the most junior or least skilled staff and, instead, counter with an offer to use inexperienced staff on a project to allow them to gain technical skills, if desired experts are also assigned.

A systematic and comprehensive team-building process requires an investment of time between the decision point to undertake a risk assessment and the point when a contractor initiates work. Within this window of time, a project manager must build a team of internal experts, termed a project team, which will help develop and oversee a project and will bring internal credibility to the final product. A project manager may be tempted to minimize team-building and plunge into "real work" of risk assessment, but it is wise to resist this impulse. Before people can effectively work together, they must become a team, and, thus:

- They must build a common problem definition and a shared project purpose.
- They must agree on procedures and roles.
- They must develop a sense of cohesiveness and trust.

Investing in team building at the onset will yield benefits throughout the project. It will improve quality of the Scope of Work, ease selecting and hiring a contractor, and heighten attention with which project team members review contractor work.

D. Organize a Project Management Team

A project manager might choose to create a "project management team," a project team subunit, to advise on nontechnical project issues. Persons with specialized administrative expertise (attorneys, financial experts, clerical managers, and personnel specialists) serve as advisors.

1. Kick-off meeting

A project team's first meeting, a project team "kick-off" meeting, officially starts team work. This meeting is a project manager's best opportunity to create a healthy team attitude. Working relationships begin at this meeting. Therefore, drafting a meeting agenda and selecting participants require care.

a. Participants

Every member of a project team and, of course, their project manager should participate in a kickoff meeting. If a risk advisor has been identified, this person should also attend. Deciding who to invite depends on meeting goals. Internal experts and managers to be intimately involved with planning, developing, and reviewing a risk assessment might need to attend. Ideally, such influential people would serve on a project team, but formal involvement is not always realistic. If these people will be informal team members, bound by identical rules as full team members, they should attend. Otherwise, participation of influential outsiders should be limited. A brief pep talk from a top manager, who then leaves, might be appropriate. Team members universally resist team-building efforts, however, if they feel they are being observed by outsiders, especially powerful managers.

b. Agenda

A specific kick-off meeting agenda depends on time and on timing. If a project has long timelines, an entire meeting might be dedicated to introducing the project manager's concept of team work. As an introduction to team work, a project manager should allot plenty of time for a project team discussion of team member roles and project goals. By exploring roles, a project manager will deal with concerns about time commitments, establish a uniform set of realistic team expectations, and allow time for project team buy-in. Discussing project goals helps establish a common team purpose, identifies areas of agreement on technical and procedural issues and sets a pattern for resolving disagreements. A project manager who facilitates such a discussion, both lays claim to a leadership role and provides a practical team work demonstration. This can help set a participatory tone that can positively influence an entire project. In addition to initiating team building, a project manager should use a kick-off meeting to orient a team to contract management (i.e., iterative review or reactive management, which are discussed in detail throughout Part I) and to orient team members to the standard four-step process of risk assessment (see Chapters 2, 3, and 5).

Team-building takes time. If a project manager and project team have worked together before, or if time is short, a project manager might choose to move a team through early stages of teamwork quickly and tackle substantive issues during the first meeting. In this case, for example, a team might draft project timelines and discuss scoping and contractor selection, in detail. A project manager might even assign team members to initiate work on scoping and on logistics of contractor recruitment. As with many aspects of project management, this choice is a professional judgement call.

E. Document Project Expectations

Next, a project manager and project team collaborate to establish project expectations. These expectations form a basis for performance standard development. Consequently, clearly articulated expectations are essential because they must be linked to contractor performance, or a project cannot succeed. In addition, a project manager must identify and resolve conflicting expectations.

A project manager can identify expectations through a series of interviews. An interview might begin by sharing a preliminary list of expectations, simply as a starting point for discussion. A project manager should develop this list with input from project team members to ensure it is realistic and to promote a sense of team ownership. Depending on the nature of a project, a project manager might choose to track frequency of each expectation being raised, as a rough gauge of importance. A project manager certainly should note who raises certain expectations, paying particular attention to opinions of individuals who control funds, pronounce project success or failure, and who will actually use risk findings.

A project manager then organizes a list of project team expectations into a preliminary expectations list. Some items will deal with process (how a project proceeds), while others relate to work products. Items should be grouped by their relevance to process or product characteristics. These groups of preliminary expectations eventually become formalized into project performance standards, i.e., process standards and product standards. Process standards and product standards, jointly referred to as performance standards, are defined and discussed in detail in Chapter 1, Section VI.

A project manager reviews preliminary information, analyses patterns, and notes any apparent conflicts, in preparation for another series of interviews with influential individuals. Influential individuals include anyone who will evaluate project success (either informally as end-users of a risk assessment report, or formally as reviewers of report technical or regulatory adequacy). A project manager should meet with as many influential individuals as possible to learn their expectations. Understanding end-user expectations and formal review requirements is essential, because these people judge project success.

End-users are individuals within an organization who will use a risk assessment to make important decisions. They are probably top-level managers, risk managers, or senior staff who counsel top management. In meeting with end-users, a project manager should explore why they feel a project is necessary, how they expect to use report findings, and what secondary benefits they hope to gain from a project.

This might be an opportunity for a project manager to understand and resolve conflicting expectations. It is not a time to make promises, but rather to listen and try to understand what these influential people need from a project. A written record should be kept of each interview. Those interviewed should be invited to review a synopsis of interview results to ensure that all important points are captured.

Regulators review a risk assessment report's adequacy for regulatory decisions (e.g., acceptable or unacceptable risks) and, therefore, must also be consulted about expectations, if a risk assessment is undertaken to meet regulatory requirements. A project manager should meet with regulatory staff who will review a project or, at least, review a written description of agency review standards and risk assessment requirements. Regulators should have a written policy on environmental risk assessment, or risk assessment technical guidance, articulating agency requirements. If an agency does not have an official policy in writing, agency expectations may be gleaned from a review of previously accepted risk assessment reports, assuming of course that agency policy, staff, or leadership have not radically changed.

Meeting with regulatory review staff is ideal, especially if their review determines the adequacy of the completed report and validity of the risk findings. It offers a project manager a tremendous learning opportunity. Regulatory review staff often have experience as risk assessors and project managers, as well as reviewers (perhaps more professional experience than any other technical resource). Better still, regulators' assistance is usually available at little, or no, cost. Astute project managers recognize regulators as a resource worth cultivating.

Clashes in expectations with regulators must be resolved early in scoping. A project manager should carefully integrate reviewers' expectations into a Scope of Work because they represent definitive technical oversight.

F. Identify Project Limitations

Limitations must be identified as early as possible so they do not derail a project. After a project manager has identified report expectations, a project manager and project team must identify project limitations. Time and money are two likely constraints. Limits of either will influence all other aspects of a project. Highly rigorous projects are generally more costly and lengthy than less rigorous projects. When timelines are compressed, either technical rigor suffers or costs increase dramatically, or both. On the other hand, a break point exists in the relationship between time, money, and rigor. For example, it may simply be impossible to perform at a high degree of rigor, if project timelines are unreasonably short, no matter how much funding is available.

Expertise is another significant limitation. Persons with key technical skills and significant professional experience can be hard to locate or impossible to hire because they are in high demand or short supply. Successful consultants have on-going professional commitments. Projects are scheduled far in advance. Conflicts of interest may further limit available qualified scientists. For example, some consultants work only for a particular type of industry, or only for government. They strive for on-going relationships and assiduously avoid any project that might impinge on their potential for future income from a long-term client. Thus, an organization may learn that its technical requirements, budget, time frame, geographical preferences, or other standards are unrealistic. If so, it will need to adjust project expectations.

After selecting and organizing team members and delineating project expectations and limitations, a project team generates its first major work product, a detailed Scope of Work, which will serve as a guide for the entire project.

G. Scoping a Project

1. Organizing Information

A risk assessment project team develops a Scope of Work to identify project needs and to select methods to meet them as efficiently as possible. A risk assessment Scope of Work covers all aspects of a risk assessment report. It presents a project as a series of deliverables (interim and final work products) and states performance standards (process and product standards) that guide development and evaluation of each deliverable. Work scopes can also specify important project-specific requirements, such as need for:

- Specialized work tasks
- Staff with specific education, experience, or skills qualifications
- Compliance with technical requirements of regulatory and governmental agencies
- Specialized facilities and equipment
- Willingness to cooperate with iterative review, comment and approval of interim and final deliverables

Typically, a Scope of Work also specifies a closing date, format requirements, and page limitations for contractors' bids. A clear, specific, and thorough Scope of Work is a worthwhile undertaking. It will guide all subsequent project work, keep contractor work properly focused, and avoid project delays and cost overruns. A systematic approach to developing a Scope of Work helps ensure that nothing is missed in project planning.

The following sections elaborate on how best to prepare to write a Scope of Work. First, data (and other information) is gathered and organized. Next, although not strictly required, a briefing document is prepared. Writing a briefing document helps a project manager and team evaluate adequacy of information they have amassed for drafting a Scope of Work. If available information suffices, a Scope of Work is prepared. If severe information gaps exist, they must somehow be addressed, prior to proceeding with a risk assessment.

a. Locate Existing Information for Use in a Scope of Work

In preparing a Scope of Work, a project manager and a project team become familiar with subject and circumstances of a risk assessment project. They gather sufficient information to understand potential problems associated with a site, activity, or facility of concern and begin to analyze this information. A project manager is typically responsible for collecting existing data for review and analysis by a technical support team.* Available information might include:

- Chemical contamination or release data
- Physical data
- Previous studies
- Risk assessment process requirements
- Report data requirements
- Regulatory requirements

This information will serve as a factual basis for scoping discussions and for drafting of a Scope of Work. It also helps a project team get a sense of what types of studies already exist and those that a contractor must perform to adequately characterize and assess risk associated with a project. Types of required data sets can vary. Project managers should verify current requirements with appropriate regulatory agencies. Current requirements should be incorporated into Scope of Work data needs and may need to be reflected in other scoping decisions as well.

Information pertaining to a risk assessment project can come from a variety of sources. Creativity pays dividends at this point. Information sources to consider include: government agency files, site owner/operator records, professional trade organizations, libraries and other collections, and files of knowledgeable individuals, each addressed briefly below. There are many other possible resources, of course, including scientific journals, magazines, newspapers, commercial databases, government agency bibliographies, and circulars. One useful, and often overlooked information source, is comprised of private companies that produce fire insurance maps of property uses.

Government files — Government files are sources of obvious value. U.S. EPA is a good source of environmental information on a site, facility, or process. U.S. Geological Service produces 7.5 minute topographic maps showing elevations as well as natural and manmade features at a scale of 1:24,000 can be extremely valuable in understanding site geography and physical relationships to surrounding areas. Other map collections are kept by Department of Agriculture. The Natural Resource Conservation of USDA produces detailed maps, such as county soil survey reports. Federal Emergency Management Agency (FEMA) produces floodplain maps. Other land use information may be obtained from U.S. Forest Service, Department of the Interior, Fish and Wildlife Service, Bureau of Land Management, Bureau of Reclamation, and Army Corps of Engineers. U.S. Census Bureau keeps population records. The National Weather Service and National Oceanic and Atmospheric Administration collect information on weather patterns. Valuable information services from the federal government include Congressional Reports, hydrogeologic investigation reports, and the Ground Water Site Inventory (GWSI) database. Specialized information can be obtained from a myriad of other federal agencies,

---

* A technical support team is usually a subgroup of the project team comprised of technical experts. A technical support team may exist to focus expertise on narrow, highly technical issues or work products.

including National Marine Fisheries Service, Department of Energy, Department of Justice, and Department of Commerce.

State government agencies are even more likely to have directly relevant information on a site, facility, or activity of concern. Secretary of State's office and state health, environmental protection departments, and water boards, may possess site-specific facility records in their files, such as inspection records, permits, prior removal or cleanup activities, facilities listing (e.g., Federal CERCLIS,* NPL** RCRA), waste discharge permits, landfill or solid waste disposal lists, leaking or registered underground storage tanks, emergency response activities, hazardous materials records, and corporate ownership and officers. Do not overlook files kept by offices such as State Attorneys General, Commissioners of Insurance, and Secretary of State.

State and federal agencies are integrating much of their information into geographic information systems. These computer systems pull together data on a variety of subjects, such as environmental contamination, natural resource distribution, human demographics and distribution, physical geography, agricultural patterns, fish and wildlife habitat, industrial patterns, land uses, water resources, and air pollution patterns. Useful data may be obtained from National Well Water Association WELL-FAX Data Base, National Water Data Exchange (NAWDEX), USGS, or U.S. EPA Regional STORET Data Base. Also of value may be work of geologic, natural history, or water surveys; studies conducted by colleges, universities, or specialized conservation or environmental programs, such as Federal Aid in Fish and Wildlife Restoration Program, Anadromous Sport Fish Conservation Program, Cooperative Fish and Wildlife Units, Endangered Species Grant Program, basin commissions, Seagrant advisory service, Great Lakes Fish Commission, or national sport fishing federations.

Data from local sources is also likely to be valuable, although it is unlikely to be of uniform quality and availability. Local agencies were historically responsible for public health protection. After being overshadowed for a number of years by federal and state programs, their importance is rebounding in environmental protection, public health, land use planning, public works, traffic control, natural resource development, agriculture, and waste management and disposal. Local tax assessors offices, local court records, county records, and local and regional historical societies contain a wealth of information, although it may not be organized in an easily accessible manner. Local airports, fire or police departments, technical colleges and high schools, emergency planning offices, agricultural extension offices, well-drilling

---

* The Comprehensive Environmental Response, Compensation, and Liability Information System (CERCLIS) is the official repository for site and nonsite specific Superfund data in support of the Comprehensive Environmental Response, Compensation, and Liability Act (CERCLA). It contains information on hazardous waste site assessment and remediation from 1983 to the present.
** Sites are listed on the National Priorities List (NPL) upon completion of Hazard Ranking System (HRS) screening, public solicitation of comments about the proposed site, and final placement of the site on the NPL after all comments have been addressed. The NPL primarily serves as an information and management tool. It is a part of the Superfund cleanup process. The NPL is updated periodically. Section 105(a)(8)(B) of CERCLA, as amended, requires that the statutory criteria provided by the HRS be used to prepare a list of national priorities among the known releases or threatened releases of hazardous substances, pollutants, or contaminants throughout the U.S. This list, which is Appendix B of the National Contingency Plan (NCP), is the NPL.

RISK ASSESSMENT PROJECT PLANNING (PHASE I) 113

companies, sewage treatment plants, waste haulers or generators, energy companies, and local educational institutions can also be sources of highly specialized local information.

Federal, state, and local government programs are increasingly posted on websites and may be found through traditional finding tools, such as *United States Lawyers Reference Directory*, published by Legal Directories Publishing Company, or Carroll Publishing Company's *State Executive Directory Annual and Municipal/County Executive Directory Annual*.

Unfortunately, U.S. EPA and some state and local agencies use confidential business information (CBI). Access to this data is generally severely restricted and in attempting to use it the project team can hinder development of a risk assessment project. When confidential information is used, and it's source (or other basic data descriptors) is withheld, public confidence suffers. Before using confidential business information, a project manager should make every effort to have CBI designation changed to allow full disclosure.

Site Owner/Operator Records — Records compiled by a site owner or operator might include descriptions of hazardous substance/waste management practices on-site. They should include documentation on types and volume of toxic materials on-site, such as product purchase invoices, waste manifests, permits, material safety data sheets, site safety plans, preparedness and prevention plans, and spill prevention and control plans. These documents supply information on hazardous substance and waste types, quantities and treatment, storage, and disposal practices.

Professional Trade Organizations — Trade organizations frequently publish guidance on manufacturing processes and common industry practices. They can also provide in-depth studies of a particular industrial problem or practice. Relevant trade associations can be identified by browsing in technical collections of government or university libraries, and through websites or finding tools, such as *Forensic Services Directory*, published by the National Forensic Center.

Libraries and Other Collections — Often documents that are not part of typical government files exist in private collections, government or university archives, and historical libraries. Private collections, government, college, university, and historical libraries may have historical photographs, maps, or other documentation of a site's history that provide important clues about a site. Often these materials do not circulate, but copies can be purchased.

Knowledgeable Individuals — Information can be obtained by interviewing people who are familiar with a site, activity, or facility of concern. People with special knowledge of a site's history might include retired wardens or conservation officers, agricultural program personnel, environmental protection agency staff, or local law enforcers. Neighbors can also possess important knowledge. Industrial practices might be well-known by employees who have retired from a facility. Local doctors might offer insights into health problems associated with a facility. A local fire chief might have special knowledge about on-site chemicals. Local naturalists — biology teachers, Audubon members, or local newspaper columnists — might have observed and documented important information. Interviews with these knowledgeable people should be conducted with a certain degree of structure so that information can be compared with other sources. Interviews should also be recorded, as part of project documentation. If possible, conduct an interview like a friendly deposition, asking a series of planned questions to ensure as much information is obtained as possible, and using a court reporter to produce an accurate interview transcript.

Compile Existing Information — Compiling existing data provides a good project overview, establishes current and historical knowledge of a site, facility, or activity of concern and, if carefully evaluated and presented, can save time and money by helping a contractor to write a risk assessment report without replicating data collection, literature reviews, policy or technical analysis or site characterizations that already exist. A touchstone in compiling this information, once it has been amassed, is to make it easily accessible for contractors' efficient use. This, of course, requires logical organization, full references, and documentation of all research that allows contractor staff (or anyone) to verify and validate your work. Chapter 5, Section III, and Chapters 9 and 22 provide useful techniques for ensuring transparent presentation of technical information.

2. Formulate the Problem

After a project manager compiles existing information, the next step is to formulate, or define, the problem. In this step:

- A site, activity, or facility of concern is described, as thoroughly as possible
- Issues concerning a project are identified
- Risk assessment project objectives are established

This acquaints a project manager and project team with project details and prepares them to target project resources toward evaluating key potential (or actual) chemical release pathways and exposure scenarios. Formulating a problem, building on efforts to locate and compile existing information, helps identify what is known about a project, which also helps identify information gaps that a risk assessment project must fill.

a. Optional Briefing Document

Next, a project manager may choose to write a team briefing document to use as an aid in determining additional data requirements, before writing a Scope of Work. A briefing document is a concise overview of all important aspects of a project and can be used to write a more extensive Scope of Work. Its purpose is to:

- Organize information gathered in previous steps into a workable format
- Describe the risk assessment development process
- Describe methods to be used to generate the risk assessment

- Identify critical skills needed by contractors to perform their work
- Articulate time lines, budgets, and other logistical issues
- Identify data needs for risk assessment, such as environmental sampling and analytical chemistry of biotic and abiotic samples
- Familiarize a project team with project details

A briefing document can contain a "conceptual site model," such as we present in Figure 1, to present primary sources of environmental releases or contaminants (e.g., drums, lagoons, structures), primary release mechanisms (e.g., spills, infiltration/percolation), secondary sources (e.g., soil), secondary release mechanisms (e.g., dust, volatile emissions, stormwater run-off), exposure pathways (e.g., wind, water, sediments), exposure routes (e.g., ingestion, inhalation, or dermal contact), and receptors (humans [area residents, transients] and biota [terrestrial, aquatic]). A conceptual site model can take many forms, including a flow chart or a pictogram.

Discussions evolving from dissemination and review of a briefing document can help identify additional information needs and, perhaps, reveal additional information sources to meet these needs.

b. Project Planning Tables (optional)

Table 3 presents questions a project manager can use, alone or with a project team in a brainstorming session, to generate management ideas. This approach organizes risk assessment project details as follows. First, a table is constructed with five columns, for questions: why, what, when, how, and who. A project manager and internal project team generate a list of project objectives and enter these objectives in a "why" column. For each objective, each deliverable (measurable accomplishments) required to achieve each objective is listed into a "what" column. Next, delivery date for each deliverable is entered into a "when" column. Then, a process for generating each deliverable is entered in a "how" column. Finally, responsibility for each deliverable is assigned. The duty may be assigned to a project team member, project manager, or contractor staff. The assignment is noted in a "who" column. Use of such a table will reduce potential of neglecting important tasks. It also creates an organized framework that a project manager will use again to generate other project documents.

c. Risk Assessment Project Planning Document (optional)

Producing a Scope of Work for a major risk assessment can be an organizational challenge. One tool that can be used to organize production of a Scope of Work is a "Risk Assessment Project Planning Document" (RAPPD). An RAPPD is developed in a brainstorming session involving internal project team members, the project manager, and, perhaps, other interested parties. First, the project team brainstorms, identifying as many project-related tasks as possible. Next, it discusses work products to achieve each task. More tasks may be added to the RAPPD. Finally, evaluation criteria are devised that will ensure each task is properly achieved. Information is organized in a simple chart, showing:

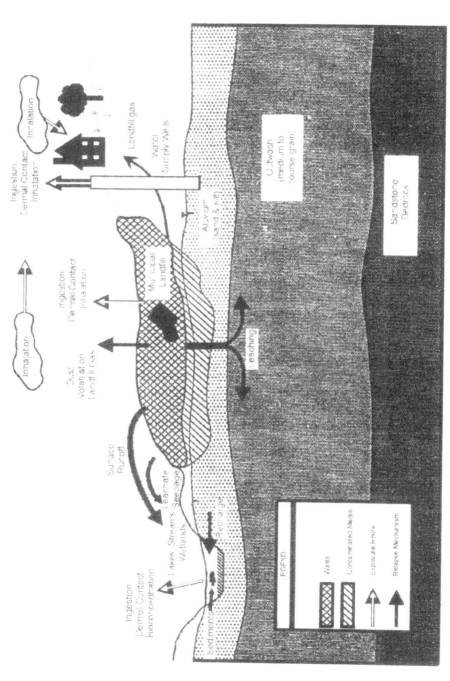

Figure 1 (Adapted from Figure 3-1, U.S. EPA, Conducting Remedial Investigations/Feasibility Studies for CERCLA Municipal Landfill Sites, Office of Emergency and Remedial Response, Washington, 1991, pg. 195.)

RISK ASSESSMENT PROJECT PLANNING (PHASE I)

Table 3   The Five Questions of Project Planning

Why: List project objectives	What: List deliverables to meet each objective	When: List due date for each deliverable	How: List process and product standards for each deliverable	Who: List person assigned to each deliverable

- Task
- Goal of task
- Work product to achieve task
- Evaluation criteria

One task, for example, might be to generate an exposure assessment report section. A goal might be to use only validated, or confirmed, mathematical models in an exposure assessment. Work products to achieve that goal might include a formal review by a project team of each interim deliverable before it is incorporated into that section. Evaluation criteria applied during review would be to verify that validated models were actually used, to check data accuracy, and to evaluate whether exposure findings generated were reasonable.

An RAPPD addressing all steps in a risk assessment process may be organized as a list of tasks, as discussed above, or as a table such as Table 4. Either way, it organizes a mass of complex information, establishes a foundation for writing a Scope of Work, and ensures that each task serves a legitimate goal and its technical sufficiency can be validated. If an RAPPD is sufficiently detailed, it may be directly incorporated into other documents, such as a Scope of Work, or RFPs and RFQs, to solicit contractor bids. If not, it should be detailed enough to serve as a complete framework for development of these documents.

3. Write a Scope of Work

When all necessary information is in hand, a Scope of Work can be written. It may contain information found in the optional briefing document, described above, or be written directly from the assembled information described earlier. A Scope of Work will be much more detailed, however. Scope of Work contents appear in Table 5.

A Scope of Work is a blueprint used by contractors to bid on a risk assessment project and is the basis for performance standards a contractor will meet to produce an acceptable risk assessment report, including scheduling and cost requirements, field, laboratory, modeling, and office work expectations, and QA/QC measures. It is also a blueprint used by a project manager and project team to develop a risk

Table 4 Sample RAPPD for Exposure Assessment Scoping

Task:	Exposure assessment.
Goal:	Estimate exposure concentrations of chemicals of concern to human and nonhuman receptors at the XYZ facility. Achieve regulatory science procedural standard.
Means to achieve goal:	Use U.S. EPA methods to evaluate all possible exposure pathways, select pathways that are likely to be completed, generate direct and indirect exposure point concentrations. Utilize both simple equations from Risk Assessment Guidance for Superfund and publicly available and validated fate and transport models.
Criteria to meet goal:	Will provide reviewers and readers with copies of all mathematical models used in the exposure assessment; will provide tables that provide all input assumptions for each variable in each equation used in the risk assessment report; will use up to date guidance documents and suggested variable values.
Task:	List and evaluate all possible routes of exposure.
Goal:	Complete analysis of potential exposure routes for local residents, site workers, transient workers, trespassers, people recreating, terrestrial and aquatic organisms. Make the analysis transparent and easily understood by reviewers.
Means to achieve goal:	Use equations found in current U.S. EPA guidance documents and validated computer models if available. Use decision criteria tables and inclusion/exclusion analysis tables.
Criteria to meet goal:	Use of tables that are easy to understand and review. If computer models are used, they will be available for use by reviewers.
Tasks:	Evaluate all reasonable exposure pathways.
Goal:	Select only those exposure pathways that have a reasonable chance of being completed.
Means to achieve goal:	Evaluate current and future land-use scenarios for exposed on- and off-site exposed populations, with an emphasis on sensitive populations.
Criteria to meet goal:	Use of clear and concise tables and figures throughout text. Use of decision criteria and inclusion/exclusion analysis tables to provide reader with reasons why each possible exposure pathway was selected or not selected for further analysis in the risk assessment.

assessment project contract and a standard against which a project manager can evaluate contractor performance.

A Scope of Work should clearly define all steps in risk assessment report development. While much of this may seem repetitive and even mundane, any one of these items, if not properly scoped, work planned, and managed, can result in cost overruns, delays in production, and even litigation. It pays to nit pick.

a. Project Limitations

In addition to information, methods, and scoping information utilized for a standard four-step risk assessment process, three project limitations must be addressed in any Scope of Work:

Table 5 Sample Scope of Work: Based on the Four-Step Human Health Risk Assessment Process

HAZARD EVALUATION
Task 1. Collect existing data: monitoring data, modeling data, surrogate data sets; information on chemicals and sources, exposure pathways, human and nonhuman receptors
Task 2. Develop conceptual model
Task 3. Define environmental modeling needs, source of chemical contaminants, data on soil, groundwater, air, surface water, and sediment conditions
Task 4. Identify background sampling needs, sampling locations, and size of samples
Task 5. Identify location of past, current, or likely, future chemical contamination, contaminated media, and contaminant identities
Task 6. Evaluate environmental media studies: soils, groundwater, surface water and sediment, air, and biota
Task 7. If needed, develop sample collection strategies: sampling types, frequency, and QA/QC measures
Task 8. Evaluate existing data (obtain new data as needed, evaluate new data for risk assessment use)
Task 9. Sort monitoring data, modeled data, and surrogate data by medium
Task 10. Determine whether existing and new data can be pooled for risk assessment use
Task 11. Evaluate methods used to gather existing data (sampling and study methods, analytical chemistry QA/QC reports, identify individual datum quality) to determine suitability for use in risk assessment report
Task 12. Develop unified data set for each chemical (statistical methods)
Task 13. Generate COPC and COC lists
EXPOSURE ASSESSMENT
Task 1. Select exposure case
Task 2. Characterize the physical setting, including climate, meteorology, geologic setting, vegetation, soil types, groundwater hydrology, and surface water
Task 3. Characterize known or potentially exposed population location, activity patterns, past, present, and possible future exposures, past, current, and future land use [residential, commercial, and recreational], sensitive subpopulations [e.g., children, infants, elderly, pregnant women, chronically ill, breeding populations, impacted populations, endangered species, and critical habitats] and their locations
Task 4. Identify exposure pathways
Task 5. Identify chemical contaminant sources
Task 6. Identify contaminated media (air, surface water, groundwater, soil, sediment, and biota)
Task 7. Evaluate fate and transport of chemical contaminants in each medium (e.g., physical and chemical parameters of each chemical [K_{oc}, K_d, K_{ow}, solubility, Henry's Law Constant, vapor pressure, diffusivity, bioconcentration, media specific half-life])
Task 8. Identify exposure points (on- and off-site) and exposure routes (dermal, inhalation, and ingestion)
Task 9. Identify exposure routes expected to be completed (groundwater, surface water, sediment, air, and food by ingestion, inhalation, or dermal exposure), and the physical phase of the chemical (vapor, particulate, absorbed, or adsorbed to soil particles, and on or in homegrown or store purchased food)
Task 10. Quantify exposure concentrations (use modeling, monitoring, and default data, simplifying assumptions, and steady or non-steady state conditions).

- Estimate exposure concentrations in each medium (soil, surface water, groundwater, indoor or outdoor air, sediment, and food, using monitoring or modeling data, and mathematical models)
- Estimate chemical intakes and uptakes from groundwater and surface water from ingestion, recreation, drinking water, and dermal contact; soil, sediment, and dust from incidental exposure, ingestion, and dermal contact; air from vapor phase and particulates; and food products |

Table 5 continued

TOXICITY ASSESSMENT
Task 1. Obtain or derive non-carcinogen and carcinogen toxicity values from U.S. or foreign government documents and databases, peer reviewed literature, or the grey literature. Task 2 Develop toxicological literature review for each chemical of concern.
RISK CHARACTERIZATION
Task 1. Organize exposure and toxicity assessment findings (tables, figures, exposure duration, absorption adjustments, and consistency checks) Task 2. Quantify pathway specific risks for each COC (cancer risk levels and non-cancer hazard quotients and hazard indexes for each pathway) Task 3. Sum risks across pathways for individuals and time frames (carcinogens and non-carcinogens by similar toxic endpoints) Task 4. Conduct uncertainty analysis (qualitative or quantitative, such as probabilistic analysis) Task 5. Summarize risk assessment results (executive summary and conclusions)

- Project budget
- Project schedule
- Quality Assurance/Quality Control measures

Effective project control depends on a project manager's ability to establish a reasonable budget and a realistic project schedule, and to then monitor progress and, when necessary, take action to halt drift away from an established budget, schedule, or workplan. Contract management involves regular monitoring of performance and periodic reviews to ensure that products are produced on schedule. Such monitoring typically relies on product/project status reports that set forth, as text or graphics, actual versus scheduled risk assessment product status, and provide a discussion of reasons for product problems (such as production delays), how production problems are to be resolved, and how to get back on schedule. In addition to the procedural aspects of contract monitoring, described above, a project manager needs to ensure that product standards are also being met. This may involve periodic review of work products against product standards, such as QA/QC plans, DQOs (see Chapter 11), or other technical measures of success.

b. Budget

Establishing and maintaining a risk assessment project budget is one of a project manager's most demanding responsibilities. Whether a projected budget has been met is a clear and most common measure of project success. However, establishing a realistic budget is difficult. Effective cost control is essential to project success. Iterative review, comment, and approval process, discussed in Chapter 1, is one method for containing costs and maintaining product quality. Next, we explain how to generate an acceptable project budget.

Report complexity influences risk assessment project costs. Many factors affect complexity, and cost, of a risk assessment including:

- Type of risk assessment needed (human health or ecological)
- Technical rigor required for each task and for overall report
- Extent and type of data collection and evaluation
- Requirements for analytical chemistry, Quality Assurance/Quality Control, and field sampling and analysis
- Extent and type of environmental fate and transport modeling
- Strategy for toxicity review and risk characterization
- Rigor of technical writing and technical review

Table 6 illustrates how a risk assessment report can be viewed as a series of factors ranging in complexity levels and costs. Table 6 is offered as an aid to help a project manager address project complexity and cost. A project manager and technical support project team should work together to refine this table and complete it to evaluate complexity and determine costs.

There are many reasons to produce such a table. Of course, it links cost to complexity. Also, it systematically creates and articulates process and product standards for a risk assessment project. It helps to define how each product (interim deliverable, task, and subtask) will be produced, how it should look when complete, associated performance standard, and level of organization for each product. Finally, it serves as one type of map of the entire project.

It is important to address each factor during scoping. Rigor of each factor affects allocation of project resources. For example, total project costs and individual task costs increase when:

- A project requires more skilled, educated, and experienced personnel
- Technical complexity increases as a result of required rigor
- Intense editing and organizational review is more demanding as a result of project implications or political climate
- Additional data are required

After initial tables are complete, they can be used to assess project costs and complexity. If complexity of individual tasks or cost is unacceptably high, table inputs can be adjusted to reduce complexity and cost. Table 6 will probably be reworked several times during contract negotiations to reconcile costs with project goals, performance standards, and technical methods. After several rounds of table input changes, project managers and technical support staff can produce a series of planning and cost projections that show resources and time required for various levels of report rigor. A final table should be part of work scopes provided in an RFP process to assist contractors in developing project proposals.

c. Schedule

Project scheduling is an iterative process. A schedule is written for a Scope of Work and finalized in a contract. In practice, however, schedule adjustments occur right up to acceptance of a final report. Scheduling is carried on through scoping and work plan development phases of a risk assessment project. Timelines are adjusted if unforeseen problems or efficiencies occur. A fortunate project manager will not

Table 6 Sample Risk Assessment Complexity Rating and Costing Scheme

Complexity Evaluation Factors				Complexity			Cost		
				Low	Medium	High	Low	Medium	High
Type	Human health	Cancer							
		Noncancer							
	Ecological	Individuals							
		Populations							
		Communities							
Scope	Contaminated media	Air							
		Water							
		Soil							
	Exposure media	Air							
		Water							
		Soil							
	Routes	Ingestion	Direct						
			Indirect						
		Inhalation							
		Dermal							
	Number of COPCs								
Scale	Local								
	Regional								
	Global								
Time	Current								
	Future								
Data	All data available								
	Major data types mostly available								
	Little or no data available								
Rigor	Scientifically defensible								
	Regulatory level								
	Planning level								

have to meet tight deadlines for interim and final products. However, few project managers enjoy such luxury as most work under constant time pressure.

While project management books offer elegant techniques for estimating project schedules and completion dates, anyone who has participated in a risk assessment project knows that start-up and completion dates are usually dictated by forces beyond their control. The art of risk assessment project management is the art of squeezing a complex project into available time frames. We suggest using simple scheduling methods that will adjust for deadline slippage and scheduling adjustments.

If choosing between iterative review and reactive project management, at first glance reactive management may appear easier. Iterative review demands careful management of draft work product delivery and review, as contractor deliverables undergo potentially multiple reviews. However, establishing realistic time frames and maintaining schedules is problematic under a reactive project management approach. When all review occurs at, or near the end of the project time frame, as with reactive project management, time required for adequate technical review is nearly impossible to predict. The technical advantages of iterative review make it a clearly superior choice, we believe, despite potentially arduous management burdens it places on a project manager and project team.

When project managers have freedom to create timelines, they begin by listing major tasks and perhaps subtasks in planning and implementation phases of a risk assessment project. Start and finish times are assigned for each task. Using these estimates, a project manager can establish a project start time (time zero) and then project an expected completion date. If deadlines are beyond a project manager's control, scheduling is a process of back calculating from a project deadline to allocate available time among essential tasks.

It is especially helpful for project managers to identify "project float." Project float relates to tasks that can be performed any time during a risk assessment project and also to tasks that can be delayed without stalling other parts of a project. Finally, a project's "drop dead" dates, deadlines that cannot be missed under any circumstances, should be clearly stated.

Computer software or hand-drawn figures can be used for project scheduling. A project manager must be prepared to shift start and completion dates to match actual product generation and review schedules.

d. Quality Assurance/Quality Control

A Scope of Work defines quality assurance procedures to ensure quality (data precision, representativeness, completeness, and comparability) of field data, laboratory data, data from literature, and desktop-derived data meets DQOs* (see Chapter 11). Standard quality assurance methods are available from professional literature.

e. Staffing

A Scope of Work defines who will perform work tasks and their required skills. This limits problems that sometimes occur when contracts are awarded based on stellar resumes of senior risk assessors who, after contract signing, may delegate project work to very junior staff.

4. Scoping Based on Report Rigor and Performance Standards

a. Appropriate Risk Assessment Rigor

Little guidance exists on how an organization develops a scoping document and manages the process of defining proper technical rigor for a risk assessment report. This is startling, considering number, cost, and societal implications (economic, legal, and health and welfare) of environmental risk assessment projects.

---

* DQO can be defined as quantitative and qualitative statements about the level of scientific and mathematical rigor that data used in a risk assessment must possess in order for it to meet the needs of decision makers.

In determining proper technical rigor, risk managers and their teams balance available resources against demand for credible risk assessment findings. We know of no mathematical formula in common use to guide this cost/benefit analysis. Rather, determining proper rigor reflects a sense of expected level of scrutiny, especially on risk assessments for public risk management decisions. Projects where little, or no, public opposition, oversight, or interest exists seem to receive fewer resources than high profile projects.

While there is rough logic to this practice (e.g., projects with potential to release harmful amounts of toxic chemicals often attract significant public interest), this practice can easily lead to policy blunders. For instance, gross overinvestments of risk assessment resources may be squandered on politically contentious projects that are, otherwise, benign. Even worse, serious underinvestments may be made in projects which pose risks of a type or magnitude of risks unrecognized by technical experts or the public.

There is a better way, however, to gauge appropriate degree of technical rigor for a risk assessment, based on project purpose. In an early paper on this subject, Belluck, et al. (1992) presented a continuum of technical report rigor. Three levels of rigor exist within the continuum: "scientifically defensible," "regulatory science" and "planning." A level of rigor is selected for an entire risk assessment report, as well as for individual tasks within a report, in order to best achieve project goals.

Scientifically-defensible level is the highest level of technical rigor. At this rigor, all information used in a risk assessment must be verified and validated; all methods and data withstand professional peer review; no default assumptions are employed; and data are only used if amenable to statistical inference and hypothesis testing. A highly rigorous site-specific, quantitative risk assessment results. This level of rigor is appropriate for a risk assessment that will face extreme scrutiny, perhaps for research science, litigation, or a proposed project involving potentially catastrophic harm. There are few technical areas where such knowledge and data exist, however, making it almost impossible to perform an entire risk assessment at a scientifically defensible level. Instead, this level of rigor usually applies only to critical components of risk assessment.

Regulatory science is an intermediate level of rigor. This level of rigor reflects a practical reality of producing risk assessment reports. At this level, a report uses a combination of verified and validated data, default values, and simplifying assumptions to produce a site-specific, semi-quantitative risk assessment. Best-available, peer-reviewed science and data are used, where practical, but limited time, financial resources, and expertise make it necessary to use data and models of lesser quality, as well. Most risk assessments fall within this category.

Planning level, the least rigorous level, employs a combination of site-specific data, possibly verified and validated data, and qualitative discussion of potential site risks. It results in a generic, qualitative risk assessment. A planning level risk assessment offers minimal quantitative insight concerning risk levels. It is rarely desirable to conduct an entire risk assessment to this low level of rigor, unless your purpose is merely to generate a preliminary estimate of risk for internal use. It is more often applied, however, to less crucial aspects of a risk assessment project where a high degree of certainty is not required.

One practical effect of establishing technical rigor for an entire report, as well as for each component, is to clarify purpose of the work. Clearly, most risk assessment reports are somewhere within a range of regulatory-level rigor. Whether they fall closer to achieving scientifically-defensible rigor or planning-level rigor depends on rigor achieved in each critical project component.

In addition to reflecting project purpose, rigor required in a risk assessment also affects credibility of risk management decisions to be based upon risk assessment findings. For example, planning-level rigor uses relatively simple, often qualitative, data to conduct a risk analysis. Any risk management decisions resulting from this level of analysis must recognize its inherent limitations. It is likely that additional studies will be required, prior to making any important risk management decisions.

Usability of regulatory-science level of rigor depends on project specifics. Most risk assessments of this type use a mixture of quantitative, semi-qualitative, and qualitative elements to generate risk findings. On a given project, this could result in risk estimates sufficient for risk management decision-making, or require additional research and analysis. Credible risk management decisions can be based on such analyses, if data limitations are made explicit.

If the ultimate use of a risk assessment is uncertain, it is wise to press for the highest possible rigor, given project resources. Doing so will generate risk findings with greater immediate utility and reduce opportunities for risk estimates of limited probative value to be misused.

b. Enforcing Rigor through Performance Standards

After decisions on overall report rigor are made, level of rigor (for each project phase and every task) is then translated into specific and measurable project performance standards. Use of performance standards improves risk assessment credibility by providing a clear measure of project success. If each standard is achieved, for every task and all phases, by definition the report is acceptable.

In contrast, if sufficient performance standards are not met, a report loses credibility and, by definition, is unacceptable. The utility of this definition of acceptability transcends simple comfort a project manager will derive from an algebraic adherence to preestablished standards of success. In practical terms, if performance standards exist for sensible reasons, failure to meet one or more standards should warn of real project inadequacies. For example, a simple performance standard states that all mathematics must be correct (and easy to check). If not, this performance standard has not been met. In a particular case of agency risk assessment reports, failure to meet performance standards may also trigger court challenges to a risk management decision, as based on arguably flawed risk findings.

Defining acceptability in terms of performance standards also helps contractors gauge resources and time required to meet client project expectations. This can avoid misunderstandings and conflict. Performance standards are delineated in the Scope of Work and contract.

Project manager and consultant both benefit from a clearly defined working relationship. In reality, many contracting organizations have difficulty articulating performance standards because no one who understands contracting also understands risk assessment. Under these circumstances, a contract might (essentially) state "We, the contracting organization, trust you, the risk assessment contractor, to deliver a document that we can live with and defend, based on the generic work plan you provided, which is appended to the contract." Both parties to this type of contract should expect trouble.

Contracting for risk assessment services without articulating performance standards is akin to three blind men and an elephant — one claims it's a tree, one asserts that it's a snake, and another thinks it is a rope. All are wrong because they have no definition of an elephant with which to unify these seemingly disparate parts. Like an elephant, risk assessment is an unwieldy beast with many fascinating parts. It is so complex that it is difficult to grasp in its entirety. It is not just modeling, toxicity testing, or report writing. Like blind men with an elephant, we need a way to unify many parts of a risk assessment project into an "acceptable" whole. Performance standards provide the way.

c. Scope of Work as Defined through Performance Standards and Report Rigor

Scoping defines two types of performance standards. "Process standards" articulate how an organization expects to work with its contractors. "Product standards" mandate required characteristics of work products. Process and product standards serve as a basis for contractor selection and performance evaluation. These standards also drive requirements for staff training, experience, and technical disciplines required among experts who produce a risk assessment report. Linking each step in risk assessment to performance steps also links planning to required actions and work products.

H. Fund the Project

A project manager must next ensure project funding. Overall project funding and funding allocation among project phases must be proposed, approved, and segregated in a project fund. Depending on the nature of funding sources, funding may even be encumbered at this point.

Proposing funding levels and allocating funds is challenging. Costs are difficult to estimate. Rough estimates can be developed, however, by adding a cost estimate to an RAPPD table. Technical rigor required for each task should be identified and its cost, based on rigor, should be estimated. Informal discussions with members of the risk assessment community may be helpful. Experienced consultants and project managers can provide insights into costs for most risk assessment project tasks. Later, cost estimates will be compared to item-specific costs provided in contractors' bids and may need to be aligned with bids and project demands (or vice versa).

After producing a Scope of Work, a project manager is ready to solicit bids from potential risk assessment contractors.

I. Solicit Contractor Qualifications or Proposals

Typically, government organizations have more ponderous contracting procedures than private sector organizations. After a Scope of Work is completed, a strategy is needed for soliciting and evaluating project bids, and then selecting the most qualified contractor. There are several possible approaches. The choice depends on how much effort can be devoted to this process.

For example, a project manager can develop either an RFQ (see Table 7) or an RFP. RFQs are used to determine what firms are interested in bidding on a contract and their risk assessment qualifications. RFQs focus on a firm's qualifications. In contrast, RFPs ask for more information (such as how a firm will perform a risk assessment, associated costs, project staff qualifications, and project management philosophy).

An RFP, in contrast to an RFQ, asks bidders to provide both their qualifications statement and a detailed proposal in response to an RFP announcement and its associated Scope of Work (generated by an organization seeking to establish a contract for services with a risk assessment contractor) (see Table 8).

In some cases, consultants responding to an RFQ are determined to be outstanding candidates and as a result, no RFP is issued. RFQ respondents are provided with a Scope of Work and asked to bid directly on a project, foregoing any RFP requirement.

A properly scoped, well-written bid solicitation package will clearly articulate all performance standards. Developing a clear solicitation package takes time, but ultimately it will improve project efficiency by attracting contractors that can provide all required services. If performance standards are not made clear until contract negotiations are underway, a project manager may waste time negotiating with a contractor that is unwilling or unable to work as required. If contract negotiations break down, an organization may need to reopen the bidding process or repeat contractor selection steps.

Certain decisions influence how a project will be advertised and these decisions must be made prior to issuing an RFP or RFQ. For example, does it matter if a contractor is local, or can a firm from another region do the work? Advantages to having a local contractor include reduced meeting costs (e.g., plane travel, hotels, per diem, etc.) and greater opportunities for emergency meetings on important issues. Determining whether to rely on a local contractor can be a matter of politics. Many government agencies use local contractors to avoid criticism for spending tax money outside of the political community. If this is a major constraint, it can be addressed by subcontracting through a local vendor with national firm connections or by nonlocal contractors opening local project offices to manage the contract, although technical work is performed elsewhere. Certain regions seem to attract highly qualified risk assessment contractors, who participate in national debate on risk assessment issues and stay up-to-date on all aspects of risk assessment report requirements. In other areas, risk assessment contractors are scarce or uninvolved in cutting-edge risk assessment techniques. If local contractors are less technically qualified, solicitation should be national.

Table 7   Examples of RFQ Solicitation

RISK WRITERS, LTD. REQUEST FOR STATEMENTS OF QUALIFICATIONS OF CONSULTANTS TO ASSIST IN THE PREPARATION OF A HUMAN HEALTH AND ECOLOGICAL RISK ASSESSMENT FOR THE PROPOSED SQUARE WHEEL DOUGHNUT FACTORY
Risk Writers, Ltd. plans to retain one or more contractors to conduct a human health and ecological risk assessment at the Square Wheel Doughnut Factory to be located adjacent to an ICBM field and wastewater treatment plant at the junction of county Road A and C. When completed, the factory will produce emissions normally associated with baking activities. Risk assessment contractors hired under this contract will be expected to quantitatively evaluate, where possible, the effects of doughnut production on surrounding human and non-human populations. Successful contractors will have significant demonstrable experience in human health and ecological risk assessment, risk assessment project management, and appropriately qualified staff. Responses to this request should focus on the contractor's demonstrable abilities to conduct human health and ecological risk assessments for the proposed doughnut factory. Persons having any questions about project details should contact: David A. Belluck Risk Writers, Ltd. 3108 46th Ave. S. Minneapolis, MN 55406 612-721-1809

Table 8   Examples of RFP Solicitation

RISK WRITERS, LTD. REQUEST FOR PROPOSALS OF CONSULTANTS TO ASSIST IN THE PREPARATION OF A HUMAN HEALTH AND ECOLOGICAL RISK ASSESSMENT FOR THE PROPOSED SQUARE WHEEL DOUGHNUT FACTORY
Risk Writers, Ltd. plans to retain one or more contractors to conduct a human health and ecological risk assessment at the Square Wheel Doughnut Factory to be located adjacent to an ICBM field and wastewater treatment plant at the junction of county Road A and C. When completed, the factory will produce emissions normally associated with baking activities. Risk assessment contractors hired under this contract will be expected to quantitatively evaluate, where possible, the effects of doughnut production on surrounding human and non-human populations. Successful contractors will have significant demonstrable experience in human health and ecological risk assessment, risk assessment project management, and appropriately qualified staff. Responses to this request should focus on the contractors demonstrable abilities to conduct human health and ecological risk assessments for the proposed doughnut factory. Responses to questions should be limited as described in the scope of work. Responses to this solicitation should also include a brief workplan. The scope of work for this project can be obtained by calling Risk Writers, Ltd. Persons having any questions about project details should contact: Sally L. Benjamin or David A. Belluck Risk Writers, Ltd. 3108 46th Ave. S. Minneapolis, MN 55406 612-721-1809

Bid solicitation packages vary in complexity. A simple one-page announcement with a general description of a proposed project may suffice for one project, while another requires an extensive information packet.

RFPs, RFQs and bid solicitation packages can contain a variety of documents. Tables 7 and 8, for example, present examples of RFQ and RFP solicitations and Table 9 shows a risk assessment Scope of Work. Solicitation packages may include more than an RFQ or RFP and Scope of Work, however, depending on the needs of the project. Providing detailed information about the project improves the chance of getting useful responses from qualified contractors.

When planning and evaluating risk assessment needs, a project manager and project team need to generate a list of contractor services, capabilities, and experiences that are either obligatory or optional. See Table 10 for an extensive list of contractor services, capabilities, and experiences. This table can be used in several ways. It can serve as a checklist of information that a project manager might consider seeking from prospective contractors when drafting documents for solicitation packages. It might also serve as a score sheet when evaluating contractor submissions. While not comprehensive, Table 10 provides space for note-taking during the evaluation process, and has room for additional attributes.

After assembling a bid solicitation package, a project manager should obtain permission to solicit bids. Bids are solicited by distributing a solicitation package, publishing a solicitation announcement, or both. At this point, risk assessment contractor selection becomes a public process. It is crucial, therefore, that documents in a solicitation package communicate precisely what a contracting organization wants to communicate in public, and communicates nothing that should not be made public.

Mechanics of soliciting bids is another consideration. Solicitation packages can be mailed directly to consulting firms or they can be made available upon request through announcements in government publications, newspaper advertising, or posting on websites. Only official contact persons should distribute information. An official contact must keep records of all information requests, materials provided, and other related communications.

After a reasonable amount of time has passed to allow potential contractors to digest solicitation documents, an optional meeting, a "kick-off meeting," can take place with parties interested in bidding.

J. Host a Kick-off Meeting for Potential Contractors

A project manager has an option to hold a kickoff meeting to answer questions before interested contractors submit project bids. It might be limited to organizations or persons who have been selected through the RFQ process for further consideration, or it could be an open meeting for all interested parties. By allowing a project team to clarify its needs with potential contractors, a kick-off meeting can result in clear, succinct bids that a team can easily use to identify qualified candidates, and it can vastly improve contractor work. Such a meeting can be managed in a variety of ways.

Table 9 Example of a Risk Assessment Scope of Work

RISK WRITERS, LTD. **SCOPE OF WORK FOR CONSULTANTS RESPONDING TO AN RFP TO ASSIST IN THE PREPARATION OF A HUMAN HEALTH AND ECOLOGICAL RISK ASSESSMENT FOR THE PROPOSED SQUARE WHEEL DOUGHNUT FACTORY**
Risk Writers, Ltd. plans to retain one or more contractors to conduct a human health and ecological risk assessment at the Square Wheel Doughnut Factory to be located adjacent to an ICBM field and wastewater treatment plant at the junction of county Road A and C. When completed, the factory will produce emissions normally associated with baking activities. Risk assessment contractors hired under this contract will be expected to quantitatively evaluate, where possible, the effects of doughnut production on surrounding human and nonhuman populations. Successful contractors will have significant demonstrable experience in human health and ecological risk assessment, risk assessment project management, and appropriately qualified staff.
DETAILED SITE INFORMATION
The proposed Square Wheel Doughnut Factory will be located on 200 acres of land previously used by the county road department to store cadmium based paints and PCB wastes. There are several small streams and a wildlife refuge located within 1/4 mile of the proposed plant. Subsistence farming occurs in the area. The proposed Square Wheel Doughnut Factory will produce specialty doughnuts for specialty doughnut vendors. Large quantities of flour, oils, sugar, preserves, and spices will be brought into the factory via County Road A and finished product will leave via County Road C. An estimated 100,000 doughnut units (absent holes) will be produced daily at this facility. Off-spec doughnuts will be sold at a factory store attached to the proposed facility. Significant vehicle traffic will occur on unpaved roads. Large amounts of volatile solvent cleaners will be stored and used on property. Large amounts of waste water will be generated. Of special significance is the threat of ingredient spills from site storage facilities into the surrounding environment. Contractors will need to develop realistic exposure cases and scenarios to meet this special need.
PROJECT TIME LINES
One or more contractors will be hired within 3 months of the issuance of this solicitation. Work is expected to begin by August 1. Report development is expected to last 4 months. Interim deliverables will be delivered at times specified in the contract.
DEMONSTRABLE COMPETENCE
Contractors responding to this solicitation must demonstrate competence in the following areas of human health risk assessment (HHRA):
a. Generic human health risk assessments (HHRA) b. Food production facility HHRA. c. Doughnut factory HHRA d. Hazard evaluation for HHRA e. Exposure assessment for HHRA f. Toxicity assessment for HHRA g. Risk characterization for HHRA h. Risk assessment report QA/QC procedures for HHRA i. Peer review of HHRAs j. Multi-pathway analysis for HHRAs

RISK ASSESSMENT PROJECT PLANNING (PHASE I)

Table 9 continued

k. Current U.S. EPA methods for HHRAs
l. Ability to use complex HHRA computer models
m. Ability to perform quantitative uncertainty analyses
n. Formal training in HHRA project management
o. Experience in HHRA project management
p. Experience in formal communications protocols
q. Experience in the development and maintenance of project finances and timelines
r. Experience in writing and editing large risk assessment reports
s. Experience working closely with clients in the development of a risk assessment report

Contractors responding to this solicitation must demonstrate competence in the following areas of ecological risk assessment (ERA):

a. Generic risk assessments (ERA)
b. Food production facility ERA
c. Doughnut factory ERA
d. Hazard evaluation for ERA
e. Exposure assessment for ERA
f. Toxicity assessment for ERA
g. Risk characterization for ERA
h. Risk assessment report QA/QC procedures for ERA
i. Peer review of ERAs
j. Multi-pathway analysis for ERAs
k. Current U.S. EPA methods for ERAs
l. Ability to use complex ERA computer models
m. Ability to perform quantitative uncertainty analyses
n. Formal training in ERA project management
o. Experience in ERA project management
p. Experience in formal communications protocols
q. Experience in the development and maintenance of project finances and timelines
r. Experience in writing and editing large risk assessment reports
s. Experience working closely with clients in the development of a risk assessment report
Estimated funding available for this project range from $100,000 to $200,000

Total response length for the entire submittal should not exceed 25 pages. Responses to each question will be graded using a point system. Points given to each question will reflect the substance of each question answered by the contractor in their submission. Responses should appear in the order asked.

Contractors should provide copies of relevant project reports to demonstrate their report writing competence. Please provide three references that can be contacted to verify statements in your submittal. A draft work plan and budget should accompany the RFP. Costs should be shown down to the task level indicating the number of hours to complete the task, the person and their qualifications to perform a given task, and their billing rate, with and without indirect costs.

SELECTION PROCESS

RFPs will be reviewed and ranked by a Risk Writers, Ltd. selection panel. The top three candidates will be invited to submit formal proposals. Separate firms may be hired to produce the ERA and HHRA should a single firm not rank first in both categories based on their submittals.
Each submission will be graded.

DISCLAIMER

This notice does not obligate Risk Writers, Ltd. to enter into contract for any services, or to otherwise reimburse any party for services or products provided. Risk Writers, Ltd. reserves the right to reject any and all submittals.

Table 9 continued

CONTACT PERSON
Persons having any questions about project details should contact: David A. Belluck Risk Writers, Ltd. 3108 46th Ave. S. Minneapolis, MN 55406 612-721-1809

K. Evaluate Bids

This step is one of the least enjoyable for many people. It requires close inspection of project proposals and firm qualifications. A few simple precautions can reduce stress involved in this step and improve efficiency.

There are actually several steps to bid evaluation. First, an "evaluation" team must be assembled. Members may be drawn from an internal project team or they may be recruited only to work on this part of the project. Second, a project manager and project team should prepare a scoring sheet and a rating system. Third, all RFPs must be assembled and assessed to determine whether they meet minimum requirements.

A prudent project manager will establish minimum requirements and will reject bids that do not conform. The RFP may strictly limit page length of bids and type of information they may contain. Such limits make bid content comparable, and discourage applicants from padding submissions with extraneous information. If, on the other hand, the RFP does not limit and focus potential contractor responses, or contractors fail to follow submission instructions, project proposal review can become a nightmare. Review, rating, and ranking of candidates can proceed quickly, if the project team has carefully drafted an RFP solicitation, and if proposals are organized in a prescribed format.

A reasonable balance must be struck. Limits that are too stringent or are too strictly enforced might sharply reduce numbers of applicants or eliminate proposals from qualified contractors. In that case, or if all bids are deficient, a project manager has a choice to either relax bid requirements or request supplemental submissions.

Bids that meet minimum requirements are reviewed and ranked by a review team according to a set of uniform evaluation standards. These standards are adapted from a Scope of Work or from criteria set forth in a solicitation packet. It is important to have a standard mechanism for evaluating RFPs. Table 11 provides a sample RFP evaluation form. This table could be provided to prospective contractors to illustrate how their proposals will be evaluated. Contractors that submit top-ranking proposals or qualifications are considered in our next step.

L. Select a Contractor

In order to select a contractor, a project manager organizes a team of interviewers and conducts an interview process. Involving a wide mix of professionals, all with some relevant technical qualifications, improves a team's ability to evaluate breadth and depth of contractor credentials in technical and nontechnical areas. Interviewers

RISK ASSESSMENT PROJECT PLANNING (PHASE I) 133

Table 10 Examples of Information to Consider for Inclusion on a Contractor RFP Response

Overview of consulting firm	Brief history of firm		
	Project management experience		
	Organizational structure		
	Staff capabilities	Engineering	Environment
			Chemical
			Geotechnical
			Geologic
			Hydraulic
			Hydrologic
			Water resources
			Structural
			General civil
			Mechanical
			Computer sciences
			Electrical
			Mechanical
		Sciences	Atmospheric
			General Chemistry
			Toxicology
			Aquatic
			Terrestrial
			Soils
			Geochemistry
			Hydrogeology
			Natural resources damages
			Geophysics
			Combustion chemistry
			Environmental fate modeling
			Forestry
			Data QA/QC
			Analytical chemistry
			Environmental sampling
Overview of consulting firm	Staff capabilities	Management services	Risk Assessment Project Management
			Data management

Table 10 continued

				Team management	
				Analytical Data QA/QC	
				Air Dispersion modeling	
				Hazard evaluation	
				Fate and transport	
				Exposure assessment	
				Toxicity assessment	
				Risk characterization	
				Uncertainty Analysis	
				Report QA/QC	Report format compliance
					Report technical review
					Report technical editing
				Liaison services	Technical/ Regulatory agencies relations
					Community relations
					Media relations
				Litigation support on risk assessment issues	
				Court testimony on risk assessment issues	
				Project accounting services	
				Risk assessment subcontractors	
				Simultaneous risk assessments for a single site, activity or facility	
			Analytical chemistry services	In-house services	
				Subcontracted services	
			In-house library services	Books	
				Journals	
				CDs	
				Data-base access, certifications and training	
				Electronic bulletin boards	

RISK ASSESSMENT PROJECT PLANNING (PHASE I)

Table 10 continued

Overview of consulting firm	Staff capabilities	Support services	Computer science	
			Laboratory operations	
			Field operations	
			Surveying	
			Drafting	
			Graphics	
			Technical writing	
			Public relations	
			Word processing	
			Accounting	
			Economic analyses	
			Alternatives analyses	
		Environmental permitting services	Air emissions	
			Solid waste	
			Water rights	
			Wastewater	
			Noise	
			Land use	
			Stormwater	
			Wetlands (404)	
			Water quality (401)	
			FERC licensing	
	Computer capabilities	Programs	Word processing	
			Spreadsheets	
			Project management	
			Graphics	
			Toxicology	
			Risk assessment	
			Air dispersion modeling	
			Environmental fate	
			Analytical chemistry QA/QC	
			Statistical	
			Word processing	
			Spreadsheets	
			Project management	
			Graphics	
			Toxicology	

Table 10 continued

			Risk assessment	
			Air dispersion modeling	
			Environmental fate	
			Analytical chemistry QA/QC	
			Statistical	
		Hardware		
		Internal networking		
		Client networking		
	Field equipment	GIS		
		Analytical Hardware		
		Sampling Hardware		
		Computer Hardware		
	Miscellaneous Staff Capabilities	Engineering and Modeling for Permit Applications		
		Project Inter- and Intra-agency Coordination		
		Project Strategy Planning		
		Risk Management Services		
		Risk Communication Services		
		Environmental Audits (Compliance and Hazard)		
		Site Safety Plans		
		Endangerment Analysis		
	Location of Personnel	Local Office	Contractor project manager	
			Risk assessment staff	
			Contractor fiscal services staff	
			Contractor modeling staff	
			Other contractor technical staff	
		Regional Office	Contractor project manager	
			Risk assessment staff	
			Contractor fiscal services staff	
			Contractor modeling staff	
			Other contractor Technical Staff	

RISK ASSESSMENT PROJECT PLANNING (PHASE I) 137

Table 10 continued

		Corporate Office	Contractor project manager	
			Risk assessment staff	
			Contractor fiscal services staff	
			Contractor modeling staff	
			Other contractor	
			Technical staff	
Air quality services	Technical services	Air emission permit applications	Compilation of air quality regulatory requirements	
			Dispersion modeling analysis	Number and type of sources to be evaluated
				Data values defining worst case
				Emission inventories
				Stack parameters
				Surface and upper air meteorological data
				Information on terrain
				Data on building downwash and cavity effects
				Location of receptors
				Modeling QA/QC analysis
				Report documentation
				Dispersion
				Coefficients
				Receptor grids
				Input variable values
				Pollutant concentrations at census tract centroids (Chronic risk assessment)

Table 10 continued

					Fault tree analysis (Acute risk assessment)
					Meteorological monitoring
					Visibility analysis
					Application of emission factors
					Compilation of emission Inventories
					Statistical analyses of aerometric and meteorological data
					BACT demonstrations
					Regulatory applicability analyses
					BACT analyses
					LAER analyses
					NSPS analyses
					NESHAPS analyses
					RACT analyses
				Elevation	
				Non-attainment	
				New source review	
				Prevention of significant deterioration	
				Construction and operating permits	
				Regulations review	
				Agency negotiations	
				Emission factor development	
				Technical/ economic control technology review	

Table 10 continued

				Modeling protocol preparation	
				Dispersion modeling	Criteria/toxic pollutants
					Building wake
					Effects
					Complex terrain
					Property line impacts
					NAAQS and PSD compliance
					Model evaluations
					Accidental releases
					Control technology option evaluations
					Health and ecological risk assessment
					Congeneration facilities
					Chemical manufacturers
					Printing operations
					Coating operations
			Emissions monitoring	Power and steam generators	
			Air emission permit applications	Dispersion modeling	Pulp and paper operations
					Municipal and hazardous waste facilities
					Thermal waste treatment systems
					Petroleum refineries
					Sludge composting operations
				Air sampling	

Table 10 continued

			Air sample analytical support services	
		Toxic air emission inventories	Petroleum manufacturing	
			Chemical manufacturing	
			Painting and coating	
			Fiberglass boat manufacturing	
			Flexible circuit manufacturing	
			Landfills	
			Boilers	
			Mobile sources	
			Medical products manufacturing	
			Tannery	
			Incineration	
			Munitions disposal	
			Grain handling	
			Mineral processing	
Air quality services	Technical services	Dispersion modeling	Identification of emission sources	
			Verification of emissions	
			Preliminary screening analysis	
			Dispersion modeling analysis	
		Air toxics review		
		Risk assessment		
		Stack testing		
		Ambient air monitoring		
		Air pollution control evaluation and design		
		Fugitive air emission monitoring		

RISK ASSESSMENT PROJECT PLANNING (PHASE I)

Table 10 continued

		Environmental compliance support	Cost-effective problem definition	
			Regulations review	
			Emission estimate calculations	
			VOC/RACT determinations	
			NAAQS/PSD/air toxics analysis	
			SARA Title III compliance	Emergency planning for toxic releases
				Emergency notification
				Community right-to-know
				Emission testing
				Ambient monitoring
				Dispersion modeling
				Multiple linear regression analyses
				Model results reconciliation
Air quality services	Technical services	Process engineering and design		
		Control equipment and design and fabrication		
		Ambient air monitoring program auditing		
		Source air testing		
		SARA Title III reporting		
		Air pathway analysis		
		Rules interpretation		
		Environmental impact assessments		
		Fatal flaw analyses		
		Site selection		
		Hazardous indices calculations		
		Process engineering reviews		

Table 10 continued

		Ambient concentration predication and evaluation		
		Landfill gas monitoring and modeling		
		Odor/noise evaluation assessments		
		Source reduction design		
		Control equipment evaluation		
		Control equipment evaluation		
		Existing source compliance		
		Control technology analysis		
		Tracer studies		
		Modeling		
		Strategic planning		
	Assistance with regulatory requirements	State and federal air emission permit applications		
		Prevention of significant deterioration permit applications		
		Environment assessments		
		Environmental impact assessments		
		Pollution prevention planning		
		Regulatory compliance audits		
		Prevention of accidental release planning		
		Emission factor development		
		New methods development		
Air quality services	Data management services	SARA Title III		
		Routine permit compliance records, data management and reporting		
		Continuous emission monitoring system automation		
		User requirement interviews and analysis		
		System logical design		
		System physical design		
		System program specification		
		Program development		
	Representative projects directly related to SOQ or RFP	Type or site, facility or activity		
		Statement of project problems		
		Statement of project goals		

RISK ASSESSMENT PROJECT PLANNING (PHASE I)

Table 10 continued

		Statement of activities to achieve goals		
		Project deliverables		
		Outcomes		
		Contractee reference		
Risk assessment services	Targeted services	Superfund		
		RCRA		
		Property transfer		
		Leaking underground storage tanks		
		Human health risk assessment		
		Ecological risk assessment		
		Natural resource damages		
		Environment sampling		
		Faunistic surveys		
		Floristic surveys		
		Soil surveys		
		Incremental risk assessment		
		Cumulative risk assessment		
		Comparative risk assessment		
		Human toxicology		
		Environmental toxicology		
Risk assessment services	Targeted services	Superfund		
		Environmental chemistry		
		Combustion chemistry		
		Epidemiology		
		Biology		
		Public Health		
		Chemical fate and behavior		
		Air emission modeling		
		Dispersion modeling		
		Hydrogeology		
		Toxicology		
		Database searches (readily accessible by risk assessment team)		
		Data collection and evaluation		
		Toxicity assessment		
		Exposure assessment		
		Risk characterization		

Table 10 continued

		Uncertainty analysis		
		Sensitivity analysis		
		Chemical criteria development		
		IRIS database usage		
		HEAST database usage		
		Soil cleanup		
		Goal calculation		
		Hazard ranking scheme usage		
		Risk assessment litigation support		
		Chain of custody usage		
		Pesticide risk assessment		
		Organochlorine risk assessment		
		Metals risk assessment		
		Volatiles risk assessment		
		Semi-volatiles risk assessment		
		Inorganics risk assessment		
		Organics risk assessment		
Risk assessment services	Targeted services	Dioxin risk assessment		
		Furan risk assessment		
		Lead risk assessment		
		Mercury risk assessment		
		Insitu bioassays		
		Ex-site bioassays		
		Waste reutilization risk assessment		
		Aquatic risk assessment		
		Terrestrial risk assessment		
		Cross-media risk assessment		
		Bioaccumulative substance risk assessment		
		Carcinogen risk assessment		
		Non-carcinogen risk assessment		
		Chemical mixtures risk assessment		
		RfD development		
		RfC development		
		Cancer potency factor (Q1*) development		

Table 10 continued

		Health advisory development	
		Qualitative mass balance analysis	
		Quantitative mass balance analysis	
		Incinerator risk assessment	
		Industrial facility risk assessments	
		Inhalation risk assessment	
		Dermal risk assessment	
		Ingestion risk assessment	Crop ingestion	Direct deposition
				Root uptake
			Mothers milk	
			Fish ingestion	
			Soil ingestion	
			Meat ingestion	
		Hazardous air Pollutant risk assessment		
		Risk-based cleanup criteria		
Risk assessment services	Targeted services	Regulatory toxicology		
	Electromagnetic field toxicology			
	Pharmacology			
		Expert testimony		
		Food chain modeling		
		Ecotoxicology	Marine	
			Freshwater	
			Estuarine	
			Terrestrial	
			Vertebrate	
			Invertebrate	
		Field and laboratory organism identification		
		Environmental chemistry		
		Biostatistics		
		Oceanography		

Table 10 continued

		GIS mapping		
		Field sampling design		
		Data management and analysis		
		Hypothesis testing		
		Probability modeling		
		Monte Carlo simulations		
		Graphics for quantitative information		
		Risk based remediation		
		Geostatistics		
		Multiple pathway/multiple contaminant risk assessment		
		Water quality modeling (surface and groundwater)		
		Wetlands delineation and functional analysis		
		Habitat evaluation		
		Veterinary pathology		
		Archaeology		
		Botany		
Risk assessment services	Targeted services	Ecology	Industrial hygiene	
				Meteorology
		Zoology		
		Indoor air risk assessment		
		OSHA compliance		
		Mammalian toxicology		
		Model intercomparisons		

Table 10 continued

		Due diligence risk assessment		
		Toxicology research		
		MSDS preparation		
		Product safety and liability		
		Liability assessment		
		Process safety management		
		Physiologically-based pharmacokinetic modeling		
		Qualitative and quantitative analysis of trace odor constituents		
		Representative project	Type of site, facility or activity	
		Directly related to RFQ or RFP		
			Statement of project problems	
Statement of project goals				
Statement of activities to achieve goals				
Project deliverables				
Outcomes				
Contractee reference				
Contract cost				
EIS services	Study design	Characterization of environmental compartments		
		Reference site identification		
		Identification of risk assessment type		
		Regulatory compliance issues		
		Environmental modeling		
EIS services	Field sampling management			
	Laboratory testing			

Table 10 continued

	Environmental assessment			
	Mitigative measures formulation			
	Mitigative measures reporting			
References	Reference vital information	Project type		
		Name of reference		
		Project responsibility		
		Title		
		Address		
		Telephone number		
		Project summary		
Statement of possible conflicts of interest for this project				
Proposed project staff	Organization chart	Contractor principal in charge of project		
		Contractor project manager (contract coordinator)		
		Data analysis component	Data analysis	Task manager
				Staff
			Computer services	Task manager
				Staff
		Technical services component	Risk assessment modeling	Task manager
				Staff
			Miscellaneous technical services	Task manager
				Staff
			Permits	Task manager
				Staff
			Risk assessment	Task manager
				Staff
			Estimating ambient concentrations	Task manager

RISK ASSESSMENT PROJECT PLANNING (PHASE I)

Table 10 continued

					Staff
				Multimedia environmental assessment	Task manager
					Staff
Proposed project staff	Organization chart	Technical services component		Control technologies	Task manager
					Staff
				Water quality	Task manager
					Staff
				Site assessment	Task manager
					Staff
				Air modeling	Task manager
					Staff
				Air pollutant emission estimates	Task manager
					Staff
				QA/QC coordinating committee	Task manager
					Staff
				Air emission control technologies	Task manager
					Staff
				QA/QC data validation	Task manager
					Staff
				Emissions inventory/ estimates	Task manager
					Staff
			Communications component	Technical writing	Task manager
					Staff
				Media communications	Task manager
					Staff
				Technical editing	Task manager
					Staff
				Report format	Task manager
					Staff
				Report graphics	Task manager
					Staff

Table 10 continued

			Community relations	Task manager
				Staff
Proposed project staff	Organization chart	Coordination component	Task manager for report coordination	Staff
		Local liaison component	Task manager for liaison with contractee	Staff
		Human health risk assessment liaison component	Task manager for risk assessment liaison with contractee	Local office staff
			Location housing risk assessment team	Staff
	Brief resumes of all proposed staff	Project title		
		Project function	Education	General
				Directly relevant to project
		Directly relevant experience		
		Specialty areas (be very specific)		
		Publications		
		Communications training and experience		
		Managers, supervisors and coordinators	Project management experience	
			Project management training	
		Risk assessment staff experience		
		Continuing education		
		Percent of time for RFP listed tasks		

Table 10 continued

Subcontractors	Name(s) and EIN(s) of proposed sub-contractor(s)			
	Proposed services			
	Contact name			
	Contact address			
	Contact telephone number			
	Summary of directly relevant experience			
	Personnel experience profiles as above			

can be recruited from project team members, the evaluation team (discussed in Evaluate Bids, above), from other internal staff, or from outside organizations. A project manager will be tempted to use the same people repeatedly. If certain staff are freely available, it might make sense to involve them in several aspects of project organization and management, because they already understand the project and their participation can provide continuity. However, it may not be possible. Most workers have a limited amount of time to share. If an expert's availability is limited, it should not be wasted on tasks that others can perform.

If an RFQ is used, contractors that receive the highest RFQ scores are usually invited to complete an RFP and proceed through bidding, scoring, interview, and selection process. If only RFPs are issued, top-scoring firms are interviewed and the highest scoring firm is usually offered an opportunity to negotiate a contract.

Interviews should be conducted in a standardized manner for each bidder, to ensure that all contractors have a fair and equal opportunity to respond to questions. During a time set for each interview (e.g., 45 minutes) a selection team* asks each potential contractor identical questions in identical order. A record is kept of answers. After all standard questions are asked and answered, time may remain for free-form discussion. After the interview, each contractor is thanked and dismissed from the interview room.

Then, responses to each question are discussed and scored. Each interviewer should calculate a score and provide a qualitative evaluation for the candidate. A contractor's final score should be a sum of all interview team members' scores,

---

* A selection team may be comprised of project team members, or not. The decision depends on staff availability and on specific skills required to evaluate contractor proposals. For example, staff with deep, but narrow technical expertise might serve well on a project team, but lack sufficient breadth to evaluate contractor proposals.

Table 11 Sample RFP Evaluation Factors Exhibit

EXHIBIT 1	
RFP EVALUATION FACTORS FOR SELECTING CONSULTANT TO PERFORM A HUMAN HEALTH RISK ASSESSMENT FOR THE FLYING LEAP LANDFILL.	
BASIS FOR SELECTION	
The successful Contractor will be responsible for conducting a scientifically defensible human health risk assessment at the Flying Leap Landfill. The successful Contractor will scrupulously follow all local, state, and federal laws and guidance concerning risk assessments and will adhere to all provisions of any attachments to the contract including the contract management protocol. In all cases where there is uncertainty in interpretation of contract provisions or implementation, final decision authority rests with the contractee Project Manager.	
A point system will be used to evaluate all Contractor proposals received on or before the RFP receipt of bids closing date. Contractors may be asked by the contractee to provide additional supporting material to support or clarify their RFP at the discretion of the contractee Project Manager.	
Evaluation points will be awarded on the following basis:	
Contract Price	50 points
General Human Health Risk Assessment Experience	25 points
Landfill Human Health Risk Assessment Experience	50 points
References	50 points
Quality of Technical Writing and Editing in Previous Risk Assessment Reports	25 points
Technical Qualifications and Experience of Contractor Project Manager and Technical Staff	50 points
Technical Qualifications and Experience of Contractor Technical Editors and Writers	25 points
Total Possible Points	275 points

tempered if necessary by a team's qualitative evaluation. This approach provides comparable information to a review team and can help prevent claims of bias. Yet, it also allows team members to exercise some discretion.

Using this process can result in an efficient selection of the best contractor and provides a record to demonstrate that the process was open, fair, and uniform. A written record of this kind helps protect an organization from litigation.

Scoping may end with contractor selection, or it may continue through development of draft contract language, discussed below.

M. Negotiate a Contract

The top scoring firm is notified of their ranking and invited to negotiate a contract. If a firm is interested in proceeding, contract negotiations begin. If not, the second-ranked firm is offered the opportunity.

Effective risk assessment contracts protect financial and legal interests of both contractors and an organization who hires contractors. A contract should define all terms and clearly state all obligations, including performance standards, payment terms, bonus and penalty provisions, and contract dispute resolution procedures.

Comprehensive contracts help ensure that all parties understand their roles and provide clear guidelines on how to reach a successful project conclusion. Chapter 8 provides a detailed discussion of contracting.

Contract negotiations can be simple or complex. Many organizations start with a set of standard contract terms and modify these terms, as necessary. Some start with a blank piece of paper and work through all contract issues to a desired endpoint. Regardless of your approach, a good contract will clearly articulate obligations of both parties. Specifically, a contract for risk assessment services should address all performance standards, either directly or by reference. For example, contract provisions deal with project timing, payment (including reward and penalty provisions), risk allocation (regarding errors or omissions), project staff qualifications, dispute resolution, and contract management method (either reactive management or iterative review process (see Chapter 5, Section II).

A contract may articulate all project performance standards, or incorporate them by reference to statutes, organizational policy documents, Scope of Work, risk assessment workplan (discussed in the next section), or other relevant documents. For example, a work plan stipulation may be attached to a contract. Agreeing in writing to abide by these performance standards reinforces their importance and may help prevent future disagreements.

N. Negotiate a Work Plan

A work plan is a contractor-prepared document, with oversight by a project manager on behalf of the contracting organization. A work plan defines all work to be performed by a contractor during a risk assessment project. Contractors are usually not permitted to perform billable work until they prepare an acceptable work plan, usually to be appended to their contract for services. In some instances, billable work is permitted to help develop a workplan. Billable work planning is justifiable in cases where little is known about a site, facility, or activity of concern. In such cases, limited billable planning activity makes sense. Otherwise, a contractor should not initiate work, until an acceptable work plan is incorporated into a final contract.

A work plan serves two important functions. First, it presents a detailed description of work a contractor commits to perform under a contract. Second, work planning, the process of developing a work plan, forces a project manager and contractor to discuss, and to agree on, every aspect of a project. Work plans are crucial to managing a contractor, evaluating work products, and resolving disputes. If project resources are inadequate to complete a project as originally conceived, work planning is an opportunity to set priorities and avoid disputes.

Work plan development can occur after contract signing. However, documenting work to be done is best served if work planning is completed before a contract is signed. This allows a work plan to be appended to the contract, making it legally binding and showing exactly what work contracting parties agreed must be performed.

The function of clarifying and reaching agreement on details of a project is best achieved through an iterative approach. In an iterative approach a project manager and contractor work to develop a series of work plan drafts, only finalizing a work plan when all issues have been identified and addressed.

Both parties benefit from this approach. It allows a project manager to explain performance standards, identify any standards that are not met by a contractor's proposals, and ensure shortcomings are addressed. A contractor benefits from an opportunity to work closely with the client, to discuss project details, and to express concerns. This will improve project profitability. It also helps minimize debates about whether a task is "in-scope" or "out-of-scope"* and, thus, to avoid potential delays and costs associated with dispute resolution or litigation.

It is important to write a clear, logical work plan to ensure complete agreement on what work will be done, who will perform each task, what it will cost, and when it will be completed. Every work plan should include certain elements: a statement of project purpose; a description of a project technical approach; a list of deliverables and a description of each; a project schedule, with delivery dates for each deliverable, completion dates for each phase of a project, and time lines for other milestones; a contractor project staff list, including subcontractors, and staff qualifications; a description of costs by each deliverable; and an explanation of how data quality objectives were used in work plan development. In addition, a contractor should certify that the work plan is consistent with the Scope of Work, or should note and explain inconsistencies.

These elements of a work plan might seem obvious, but they need to be articulated because work plans are often surprisingly vague. A poorly written work plan invites a contractor and a project manager to interpret ambiguous terms to their advantage. This leads to conflicting views and can rapidly undermine performance.

1. Drafting a Risk Assessment Work Plan

It is advisable to compensate a contractor for extensive work plan development. As discussed above, developing work plans can require several iterations before all parties are satisfied. Generic work plans are a good starting point for developing detailed risk assessment work plans. Table 12 presents an example of such a generic plan. Special considerations for an ERA work plan are presented in Table 13. Many elements presented in Table 12 can be merged with Table 13 data elements.

The level of detail appropriate in evaluating workplan proposals is a matter of professional judgement. We present two extremes as tables at the end of this chapter: Table 14 is an extensive table of data elements that a project manager can use in RFP or contracting process to gauge adequacy of a contractor's proposed workplan. Table 15 presents an abbreviated evaluation form.

Contractors often develop complex work plans with little or no compensation because work plan development is considered a cost of obtaining an environmental risk assessment contract.

---

* In-scope work is work that a contractor agreed to perform at cost bid for a project. In-scope work is performed without additional compensation. Out-of-scope work is work that was not included in project scoping, workplan, or contract terms or conditions. The contractor receives additional compensation for approved out-of-scope work.

RISK ASSESSMENT PROJECT PLANNING (PHASE I)

Table 12 Example of a Human Health Risk Assessment Contractor Work Plan Outline

INTRODUCTION
____ Project description, purpose and objectives   ____ Technical approach to produce risk assessment report   ____ Project work to date, if any   ____ List of documents used to generate work plan   ____ Description of tasks performed and products generated in previous years that are used to generate work plan   ____ Description of work plan contents   ____ Applicable laws and rules   ____ Statement that work plan is consistent with client organization approved Scope of Work   ____ Comparison of Scope of Work with proposed work plan   ____ Discussion of how data quality objectives were used to develop work plan   ____ Project meetings reporting and communication requirements   ____ Staffing for report   ____ Report oversight
HAZARD ASSESSMENT
____ Data collection   - Site visit   - Data Quality Objectives   - Quality Assurance/Quality Control Plan   - Sampling and Analysis Plan (i.e. Field Sampling Plan and Quality Assurance Project Plan) for new field work   - Other plans (e.g., Community Relations Plan, Data Management and Data Validation Plan, Laboratory Analysis Plan)   - Analytical chemistry and other laboratories to be used in data collection and evaluation phase   - Analytical chemistry and other laboratory methods to be used in data collection and evaluation phase   - Precision of analytical techniques   - Existing and needed sampling data   - Chemicals of potential concern list   - How nutrients and background chemical concentrations will be handled   - Mathematical methods to calculate chemical concentrations   - Uncertainty analysis
____ Environmental setting   - Site description and history   - Demographics   - Physical description (e.g., physiography, topography, climatology, meteorology, biology, geology, soils, hydrology (i.e. surface and groundwater)   - Ecological resources (e.g., on-site and near-site habitats, vegetation and animals), aquatic resources, wildlife   - Summary Section
____ Data evaluation   - Previous and recent sampling efforts   - Data quality and data representativeness: Listing of data set sources considered of adequate quality for use in a risk assessment (criteria include source and recentness of data, sampling locations, adequacy of documentation, data validation results, adequacy of analytical methods, detection limits, completeness, and comparability)   - Adequacy and representativeness of database for calculation of exposure point concentrations for each contaminated medium (e.g., surface soil, subsurface soil, groundwater, surface water, sediment, and ambient air)

Table 12 continued

- Data useful for calculating background concentrations of chemicals of potential concern (COPC) and chemicals of concern (COC)
- Data quality designations (e.g., usable for screening level or enforcement level activities)
- Listing of COPCs (e.g., criteria include evaluation of contaminant concentrations in environmental media, comparison with background concentrations, and toxicity evaluation
- Data base uncertainties (e.g., lack of or limited monitoring data, bioavailability data, background concentration data, seasonal groundwater and surface water data, sediment and soil chemistry data, and soil quality data)
- Selection of chemicals of concern
- Summary section

EXPOSURE ASSESSMENT

____ List of guidance documents to be used to generate exposure assessment
____ Goals of exposure assessment (e.g., exposure cases to be used)
____ Description of current or future land use scenarios (e.g., residential, commercial, recreational)
____ Develop exposure conceptual model (i.e., link source of contamination to transport or release
____ Exposure case(s)
____ Computer models and mathematical equations selected for exposure assessment
____ Fate and transport modeling
____ Populations and individuals to be modeled or studied
____ Desktop and field evaluation techniques
____ Sources of data for use in exposure models and equations, mechanism, exposure point, and route of exposure)
____ Exposure pathway analysis to determine which pathways are to be retained for quantitative evaluation (e.g., evaluation criteria include likelihood of pathway completion, relative importance of pathway to total exposures, size of potentially exposed population, and appropriateness of pathway for a given location)
____ Discussion of potential receptors
____ Discussion of potential exposure routes
____ Discussion of exposure assumption (e.g., use of standard EPA default values when case specific information is not available; use of case specific data such as exposure frequency, exposure duration, land use, and bioavailability)
____ Define mathematical methods to calculate exposure point concentrations
____ Define mathematical methods to calculate chronic daily intakes
____ Discuss major uncertainties with exposure assessment (e.g., adequacy of chemical databases, exposure pathways and receptors, general exposure assumptions, and pathway specific exposure assumptions)
____ Summary section

TOXICITY ASSESSMENT

____ Define hierarchy of data sources
____ Methods to be used when no regulatory agency acceptable toxicity factors are available for use in the risk assessment report
____ Define sources of toxicity data
____ Carcinogen toxicity values and findings
____ Noncarcinogen toxicity values and findings
____ Chemical bioavailability
____ Uncertainties in toxicity assessment (e.g., derivation of toxicity criteria)
____ Summary section

RISK ASSESSMENT PROJECT PLANNING (PHASE I)

Table 12 continued

RISK CHARACTERIZATION
_____ Description of methods used to combine toxicity and exposure data _____ Chemical specific carcinogen risk estimates _____ Chemical specific noncarcinogen risk estimates _____ How risks will be calculated _____ Multiple chemical risk characterization methods _____ Single chemical risk characterization methods _____ Use of acute, subchronic and chronic exposure risk characterization methods _____ Less than lifetime carcinogenic risk assessment techniques _____ Methods to sum risks across pathways and exposure routes, as appropriate _____ Type of uncertainty analysis to be used (e.g., Monte Carlo techniques or qualitative techniques) _____ Summary section
PROJECT SCHEDULE
_____ Start and completion dates for all sub-tasks, tasks, interim deliverables, draft final report, and final report
ORGANIZATION AND RESPONSIBILITY
_____ Responsibilities and qualifications of key personnel (e.g., project manager, risk assessment manager, risk assessment technical advisor, QA/QC officer, health and safety officer, and project staff)

Note: *This is not an exhaustive list.

Table 13 Outline of Ecological Risk Assessment Report Work Plan for Ecological Effects Evaluation

1. Describe qualitative, semi-quantitative, and quantitative surveys of flora and fauna in potentially exposed habitats and reference sites
2. Describe chemical sampling of media and biota in potentially exposed habitats and reference sites
3. Describe laboratory and on-site toxicity testing
4. Describe tissue analyses, enzyme studies, and bioaccumulation studies
5. Describe fate and transport modeling studies
6. For each of the proposed studies above, the following details are provided: • Study objectives • Effects to be measured • Relevance of studies to ERA • Proposed field and laboratory methods • Risk based detection limits of laboratory methods • Sources of methods • Sampling criteria and plans • Benchmark values • Background values • Statistical methods • QA/QC practices
7. Additional ERA paradigm requirements

Note: This is not an exhaustive list.

Table 14 Example of a Detailed Work Plan Evaluation Table for Use by Project Manager to Judge Work Plan Adequacy

Item	Essential/ Optional/Not Needed (E/O/N)	Discussed in work plan (Y/N)	Cost	Acceptable (Y/N)
Management style				
Proactive Style				
Reactive Style				
Team-Approach				
Internal staff costs				
Charge-Back for Staff Hours				
Other:				
Contractor Costs				
Client visits				
Work Plan Development				
Overhead				
Profit Margin				
Specialists' Billable Rates				
Administrative Costs				
Internal Meeting Costs				
Technical Documents				
Response to Client Telephone				
Response to Client Written				
Internal QA/QC Costs				
Travel Expenses				
Computer Time				
Computer Software				
Computer Hardware				
Contractor On-Site Costs				
Staffing				
Housing				
Supplies and Equipment				
Risk Assessment Report Costs				

RISK ASSESSMENT PROJECT PLANNING (PHASE I)

Table 14 continued

Item	Essential/ Optional/Not Needed (E/O/N)	Discussed in work plan (Y/N)	Cost	Acceptable (Y/N)
Hazard Identification				
Exposure Assessment				
Toxicity Assessment				
Risk Characterization				
Uncertainty Analysis				
Sensitivity Analysis				
Response to Public Comments				
Response to Client Comments				
Number of Drafts of Interim				
Number of Drafts of Final Report				
Size of Executive Summary				
Rigor of Executive Summary				
Use of Separate Technical Reports				
Generation of Summary Document from Technical Reports				
Extent of Modeling				
Scientific Rigor Requirements				
Mathematical Rigor Requirements				
Use of EPA Toxicity Values				
Derivation of Toxicological Values				
Use of Default Values				
Target Audience				

Table 14 continued

Item	Essential/ Optional/Not Needed (E/O/N)	Discussed in work plan (Y/N)	Cost	Acceptable (Y/N)
Develop Cutting-Edge Science				
Use Existing Science				
Number of Chemicals Fully				
Rigor of QA/QC				
Quality of Analytics				
Number of Exposure Scenarios				
Emphasis on Text				
Emphasis on Graphics				
Rigor of Toxicity Profiles				
Use Case Specific Field-Collected				
Use Surrogate Data Sets				
Use Default Assumptions				
Pharacokinetic Modeling				
Chemical Mixtures Risk				
Bioassay Use				
Report Production				
Layout and Design				
Standard Formats				
Unique Formats				
Required Formats				
Editing				
Technical Edit by Scientists				
Edit by Technical Writer				
Proofing Text and Graphics				
Graphics Versus Large Text Blocks				
Maps				

RISK ASSESSMENT PROJECT PLANNING (PHASE I)

Table 14 continued

Item	Essential/ Optional/Not Needed (E/O/N)	Discussed in work plan (Y/N)	Cost	Acceptable (Y/N)
Photos				
Figures				
Tables				
Cover Art				
Line Art				
Report Reproduction				
Typesetting				
Printing				
Binding				
Mailing				
Training				
OSHA 40-Hour Training				
Risk Assessment Fundamentals				
Advanced Risk Assessment				
Field Work				
Initial Site Walk and Evaluation				
Epidemiological Study				
Transect Analysis				
Faunistic Study				
Floristic Study				
Site Records Search				
Environmental Media				
Sampling				
Bioassays				
Basic and Applied Research				
Laboratory Costs				
Analytical Chemistry Analysis of				
Chemicals				
Hardware				
Software				
Personnel				

Table 14 continued

Item	Essential/ Optional/Not Needed (E/O/N)	Discussed in work plan (Y/N)	Cost	Acceptable (Y/N)
Space				
Methods Development				
Miscellaneous Supplies				
Communications Requirements				
Public Information				
Public Involvement				
Technical Writer On-Staff				
Public Relations				
Legal Considerations				
Defining In-Scope/Out-of-Scope				
Contract Negotiation and Drafting				
Contracting (Time and Materials, Firm, Fixed Cost)				
Regulatory Compliance				
In-House Counsel				
Contingency Costs				
Additions to Scope of Work				
Additions to Work Plan				
Additions to Contract				
Additional QA/QC of Mathematics				
Additional QA/QC of Science				
Additional QA/QC of Writing				
Contractor Staff Inexperience				
Contracting Organization				

Table 14 continued

Item	Essential/ Optional/Not Needed (E/O/N)	Discussed in work plan (Y/N)	Cost	Acceptable (Y/N)
Inadequate Knowledge of Current Regulations				
Inadequate Knowledge of Law				
Inadequate Knowledge of Guidance Documents				
Inadequate Communications				
New Regulatory Requirements Between Contract Date and Report				

Table 15 Example of Work Plan Evaluation Criteria

___ Does the work plan address all Scope of Work, RFP, other requirements?
___ Does the work plan include any work not required by the Scope of Work?
___ Are all budget items acceptable?
___ Are health and safety provisions acceptable?
___ Does the work plan contain language that specifically acknowledges acceptance of a contracting organization's contract management methods and acceptance of all provisions associated with its use?
___ Does the work plan contain adequate project control mechanisms?
___ Does the work plan contain acceptable deliverable timelines?
___ Are work plan product and process standards acceptable?
___ Are project QA/QC methods acceptable?
___ Is contractor management structure for the risk assessment sound?
___ Are sub-contractors and their activities proposed acceptable and appropriate?
___ Are data management methods acceptable?
___ Are communication protocols with the contractor project manager acceptable?

Paying a contractor a reasonable sum has at least three project benefits. First, it gets a contractor's attention. Contractors are usually under high pressure to make billable hour goals. Work that is not billable, such as developing a work plan for free, tends to receive less attention than billable work. Second, paying a contractor puts the working relationship on a professional level. Third, work planning provides a project manager with a chance to judge a contractor's professional style before signing a contract. If a contractor who is being paid for work plan development behaves in an unprofessional manner, a project manager can reconsider offering a contract or can tighten terms to help ensure professional performance.

One approach to work plan compensation is to accept a contractor's general work plan, sign a contract, and then develop a detailed work plan as a first part of a risk assessment project. This process is favorable to contractors, but less so to clients who might find themselves contractually bound to work with a contractor who is, in some way, unable to perform a risk assessment in a manner that meets their needs. Another approach is to insert a contract clause that declares a contract void if a contractor fails to produce an acceptable work plan.

2. Assessing Work Plan Acceptability

What constitutes an acceptable work plan depends on an agreement between a project manager and contractor. This agreement, in turn, depends on performance standards established for a project. These standards, in turn, depend on expected uses of a risk assessment report and risk estimates. Chapter 7 provides a summary of legal context of risk assessments which may drive performance standards.

Work plans for environmental risk assessments should describe work to be performed for each step in a risk assessment process (hazard assessment, exposure assessment, toxicity assessment, and risk characterization). See Tables 11 and 12 for ideas of elements that should appear within a work plan. Within each step, a contractor should consider all possible work. For example, if data are adequate for a site, additional sampling will not be required. On the other hand, existing data may need to be supplemented. If sampling is required, additional work will include: setting and achieving data quality objectives; establishing and implementing plans for QA/QC sampling and analysis; identification of appropriate analytical chemistry laboratories, methods, and techniques; and assignment of qualified personnel to perform associated tasks.

In addition, ERA work plans should address certain studies not required in an HHRA, including:

- Qualitative, semi-quantitative, and quantitative surveys of flora and fauna in potentially exposed habitats and reference sites
- Chemical sampling of media and biota in potentially exposed habitats and reference sites
- Laboratory and on-site toxicity testing
- Tissue analyses, enzyme studies, and bioaccumulation studies

For each of these studies, an ERA work plan must state study objectives; identify effects to be measured; explain relevance of the study to ERA proposed field and laboratory methods, risk based detection limits of laboratory methods; sources of methods; sampling criteria and plans; benchmark values; background values; statistical methods; and QA/QC practices.

A work plan will demonstrate a contractor's understanding of a project. Proposed product timelines must not be too long or too short. A project budget must be realistic. Cost proposals must state number of hours to complete delineated work and any associated hourly costs. Indirect cost rates must comply with rates agreed to in the project contract. All project costs must be well justified and should match project costs outlined in the Scope of Work and contract. Discrepancies should be explained.

A work plan will also indicate whether a contractor can deliver promised services. A staffing plan must list only qualified personnel. If subcontractors will provide some services, the general contractor must articulate how control will be maintained over subcontractor work quality.

3. Work Plan Dispute Resolution

It should come as no surprise that a project manager and contractor, as well as every other participant in a risk assessment project, have different interests. Understandably, a project manager wants to obtain the best possible work for the least cost, whereas, contractors want to generate high quality products and maximize profits. Work plans and contracts balance these competing interests.

Work plan development always involves some disagreements over performance standards. Sometimes project needs exceed available funding. Goals may be inconsistent or may run afoul of standard methods. Resolving disputes is essential, and possible, but it requires a cooperative stance in which disputing parties focus on problem-solving, rather than on winning.

A key to successful problem-solving is to determine interests of each party to a dispute, avoiding staking out positions and, instead, undertake creative problem-solving to find a mutually satisfactory resolution. This does not come naturally to most people. Preparation can help.

When a dispute arises, a project manager and contractor typically meet to attempt to find an informal solution. Both should prepare to meet by identifying their concerns and interests and then asking themselves "Why is this an issue?" Rather than focusing on what is required, getting to *why* it is required can point to a path toward resolution. It is wise to be thoroughly familiar with documents that control the working relationship — a contract (if it has been signed), Scope of Work, and any performance standards. In meeting, both parties will benefit by being honest about their needs and their reasoning and motivations, and by seeking creative ways to accommodate these needs. If workable changes are identified to resolve a problem, it may be wise to have a project attorney document agreement in a legal manner. Until an agreement is reached or it becomes clear that agreement will not be reached, parties' attorneys should probably not be involved in problem-solving.

If a contract has been signed, check to see if contract terms address dispute or mandate use of Alternate Dispute Resolution (ADR). If a contract does not address the specific situation, and the project grinds to a halt, either party might claim damages for breach of contract. A contractor might raise a claim if a project manager attempts to hire a different contractor to perform a risk assessment; a project manager might raise one, if a contractor has refused to perform. Since a breaching party can be held liable for damages to the other party, both parties benefit from resolving disputes.

It is prudent to make a final work plan part of the contract. This may be done by stating the terms of a work plan within the contract text, by incorporating it by reference, or by stipulation. A stipulation states that the contractor is aware of project performance standards, understands and agrees to meet them and to inform a project manager, if compliance with a performance standard is not possible. It may also state that the contractor agrees that failure to comply with all performance standards constitutes a breach of contract and that, at a project manager's discretion, noncompliance may trigger contract penalties, including a suit for damages and liability for the contractee's court costs and reasonable attorneys fees. Work plan stipulations reaffirm and highlight the importance and binding effect of performance standards. Of course, like all contract terms, such stipulations are open to negotiation and may be drafted to benefit both parties.

O. Hire the Contractor

If a contractor and project manager agree on contract terms and work plan details, the contract is finalized, approved by organization management, and signed by both parties. At this point, a contractor is officially hired and project work can begin.

III. CONCLUSION

After completing Phase I of a risk assessment, summarized in Table 16, a project manager and contractor are poised to undertake "real" risk assessment work in Phase II. It is important to note, however, that efficiency in Phase II is likely to be directly proportionate to the degree of care taken in these initial project planning steps.

REFERENCES

Belluck, D.A., et al., Breaking the reactive paradigm: a proactive approach to risk assessment management, *Tot. Qual. Environ. Manage.,* 1(3), 253, 1992.

Gilbreath, R.D., *Winning at Project Management: What Works, What Fails and Why,* John Wiley & Sons, New York, 1986.

Kerzner, H. *Project Management: A Systems Approach to Planning, Scheduling and Controlling,* 3rd ed., Van Nostrand Reinhold, New York, 1989.

Lucero, G.A., *Endangerment Assessment Handbook,* Office of Waste Programs, Washington, 1985.

Randolph, W.A. and Posner, B.Z., *Getting the Job Done: Managing Project Teams and Task Forces for Success, revised ed.,* Prentice Hall, Engleswood Cliffs, New Jersey, 1992.

U.S. Environmental Protection Agency, *Conducting Remedial Investigations/Feasibility Studies for CERCLA Municipal Landfill Sites,* Office of Emergency and Remedial Response, Washington, D.C., 1991.

U.S. Environmental Protection Agency, *Getting Ready, Scoping the RI/FS, Solid Waste and Emergency Response* (OS-220), Washington, D.C., 1989.

U.S. Environmental Protection Agency, *Guidance for Conducting Remedial Investigations and Feasibility Studies Under CERCLA, Final,* Office of Remedial Response, Washington, D.C., 1988.

U.S. Environmental Protection Agency, *Proposed Guidelines for Carcinogen Risk Assessment,* National Center for Environmental Assessment, Washington, D.C., 1996.

U.S. Environmental Protection Agency, *Risk Assessment Guidance for Superfund Human Health Evaluation Manual,* Office of Emergency and Remedial Response, Washington, D.C., 1998.

RISK ASSESSMENT PROJECT PLANNING (PHASE I)

Table 16 Critical Elements for Planning a Risk Assessment Project

Elements in Phase I: Planning A Risk Assessment Project	Applicable? (Y/N)	Addressed in Planning? (Y/N)	Sufficient? (Y/N)	Complete? Pass?	Complete? Fail?
STEP 1. DETERMINE WHETHER A RISK ASSESSMENT IS NEEDED					
Risk Assessment Mandated? • Applicable law • Organizational policy • Other					
Risk Assessment Discretionary? • Use existing decision matrix to determine need for risk assessment • Develop decision matrix to determine need for risk assessment					
STEP 2. SELECT A PROJECT MANAGER					
Qualifications of Project Manager • Experienced • Technical qualifications • People-skills					
STEP 3. BUILD A RISK ASSESSMENT TEAM					
Team building considerations • Purpose • Team Process • Internal Project Team (technical experts, other professionals) • Risk Advisor					

Table 16 continued

Elements in Phase I: Planning A Risk Assessment Project	Applicable? (Y/N)	Addressed in Planning? (Y/N)	Sufficient? (Y/N)	Complete?	
				Pass?	Fail?
STEP 4. ORGANIZE A PROJECT MANAGEMENT TEAM (OPTIONAL)					
Team building considerations • Purpose • Team Process • Team members					
STEP 5. DOCUMENT PROJECT EXPECTATIONS					
Legal, regulatory requirements					
Policies or strategies of organization					
Reviewers' standards					
Informal or unstated expectations of influential					
Concerns of the general public					
STEP 6. IDENTIFY PROJECT LIMITATIONS					
Budget					
Timeline					
Expertise					
STEP 7. SCOPING THE PROJECT					
Prepare a Risk Assessment Project Planning Document					
Prepare an RFP/RFQ Solicitation					
Name the project					
Project location					
Type of site, activity or facility					

RISK ASSESSMENT PROJECT PLANNING (PHASE I)

Table 16 continued

Elements in Phase I: Planning A Risk Assessment Project	Applicable? (Y/N)	Addressed in Planning? (Y/N)	Sufficient? (Y/N)	Complete?	
				Pass?	Fail?
Funding level if available (in some cases no funding level will be provided in a scope, RFQ, or RFP).					
Contractor personnel training (e.g., university and graduate classes and degrees, research, project training) and experience levels (e.g., years as a professional, types of project experience, role in project teams) and publications experience (e.g., report writing, book chapters, peer reviewed papers) required to qualify a firm for a risk assessment contract.					
Contractor project manager project management training (e.g., university, training courses) and experience levels (e.g., number and types of risk assessment projects and role in each project) required to qualify firm for a risk assessment contract.					
Process standards (e.g., how the contractor will work with the contractee organization, use of a linear deliverable review, comment and approval process or reactive contract management process).					
Product standards (e.g., what level of rigor, QA/QC, readability, and review each deliverable must achieve to be acceptable).					
Details on how each listed RFQ/RFP question is to be answered and presented (e.g., response limited to 1 page for each question for a total of no more than 25 pages).					
How each question will be scored by the contractee to create a final list of bidders.					
How each question will be evaluated for responsiveness.					
Following scoring, how will a contractor perform the risk assessment be selected.					
Evaluation and project timelines.					
Description of how the RFQ/RFP process will be managed (e.g., RFQ only, RFQ then RFP, RFP only).					
Contact names and availability of the contractee scoping document for further information.					

Table 16 continued

Elements in Phase I: Planning A Risk Assessment Project	Applicable? (Y/N)	Addressed in Planning? (Y/N)	Sufficient? (Y/N)	Complete?	
				Pass?	Fail?
STEP 8. FUND THE PROJECT					
• Adequate • Encumbered or reliable					
STEP 9. SOLICIT CONTRACTOR QUALIFICATIONS OR PROPOSALS					
• Proper geographic area • Proper types of firms • Detailed RFQ/ RFP requirements • Linked to project expectations identified in scoping					
STEP 10. HOST A KICK-OFF MEETING					
Agenda					
Attendance					
STEP 11. EVALUATE BIDS					
• Scoring system linked to expectations identified in scoping • Evaluation and selection team • Formal process (timing of presentations and set of questions)					
STEP 12. SELECT A CONTRACTOR					
STEP 13. NEGOTIATE A CONTRACT					
STEP 14. NEGOTIATE A WORK PLAN					
STEP 15. HIRE A CONTRACTOR					

U.S. Environmental Protection Agency, *Risk Assessment Guidance for Superfund, Vol. I, Human Health Evaluation Manual (Part A), Interim Final,* Office of Emergency and Remedial Response, Washington, D.C., 1989.

U.S. Environmental Protection Agency, *Risk Assessment Guidance for Superfund, Vol. I: Human Health Evaluation Manual (Part B, Development of Risk-based Preliminary Remediation Goals), Interim Final,* Office of Emergency and Remedial Response, Washington, D.C., 1991.

U.S. Environmental Protection Agency, *Risk Assessment Guidance for Superfund, Vol. I: Human Health Evaluation Manual (Part C, Risk Evaluation of Remedial Alternatives), Interim Final,* Office of Emergency and Remedial Response, Washington, D.C., 1992.

U.S. Environmental Protection Agency, *Risk Assessment Guidance for Superfund: Vol. 1. Human Health Evaluation Manual (Part D, Standardized Planning, Reporting and Review of Superfund Risk Assessments) Interim,* Office of Emergency and Remedial Response, Washington, D.C., 1998.

U.S. Environmental Protection Agency, *Survey Management Handbook, Vol. I: Guidelines for Planning and Managing a Statistical Survey,* Office of Policy, Planning, and Evaluation, Washington, D.C., 1983.

U.S. Environmental Protection Agency, *Scoper's notes, an RI/FS Costing Guide*, Solid Waste and Emergency Response, Washington, D.C., 1990.

U.S. Environmental Protection Agency, *Guidance for Performing Preliminary Assessments Under CERCLA,* Office of Solid Waste and Emergency Response, Washington, D.C., 1991.

U.S. Environmental Protection Agency, *Eco Update, Developing a Work Scope for Ecological Assessments,* Office of Solid Waste and Emergency Response, Washington, I,4, D.C., 1992.

CHAPTER 5

Managing Risk Assessment Report Development (Phase II)

David A. Belluck and Sally L. Benjamin

CONTENTS

I. Introduction 173
II. Managing a Project 174
    A. Implement Iterative Review, Comment,
        and Approval of Interim Deliverables 189
    B. Conduct Hazard Assessment 196
    C. Conduct Exposure Assessment 202
    D. Conduct Toxicity Assessment 208
    E. Conduct Risk Characterization 208
    F. Conduct a Final Review of the Draft Risk Assessment Report .. 209
III. Human Health Risk Assessment Review 211
    A. Introduction 211
    B. Peer Review of a Human Health Risk Assessment 214
    C. Risk Assessment Report Checklists 216
    D. Input/Output Analysis: Risk Assessment Review Accounting 217
IV. Concluding HHRA Review 220
V. Conclusion of Phase II 220

I. INTRODUCTION

Our previous chapter presented fifteen steps of risk assessment project planning (Chapter 4, Table 1). Here, we address Phase II and the seven steps of risk assessment project management, in which the risk assessment contractor undertakes the "real" work of performing a risk assessment.

II. MANAGING A PROJECT

This chapter assumes a project will be managed by "iterative review," which is simply a process of verifying that contractor work meets relevant performance standards prior to allowing its incorporation into subsequent work products. In iterative review, performance standards developed in Phase I will guide project work in Phase II. There are two general types of performance standards: (1) process standards guide how work is to be done, (2) product standards specify a final product's attributes.

An alternative to iterative review is to merely react to contractor work at the end of a project. We call this management approach "reactive management." In reactive management, an organization hires a contractor, establishes few, if any, standards to guide project work and trusts a contractor will produce an acceptable final product. Technical review of work does not occur until after a contractor produces a draft risk assessment report. Then, an organization's experts (or those of a regulatory agency) review the report, make note of inadequacies, and negotiations begin on how to correct shortcomings within its remaining budget and time frame. A contractor may generate a draft report more quickly without iterative review, but time saved typically drains away as contractor and project manager rework portions of the report where mistakes were integrated into subsequent work.

Reactive management is widespread within the world of risk assessment project management and is a source of problems wrongly attributed to shortcomings of technical and scientific risk assessment disciplines. Consider a project manager who discovers problems in a draft final risk assessment report. This person faces a disagreeable choice. If a project manager tries to force a contractor to fix problems to achieve desired technical quality, the project will probably be delayed, it may exceed its budget, and a contractor might refuse, either because there is little incentive to rework flawed sections as most of project payments have been received, or because projects are scheduled to begin for other paying clients. If problems are not corrected, a potentially fatally flawed report will be accepted.

Reactive management also places a contractor in a difficult position. Many decisions involving professional judgement take place during an environmental risk assessment project. A contractor is unlikely to make "right" decisions every time without feedback from those who will use a report. If a client does not participate in a risk assessment process, only an extraordinarily lucky contractor (or a mind reader) will guess right at each decision point. When a client finally reviews a draft final risk assessment report, a contractor is forced to justify and defend work, and may even be tempted to discount problems and advocate for no substantive changes.

Not surprisingly, risk assessments prepared under a reactive management style frequently encounter delays, run over budget, or are accepted despite technical flaws. For these reasons, we advocate use of iterative review, a process much more likely to yield a final draft report that contains few surprises and requires minimal correction. See Table 1 for a list of critical elements to consider when managing a risk assessment.

MANAGING RISK ASSESSMENT REPORT DEVELOPMENT (PHASE II) 175

Table 1 Examples of Critical Elements for Managing a Risk Assessment Report

ELEMENTS IN PHASE II: RISK ASSESSMENT REPORT DEVELOPMENT	Element (Applicable ? Y/N)	Present in Report? (Y/N)	Math/Science Correct and of Adequate Rigor? (Y/N)	Review of work product complete?	
				Pass?	Fail?
STEP 16 BEGIN CONTRACTOR WORK (see Chapter 5)					
STEP 17 IMPLEMENT ITERATIVE REVIEW, COMMENT, AND APPROVAL OF INTERIM DELIVERABLES (see Chapters 4 and 5)					
STEP 18 HAZARD ASSESSMENT					
Section properly formatted					
Sufficient tables to support text					
Sufficient figures to support text					
Comprehensive data collection					
Appropriate number of significant figures used in mathematical expressions					
Site-specific data from existing sources (e.g., physical, chemical, and biological parameters for all media) • Appropriate use of field data • Appropriate use of statistical methods • Appropriate use of pooling methods					
Methods to address data gaps					
Quantitative uncertainty analysis presented					
Qualitative uncertainty analysis presented					
Data summary tables in the text of this section					
Raw data provided used for QA/QC and data validation process and generation of risk assessment concentrations (for use in fate and transport and exposure equations)					

Table 1 continued

Comprehensive data evaluation							
Analysis of all data considered for use in the risk assessment report to ensure of usable quality							
Definition of usable data for the risk assessment report (e.g., meets all DQO, meets state requirements for data usability)							
Data sets meet data quality objectives							
Quality rating for each data set (e.g., adequate, marginal, inadequate) and usability in a risk assessment report							
Analytical data quality ranking							
Statistical analysis of raw analytical data							
Surrogate data sets • Appropriate usage • Evaluate surrogate data sets for data quality and ability to be pooled with other data							
Chemical selection process for assessment • A formal chemical selection process unnecessary if all chemicals found at a site are to be evaluated quantitatively in the risk assessment report • No chemicals removed from the list of COCCs without the written concurrence of the contracting organization • At least 99% of potential toxicity must be represented by chemicals listed for further evaluation in the risk assessment report when using concentration toxicity screens • Separate chemical selection processes should be employed for human health and ecological risk assessment • Toxicity values derived, where possible, from readily available reputable sources (e.g., IRIS, HEAST) for chemicals without readily available toxicity values • A route-to-route extrapolation (e.g., modifying inhalation toxicity values for use as ingestion toxicity values) evaluated by contractor, if EPA derived toxicity values are unavailable for a given chemical by a given route • Unless agreed to in writing by the contractee, a fully quantitative chemical selection process will be done for all identified site chemicals.							

Table 1 continued

Decision criteria table in chemical selection section lists:									
• Chemical name • Toxicity value (oral: Cancer Potency Factor, Carcinogenicity Group, Ingestion Reference Dose; Inhalation: Carcinogen Potency Factor, Inhalation Reference Dose) • Relative Toxicity Rating (e.g., high, medium, low and unknown with their associated definitions) • Sampling location and media • Persistence and mobility in the environment and associated definitions • Number of sites with adequate analytical data (determined by QA/QC) • Reasons for including or excluding a chemical from a quantitative risk analysis (inclusion/exclusion analysis)									
Chemicals expected to be found at a site, but not found, will be given a default value (e.g., 1/2 its quantitation limit)									
The contractor will provide a discussion in the body of an interim deliverable whether low concentration/high toxicity chemicals should be included in the risk assessment in spite of the relatively low numerical risk factor they receive in the toxicity/concentration screen									
Additional data obtained by proper sampling methods or appropriate use of surrogate data sets									
Statistical validity of environmental media samples (e.g., background and contaminated media soil, water, sediment, air, and biota samples over time) will be determined									
Source concentration data from actual media sampling or surrogate data sets									
Concentrations of analytes measurable at levels of human health or ecological concern									
Site concentrations									
QA/QC to assure required data quality									
Sample validity verification (e.g., QA/QC of samples)									

Table 1 continued

QA/QC report on all analytical data											
Comprehensive characterization of the site, activity, or facility											
Conceptual evaluation model presents sources of chemical contamination and concentrations, time, and location											
Conceptual evaluation model presents release pathways (e.g., media, rate of release)											
Conceptual-evaluation model presents all significant release mechanisms evaluated											
Conceptual-evaluation model presents all significant contaminant transport pathways evaluated											
Conceptual-evaluation model presents all significant cross-media transfer effects evaluated											
Conceptual-evaluation model presents all significant contaminated media evaluated											
Conceptual-evaluation model presents all significant site-specific characteristics evaluated											
Conceptual-evaluation model presents all significant spatial relationships evaluated											
Conceptual-evaluation model presents receptors (e.g., types, sensitivities, numbers, locations)											
Conceptual-evaluation model presents all significant land use scenarios evaluated											
All data usable in a risk assessment organized into a report or a summary table											
Decision criteria tables will be used by the contractor to demonstrate their train of logic in determining what constitutes an acceptable data set for risk assessment purposes											
Inclusion/exclusion analysis tables will be used by the contractor to demonstrate their train of logic in determining what data sets will be included or excluded from use in their risk assessment purposes											

MANAGING RISK ASSESSMENT REPORT DEVELOPMENT (PHASE II)

Table 1 continued

Summary table of data collected for a site, activity, or facility (chemical name, range of sample analytical quantitation limits, range of detected concentrations, and background levels for *each* medium)										
Summary table of chemical concentrations in all media sampled and minimally contains the chemical name, and the range of concentrations in each medium (e.g., soil, groundwater, surface water, sediments, and air)										
Summary table comparing background concentrations to chemical concentrations in suspected contaminated media										
Certifications that a rigorous technical review was carried out by qualified persons in the field of analytical chemistry QA/QC										
Certifications that all chemicals and their concentrations found in environmental samples will exhibited in tables until it has been determined, in concert with the contractee, that they will no longer be evaluated in the risk assessment										
Certifications that agreed to methodologies were used in screening COPCs to arrive at a final list of COC — this includes QA/QC methodologies										
Certifications that prior to delivery of any deliverable, that a rigorous mathematics, science, and writing review have been done and that the deliverables meet the quality requirements as stated in the contract or work plan										
STEP 19 EXPOSURE ASSESSMENT										
Description of the mathematical models used to generate the chemical concentration used in the exposure equations and justification of their use										
Sufficient tables to support text										
Sufficient figures to support text										
Section properly formatted										
Appropriate number of significant figures used in mathematical expressions										
Appropriate equations										
Appropriate variables										

Table 1 continued

Appropriate statistical methods								
Appropriate mathematical approaches								
Appropriate default values								
Adequate descriptions and justifications								
Groundwater Pathway Analysis — nature of chemical containment • Leaks • Spills • Ground spreading, underground burial of materials • Deposition of chemicals in trenches or impoundments in permeable soils • Incomplete containment								
Groundwater Pathway Analysis — Groundwater contamination likely • Landfills • Underground storage tanks • Surface impoundments • Lagoons • Open dumps								
Groundwater Pathway Analysis — quantity of releasable chemicals								
Groundwater Pathway Analysis — precipitation types and amounts								
Groundwater Pathway Analysis — infiltration rate								
Groundwater Pathway Analysis — subsurface conditions • Karst formations • Sands • Clay • Gravels								
Groundwater Pathway Analysis — groundwater mobility • Rapid • Slow								
Goundwater Pathway Analysis — most likely groundwater use scenarios								

Table 1 continued

Groundwater Pathway Analysis — analytical data on chemical concentrations • High • Medium • Low												
Groundwater Pathway Analysis — sources of groundwater contaminant exposures • Municipal • Private drinking water • Irrigation • Food processing • Aquaculture • Swimming pools • Recreation • Animal watering												
Groundwater Pathway Analysis — sensitive populations and environments												
Surface Water Pathway Analysis — distance between source of chemicals and surface water bodies												
Surface Water Pathway Analysis — quantity of releasable chemicals												
Surface Water Pathway Analysis — drainage area feeding into surface water body												
Surface Water Pathway Analysis — precipitation types and amounts												
Surface Water Pathway Analysis — infiltration rate												
Surface Water Pathway Analysis — likelihood of significant runoff or flooding												
Surface Water Pathway Analysis — nature of chemical containment												
Surface Water Pathway Analysis — presence of stressed vegetation												
Surface Water Pathway Analysis — fish and wildlife are unnaturally absent												
Surface Water Pathway Analysis — chemical releases have reached surface water bodies												

Table 1 continued

Surface Water Pathway Analysis — contaminated ground discharged to surface waters											
Surface Water Pathway Analysis — surface water contaminated											
Surface Water Pathway Analysis — drinking water intakes threatened											
Surface Water Pathway Analysis — human food chain impacts of contaminant releases											
Surface Water Pathway Analysis — sensitive populations and environments											
Surface Water Pathway Analysis — sources of surface water contaminant exposures • Municipal drinking water • Private drinking water											
Surface Water Pathway Analysis — water uses (irrigation, food processing, aquaculture, swimming pools, recreation, animal watering, in relation to local population exposures)											
Surface Water Pathway Analysis — surface water types present (rivers, wetlands, coastal tidal waters, disappearing rivers, ditches, intermittent flowing water, lakes, ponds, static water channels, oxbow lakes, embayments, harbors, sounds, estuaries, streams, back bays, lagoons, intermittent ponds and lakes, watersheds, drainage areas)											
Soil Pathway Analysis — nearby resident populations											
Soil Pathway Analysis — transient populations											
Soil Pathway Analysis — resident populations											
Soil Pathway Analysis — movement off-site as dust											
Soil Pathway Analysis — movement off-site as particles in soil runoff											
Soil Pathway Analysis — movement on vehicles											
Soil Pathway Analysis — movement to structures with transient or resident populations											

MANAGING RISK ASSESSMENT REPORT DEVELOPMENT (PHASE II) 183

Table 1 continued

Soil Pathway Analysis — occupational exposures				
Soil Pathway Analysis — sensitive populations and environments				
Soil Pathway Analysis — soil exposures to farm animals				
Soil Pathway Analysis — food chain impacts				
Air Pathway Analysis — population exposures in areas adjacent or nearby release source				
Air Pathway Analysis — unusual odors				
Air Pathway Analysis — actual releases to the environment				
Air Pathway Analysis — populations with adverse health effects from air exposures				
Air Pathway Analysis — circumstantial evidence of releases to the air				
Air Pathway Analysis — transient populations				
Air Pathway Analysis — resident populations				
Air Pathway Analysis — sensitive populations and environments				
Graphics • exposure routes of concern • exposure point concentrations • calculated risks at the appropriate exposure points				
Site Conceptual Model • Illustrates all potential contaminant transport pathways (e.g., direct air transport downwind, diffusion in surface water, ground water flow, and soil gas migration) • Illustrates all relevant cross media transfers (e.g., volatilization to air, wet deposition, dry deposition, ground water discharge to surface water, etc.) • Highlights exposure routes to be quantified in the text and identifies where they are discussed in the text				
Chemical fate and transport models appropriate to the physical setting (e.g., using a dispersion model designed for flat terrain for mountainous areas)				
Methods to address data gaps				

Table 1 continued

Quantitative uncertainty analysis presented												
Qualitative uncertainty analysis presented												
Include inclusion/exclusion and decision criteria tables list • All reasonable exposure pathways • Which pathways are included and excluded for further analysis • Reasoning used to include or exclude a pathway												
Selection of appropriate exposure case(s)												
Contaminant concentration isopleths presented by individual chemical and medium												
Fate and transport models • Chemical fate and transport models are appropriate to the physical setting • Wet and dry deposition modeling are performed for air toxics risk assessments • Models consider current and future risk scenario vapor-particle partitioning, groundwater to surface water discharges, and partitioning processes in soil, ground water, and surface water												
Evaluation of highly exposed and of susceptible groups and individuals												
Tabular presentation of equations used to generate exposure levels, all variables and their definitions, variable values and calculated intakes.												
Variable values are presented properly												
Complete variable selection tables are provided for each exposure equation used in the risk assessment												
Variable values in the body of the report correspond exactly with those approved in variable submission tables												
Exposure equations in the body of the report correspond exactly with those approved in variable submission tables												
Exposure pathways are properly illustrated												
Exposure routes are shown												
Exposure point concentrations are noted												

Table 1 continued

Calculated risks are indicated at the appropriate exposure points								
All assumptions used in exposure analysis were adequately justified								
Exposure equations and variable values in the main body of the report, text, and appendices correspond exactly to those approved in variable submission tables								
Appropriate equations								
Appropriate variables								
Appropriate statistical methods								
Appropriate mathematical approaches								
Appropriate default values								
Adequate descriptions and justifications								
Chemical fate and transport models appropriate to the physical setting (e.g., using a dispersion model designed for flat terrain for mountainous areas)								
Both wet and dry deposition modeling performed when air is the source of contaminant distribution to determine route-specific risks.								
Both current and future risk scenarios have been modeled unless otherwise specified by the contract								
Human health and ecological risk calculations based on well water withdrawals and discharges when ground water is the source of contaminant distribution								
Vapor-particle partitioning quantitatively addressed for contaminants that are determined to be transported through the air; partitioning processes in soil, ground water, and surface water also been quantitatively addressed								
Tables and text in the exposure analysis section provide the basis for selecting values and input parameters to exposure equations or models (when based on data, information is presented on the quality, purpose, and representativeness of the database; when based on assumptions, the source and general logic used to develop the assumptions [e.g., monitoring, modeling, analogy, professional judgement] is discussed)								

Table 1 continued

Discuss individual risks under various exposure cases					
Discuss important subgroups of the population such as highly exposed or highly susceptible groups or individuals					
Discuss population risks					
Comprehensive list of COPCs					
Appropriate number of COPCs					
Comparison of background chemical concentrations to contaminated media					
Analytical methods at or below levels of health concern					
STEP 20 TOXICITY ASSESSMENT					
Appropriate number of significant figures used in mathematical expressions					
Section properly formatted					
Sufficient tables to support text					
Sufficient figures to support text					
Quantitative uncertainty analysis presented					
Qualitative uncertainty analysis presented					
Source, calculation method, and reliability of toxicity values are discussed					
Methods to address data gaps					

Table 1 continued

Hierarchy of data sources to derive toxicological values and information • Integrated Risk Information System (U.S. EPA) • Health Effects Assessment Summary Tables (U.S. EPA) • Other EPA publications, directives, etc • Agency for Toxic Substances and Disease Registry reports • Other government agency publications or databases (e.g., California and New York), or federal agency (e.g., U.S. Fish and Wildlife Service) • International environmental or health agencies (e.g., International Agency for Research on Cancer, World Health Organization) • Professional organizations • Peer-reviewed computer databases • Peer-reviewed journals • Nonpeer-reviewed computer databases • Nonpeer-reviewed journals					
Relationship of extrapolation model selected and available information on how a chemical causes observed toxicity					
Regulatory concentrations for each COC					
Influence of toxicity data sets selected on numerical potencies of a chemical					
Discussion of use of animal-based toxicity values in HHRA					
Similarity between expected routes of exposure and those used in the risk analysis					
Nature, reliability, and consistency of the particular studies in humans and laboratory to risk assessment uncertainty					
Mechanistic basis for a chemicals toxicity					
Current toxicity values are used (current values verified less than one month prior to publication)					
Experimental animal responses to chemical exposures and their relevance to human illness or death					
Toxicity equations and variable values in the main body of the report must correspond exactly to those approved in variable selection tables					

Table 1 continued

Chemical mixtures toxicology data discussed					
Chemical mixtures risk assessment performed for carcinogens and noncarcinogens in accordance with EPA's 1986 guidelines and subsequent documents					
Group A, B, and C carcinogens evaluated using a current EPA CPF					
Group A, B, C, D, and E carcinogens evaluated using EPA RfDs and RfCs					
Chemical toxicology profiles are up-to-date (e.g., IRIS data from data search conducted within one month of report submittal)					
STEP 21 RISK CHARACTERIZATION					
Section properly formatted					
Sufficient tables to support text					
Sufficient figures to support text					
Methods to address data gaps					
Appropriate number of significant figures used in mathematical expressions					
Provides tables showing chemical risks by toxic and by exposure scenario					
Provides tables showing incremental risks					
Provides tables showing cumulative risks					
Carcinogenic risks listed for Group A, B, and C carcinogens					
Clear statement of the acceptable/unacceptable risk threshold					
Quantitative uncertainty analysis presented					
Qualitative uncertainty analysis presented					
STEP 22 CONDUCT A FINAL REVIEW OF THE DRAFT RISK ASSESSMENT REPORT					

Note: This table assumes use of iterative review and comment contract management approach, rather than reactive management.

Y = Yes
N = No
PNA = Potentially Not Applicable

A. Implement Iterative Review, Comment, and Approval of Interim Deliverables

As discussed above, in iterative review, a contractor generates all work products (referred to as "interim deliverables" until all project work has passed final review) in accordance with performance standards (see Tables 2 and 3 for examples of product and process components for risk assessment report review listed by report section). A contractor completes an interim deliverable and delivers it to a project manager, who conducts a brief review of the work. If there are no obvious problems, it is passed on to a project team for an in-depth review. If work meets all performance standards, it is approved. If not, it is returned to the contractor with an explanation of its shortcomings. An interim deliverable may be integrated into other work products only after it meets all performance standards and is approved.

It takes time to establish performance standards and to review interim work. Even so, iterative review probably improves overall project efficiency because it ensures that a consultant's work never strays far from performance requirements. The result is delivery of a report that is very nearly perfect as a project draws to a close.

Managing production of technical documents, in this case, risk assessment interim deliverables, can be a very intensive task. During this phase of report development, a project manager must monitor timelines, budgets, document development, and product technical quality. Many project managers run into trouble during this phase of report development by trying to replace, supplant, or bully their technical experts rather than use or guide their technical experts to tell them if an interim deliverable is technically credible.

The number of interim deliverables that will be generated during this phase of report generation depends on decisions of the project manager and project team. Some risk assessment teams prefer a limited number of interim deliverables (e.g., four, based on the HHRA paradigm), in order to speed delivery and review time. Others prefer to break a report into a large number of interim deliverables, hoping to prevent a major error in an interim deliverable from being propagated in later parts of the report. The smaller the interim deliverable, the less damage an error can cause. A decision on whether or not to use standard tables and formats, and if so which ones, will need to be made by a project manager.

Interim deliverables can be managed in two basic ways. One approach calls for an interim deliverable listing all methods, data, equations, etc., to be approved before calculations are allowed to proceed, and a second interim deliverable that provides results of calculations. Another approach uses only one interim deliverable that provides both inputs and outputs used in calculations, a more streamlined approach to interim deliverables. As with all interim deliverables, subsequent deliverables cannot be started by a contractor until an interim deliverable under review is formally reviewed and approved by a client (i.e., the organization hiring risk assessment services from a contractor).

Once again, a project manager is faced with a balancing decision — time restrictions vs. error and quality control; paired vs. single interim deliverables; extensive

Table 2 Examples of Product Components for Risk Assessment Report Review Listed by Report Section

Section	Components	
	General	**Specific**
Executive Summary	Risk findings	
	Summary tables and figures	
	Summary of risk assessment process	
	Summary of assumptions	
	Report uncertainties and effect on risk findings	
Introduction	Report scope	
	Report organization	
Hazard Evaluation	Background data	Site description (map)
		Site geography
		Sampling locations
		Sample media
Exposure Assessment	COCs	Narrowing of COPC list to COC list
	Potential exposure pathways	
	Data needs	Background sampling
		Sampling locations
		Sampling media
		QA/QC methods
	Data evaluation	Analytical methods
		Quantitation limits
		Qualified and codified data
		Use of blanks
		Tentatively identified compounds
		Chemical concentration calculations
		Comparison of concentrations to background
		Concentration-toxicity screen
		Data gaps and limitations
	Analyze uncertainty	
	Characterize exposure setting	Physical setting
		Potentially exposed populations
Exposure Assessment	Identify and describe exposure pathways	Source of receiving media
		Fate and transport of chemicals
		Exposure points
		Exposure routes
		Integration of sources
		Integration of releases
		Integration of fate and transport mechanisms

Table 2 continued

Section	Components General	Components Specific
		Integration of exposure points
		Integration of exposure route
		Summary of exposure pathways to quantify in assessment
	Quantify exposure	Exposure concentrations
		Chemical intakes/uptakes for each pathway
	Analyze uncertainty	
Toxicity Assessment	Noncarcinogenic	Toxicity values
		Regulatory concentrations
		Appropriate exposure durations
	Carcinogenic	Toxicity values
		Weight of evidence classification
	Chemicals lacking toxicity values	Qualitative evaluation
	Analyze uncertainty	
Risk Characterization	Noncarcinogenic	Single and multiple pathway
		Individual chemical risk and summation of total carcinogenic risk
	Carcinogenic	Single and multiple pathways
		Acute HQ for individual substances
		Subchronic HQ for individual substances
		Chronic HQ for individual substances
		Summation of HQs into hazard indexes by similar toxic endpoints
Analysis of uncertainties (quantitative or qualitative)		
Statement of findings		
Editorial section	A formal report section where project proponents (such as permitees) or responsible parties can provide editorial comments on the risk assessment report	Discussion of risk conservatism in the report and its effects on risk findings
		Biases associated with chemical selection
		Uncertainties associated with toxicity values
Appendices		

Table 3 Examples of Process Components for Risk Assessment Report Review Listed by Report Section

Element	General Attribute	Examples
Writing Style	Proper level for intended audience	
	Strong topic sentences	
	Clear, concise, and comprehensive	
	Risk neutral language	Presentation factual
		No editorializing, loaded, or biased terms (just the facts)
		Terms used accurately
	Clear linkages betwen text sections	
	Standard format	Consistent use of scientific notation (E5, 1×10^{-5}, etc.)
		Headings
		Type style
	Technical rigor appropriate to intended use of report	
Tables	Source of information identified (in each table)	
	Explain utility of each table	
	Clearly linked to related text, figures, tables, and appendices	
	Decision criteria tables present	
	Inclusion/Exclusion analysis	Shows all options evaluated
		Explains why some were selected for further analysis
	Risks tabulated	Individual exposure routes
		Individual exposure pathways (indirect and direct)
		Across exposure routes and pathways
	Compare risks to appropriate risk standards for each endpoint	
	U.S. EPA Standard Tables used, or adapted	
Figures	Identity source for each figure	
	Explain utility of each figure	Illustrate pathway analysis and exposure analysis
		Conceptual model of site, activity, or facility of concern
	Can stand alone (self-explanatory)	
	Clear, accurate, and precise	
	Clearly linked to related text, figures, tables, and appendices	
Technical Concerns	All relevant exposure pathways evaluated	Direct exposure
		Indirect exposure
		Uses comprehensive COC list, or justified exclusion of COPCs
	Specifies exposure case	RME, MEI, 95th percentile, 99th pecentile
	Specifies evaluation of risk to individual or populations	

Table 3 continued

Element	General Attribute	Examples
	Identifies guidance documents used (U.S. EPA risk assessment guidelines and technical publications, other federal or state agency guidance or requirements)	
	Explains how guidance documents were selected	
	Specifies types of data and models used	Verified data and models
		Comprehensive COC list
	Specifies how risk estimates generated	Single, complex computer model only
		Combined mathematical models and desktop methods
	Addresses level of rigor	All calculations are mathematically correct
	Addresses level of scientific analysis	Cumulative or incremental risk assessment selected and used throughout report
	Risk case achieved by use of proper input variables	Selection of toxicity values
		Selection of COCs
	Risk case achieved by use of proper input variables and selection of data for release quantification and modeling	Selection of toxic endpoints
		Selection of fate and transport models
		Selection of risk characterization methods
		Selection of uncertainty analysis methods
		Selection of sensitivity analysis methods
		Selection of study area
		Selection of exposure equations/models
		Selection of environmental conditions to model
		Distinguishes occupational from nonoccupational exposure
		Provides a site reconnaissance report
	Identify	Alternate methods to calculate environmental releases
		Statistics used to pool sampling data
		"Minor components" in chemical releases
		Fate and transport models verified and validated

Table 3 continued

Element	General Attribute	Examples
Technical Concerns	Identify	Quality of exposure models (for example, verified and validated)
		COC body burdens in exposed populations and individuals
		Use of default values, unvalidated assumptions, policy values, etc.
		Risk yardsticks (precise numerical value, a range of numerical values, narrative standard or guidance, or a numerical increment)
		Risk assessment case
		Outstanding issues that need further study
		Potentially fatal flaws and flaws serious enough to reduce usefulness of risk assessment findings (no confidence in environmental sampling data)
		Public concerns and project response (responsiveness summary)
	Define	How risk assessment case is achieved
		Meaning and use of confidence intervals and confidence levels
		And describe statistics used in risk assessment and their use and meaning
		Criteria for selection of COCs to be quantitatively or qualitatively evaluated in risk assessment criteria for total elimination of chemicals from quantitative or qualitative evaluation
		Quantifiable carcinogen (such as, U.S. EPA Group A, B, or C)
		Information sources used to obtain data used in risk assessment
	Spreadsheets allow all math to be checked from start to finish of risk assessment without gaps	e.g., inputs of Equation A yield outputs that are then used as inputs for Equation B that yields outputs that become inputs for Equation C and so on
	Summarize	Risk findings in table form with comparison to risk yardsticks
		Major uncertainties

MANAGING RISK ASSESSMENT REPORT DEVELOPMENT (PHASE II) 195

Table 3 continued

Element	General Attribute	Examples
		Variability in environmental sampling and effects on exposure and calculated risks
		Data quality used for each part of risk assessment
Regulatory/ Legal	All relevant statutes and administrative rules under which the risk assessment has been performed identified and requirements met	
	Applicable Risk Yardsticks	Legal or regulatory basis identified
		Clearly stated
Transparent	Discussion of risk assessment assumption, model, data, and findings in terms of accuracy, representiveness, completeness, precision, and relevance to known or project conditions	
	Reasons for selecting risk assessment type, chemicals of concern, exposure scenarios, toxicity values, report input variables, and models	
Uncertainties	Sources of risk over- and under-prediction discussed in text or tables.	
Review	Type used	Iterative: comment and approval of draft interim and final deliverables completed
		Reactive: only final draft and final deliverables
Source of Reviewers	External (stakeholder groups, government agencies, proponents and opponents)	Internal (staff)
	Response	Issues answered, addressed, denied, ignored

vs. intensive review procedures. As with many decisions made during a risk assessment, there is no one right answer.

This chapter will present a detailed description of many interim deliverable products that can be generated by a risk assessment contractor. The risk assessment project manager and project team must determine appropriate numbers of interim deliverables and their technical complexity. Each interim deliverable can contain text, tables, and figures that are to be linked to previous and future interim deliverables. When all interim deliverables are linked together to form a final report, they should demonstrate an unbroken, transparent, logical, and technically compelling argument that data, methods, and risk findings are reasonable and clearly articulated. The interim product approach helps to ensure that these goals are achieved.

In this chapter, interim deliverables are suggested for each of four steps of an HHRA. A short discussion of each interim deliverable is provided to explain the need for such a document. However, this discussion in no way replaces technical discussions provided by guidance and technical documents that fully discuss each technical step in risk assessment report generation.

B. Conduct Hazard Assessment

Hazard assessment, also referred to as "hazard evaluation" or "data collection and evaluation," involves collection and evaluation of data to ensure that adequate information exists to identify, examine, and to fully characterize all exposure pathways (see Table 4).

Data amassed during hazard assessment include both new and existing data about a site, activity, or facility of concern. Site sampling provides new data directly related to the site, activity, or facility under evaluation. Surrogate data sets are also valuable, however, especially to evaluate potential risks from proposed facilities or activities. Interim deliverables produced during hazard assessment are intended to generate reliable data on chemical release and exposure, for use in the exposure assessment phase of a risk assessment report.

Hazard assessment can generate voluminous data collections of variable quality. It is important, therefore, to organize and categorize data by quality or "useability" in a risk assessment. A contractor can achieve this by developing four types of reports:

- Site Sampling and Analysis Plan (or an Existing Data Analysis Plan)
- Site Sampling Analytical Chemistry Data Report
- Analytical QA/QC Data Validation Report
- Chemical Selection Report

1. Site Sampling and Analysis Plan

A Site Sampling and Analysis Plan (or Existing Data Analysis Plan) presents a work plan for data sampling and analysis to generate statistically and biologically credible contaminant data. Either surrogate data sets or monitoring data may be used, depending on project requirements. Monitoring data sets provide the best information about a specific site, activity, or process. However, use of surrogate data may be justified either because of cost or because monitoring is impossible (e.g., for a project involving a proposed facility).

During data collection and evaluation, risk assessors consult with chemists and statisticians. Chemists help determine what contaminants are present or expected, and at what levels. Statisticians can help determine whether on-site contaminant concentrations differ significantly from background concentrations, and the appropriate numbers of samples required to make that determination with a degree of scientific certainty. These experts may also help with such deliberations as:

- Whether the proposed sampling methods are adequate to measure COPCs at concentrations of human health or ecological concern, given the nature of potential exposures and risks
- The characteristics of the site, facility, or activity
- Whether sampling results will be representative of important characteristics
- Whether it is appropriate to combine data gathered at different times, given the level of precision and accuracy of the reported sample concentrations

MANAGING RISK ASSESSMENT REPORT DEVELOPMENT (PHASE II) 197

Table 4 Examples of Concerns for Review of a Human Health Risk Assessment Hazard Evaluation

Concern	General Concern	Examples
Figures	Conceptual model	
	Habitat/land use map	
	Food web	
	Isopleth maps of actual or calculated/modeled contaminant concentrations	
Technical	Sufficient literature and database searches to support analysis	
	Reasonable data usability hierarchy	e.g., sampling data from existing facility that is very similar to proposed facility, sampling data from facility somewhat similar to proposed facility, sampling data from facility not closely related to proposed facility, sampling data generated using models and proposed facility inputs and outputs
	Data sufficiency	
	Samples	Types of samples (e.g., site specific, current, old, surrogate samples from similar facilities)
		Sufficiency
		Relevance
		Analytical chemistry QA/QC
		Usability
	Sampling statistics	Representativeness of sampling
	Appropriate use of fate and transport analysis	
	Comprehensive list of COPCs	
	Appropriate methods used and presented to generate final COC list	
	Receptor location physical properties fully described	
	Credible mass balance	
	Derivation of bioaccumulation factors (BAFs) and bioconcentration factors (BCFs)	
	Terminology	Correct scientific names for organisms
		Correct technical terms for abiotic components of ecosystem
	Receptor location	
	Site inspection	
Report Transparency	All assumptions and decisions employed are fully discussed	
	All information and numerical data referenced.	
Uncertainties	Data sources	Samples
		Surrogate data sets
		Other data
		Data sufficiency
	Data accuracy	Emission rate data
		Pollution control equipment efficiency

Table 4 continued

Concern	General Concern	Examples
	Does data adequately represent site, activity, or facility characteristics	Ability to maintain pollution control equipment efficiency over time
		Effects of catastrophic releases on total release estimates
		Point estimates do not reflect data distribution
		Assumed vs. absolute and operational ability of pollution control devices or methods to function at design control efficiencies
		Releases caused by decreased release control efficiencies and catastrophic releases
		Identities of chemical species released into the environment
		Physical form of released chemicals (such as, particulate sizes and distribution, chemical distribution, and quantities on particles of different size)
Uncertainty Analysis	Exclusion (or inclusion) of chemicals from quantitative analysis	e.g., does report only quantitatively evaluate chemicals considered to have a high potential for release, high release rates, or readily available toxicity values
	Confusion of uncertainty and conservatism	
	Conservative (non-conservative) assumptions	Regarding percentage of total release attributed to each chemical of concern
	Assumed vs. absolute and operational ability of pollution control devices or methods to function at design control efficiencies	
	Releases caused by decreased release control efficiencies and catastrophic releases	
	Physical form of released chemicals	e.g., particulate sizes and distribution, chemical distribution, and quantities on particles of different size
	Model quality	Validated or unvalidated
	Release Model quality	Identification of all release sources
		Quantification of releases from all sources
		Release rates
		Release composition
		Fugitive release estimates
		Plume/release depletion/attenuation evaluation
		Emission size distribution estimates (such as, particulates)
		Effects of transport and dilution on receptor point concentrations of released substances

MANAGING RISK ASSESSMENT REPORT DEVELOPMENT (PHASE II)

Table 4   continued

Concern	General Concern	Examples
		Effects of loading and handling on releases
	Facility Model quality	Facility lifetime
		Input/output rates from facility
		Accuracy of mass balance calculations
	Exposure case selection and implementation	Use of incremental vs. cumulative risk scenario
		Fate and transport model inputs and outputs
		Location of release sources and receptors
		Conservatism of each input to each other and risk assessment exposure case
		Wet and dry deposition rates
		Discharge rates
		Runoff rates
—		Meteorological conditions
		Water body parameters
		Removal efficiencies
		Mixing rates
		Soil density
		Movement of COCs within and between environmental compartments
		Days per year and hours per day that environmental releases are expected
		Changes in release characteristics
		Release rates from handling or transshipping chemicals of concern
		Fugitive releases
		Validation of data and models.

Answers to these questions determine how data can be used in the risk assessment (in a qualitative, semiquantitative, or quantitative manner) and overall report rigor.

Careful evaluation, as described above, reduces uncertainties associated with measured chemical concentrations and improving overall report rigor. In contrast, poor sampling and data analysis can generate false negatives (indicating no problem where a problem exists) or false positives (indicating a problem where there is none). Failure to achieve minimum data quality requirements for environmental sampling and analytical chemistry will undermine an entire risk assessment process and make it impossible to reliably assess exposure pathways or to even establish contaminant source concentrations. U.S. EPA guidance and many other technical documents provide detailed guidance on the proper techniques for sampling soil, groundwater, air, surface water, sediments, food, and human tissue.

In some cases existing environmental sampling is adequate for a risk assessment project's needs. Data sets must simply be analyzed to determine whether they can be used alone, or in combination with other data. An analytical chemistry QA/QC process evaluates quality of all data considered for use. This is especially important when data sets originate from different days, locations, or laboratories, or if surrogate data sets are only available from situations that do not exactly match the proposed facility. A qualified expert in QA/QC appraises usability of existing data sets by reviewing source of data and all available documentation on how it was collected and analyzed. QA/QC performance standards are used to guide a contractor performing data usability review (see Chapter 11).

In other cases, site sampling is not possible. Site sampling may be impossible, for example, if a risk assessment is for a proposed facility, i.e., a facility that has not yet been built. If so, surrogate data sets are obtained from technical literature and are evaluated to determine quality of each data point and to decide whether data sets can be combined. These data sets must be chosen and analyzed with care to ensure their quality and applicability to the current project.

Data sets deemed usable by systematic evaluation are organized into a data summary. This is a report, table, or list summarizing data in one of two possible formats. Data are presented either as chemical concentrations in a specific environmental medium or as chemical concentrations in all environmental media. Data summaries organize data for easy use and efficient review.

A contractor uses a data summary to determine whether chemical concentrations at a site are less than, equal to, or greater than background concentrations. Chemicals at concentrations exceeding background concentrations will probably undergo quantitative evaluation, whereas a contractor might perform a less rigorous evaluation of risks of on-site chemicals at levels below or equal to background concentrations. Finally, chemicals strongly suspected to be on-site may be listed in a data summary, although sampling failed to detect them. Since actual concentration is unknown, chemicals suspected to be present are assigned a theoretical concentration, such as one-half the analytical detection level for that chemical.

In addition to presenting a work plan, a Site Sampling and Analysis (or Existing Data Analysis) Plan sets forth the following information:

- Risk assessment data needs (e.g., what media must be sampled to generate data for use in risk assessment)
- DQO (qualitative and quantitative statements that ensure that data of known and documented quality are obtained during site sampling and analysis);
- A site conceptual evaluation model of all potential or suspected sources of contamination, their identities, types, and concentrations, potentially contaminated media, potential exposure pathways, and probable completed exposure pathways
- Fate and transport models and exposure variable data needed in exposure assessment

After a project manager approves the Site Sampling and Analysis Plan, a consultant can begin environmental sampling, data compilation, and analysis.

2. Site Sampling Analytical Chemistry Data Report

A Site Sampling Analytical Chemistry Data Report, a second type of report that can be produced by a contractor, discusses data usability and organizes project data into tables. This will help determine whether there is sufficient data of required quality to meet project needs. It also aids reviewers in evaluating data usefulness. Since data quality affects credibility and uncertainties of a risk assessment report, data sets of sufficient quality must be selected to match intended risk assessment rigor.

3. Analytical QA/QC Data Validation Report

Analytical chemistry data for a site, activity, or facility usually is collected on different dates, from different locations, using different sampling and analytical methods. In a third type of report, an Analytical QA/QC Data Validation Report, a contractor discusses whether available data sets for a site, activity, or facility meet minimum data quality requirements and, therefore, can be combined to generate a single numerical chemical concentration. It is not unusual for significant portions of an analytical chemistry database to be of such poor quality as to be unusable. An Analytical QA/QC Data Validation Report presents each element of a chemical database in a table with a quality rating (e.g., adequate, marginal, inadequate) assigned to each data point. A report may also compare environmental sample concentrations with natural and anthropogenic background concentrations and compare risks from background chemical concentrations to risks from a site, facility, or activity of concern. This table helps risk assessors determine quality of available data, as a group, as subsets, and as individual data points.

4. Chemical Selection Report

A site, facility, or activity can release hundreds (or even thousands) of chemical compounds, COPCs. Costs, time, and data limitations may preclude a quantitative evaluation of every chemical listed in data summaries. Screening out chemicals from full, quantitative assessment (i.e., reducing a COPC list to just a few COCs), has been justified due to cost and efficiency. Such reasoning has been undermined by significant improvements in risk assessment tools. Desktop computing capabilities, dedicated risk assessment software, and widespread availability of appropriate data sets all belie any need to screen out COPCs for efficiency. It is now possible for any qualified risk assessment contractor to quantitatively evaluate all chemicals for which concentration (or emission levels) and toxicity factors exist or can be derived. However, chemical screening continues.

In a fourth type of report, a Chemical Selection Report, a contractor selects a list of COCs, chemicals slated for full, quantitative assessment, from a comprehensive COPC list. Although not justifiable in most cases, methods such as Concentration-Toxicity Screens are used to identify a subset of chemicals that represent a majority of risks (e.g., >95%) in an environmental medium. This subset is considered COCs, and only chemicals in this subset undergo full quantitative evaluation.

In a Chemical Selection Report, a contractor summarizes data on COPCs / COCs, arranged either by specific environmental media (e.g., for soil list: chemical name, frequency of detection, range of sample quantitation limits, range of detected concentrations, and range of background levels) or for all sampled media (e.g., list each chemical and its concentration range in each medium).

C. Conduct Exposure Assessment

In the exposure assessment phase, a contractor evaluates magnitude, frequency, duration, and route of chemical exposures of receptors. This evaluation may be qualitative or quantitative; it may assess past, current, or future exposures, and it may consider exposure of human or non-human receptors. Table 5 presents examples of concerns for review of an HHRA exposure assessment.

First, a contractor characterizes the exposure setting, gathering all relevant information on physical setting, such as meteorologic patterns, geographic features, and social factors; land use (past, current, and future); population density and demographics; and behavior of nearby populations. Next, a contractor identifies exposure pathways by identifying chemical sources, mechanisms of chemical releases, and environmental media capable of transporting chemicals to locations of exposed organisms. Figure 1 illustrates possible fate and transport paths. The contractor then evaluates possible exposure pathways and identifies those most likely to be completed.

Completed exposure pathways, those where a release results in known or probable exposures, are analyzed in detail in the exposure section of a risk assessment report. An exposure pathway is completed when a chemical moves away from its source through the environment to a location where an organism is directly or indirectly exposed. Figure 2 illustrates exposure pathways. Movement from a chemical's source to another location is termed its "fate and transport." Chemicals may move through one or several environmental media, including water, soil, air, sediment, terrestrial or aquatic plants, and terrestrial or aquatic animals. The environmental medium by which a chemical encounters an organism and exposure occurs is termed "exposure medium." An exposed organism is termed a "receptor."[*,**] After an exposure occurs, contaminants from the exposure medium enter a receptor by three possible "exposure routes": inhalation, ingestion, and dermal absorption.

The goal of interim deliverables generated during exposure assessment is to estimate type and magnitude of exposure to COCs that are present or migrating from a site, activity, or process. Toward this end, contractors can produce three types of reports:

- Fate and transport modeling recommendations report
- Exposure point concentrations report
- Variable selection table report

---

[*] Human receptors include: adult resident, subsistence farmer, worker, home gardener, subsistence fisher, child resident, and swimmer.
[**] Ecological receptors include: mammals (subdivided into predators or nonpredators), fish, benthos, birds (subdivided into predators or nonpredators), plants, insects, crustacea, and soil fauna.

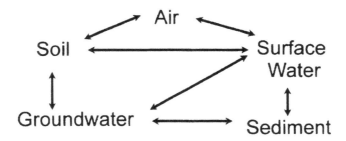

Figure 1 Example of fate and and transport diagram. (Adapted from U.S. EPA, 1995, Development of Human Health Based and Ecologically Based Exit Criteria for the Hazardous Waste Identification Project, Vol. 1, Figure 1-1, page 1–6.)

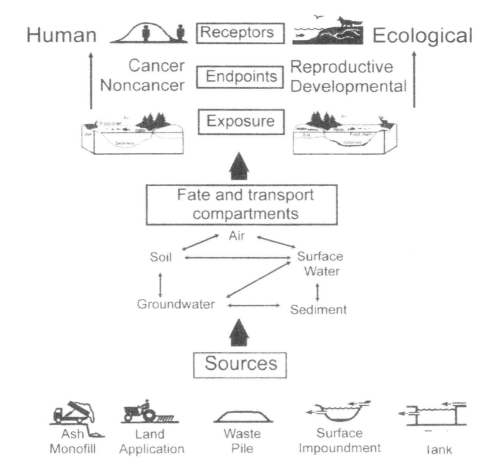

Figure 2 Example of exposure pathway analysis.(Adapted from U.S. EPA, 1995, Development of Human Health Based and Ecologically Based Exit Criteria for the Hazardous Waste Identification Project, Vol. 1, Figure 1-1, page 1–6.)

Table 5 Concerns for Review of a Human Health Risk Assessment Exposure Assessment

Concern	General Concern	Examples
Tables	Decision criteria tables	
	Variable selection tables	
	Tables with assumptions	
	Input/Output tables	
	Exposure pathways	
	Routes of exposure	
	Direct exposure	
	Indirect exposures	
Figures	Exposure pathways	
	Routes of exposure	
	Direct exposure	
	Indirect exposures	
Technical	Proposed time span of a facility, activity or process to be modeled in the risk assessment	e.g., industrial facility operational lifetime
	Effects of local meteorological during a typical year	On environmental releases
		On effectiveness of emission controls
	Typical lifetimes for similar sites, activities or facilities	
	Breakdown of lifetime exposure	e.g., simple 70 year assumption or division into years of exposure
	Exposure breakdown by body weights by age class	e.g., infant, child, young adult, adult
	Exposure scenarios	Types of exposure scenarios evaluated quantitatively
	Routes of exposure quantitatively evaluated	
	Receptor types quantitatively evaluated	e.g., average person, 95th percentile individual, 99th percentile individual, etc.
	Routes of exposure and receptor types qualitatively evaluated	
	Handling of incidental exposures	
	Selection and correct scientific name of nonhuman species for food web analysis	
Regulatory/ Legal	Standard default values	Values identified
		Source of values identified
		Used correctly
Report Transparency	Models fully justified	
	Models fully referenced	
	Input values fully justified	
	Input values fully referenced	

MANAGING RISK ASSESSMENT REPORT DEVELOPMENT (PHASE II)

Table 5 continued

Concern	General Concern	Examples
Uncertainties	Selection of appropriate models	
	Selection of appropriate input values	Bioconcentration factors
		Bioaccumulation factors
		Trophic/food web structure
	Rigor of analysis	Ingestion exposure analysis
		Inhalation exposure analysis
		Dermal exposure analysis
	Depth of contamination in sediments and soils	
	Distribution, concentration and chemical speciation in food items	
	Behavioral characteristics of potentially exposed population	
	Exposure duration	
	Exposure scenarios	
	Indirect exposure pathway completion	
	Direct exposure pathway completion	
	Ingestion rates	
	Exposure frequency	
	Bioavailability of COC	
	Exposure averaging time	
	Sources of exposure	Soil location and mixing depth
	Sediment/Soil mixing depths	
	Water vs. food as source of biocumulative chemicals in aquatic organisms such as fish	
	Percent chemical and chemical species in a particular phase	e.g., particulate versus dissolved
	Determination of COC background concentrations	
	Handling of independent and dependent variables	
	Change in environmental/exposure conditions between time hard data collected and time period risk assessment models	
	Intermittent exposures	
	Background exposures and body burdens	

1. Fate and Transport Modeling Recommendations Report

A fate and transport modeling recommendations report discusses how a risk assessment will calculate concentrations at receptor points, either for a single numerical concentration or a range of concentrations for use in intake/uptake equations. Fate and transport models allow a contractor to estimate drop in concentration of a chemical as it moves from a point of release to a receptor point.

In theory, concentration at an exposure point can be monitored. In practical terms, however, monitoring is difficult. There are spatial problems — unless a receptor wears monitoring equipment, exposure location will differ from monitoring location. There are temporal problems — exposure data are collected at discrete points-in-time, although exposure is likely to be continuous. There are analytical limitations — exposure point concentrations may fall below current analytical capabilities. Even if exposure concentration can be measured with a degree of certainty at a receptor point, it is very difficult to monitor intake/uptake of a chemical by individual receptors.

Models offer a way around spatial, temporal, and analytical difficulties that impede monitoring data collection. Although models merely approximate reality, they play an expanding role in environmental risk assessment. Risk assessors use fate and transport models to represent how chemical contaminants move through environmental compartments and to estimate chemical concentrations at points along these paths, including concentrations at exposure points. Intake/uptake equations, another type of model, estimate the amount of chemical that passes from the environment into a receptor at a point of exposure.

In order to ensure proper use of models, a project manager can require a fate and transport modeling recommendations report. This report is reviewed and, if acceptable, approved by a project manager.

Fate and transport modeling recommendations reports present estimates of concentrations for each environmental medium where COCs may reside (e.g., downstream in surface water, down gradient in groundwater, as particulates or vapor in air). These estimates are obtained either from monitoring data (e.g., water sampling) or from modeling.

Estimates are modeled by entering source concentrations (obtained from data collected and evaluated in hazard assessment) into environmental fate and transport models. Whereas concentration at a source is known, concentration of a contaminant at some distance from its source must be calculated. Either mathematical or computer simulation models can be used. These models assume that environmental conditions will change concentration of a contaminant as it moves through an environmental medium (e.g., air, surface, water, groundwater) from its point of origin to a receptor. Assuming no input of contaminant from other sources, a contaminant concentration should drop as a function of distance from release point. However, rate of decrease may be significantly altered if a chemical transforms physically (e.g., from solid to gas, solubilized in precipitation), chemically (e.g., photolysis, hydrolysis, oxidation, reduction), or biologically (e.g., biodegradation, formation of metabolites), or if it accumulates because of an affinity of certain types of chemicals for certain environmental media. Analysis should account for these possibilities. Depending on risk assessment project purpose, the highest receptor point concentration is determined, either for each chemical or for the highest overall exposure concentrations for all chemicals.

Choice of models is important. Modeling results are not reality. They are mathematical representations of reality. Accuracy and precision of this representation depends on validity of model structure and inputs. It is extremely important, therefore, to ensure accurate representation of physical, chemical, and biological processes by using models that have been calibrated and validated or confirmed for a given situation. Unfortunately, many models commonly used for risk assessment have not been calibrated or validated. So, there is no way to ascertain how well they represent reality. In addition, there are many different types of environmental fate models, each with strengths and weaknesses. Models may be specific to one medium or may model movement of contaminants through several media.

Exposure pathways with the highest likelihood of completion are carried into exposure quantification step of exposure analysis.

2. Variable Selection Table Report

Before exposure can be quantified, however, a contractor must select, or calculate variable values for use in intake/uptake concentrations and equations, discussed in our next section. A contractor prepares a variable selection table report to aid variable selection review by a project manager and project team.

A variable selection table report organizes intake/uptake concentrations, equations, and input variable values in tables to allow reviewers to easily determine whether values selected represent the required level of risk conservatism. These tables present each equation used in exposure quantification (see below) and show the value selected for each variable in an equation. In addition, each variable is defined, a range is presented of values that could have been selected for this variable, and the value selected is identified on a distribution curve of all possible values (e.g., mean, mode, 95th percentile). There is a section stating reasons for selecting the value and a citation is given for the source of each variable value. Variable selection tables permit very efficient review of numerical values proposed for use in exposure calculations in exposure assessment.

3. Exposure Point Concentrations Report

Finally, a contractor quantifies exposure by calculating chemical intakes and uptakes by receptors, using equations that are route-specific and media-specific. A contractor presents these calculations for review in an EPC report. This report is a series of tables that list numerical concentrations to be used, and intake/uptake equations employed to model direct and indirect exposure. Whether they are simple calculations or complex computer programs, all exposure equations use the same basic types of inputs, including variables related to the chemical (concentration), exposed population (contact rate with medium, exposure frequency and duration, and body weight), and choice of risk assessment techniques (e.g., averaging time for exposure period). Presenting equations and their inputs in tables allows for easy review. Appendices to an EPCs report may explain strengths and weaknesses of fate and transport models used to calculate intake/uptake concentrations, as well as reasons for selecting variables that were used in each equation in a model.

4. Precautions in Exposure Assessment

Certain precautions should be observed in using exposure equations. First, it is important to recognize that a given intake/uptake equation can dramatically alter calculated exposure level. Choice of exposure equation may make a difference between an acceptable risk estimate and one that indicates unacceptable risk. For this reason, it is essential that an exposure equation be a reasonable mathematical representation and that it represent correct exposure cases. Second, variable values must be selected to represent appropriate predetermined levels of conservatism (e.g., worst case), alone, and in concert with other variables. Finally, data must be of sufficient quality to support quantitation (otherwise, exposure quantification may be inappropriate).

D. Conduct Toxicity Assessment

Toxicity assessment is next. Toxicity assessments evaluate potential for environmental contaminants to cause adverse effects in exposed individuals and populations and, if possible, determine relationship between exposure levels and increased likelihood or severity of adverse effects. During toxicity assessment, a risk assessment contractor gathers toxicity information (both qualitative and quantitative) for substances under evaluation; identifies exposure periods (acute, subchronic, chronic) for which toxicity values are needed; determines carcinogenic and noncarcinogenic (systemic) effects on test organisms; and summarizes toxicity information (see Table 6).

A Toxicology Data Report presents a contractor's toxicity assessment work for review. This report presents toxicological values which a contractor will combine with exposure concentrations (determined in exposure assessment step) to generate risk estimates. If possible, a contractor obtains information on toxicity (relationship between dose and effect) from standard toxicity databases, such as U.S. EPA's IRIS, HEAST, or California EPA toxicity values. Chemical-specific toxicity values that may be obtained from these sources include RfDs, RfCs, CPFs, and benchmark values. If toxicology values are not readily available from such standard sources, a risk assessment contractor should generate them from studies published in peer-reviewed journals, if possible. If peer-reviewed journals do not provide necessary information, toxicology values may be generated from nonpeer-reviewed literature. Thus, it should be clear that some toxicity values are more reliable than others because they derive from higher quality data sources.

A contractor organizes toxicity values into a series of summary tables. Separate tables are generated for noncarcinogens and carcinogens. Summary tables of noncarcinogenic toxicity data present RfDs and RfCs for all applicable exposure durations, confidence levels for toxicity data, critical toxic effect used to generate a NOAEL or LOAEL, route of toxicant administration for deriving an RfD or RfC, source of RfD or RfC, and numerical values used to generate a total UF or modifying factor. Summary tables of carcinogenic toxicity data present route of exposure, numerical SF, letter weight-of-evidence classification, type of cancer induced, route of toxicant administration and source of the SF.

E. Conduct Risk Characterization

The final step is risk characterization (see Table 7). This step uses exposure and toxicity data, gathered in prior steps, to calculate risks for all Group A and B carcinogens and for noncarcinogens by toxic endpoint and exposure duration, or to generate qualitative expressions of risk, if data is insufficient for quantitative analysis.

Table 6 Examples of Concerns for Review of a Human Health Risk Assessment Toxicity Assessment

Concern	General Concern	Examples
Technical	Source of toxicity values	e.g., IRIS, HEAST, IARC, WHO
	Hierarchy of data sources	
	Methods to generate toxicity values when such values are not readily available	
	Match of chemical species	
	Bioavailability and toxicity related to local environmental conditions	
Uncertainties	Toxicity values	Methods used to calculate toxicity values
		Data used to calculate toxicity value
		Weight of evidence classifications
	Noncarcinogenic effects of carcinogens	
	Interactive effects of chemicals in receptors	Synergistic effects
		Antagonistic effects
	Chemical form selected to represent all exposure to that chemical	Metallic, organic, inorganic
	Chemical species	Percent of chemical exposure attributed to a given chemical species (such as, trivalent versus hexavalent)
	Linearity of toxicity	
	Relative source contributions	
Review Process	Use of toxicology values	Correct in text
Correct in Tables		

Information used in this step should have already been reviewed and approved, under iterative review process, so a contractor can be confident using this information in standard equations for calculating risk. A risk characterization report, produced in this step, presents input data and risk calculations in a tabular format.

F. Conduct a Final Review of the Draft Risk Assessment Report

A contractor now assembles all approved interim deliverables into a final draft risk assessment report and submits it for review and approval by a project manager. Although it is likely that every draft final report will have some problems, their number and severity will diminish with review rigor if a report was developed through an iterative review process.

Before a project manager distributes a draft final report to a project team, it should undergo review by project manager and, if possible, risk advisor. This will ensure that a contractor has met minimum report standards regarding its content, format, and QA/QC standards. If a report does not meet minimum standards, it is returned for revisions. If a report meets minimum standards, a project manager distributes it to the project team for review, comment, and approval.

Table 7 Concerns Specific to Review of a Human Health Risk Assessment Risk Characterizaion

Element	General Attribute	Examples
Writing	Risk neutral language used	
Tables	Individual exposures pathway risks	
	Individual routes of exposure risks	
	Summed risks across exposure routes and pathways	
	Carcinogen risks	e.g., individual and summed
	Noncarcinogen risks	e.g., individual and summed
	Comparison of individual and summed risks to risk yardsticks	
Technical	Individual exposure pathway risks tabularized	
	Individual routes of exposure risks tabularized	
	Summed risks across exposure routes and pathways	
	Noncarcinogen risks	Individual
		Summed
	Comparison of individual and summed risks to risk yardsticks	
	Noncarcinogen risks presented for each substance by its critical toxic effect using HQ and HI approach	
	Screening level hazard index	All noncarcinogen HQs summed
	Carcinogenic risks presented for each substance	Sum risks of U.S. EPA Group A, B, and maybe C carcinogens
	Relative source contributions used to account for different sources of exposure being less than 100% of all exposures	Drinking water is often allocated 20% of total exposure, for example.
	All noncarcinogen HQs summed for use as a risk screening tool	
	Risk yardsticks defined (as single numerical values, range of values, or increments with error bars)	
	Appropriate uncertainty analysis	Qualitative uncertainty analysis, quantitative uncertainty analysis, probabilistic uncertainty analysis
	Tabular presentation of all assumptions, in order of appearance in text, and effects of uncertainties on risk assessment numerical findings	Impact on risk assessment conservatism, numerical contribution of each assumption to final risk estimates
	Sensitivity analysis	
Uncertainties	Dose-response curves	
	Dose-response model	
	Risk yardsticks	
	Synergistic and antagonistic effects data base	
	Chemical mixture data base	
	Additivity assumptions	
	Linearity of dose and effect relationship	

MANAGING RISK ASSESSMENT REPORT DEVELOPMENT (PHASE II)

Table 7 continued

Element	General Attribute	Examples
	Derivation of toxicity values (may be greatest uncertainty in risk assessment)	
Applicability of Regulatory Concentrations		
Slope of Dose Response Curve		

Performance standards* developed during risk assessment report planning will be used to evaluate the draft risk assessment report. While earlier reviews of each interim deliverable should eliminate need for an intensive review of a draft final risk assessment report, this is not always true. Therefore, it is necessary to perform a thorough review of a draft final document. HHRA review is discussed in detail in the next section. (ERA review is presented in Chapter 9.)

III. HUMAN HEALTH RISK ASSESSMENT REVIEW

A. Introduction

HHRAs evaluate potential risks and impacts posed by a site, activity, or facility on human morbidity or mortality. Chapter 2 describes the HHRA process in detail. Review of contractor-produced HHRA deliverables will ensure that science is consistent with current standards, that calculations are verifiable, and all product and performance standards have been met. Review of interim deliverables ensures that each deliverable meets project performance standards and ensures report quality. This will speed review and approval of the full report and prevent compounding of errors from flawed interim work products being integrated into subsequent report sections. This chapter offers tools for conducting a critical review of interim and final risk assessment deliverables.

Review of draft publications, including risk assessments, is standard practice for professional scientists, intended to detect and correct flaws before a technical document is published. Unfortunately, criticism is rarely welcome. During review, everyone involved must remember that critical comments are directed at problems, not people. Comments should be offered and received in this light. With this single caveat, in our opinion, critical reviewers cannot be too critical. Reviewers who soften their comments in interest of civility may fail to identify or clearly communicate problems.

It is a critical reviewer's job to check contractor-generated products to ensure that all product and performance standards have been met for a risk assessment report. In a properly scoped and contracted risk assessment project, performance standards are contract terms. Thus, in effect, critical review is a check to ensure compliance with terms of the contract for service. Critical review also safeguards credibility and protects against potential liability. If errors become public, they can

---

* Performance standards were developed early in the risk assessment project based on expectations of those expected to use and review the final risk assessment report. They were also integrated into the project Scope of Work, contract terms, work plan details, and iterative review standards.

be used to undermine a project or to embarrass, sue, or bring undue pressure to bear on an organization, risk assessment team, and its contractor. These attacks could seriously skew a risk assessment outcome.

Risk assessment reports can be huge documents. Reviewing them takes considerable professional time and effort. They typically involve hundreds of pages of text, supported by scores of figures and spreadsheets presenting mathematical model inputs and outputs for each step in calculating project risk estimates. Each piece of a risk assessment builds on previous pieces. One erroneous calculation can undermine all subsequent calculations and even a relatively small error in a critical calculation may give rise to a fatally flawed risk estimate. Thus, each step and every calculation of a risk assessment must be critically reviewed, both alone and in conjunction with all related steps. Each step and calculation must be clearly presented, if a risk assessment is to be easily reviewed.

In conducting a critical review, the quality of each technical decision must be considered, as well as its impact on other decisions in the risk assessment . The potential for a decision to influence other decisions depends on whether it is "independent" or "dependent." A decision is independent, if changing it will not directly affect other decisions. For example, if changing a value selected for one exposure variable alters no other values, exposure variable selection is an independent decision. A decision is dependent, if changing it affects other decisions. Due to their far-reaching effect, dependent decisions must be made in a reasonable manner, must be fully explained, and must be carefully reviewed.

It is rare for an individual to possess sufficient expertise to conduct an adequate review of an entire risk assessment report. The multidisciplinary nature of risk assessment means that a report reviewer, well qualified to review work in some disciplines, is probably less qualified to review work in others. An engineer is not a toxicologist, for example, and is rarely technically trained and qualified to review the technical accuracy and professional decisions related to toxicology. Similarly, toxicologists are not hydrologists and cannot evaluate placement and depth of monitoring wells. Team work solves this review problem.

A reviewer should have experience, as well as education in a proper discipline. Professional experience will help a reviewer appreciate both the general principles and the case-specific issues of the situation under review. A reviewer likely to have such appreciation will have extensive field, laboratory, and desktop risk assessment experience. Finally, certain personality traits, such as intuition, common sense, tact, and an analytical mind are desirable in a reviewer. The best reviewers seem to have a sixth sense for where problems will pop up. Some are so grounded in reality that they can immediately spot problematic risk assessment assumptions or calculations. Other excellent reviewers spot problems simply by thoroughly checking every equation, as well as every input used and output presented in the risk assessment report.

Since it is a labor-intensive undertaking, review should not begin until a contractor's staff scientists have completed their review, made all necessary modifications, and are ready to stand behind the accuracy and thoroughness of the work. This applies to interim work products, as well as to the final deliverable. In fact, careful review of interim deliverables is arguably more advantageous than review of final work if a risk assessment project uses the iterative review process.

Under iterative review, a contractor must meet performance standards before delivering a work product, whether the work is an interim deliverable or a final report. A contractor completes an interim product, checks it against all performance standards and, only then, sends it to the client for review by a project manager and project team. If a contractor complies with this contractual obligation (and appropriate performance standards were developed), a project manager and the internal project team should have only minor corrections and comments on each interim deliverable. As work proceeds, each deliverable undergoes similar scrutiny by both contractor and client, until all interim work is complete. It is then integrated into a final draft risk assessment report. At this point, review of the entire product occurs first by the contractor, and then by the client team and project manager.

If a contractor fails to meet performance standards and fails to adequately review work products before delivery, careful review by the project manager and project team becomes crucial. Such review can help clarify whether a performance standard was impossible to meet and it can serve as a means to enforce contract terms and ensure an acceptable product. Relying on a client to catch problems is poor practice, and should embarrass any professional contractor. It will undermine a client's trust and it may even lead to legal remedies under the contract, such as requiring a contractor to bear costs of fixing problems or a loss of financial incentives offered under the contract (see Chapter 8).

Simple, but effective, mechanisms exist to prevent errors. First, contractors should review all product and process standards in the project contract, as well as any contract language concerning their responsibility to conduct peer review. Contractors should check whether applicable performance standards are being met, especially if the contract requires a warranty that all process and product standards were met prior to product delivery. Tables 3 and 4 present these two types of performance standards for risk assessment report review.

Second, contractors should build critical review into the production process. In addition to review by technical peers, a review by a technical writer or editor is wise. A thorough review by a technical writer or editor can greatly enhance the clarity of the deliverable, thus improving the client's ability to appreciate the quality of the technical work (see Chapter 22).

Third, use of "input/output" analysis, discussed below, is a highly effective tool to aid review for tables, which comprise the technical core of most risk assessments, and also for text. The basis of input/output analysis is recognition that every piece of data must have a source and a purpose in a risk assessment. In other words, data comes from somewhere and goes somewhere. Starting with the first table in a risk assessment each data point is examined. What is its source and what is its application? Data may, for example, be an input to an equation in another part of the same table, or it might be carried elsewhere in the risk assessment report as input to another table. Properly prepared tables will make this clear. By following data in each table's inputs (inputs from referenced sources or earlier tables or calculations) and outputs (to equations elsewhere in the report), a reviewer can verify that data originates from legitimate sources and is properly used (see Table 8).

Finally, a project manager can further streamline review of contractor interim deliverables by distributing them to a subset of the project team to determine whether

Table 8 Examples of Basic Methods for Checking Inputs, Outputs, Logic Train, Seamlessness and Mathematical Correctness in a Risk Assessment.

Data Source (reference or page number in report)	Input (units)	Equation	Purpose of Equation	Output (units)	Application of Output (page number where it will be used; description)	Comments

Notes: Explanation of Table — This systematic review of risk assessment inputs and outputs ensures that all risk assessment inputs, outputs, and calculations are properly linked. Comments may include notes on whether math is correct and verified, gaps in logic, and identification of missing or dangling inputs or outputs.

the product meets minimum standards prior to distributing the document for full review by all team members and other interested parties.

B. Peer Review of a Human Health Risk Assessment

1. Importance of Peer Review

A risk assessment review team ensures that all important process and product standards were met in the risk assessment report. Review is the only way to ensure that the report will comply with organizational policies and strategies, have the proper level of technical rigor, and will be written and presented in a manner that can be understood by its intended audiences. Interim reviews help keep the project on track. The final report review represents the last chance to identify and fix mistakes and to integrate missing pieces into the report.

2. Organizing Peer Review

a. Building Immediate Review Capacity

Reviewing risk assessment reports takes time. It is tedious and may even be mind-numbing. In this age of downsizing and technical specialization, there are probably only a few persons in an organization who are qualified to critically review part, or all, of a risk assessment. If qualified peer reviewers exist, they may not have time to perform a thorough review. How can organizations receiving risk assessment deliverables from contractors respond to the need for prompt critical review?

Immediate review capacity can be achieved by hiring a qualified contractor to provide peer review services. Risk assessors and allied technical specialists are available across the country who regularly provide such services, often they are available at short notice to meet tight time lines. Obtaining their services may be expensive. Compared to the overall costs of the risk assessment and the importance of meeting deadlines, however, they are probably a cost effective option.

b. Tracking performance against established standards

Performance standards are the key to reviewing a risk assessment report. To approve of a risk assessment report's contents, the reviewer must be able to check the work against the performance standards. This means that the consultant must provide sufficient documentation to allow the reviewer to understand what work was done and verify that it was conducted properly. Ideally, the review will be based on performance standards that are supported by project contract terms. If a generic contract was used, however, performance standards might not exist. The reviewer must turn to generic checklists, such as those presented here. When performance standards are achieved, peer review is simple and generates few critical comments.

Risk assessment review can parallel the report's logical components, either in phases (such as Data Collection and Evaluation/Hazard Assessment, Exposure Assessment, Toxicity Assessment, and Risk Characterization [see Tables 5, 6, 7, and 8]) or its deliverables (such as each individual interim deliverable, or the draft and final risk assessment report). Regardless of how the risk assessment is produced or the review process is parsed, the following are essential considerations in every risk assessment review:

- Has the risk assessment captured the essential attributes of the site, activity, or facility?
- Are all critical elements of a generic risk assessment present in the report?
- Is all math and science correct?
- Were appropriate media evaluated?
- Were appropriate populations investigated?

The following sections summarize what risk assessors should consider when reviewing specific components of the risk assessment report, and present tables listing specific items for consideration. Table 1 presents an example of a critical elements review checklist. This technique follows the outline of a generic report. Table 8 approaches risk assessment review using an input/output approach. With this approach, reviewers use a pencil and calculator to follow and check all equation/model input variables and outputs. Reviewers then use table outputs as inputs to the next equation, and so on, until risk results are reported. Tables 1 and 2 provide methods for reviewing each key element in a risk assessment report and can be as general or specific as required.

c. Minimum Standards for Risk Assessment

Minimum performance standards are often contentious. Typically, the conflict centers on two questions — what are these standards and when have they been achieved? The most basic requirements are for a report to be seamless, transparent, and present an unbroken chain of logic, as called for by the U.S. EPA. Seamless reports have text, tables, figures, and appendices integrated into an unbroken whole. Transparent reports provide the reader with a clear understanding of what was done, why it was done, how it was done, where it was done, and who did the work. Reports with an unbroken chain of logic allow a reviewer to understand all logical steps in report development without having to guess or read minds.

Seamless Report Development — A seamless report integrates all text, tables, figures, and appendices into a coherent package. This is relatively simple to achieve, by careful organization and drafting. For example, an initial framework of a risk assessment might develop as a set of tables presenting all information (data, default values, assumptions, etc.) and calculations at the core of a risk assessment. U.S. EPA's standard tables, discussed in detail in Chapter 2, Section X, are a useful tool for organizing a seamless report. Next, report authors might describe these tables in text, giving the source of information, justifying use of default values, providing reasons for assumptions, and detailing relationships between tables. Additional clarification, justification, data, background information, and proofs might be offered as figures and appendices. In a seamless report, there are no loose ends. Just as a writer creates a risk assessment, a reviewer deconstructs a risk assessment, starting with an examination of all tables, moving to the text that discusses each table, and then to supporting figures and appendices (see "Input/Output Analysis: Risk Assessment Review Accounting," below).

Transparent Report Development — Transparent reports clearly present what was done, why it was done, how it was done, where it was done, and who did the work. These questions form the basis of a risk assessment review. Typically, this information appears as text, either in the body of a risk assessment or as a series of appendices (on field and laboratory methods, credentials and work assignments of researchers and technicians, and raw data, data manipulation, etc.).

Presenting an Unbroken Chain of Logic — A reviewer cannot evaluate whether a report presents an unbroken chain of logic, until it is determined to be seamless and transparent. When it is clear that all pieces tie together, and each was properly developed, a reviewer can consider how pieces fit together. Whereas previous standards deal with accounting — is everything here, is it in the right place, and was it properly created? — this standard requires analysis based in a thorough understanding of risk assessment tenets. It may even require a reviewer to challenge professional judgements made by those generating a risk assessment report. Consequently, it is vital for risk assessment authors to carefully demonstrate the complete logic flow of a report and, as much as possible, present rationale to support each step.

C. Risk Assessment Report Checklists

A critical elements review checklist follows the general outline of a risk assessment report (see Tables 3 and 4). A reviewer can use a generic checklist, deciding that a report is adequate if it addresses each critical element in the list, or a reviewer may turn to specific performance standards stated as contract terms, a work plan, or a Scope of Work. Specific performance standards will improve review rigor. Some elements apply throughout a report (proper spelling, math, and science). A few elements relate only to specific sections (see Tables 5, 6, 7, and 8).

Reviewers may make numerous comments during development of a risk assessment. Comments and questions can lead to complex answers. These, in turn, generate still more questions. Formal responsiveness summaries offer a good way to manage dialogue and ensure that a contractor addresses all reviewer comments. In its most basic form, a responsiveness summary is a series of questions, grouped by topic or chronologically, followed by answers.

D. Input/Output Analysis: Risk Assessment Review Accounting

The framework of a risk assessment is a series of interrelated tables, with data (numerical inputs) used in equations or models to generate new numerical values (numerical outputs). Numerical outputs become inputs for subsequent tables, along with data from other sources. For example, a risk assessment typically starts by gathering initial data sets (chemical identity, concentration, or emission data) and placing them in tables. These inputs generate other outputs, and so on, until quantitative risk estimates are generated and a risk assessment is complete.

Table 8 presents an input/output approach to risk assessment review, termed the "Risk Assessment Review Accounting System" (RARAS). RARAS is a systematic approach to evaluate all inputs and outputs used to generate a final quantitative risk estimate. Like any accounting method, RARAS tracks each numerical input and output, from table to table, to ensure that no transcription errors occur, that each mathematical calculation is correct, and that the numerical input values are defensible, and that the output values from each table are correct and are carried forward properly in the analysis. This ensures that data of known quality and source are used correctly to generate interim output variables and verifiable quantitative risk estimates. This system is often used as a preliminary check of accuracy, completeness, logic, transparency, and integration with figures, tables, and appendices, prior to review of assessment text.

Risk assessment reports can be difficult to review. Many are written by specialists for specialists. For ease of review and transparency, risk assessment reports can contain tables formatted to illustrate the input/output approach (see Table 9). All data in tables have a source and a use. The only table that has no further use in subsequent tables are final summary tables. Therefore, input/output tables ensure that all input data is fully referenced to its source and all output data has a clearly defined use.

There are four elements to a typical input/output table. They are:

- A column header describing the data set in the numerical value cell or the equation used to generate the value in the numerical value cell (these headers appear at the top most row of cells in the table)
- A data source stating exactly where data comes from, either elsewhere in the risk assessment report or from another fully referenced source, e.g., citation with table number, page number, column number, etc (this information appears in the second row of cells in the table)
- Statement of the numerical value (this information appears in the third row of cells in the table)
- Description of how data is used in subsequent tables (this information appears in the fourth row of cells in the table)

218   A PRACTICAL GUIDE TO ENVIRONMENTAL RISK ASSESSMENT REPORTS

Table 9 Example of an Input/Output Table Used in a Risk Assessment Report

Instructions for constructing an input/output analysis table	TABLE NUMBER TITLE OF TABLE				
	Column 1 Header	Column 2 Header	Column 3 Header	Column 4 Header	Column 5 Header
Each Column Header describes the data set in the numerical value cell or the equation used to generate the value in the numerical value cell	Input (Source):	Input (Source):	Input (Source):	Input (Source):	Input (Source):
Data Source (state exactly where data comes from — this report or another source, with full information on table number, page number, column number, etc.)	Input (Source): This report, page 9, Table 7, column 4				
State numerical value	27,000	numerical value	numerical value	numerical value	numerical value
How data is used in subsequent tables	Output (Use): Final table; this page, column 5 input to equation	Output (Use):	Output (Use):	Output (Use):	Output (Use):

Note: If Column 5 contained an equation that used data from Column 1–4, the equation would be presented in the header and the numerical value calculated would appear in the numerical value cell. The output cell would tell the reader where the output numerical value is used later in the report.

When each of these cells is properly filled, a reviewer can rapidly determine whether imported data was accurately transcribed, if calculated values are mathematically correct, and where calculated values will be used in subsequent tables in the risk assessment. While this table and its use may seem intuitive, the vast majority of risk assessments do not use this approach to their detriment. It is, therefore, not surprising that many transcription errors are not caught by reviewers and many tables have columns of information with no apparent use.

Why is this important? When columns or tables have information that is not used elsewhere in the risk assessment, reviewers and readers question why the data is present, where it was to be used, and why it was not used. Answering these questions can cost considerable time, at the contractor's expense. These types of problems can result in unease with the ability of the contractor to monitor their own work and generate a professionally credible product. Consistent use of input/output tables can forestall these problems.

An input/output accounting system provides risk assessment reviewers with a simple, powerful tool to check all logic, mathematics, sources, equations, and variable values used in the risk assessment. Using this risk review accounting system can provide a rapid determination of whether all inputs are used (e.g., several tables produce inputs and outputs that are not used for any discernable purpose), whether a series of tables and calculations are used to generate a reported result (e.g., all concentration values calculated in water do not exceed federal standards), to make sure that all necessary data have been presented in tabulated form, and that all necessary calculations have been performed and their results are mathematically correct (e.g., sometimes data in spreadsheets are incorrect, and even when correct, produce incorrect mathematical results). By ignoring nontabulated information in the risk assessment (e.g., prose and figures), the reviewer can determine whether all necessary tables are present to fully explain and justify risk assessment methods, decisions, and mathematical findings. This accounting system applies equally well to HHRAs and ERAs.

HHRA and ERA reports are comprised of innumerable decisions. Every decision receives intense scrutiny on a high profile project, like a risk assessment, including what the report omits. Careful documentation of each decision is necessary when many people are making decisions on the report development, if the project takes many months, or years, or staffing change. Decisions that cannot be explained will fuel skepticism about the quality of the risk assessment, the validity of the risk estimates, and the wisdom of the resulting risk management decisions. Although decisions are best made as part of a plan, ad hoc decisions often occur during the course of a risk assessment project. Ad hoc decisions may not be fully documented or explained in the report, unless a special effort is made to do so. Controlling ad hoc decisions, and documenting and explaining them, deserves special attention in high profile projects, like a risk assessment report. RARAs tables can help ensure that even ad hoc decisions are fully documented and justified.

IV. CONCLUDING HHRA REVIEW

Reviewing an HHRA is a team effort and it requires time and technical expertise. Review is more efficient if risk assessment reports meet comprehensibility and transparency requirements set forth in a risk assessment contract. When risk assessment reports are prepared for the public or other nonrisk assessment specialists, they must be formatted and written with this audience in mind. This often means that the report should be understandable to an educated lay person and that all math must be easily checkable (such as, with a pencil and hand calculator). If the intended audience cannot understand the text and check report math, they will not accept the risk assessment report as credible. Writing and formatting a report in a manner that an educated lay person can follow also ensures that technical reviewers will be able to understand the report and verify its train of logic from the initial inputs through to the final risk estimates.

V. CONCLUSION OF PHASE II

At the conclusion of Phase II, the risk assessment project has proceeded through the four steps of HHRA: hazard assessment, exposure assessment, toxicity assessment, and risk characterization. A similar approach can be used for ERAs. If iterative review has been used, all work will have been reviewed before being integrated into the draft final report and the technical work on the risk assessment project is essentially complete. In the next two phases of the process, the final work products will be reviewed and accepted, the contract will be concluded by the parties, and the risk assessment findings will be put to use.

CHAPTER 6

Concluding a Risk Assessment Contract (Phase III and IV)

David A. Belluck and Sally L. Benjamin

CONTENTS

I. Introduction 221
II. Conclude the Project 221
    A. Accept the Final Draft 221
    B. Close the Risk Assessment Contract 223
    C. Address Risk Management and Risk Communication 223
III. Follow-up Studies and Activities (Phase IV) 228
    A. Post-Risk Assessment Report Activities and Studies 228
IV. Conclusion 230

I. INTRODUCTION

Phases III and IV are the subject of this chapter. In these last phases of risk assessment report development, final work products are reviewed and accepted, a contract is formally concluded, final documents are delivered, payments made, and risk assessment findings are put to use. See Table 1 for an overview of Phases III and IV.

II. CONCLUDE THE PROJECT

A. Accept the Final Draft

When a contractor has produced a draft final report that meets or exceeds all performance standards, a project manager can accept it as a final report on behalf

Table 1 Critical Elements for Concluding a Risk Assessment Report

Phase III: Concluding the Risk Assessment Contract	Applicable? (Y/N)	Appropriate? (Y/N)	Required? (Y/N)	Complete?
Step 23 Accept the Final Risk Assessment Report (See Chapter 6 for discussion)				
Step 24 Close contract • Receive report and all supporting documentation • Make payment				
Step 25 Post-risk Assessment Report Activities and Studies (See Chapters 6 and 21 for discussion) • Risk management activities • Risk communication or public involvement • Further research and analysis				

Note: Y = Yes
N = No

of an organization that has hired a contractor. The contractor then produces a required number of copies of the final risk assessment report and distributes them, under contract terms. Typically, enough copies are produced to distribute them within an organization's management structure and to give copies to key staff members, regulators, political leaders, and other interested members of the general public. It may now be time to close the risk assessment contract, unless the contract provides for additional work following delivery and acceptance of a final report.

B. Close the Risk Assessment Contract

After all contractual obligations have been discharged a contract can be closed. Each organization will probably need to obtain sign-offs and complete paperwork. Prior to paying a contractor, it may be necessary to determine whether penalties or bonuses are owed under the contract.

C. Address Risk Management and Risk Communication

Although distinct from a risk assessment report generation process, risk management and risk communication tend to follow immediately upon completion of a risk assessment report.

1. Risk Management

Risk management decisions are far from scientific determinations. Rather, scientific information, embodied in risk estimates and technical risk mitigation capabilities, is integrated into societal decisions. The result is a risk management decision. The art of risk management is the art of weighing all factors involved in a case and balancing conflicting demands and data. Risk management decisions are typically trade-offs between risks and benefits or between multiple risks. For many risk managers, risk management decisions are compromises between a desire for lowest possible risks and society's demand for jobs, economic growth, and goods production.

Imagine this scenario: A year has passed since work began on a risk assessment. Close to half a million dollars have been expended on consultant services, peer review sessions, public comment meetings, and responses to public comments. So far, no serious report problems have been discovered (e.g., wrong inputs, mathematical errors, etc.). One major task remains — how to interpret risk assessment results and selection of a risk management/reduction strategy. It is a risk manager's responsibility to decide how risk findings will be interpreted by regulatory agencies and what action will be taken based on these findings.

In this case, cancer risks are estimated at three times acceptable levels set by a state health agency. If a risk manager simply compares risk estimates to an acceptable level set by a health agency, these are found to be unacceptable risks. However, risk management decisions are rarely so simple. Other considerations enter our analysis, including technical concerns, economic concerns, and social/political concerns.

Technical concerns trigger questions about risk assessment report quality and its findings. For example, how solid are these risk findings, in light of uncertainties involved in input data to a risk assessment and risk assessment methods? Is report rigor sufficient to support risk management findings? Other technical concerns center on range of options for addressing environmental risk. Are there feasible alternatives to reduce risks? Is a no-action alternative a feasible choice? Will a risk reduction solution have a domino effect and result in other environmental damage?

Economic concerns are also triggered; such as, is it feasible to spend several million dollars to achieve a threefold risk reduction merely to bring risks below regulatory yardsticks? Economic issues quickly devolve into broader social problems and associated political, regulatory, and legal issues: What is the nature and severity of identified health risks? Is a particularly deadly type of cancer in a narrow population worse, or better, than widespread effects of a nonlethal nature? With a local community in need of jobs, will a finding of unacceptable risks and costs associated with correcting these risks survive a serious political challenge? Will a decision result in social dislocation? Can a consensus be reached on risk reduction measures? Who will attack a finding of acceptable or unacceptable risk? How much discretion do applicable laws allow in making an acceptable/unacceptable risk decision? Can either decision be successfully defended in court?

Rather than ask whether a risk management decision is right or wrong, risk calculus appears to focus on whether a decision is politically survivable, socially acceptable, or economically viable. Risk management is a tough job. Unlike risk assessors, risk managers have no formal methods to follow in making their decisions. As a result, a fundamental incongruity exists between the highly formalized risk assessment process, with its standardized methods and technical peer review, and the ad hoc, values-laden risk management process.

Inconsistent rigor of risk assessment and risk management often leads to public outrage over risk management decisions that do not appear to align with risk findings. Public input to a risk assessment process seems to be unappreciated or ignored in final decisions. Until a systematic, formalized, rigorous, peer-reviewed risk management process becomes a reality, this disparity will continue to pose problems for risk professionals.

One solution to problems involving risk management non-transparency has been to develop a formal document, a so-called "Risk Management Decision Document." This document, prepared by or for risk managers, formally documents data and logic that resulted in a given risk management decision. This type of document allows regulated or interested parties to see how a risk management decision is made; how supportable is a decision based on facts presented; and weaknesses, strengths, and uncertainties associated with risk analysis, its numerical findings, and risk management determinations.

Risk management decision documents have been problematic for risk managers. In some cases, it has allowed interested parties to see exactly how a risk management decision is made. In many cases, it is based more on political and economic factors than risk factors. While risk managers may be willing to admit how they come to important risk management decisions in private, they are often unwilling to place political and economic factors ahead of health protection in public. As a result, risk management decision documents may face a difficult future.

One way that risk managers try to control risk management outcomes is to control risk assessment preparation. For example, risk managers can limit resources available to perform a risk assessment and thus limit ability of risk assessors to evaluate all appropriate exposures and risks. In other cases, nonrisk assessors may be given responsibility of developing risk assessment guidance documents in concert with regulated parties. It is not uncommon for risk managers to select risk assessment project managers based on their understanding of a desired risk outcome. While these practices may not be illegal, these approaches are at least unsavory and all too common. It is these types of manipulations that undermine public confidence in risk assessors and risk assessments. How else can some members of the general public interpret the fact that unacceptable chemical exposure risks continue to increase around the nation, while vast environmental protection bureaucracies, for almost two decades, have evaluated these risks and made risk management decisions that allow it to continue? Perhaps reversing or eliminating these poor management practices could restore confidence in the risk assessment process.

a. Separating Risk Assessment from Risk Management

Risk assessors generate risk estimates, but they do not make risk management decisions. Risk management is the purview of risk managers. It is important to keep a clear separation between these two processes. There are several good reasons for this distinction.

Risk managers make risk management decisions, decisions about how to respond to risk findings, by integrating risk estimates with other factors. Other factors typically include statutory and regulatory requirements, economic concerns, political commitments, social impacts, and technical feasibility issues, as well as a wide range of other social and technical concerns. Regulatory, statutory, economic, social, and political concerns are legitimate concerns for a risk manager. But, they are not legitimate factors within a risk assessment process. Thus, to avoid potential bias of risk estimates, these concerns must be set aside until after risk assessment is completed. This is one compelling reason to distinguish risk management and risk managers, from risk assessment and people who perform these assessments.

Yet another good reason to distinguish risk management from risk assessment is to encourage accountability. There has been an unfortunate trend toward lobbying and pressuring risk assessors to be more reasonable in their assessments of environmental risk. Such pressure on risk assessors is misplaced. It should be brought to bear on risk managers, instead. Risk managers are people with a legitimate task of determining how society deals with environmental risk. Risk managers, perhaps fearful of being held accountable for their contentious decisions, have tended to implicate risk assessors in their decision-making, pleading "the science made me do it." But, risk managers are not captives of risk assessment findings. Risk estimates are simply one important factor among many considerations in risk management decision-making. Risk managers weigh and balance costs and benefits of their risk management decisions and must accept full responsibility for their decisions.

Yet another reason for distinguishing risk management from risk assessment is to develop better risk management decisions through improved tools and better analytical methods. Just as a risk assessor relies on an accepted analytical framework to generate risk estimates, risk managers could benefit from accepted analytical methods to guide risk decision-making. The risk management field would benefit from scrutiny to improve its analysis and decision-making process. Unfortunately, a great deal of effort has been misapplied toward improving risk assessment, while risk managers have been left to muddle ahead with little or no refinement of their decision-making procedures.

A strong distinction between technical analysis, which occurs during risk assessment, and social decision-making, which occurs during risk management, must be recognized and rigorously maintained, for reasons of:

- Unbiased risk estimates
- Improved decision-maker accountability
- Advancement of methods employed in risk decision-making

4. A Systematic Approach to Risk Management

Generally, risk assessment findings are compared by risk managers to state or federal carcinogen and noncarcinogen "bright lines." Bright lines may be expressions of risk (i.e., cancer risk = E-5; noncancer HI or HQ = 1) or they may be chemical concentrations that, by law or policy, represent the upper bound of what is deemed "acceptable." If numerical risk assessment findings fall at, or below, a bright line, they are usually considered to be acceptable risks. Risk findings greater than a bright line may not be acceptable. For most government risk management programs, no guidance documents exist. Although bright lines are part of many state and federal programs dealing with environmental risk management, use of a bright line approach presents a number of problems.

First, there is a tendency to rely entirely on numeric risk findings. Given inherent uncertainties in risk assessment, blind faith in risk estimates is not justified. A better approach would help a risk manager put a risk estimate into context, both recognizing imprecision of risk findings and taking into account factors beyond the risk assessment process, such as economics and technical feasibility.

Second, the current bright line approach does not precisely delineate when a risk greater than a bright line value becomes an unacceptable risk. This imprecision leads to high transaction costs by encouraging regulated parties to generate risk assessments with risk findings at (or very close to) a bright line, and to then expend huge efforts attempting to convince a risk manager that a marginally acceptable risk estimate should not be interpreted as representing an unacceptable risk.

Third, a bright line approach results in an imbalance between technical rigor of risk assessment and risk management decision-making.

Several years ago, one state program, the Minnesota Pollution Control Agency's (MPCA) Air Quality Division, attempted to move away from a bright line approach. As discussed above, the bright line approach deems risks as "unacceptable" if a risk estimate or air monitoring data exceeds an established bright line. In a radical departure from this approach, the MPCA's planning process for use of Minnesota air quality Health Risk Values (HRVs) considered a flexible risk management strategy, termed a "Zonal Risk Management Approach." This Zonal Risk Management Approach would have allowed risk managers to place numerical risk estimates in context, to consider risk assessment quality, as well as nonrisk factors, and to respond based on case-specific or site-specific considerations.

The Zonal Risk Management Approach would have established three zones bracketing an existing bright line (see Figure 1). Immediately surrounding a bright line is a gray zone. Width of the gray zone reflects quality of risk analysis used to generate a risk estimate. A higher quality risk assessment would generate a narrower gray zone and would reduce the need for negotiation. A red zone would begin at the upper edge of the gray zone. Risk estimates that fall in the red zone would be clearly unacceptable and would not be permitted. A green zone would begin at the lower edge of the gray zone. Risk estimates in the green zone would be considered clearly acceptable and would be permitted. In certain circumstances, risk managers could permit projects in the gray zone, based on clear and compelling reasons.

The Zonal Risk Management Approach linked decision-making to the quality of the risk assessment report by recognizing that risk findings are not simple point estimates; they have a range of uncertainty around them. Correlating the size of the gray zone to the degree of uncertainty around a point risk estimate would encourage positive behavior among the regulated community and their consultants. High quality risk assessments would involve less uncertainty and, thus, should encounter a smaller gray zone and should also have a better chance of generating a risk estimate clearly within the green or red zone. Risk estimates from poor quality risk assessments would tend to fall into the grey zone. The Zonal Risk Management Approach eliminates obviously unacceptable projects, allows for efficient decisions on clearly acceptable and unacceptable projects, and rewards high-quality work by adjusting the size of the gray zone, based on certainty surrounding the risk estimate.

The idea is an example of the potential for making better use of agency resources to arrive at defensible risk management decisions, while rewarding high-quality risk assessment reports. This approach was rejected because of political considerations, but it represents an interesting alternative to the ubiquitous "bright line" approach.

2. Risk Communication

Risk managers and risk assessment project teams work with risk communication specialists to inform the public about risk findings and risk management decisions. Risk communicators identify strategies and methods to communicate clearly with the public. Risk communicators skillfully employ language skills to transfer information on the risk assessment process, risk assessment findings, and risk management decisions. Ideally, risk communication informs. It does not attempt to sell a solution, intimidate, or placate people. Risk communication should be used to provide unbiased information, not to convince people of the correctness of one risk management option over another. In other words, the goal is not to educate the public so that they agree with the organization's views. Rather it is to help them to become

Higher Risk				
	Red Zone Clearly Unacceptable risk	E-4		
	Gray Zone Risks Could be Acceptable or Unacceptable			
	BRIGHT LINE	E-5		
	Gray Zone Risks Could be Acceptable or Unacceptable			
	Green Zone Clearly Acceptable risk	E-6		
Lower Risk				

Figure 1 The Zonal Approach to Risk Management

informed and then decide for themselves upon the legitimacy of the risk assessment process and its findings (see Chapter 21).

III. FOLLOW-UP STUDIES AND ACTIVITIES (PHASE IV)

A. Post-Risk Assessment Report Activities and Studies

Persons involved in the risk assessment world often think of post-risk assessment activities being primarily comprised of risk management and communication activities, as discussed above. Numerous other activities can, however, follow close on the heels of a completed risk assessment.

Site cleanup concentrations and methods might be developed. For example, after the CERCLA or RCRA baseline risk assessments, property transfer evaluations or other specialized risk assessments. If it is determined that current site conditions represent unreasonable risks, a remediation risk assessment is conducted. The size, technical rigor, site specificity, and costs can vary greatly depending on applicable laws or regulations, hazards posed by a site, facility, or activity, and preferences of regulatory agency staff and management. Remedial risk assessments begin with the premise that some environmental medium or media must have its chemical contaminant concentrations reduced. Determining medium-specific cleanup concentrations is the key to this process. Cleanup concentrations can be mandated by federal or state legislatures, or agencies. They can be developed as risk assessment based standards or guidelines by government agencies, or on a case-specific basis.

If the risk assessment was a screening-level risk assessment, and unacceptable risks were found, the next step might be to conduct a full risk assessment. Screening-level risk assessments use conservative inputs and methods to produce conservative estimates of risk in relatively short time periods and for much less cost than full-scale risk assessments. More refined risk assessments are thought to produce lower risk estimates, since they use case-specific data and fewer conservative, generic assumptions, crude models, or default values.

Another follow-up project might involve generating Risk Reduction Tables. If a risk assessment produces unacceptable risk findings, regulators and regulated parties might want to alter the parameters and rerun the risk assessment. For example, they might select processes that release less environmental contamination or they might include equipment to control environmental releases. If chemical releases and risk are linearly related, risk levels will drop in direct proportion to reductions in environmental releases. Risk Reduction Tables are prepared that show the risk reductions that result from alterations in various project parameters.

The law might mandate the next step. Findings of unacceptable risks in certain types of screening-level risk analysis may require the preparation of an Environmental Impact Statement, a document that evaluates risks in a detailed manner. Or, litigation could be the next challenge. Parties to the risk assessment, or the public, might bring a legal challenge to how the report was produced, its findings, or the resulting risk management decisions. If so, the next steps will be governed by the rules of the court and will probably involve pleadings, discovery, case development, and perhaps settlement negotiations.

After the risk assessment is completed, an effort will be made to mend fences with those involved with the risk assessment process and the risk management decision. Risk generates animosity in the mildest of people. It can stir controversy at every level of government. Building and maintaining good relations with government officials, whose belief system or constituents were offended by a risk management decision, can be time consuming. It is also essential to the success of the next controversial project.

IV. CONCLUSION

Since each risk assessment is a customized product, there is no single way to produce a risk assessment report. Each step presented in this chapter is a suggestion to be modified to meet specific project needs. Steps should be eliminated with care, however, since each step is important. Use of the iterative review risk assessment process is highly recommended.

There are innumerable issues and technical details that must be addressed by risk assessment project managers and team members throughout the production phase of the risk assessment project. Table 1 provides a series of important principles for managing risk assessment report development. It is not exhaustive; however, it provides many fundamental principles behind the ideas presented here and discussed throughout the book.

Part II

Primers

CHAPTER 7

Legal Context of Environmental Risk Assessment

Bruce Braaten

CONTENTS

I. Introduction 233
   A. Regulatory Framework 233
   B. Expanse of Environmental Regulations 234
   C. Risk in the Environmental Regulatory Framework 235
II. How Regulations Address Risk 236
   A. Preventive Regulations: "What is Safe?" 236
   B. Reactive Regulations: "How Clean is Clean?" 237
III. Regulatory Methods for Addressing Risk 238
   A. Numerical Standards 239
   B. Technology-Based Standards 240
   C. Risk Assessment 240
IV. Conclusion 243

I. INTRODUCTION

A. Regulatory Framework

Environmental regulations generally provide the basis for conducting a risk assessment. Risk assessment is usually a specific legal requirement within an overall program of environmental regulation. The overall program significantly impinges on how risk assessment is conducted, on the process and product standards governing report acceptability, and on how risk assessment is used. Consequently, a project manager must be aware of the most recent statutes and regulations governing the risk assessment project. Unfortunately for the project manager, locating the specific legal provisions that apply to a given project can be challenging.

B. Expanse of Environmental Regulations

Since the early 1970s, environmental law has grown into a tremendous body of statutes, rules, and court decisions. Note that this chapter presents general information on the relationship between environmental law and risk assessment. Nuances of the law may be very important in specific instances, but are not presented here. Although the number of federal statutes may not appear too overwhelming, these regulations are daunting in their individual complexity and scope. A single statute may, for example, embrace hundreds of pages of detailed regulatory requirements.

A number of players have a role in developing environmental regulations. To become a federal statute, a bill must be passed by both houses of Congress. The President must then sign, or not veto, the bill. Statute compilations are codified in legal codes, such as the U.S. Code (U.S.C.).

In most instances, the statute instructs the U.S. EPA to promulgate a rule to making procedures. These procedures require that the public receive notice of the proposed rule and an opportunity to comment on its provisions. After considering public comments, the U.S. EPA promulgates the rule as final. Federal rules are published in the Code of Federal Register (C.F.R.).

Under many federal environmental statutes, states may be delegated authority to administer the federal program. To receive this authority, the state's applicable statutes and rules must be at least as stringent as the federal regulations. States which develop sufficiently stringent rules and otherwise qualify to administer and enforce the federal program, are said to have "primacy."

States that seek primacy over new federal environmental programs typically pass statutes paralleling the federal statute. Based on authority found in the state statutes, the state regulatory agency, in turn, promulgates the state rule. Thus, the overall environmental regulatory framework consists of four players: the U.S. Congress; the U.S. EPA; the state legislature; and the state environmental agency.

For example, the U.S. Congress enacted the Safe Drinking Water Act of 1974 (SDWA). Congress amended the SDWA in 1986.* Within the statute, Congress instructed U.S. EPA to establish maximum contaminant levels (MCLs) for chemicals found in public drinking water systems. The MCLs were to become the national drinking water standards. In response, U.S. EPA developed the actual concentrations for the MCLs.** These standards set the maximum allowable concentration for specific chemicals. Drinking water with chemicals above these concentrations is not considered safe. The SDWA allows the states to have primacy over their public water supply programs provided that, in part, they adopt state standards no less stringent than the MCLs.*** By adopting and enforcing drinking water standards of equal or more stringency than federal MCLs, many states have received such primacy.

---

* 42 U.S.C. §§ 300f et seq.
** 40 C.F.R. §§ 141.1 et seq.
*** 42 U.S.C. § 300g-2.

To appreciate the impact of the above regulatory framework, four concepts must be understood. First, both statutes and rules have the force and effect of law. Second, with fifty states each developing regulations, differences occur in the rules governing identical situations in different states. Third, all four layers — both state and federal statutes and rules — must be grasped to gain a complete understanding of a state regulatory program. Fourth, environmental regulations are dynamic, unlike the static laws of natural science. For environmental scientists, this can be a difficult to accept. Whereas the physical laws are constant, there is no guarantee of constancy in environmental law. Environmental law is a relatively new, continually evolving, body of law. A change by Congress can cause a whiplash effect down through the federal rule, state statute, and state rule.

In addition to statutes and rules, agency guidance and court decisions play significant roles in the environmental arena. When a regulatory agency develops guidance documents, the guidance generally provides supplemental detail on how the agency will implement or apply its regulations. Sometimes the policy is formally published as guidance documents. Sometimes it is found in agency memoranda or letters. Agency guidance is not legally binding (i.e., guidance documents do not pass through the rule making process). In practical terms, however, agency staff who review project progress, certify compliance, and enforce regulations rely heavily on applicable guidance.

Ultimately, disputes over the application of statutes or rules are resolved in court. Court decisions address a wide array of issues, such as whether a statute authorizing the rule is constitutional, whether a rule developed by an agency is within the authority provided by a statute, whether the way a rule is applied to a plaintiff is both constitutional and within the bounds of a statute and rule, and whether the interpretation of myriad technical and legal terms that comprise a rule and statute is proper. Court decisions can serve as guides for judges, attorneys, and savvy project managers to project how courts will rule on future court cases interpreting similar regulatory language or addressing similar legal issues.

C. Risk in the Environmental Regulatory Framework

A single, universal objective runs throughout the vast body of environmental regulation — to protect human health and the environment. Toward that goal, environmental regulations provide the means to protect human health and environment from a wide range of threats from toxic substances. The toxic substance of concern is typically identified in either the legal definitions, lists of parameters and their associated legal concentrations, or in terms of methods or physical criteria for determining whether a given substance poses the threat addressed by the regulation.

The objective of environmental regulation is rarely debated. Instead, debate revolves around how to protect the environment and human health from a specific situation. The debate may center on whom (human health) or what (ecology) is to be protected (i.e., the receptors). It may deal with the means of controlling the threats, through technological or policy-based solutions. Or, it may focus on the appropriate level of risk to deem "acceptable" to adequately protect receptors. Risk assessment may play a role in each type of debate.

Regulatory requirements for a risk assessment are located within the thousands of pages of environmental regulations. Regulations may require that the regulatory agency perform the risk assessment, or may allow it to be performed by the regulated party. Provisions for a risk assessment may be explicit. For example, the Superfund site investigation report must include: a description of known contaminants; a description of pathways of contaminant migration; and, an identification and description of human and environmental targets.* Or, the risk assessment requirements may be implied from the broad regulatory language. For example, a requirement may be stated as "the impact cannot adversely impact the human health and the environment."

This chapter presents an overview of how environmental regulations address risk, primarily through the use of numerical standards, technology-based standards, and risk assessment.

II. HOW REGULATIONS ADDRESS RISK

In general, environmental regulations can be categorized as being either preventive or reactive. These categories are depicted in Table 1.

A. Preventive Regulations: "What is Safe?"

Preventive regulations prevent or minimize the introduction of a given environmental contaminant into the environment. The critical issue is deciding how much of a contaminant can be "safely" introduced into the environment. The preventive regulations can be subcategorized. The first preventive subcategory places restrictions on the use or application of a product containing the substance that may become an environmental contaminant. Examples of such restrictions include The Federal Insecticide, Fungicide, and Rodenticide Act's (FIFRA) regulatory prohibition of the use of a pesticide in a manner inconsistent with its registered label or Toxic Substances Control Act's (TSCA) regulation of Polychlorinated Biphenyls (PCBs) usage.** The second preventive subcategory requires the minimization or removal of a contaminant from an emission or effluent discharge by controlling the discharge or emission rates of the material. This minimization or removal occurs after the completion of a process, but prior to its release into the environment. The Clean Air Act (CAA) or The Clean Water Act (CWA) technology-based treatment requirements are examples of restrictions on emissions or effluent discharges.***

1. Role of Environmental Impact Statements within Preventive Regulations

Within the preventive environmental regulations, the role of the National Environmental Protection Act of 1969 (NEPA) warrants discussion.**** NEPA requires that

---

* 40 C.F.R. § 300.420 (c).
** 7 U.S.C. § 136j, 15 U.S.C. § 6.
*** 42 U.S.C. § 7412 (d), 33 U.S.C. § 301.
**** 42 U.S.C. §§ 4321 et seq.

LEGAL CONTEXT OF ENVIRONMENTAL RISK ASSESSMENT

Table 1 Regulatory Approaches to Risk

	Environmental Regulations Universal Objective: Protect human health and the environment		
Type	Preventive Regulations		Reactive Regulations
Goal	Prevent or minimize contamination		Respond to contamination
How	Control use	Control discharge or emission rates	Environmental contaminations not controlled
For example	Regulate pesticide application rates (FIFRA)	Set risk levels at concentration levels or SDWA MCLs	Set risk levels using risk assessment (CERCLA, RCRA, TSD facility, and LUST corrective actions)
	PCB regulation and enforcement (TSCA)	Set risk levels at technology-based standards (CWA or CAA)	

the federal government take into account environmental impacts in the administration of their functions and programs.* To that end, NEPA contains the Environmental Impact Statement (EIS) process. ** The purpose of the EIS process is to collect, analyze, and prepare information on the potential adverse environmental impacts of a proposed action. Before a decision is made to allow an action, the EIS report is to be provided to the decision makers.

Not all preventive programs require an EIS. The threshold test for determining if an EIS is required is "whether a major federal action significantly affects the quality of the human environment."*** An attorney should be consulted on a proposed project in regard to the need for an EIS. For example, "federal actions" include issuing permits.

If a federal EIS is triggered, a report is required containing: the environmental impact of the proposed action; any adverse environmental effects which cannot be avoided, if the proposal is implemented; alternatives to the proposed action; the relationship between local short-term uses of the environment and the maintenance and enhancement of long-term productivity; and any irreversible and irretrievable commitments of resources if the proposed action is implemented.

B. Reactive Regulations: "How Clean is Clean?"

The reactive regulations reduce the concentration level after a contaminant has already been released into the environment at unacceptable levels. At issue is the level of contamination that can be "safely" left in place. In other words, "how clean is clean?" The acceptable level of risk dictates how much contamination may be left in place without posing an unacceptable threat. Examples of reactive regulation programs include Superfund;**** and RCRA - Subtitle S treatment, storage, and

---
* 42 U.S.C. § 4321.
** 42 U.S.C. § 4332.
*** 42 U.S.C. § 4332 (c).
**** 40 C.F.R. Part 300.

Table 2 Federal Cleanup Process

Superfund	RCRA	
Abandoned/Inactive Disposal Sites	Permitted TSD Facilities (Subtitle S)	Leaking Underground Storage Tanks (Subtitle I)
Site Discovery/Notification ↓	Remedial Investigation ↓	Initial Release Response ↓
Preliminary Assessment/Site Investigation ↓	Corrective Measure Study ↓	Initial Abatement Measures & Site Check ↓
HRS — II Scoring/NPL Listing ↓	Remedy Selection ↓	Initial Site Characterization ↓
Remedial Investigation/Feasibility Study ↓	Remedy Design/Remedy Implementation	Free Product Removal ↓
Remedy Selection ↓		Investigation of Soil & Groundwater ↓
Remedial Design/Remedial Action		Corrective Action Plan

disposal (TSD) facility corrective actions,* and RCRA - Subtitle I leaking underground storage tank (LUST) corrective actions.**

III. REGULATORY METHODS FOR ADDRESSING RISK

Three regulatory methods address the setting of an acceptable level of risk to a contaminant exposure. These methods are: numerical concentration standards, technology-based standards, and risk assessment (see Table 1).

Numerical concentration standards and technology-based standards are generally employed in preventive programs. One subset of the preventive regulations are those that control the discharge or emission rate; these regulations usually entail the issuance of a permit. The permit specifies the standards for emission or effluent discharge of a contaminant. Depending on the regulatory program, standards can be either numerical concentrations or technology-based. Risk assessment enters the permitting process when the agency determines how stringently to set these standards. Agencies, not the project manager, typically conduct risk assessments in these programs. There are exceptions to the general pattern of agencies conducting risk assessments under preventive regulatory programs. For example, private parties seeking some combuster permits may need to conduct a risk assessment.

The second subset of the preventive regulations are those that control the use of the chemical material. These regulations may require registration or premanufacture

---

* [12] 40 C.F.R. Part 264 (proposed July 27, 1990). At the time of editing this chapter, advance notice of proposed rulemaking was issued by U.S. EPA pertaining to this proposed rule.
** 40 C.F.R. Part 280.

notice to the U.S. EPA. In these programs, the private parties conduct the risk assessment, not the agencies. For example, manufacturers are required to perform risk assessments to register a new pesticide under FIFRA, or to produce a new chemical under TSCA.

A. Numerical Standards

The first approach to risk is the use of numerical concentrations as standards. Generally, these numerical standards are human-health based. Risk assessment is used to determine the maximum concentration levels that will not cause any adverse health effects in humans exposed for a given exposure period. These numerical concentrations provide the minimum acceptable level of human health protection. There may also be numerical standards set for ecological-based protection.

To establish a health-based standard, an agency, first, collects and evaluates data to identify the COCs. Second, an exposure assessment is made to determine the level of exposure necessary to cause adverse health impacts and to evaluate the potential exposure to the contaminants. Third, the agency conducts a toxicity assessment by gathering evidence from a variety of sources "regarding the potential for a substance to cause adverse effects (carcinogenic and noncarcinogenic) in humans. These sources may include controlled epidemiologic investigations, clinical studies, and experimental animal studies."* Within these three steps, the potential risk for adverse effects to occur is characterized. Finally, toxicity values for carcinogenic and noncarcinogenic effects are developed. As part of the development for carcinogenic values, a calculation is made using a lifetime risk level assumption.** This assumption is based on a policy decision such as 1 in 100,000 or 1 in 1,000,000 risk level.

The SDWA is an example of a preventive program that uses numerical standards. The SDWA authorized U.S. EPA to promulgate health-based drinking water standards.*** In promulgating these standards, U.S. EPA performs a risk assessment to determine the level of contamination that will not adversely impact human health. U.S. EPA then issues numerical health-based standards, as Maximum Contaminant Level Goals (MCLGs) and MCLs. MCLGs are the concentrations at which no known, or anticipated, adverse effects occur to human health, and which allow an adequate margin of safety. MCLGs are not enforceable. MCLs are set as close to MCLGs as feasible. Feasibility reflects the best available technology, including cost and treatment technology.**** MCLs are the federally enforceable public drinking water regulations.

---

* [14] Risk Assessment Guidance for Superfund, Vol. I, Human Health Evaluation Manual (Part A), EPA, 7-3, 1989.
** [15] Risk Assessment Guidance for Superfund, Vol. I, Human Health Evaluation Manual (Part A), EPA, 1-7, 1989.
*** 42 U.S.C. § 300f.
**** 42 U.S.C. § 300g-1.

B. Technology-Based Standards

The second regulatory approach to risk is the use of technology-based standards. Technology-based standards consider the effectiveness of pollution control technology applied at the "end of the pipe" to minimize or eliminate air emissions or effluent discharge. The agency setting the standards selects the most effective treatment technology available (the "best") that can reasonably remove contaminants out of the process stream. The standard for removing particulate matter from air emissions, for example, would require the use of the most effective scrubber or filter treatment technology. In selecting a particular technology, the agency determines that it represents the best that can be achieved to eliminate or minimize the release of an environmental contaminant. Under technology-based standards, as long as the required technology is used, remaining emissions or discharges are deemed acceptable.

For example, Title III of the Clean Air Act Amendments of 1990 requires the use of maximum achievable control technology (MACT) as a technology-based standard for the emission of hazardous air pollutants. "The maximum degree of reduction in emissions that is deemed achievable for new sources . . . shall not be less stringent than the emission control that is achieved by the best controlled similar source."* Thus, the standard for a new source is the pollution reduction that technology can achieve, using the best control technology. The standard considers the best performing control technology of similar sources. Factors such as cost and energy requirements may be included in the technology selection. In addition, there is a unique CAA provision for setting health-based threshold levels for hazardous air pollutants.** U.S. EPA is to report to Congress within eight years of promulgating a MACT standard, and make recommendations about the health risk remaining after application of the technology-based emission standards.***

Other examples of technology-based standards include the Clean Water Act requirements for existing point sources. These effluent limitation requirements include the use of an industry specific, best practicable control technology.**** RCRA also has requirements for the use of best demonstrated available technologies, as a treatment standard, before restricted hazardous waste can be land disposed.†

C. Risk Assessment

The third approach to risk is through the use of risk assessment, as part of regulating a particular site, process, or facility. Typically, this is the type of risk assessment you will encounter as a project manager. The technical and procedural nuances of risk assessment are discussed in detail throughout this book.

Risk assessment can be used in this third regulatory approach to react to existing contamination. It can be used to calculate cleanup concentrations at Superfund sites, RCRA TSD facility corrective actions, and LUST corrective actions.

---

* 42 U.S.C. § 7412 (d).
** 42 U.S.C. § 7412 (d)(4).
*** 42 U.S.C. § 7412 (f).
**** 33 U.S.C. § 301 (b).
† 42 U.S.C § 3004 (m).

LEGAL CONTEXT OF ENVIRONMENTAL RISK ASSESSMENT

Table 3 Comparison of RCRA Subtitle S-TSD Facility Corrective Action (proposed) to Superfund Cleanup

RCRA		SUPERFUND	
Corrective action on a permitted facility		Corrective action on an abandoned facility	
Facility Assessment			
Facility assessment		Preliminary Assessment (PA)	Site Inspection (SI)
Prior to permit issuance		After environmental release	
Actual or suspected release assessment resulting in either	If no evidence of release, no further action	PA eliminates from futher consideration, sites that pose no threat	SI eliminates, from futher consideration, sites that pose no significant threat (no levels provided)
	If evidence of release, remedial investigation with specified (Subpart S) permit (action) levels not to be exceeded		
Investigation			
Facility Investigation		Remedial Investigation	
Action levels exceeded	No, no further action	Conducted on sites receiving HRS-II score greater than 28.5	
	Yes, Corrective Measure Study		
Remedy Selection			
Conditional remedy phase-in, with conditions up to length of permit	Achieve specified (in Subpart S) media cleanup standards	Achieve ARARs	ARAR waiver

These three remediation programs follow the same general approach to site cleanup (see Tables 2 and 3). First, the site is initially assessed. If sufficient contamination is found, a remedial investigation is required. During the remedial investigation, the nature and extent of the risk from the contamination is characterized. Third, during the feasibility study, plausible cleanup technologies are identified and evaluated. Fourth, a remedy is selected to achieve the cleanup goals. Evaluation criteria are provided. The final cleanup goals are selected. Depending on the specific cleanup program and the complexity of the site, the detail required within the first four steps may vary accordingly. Finally, the remedy is implemented.

Risk assessment is conducted during the remedial investigation. For Superfund, a baseline risk assessment is required during the remedial investigation. The Superfund remedial investigation (RI) collects "data necessary to adequately characterize the site for the purpose of developing and evaluating effective remedial alternatives."*

---
* 40 C.F.R. §300.430(d).

Site characterization includes conducting field investigations and conducting a site-specific baseline risk assessment. The field investigations are to characterize the nature and extent of the contamination as well as a site's physical features.

The baseline risk assessment characterizes "the current and potential threats to human health and the environment that may be posed by contaminants migrating to groundwater or surface water, releasing to air, leaching through soil, remaining in the soil, and bioaccumulating in the food chain."* The purpose of the baseline risk assessment is to: "provide risk managers with an understanding of the actual and potential risks to human health and the environment posed by the site and any uncertainties associated with the assessment. This information may be useful in determining whether a current or potential threat to human health or the environment exists that warrants remedial action ... As a general policy ... EPA generally uses the results of the baseline risk assessment to establish the basis for taking a remedial action."**

For RCRA TSD facility corrective actions, risk assessment may be required during the remedial investigation. The RCRA TSD facility RI identifies the nature and extent of the releases. The investigation may also include:

- A hydrogeologic investigation
- A characterization of solid waste management units (SWMUs) of concern
- Descriptions of human and environmental receptors
- Information in assessing risks to human health and the environment from the releases***

The proposed subpart S rule preamble states the following in relation to the third and fourth items above. "Section 254.511(a)(4) would provide the Agency with the authority to require information that will assist the Regional Administrator in the assessment of risks to human health and the environment from releases from solid waste management units. Information collected under §264.511(a)(3) also would integrate information on exposed humans and environmental systems and information on contaminant concentrations to assess the magnitude of threats to exposed populations. The interim measures are appropriate prior to selecting the final remedy or to evaluate whether a determination is warranted so that no further action is necessary (under proposed §264.514). The permittee should refer to chapter VIII of the <u>RFI Guidance</u> for information regarding the Agency's expectations for data that may be needed to conduct a risk assessment."**** Moreover, the U.S. EPA has provided the following with regard to the use of risk assessment in RCRA corrective actions: "While some implementing agencies may require the Permittee/Respondent to conduct a risk assessment, the policy on conducting risk assessments in the

---

* 40 C.F.R. §300.430(d)(4).
** Letter from Don Clay, Assistant EPA Administrator, to EPA Division Directors (April 22, 1991) discussing the role of baseline risk assessment.
*** 40 C.F.R. §264.511 (a)(proposed).
**** 55 Fed. Reg. at 30811, July 27, 1990.

corrective action program is evolving. Currently, their use is optional at the discretion of the implementing agency and should be based on site-specific conditions."*

IV. CONCLUSION

Legal considerations drive risk assessment. It may be the requirements of CERCLA or RCRA. It may be the hope of limiting future liability or liability to third parties. Legal considerations may require action or restrict action. Either way the influence of environmental law on the risk assessment process adds an interesting dimension to human health and environmental risk assessments.

---

* RCRA Corrective Action Plan, ERA 520-R-94-004, page 52 (1994).

CHAPTER 8

Risk Assessment Contract Formation

Robert Craggs and Sally L. Benjamin

CONTENTS

I. Introduction 245
II. Contracting Philosophy 246
    A. Objectives and Assumptions 247
    B. Affected Participants 247
    C. Communication Protocols 247
    D. Types of Contracts 248
    E. Interim Work Products 248
III. Contract Components 250
    A. Scope of Services 250
    B. Schedule 251
    C. Compensation 251
    D. Standard Commercial Terms and Conditions 253
IV. Common Contracting "Pitfalls" 254
    A. Lack of a Clearly Defined Scope of Work 254
    B. Misapplication of the Compensation Terms 254
    C. Contract Amendments 255
V. Conclusion 255

I. INTRODUCTION

Often perceived as a necessary evil, contract formation occurs in any business transaction where promises are made in exchange for something of value. A risk assessment project generally involves contract formation for risk assessment services.

In order to be effective, contract management requires key terms and conditions (performance standards) to be integrated into the contract. These terms and conditions are defined before beginning contractor selection and serve to create a set of interrelated requirements that the risk assessment project manager can use to ensure completion of an acceptable risk assessment, (i.e., a risk assessment that is completed at an established cost, on schedule, and includes the required information and analysis). Thus, formation of an effective contract is essential to successful management of a risk assessment project. This chapter addresses three aspects of contract formation: contracting philosophy, contract components, and common contracting "pitfalls."

II. CONTRACTING PHILOSOPHY

Before drafting or negotiating the terms of a contract, a project manager must have a solid foundation for a contract that allows for its active management. Without effective contract management there is no guarantee that a risk assessment report will comply with the project schedule, performance standards, and budget. A project manager or project representative from a contracting firm often delegate contract negotiation with a prospective contractor to third parties in their organization, either by necessity (e.g., they are technical experts or generalists, not skilled negotiators) or because they are unwilling to undertake the formal contracting process. The third party is generally an attorney, who negotiates the terms of the contract in the most favorable light for the party they represent. However, without guidance from a project manager, an attorney is not likely to understand the technical components of a risk assessment, or even the approach a contracting organization's project manager uses to develop a project. Delegation of formal contract formation to third parties tends to break the continuity required to form an effective contract, unless the project manager and the technical lead in the contractor's organization are also involved.

Effective contract formation involves identifying performance standards from which the contract should be developed. There are a wide number of process and product standards that may need to be integrated into final contract terms (see Chapters 4–6). Those deserving special attention include:

- Project objectives and assumptions
- Achievable time lines and budget
- Key personnel
- Affected participants
- Communication channels between the contractor and the client
- Appropriate types of contracts
- Interim and final work products from the project
- Performance standards, as discussed in detail in previous chapters

If a project manager must delegate contract formation responsibility to a third party, these parameters should be communicated to a negotiator in a written document. This document will serve as a basis for successful contract negotiations. Each parameter is discussed below.

A. Objectives and Assumptions

There are many reasons to conduct a risk assessment. With most human health and ecological risk assessments, there are both obvious reasons and reasons that are unstated or unrecognized. The contracting organization's project manager should attempt to articulate all of the project's short-term and long-term goals based on their complete understanding of institutional and project needs. For example, a risk assessment may be the first in a series of risk assessments on similar projects. If so, it is likely to serve as a prototype for the approach used in future risk assessments. Articulating this as an objective will help the contract negotiator recognize the precedent setting effect of the project, and negotiate accordingly.

In addition, the underlying assumptions and expectations for the risk assessment should be stated. For example, the scope of work should describe assumptions that relate to both quantitative and qualitative analysis. This approach will help clarify the expectations of risk assessment users and define the context for which the risk assessment is designed. Contract negotiators should understand a contracting organization's assumptions and expectations early in the process.

B. Affected Participants

Before contract formation, persons and organizations affected by the risk assessment process, or its results, should be identified. The organization which needs or requires the assessment has an obvious interest. Other stakeholders may be less obvious. Identify these affected parties by envisioning the assessment process and its outcomes. Insight into who is concerned about the risk assessment, and why, should influence the content and format of the interim and final work products to make them as useful as possible. The contract may not name particular participants, aside from the contracting parties, but it should reflect their influence on work products (see Chapter 31).

C. Communication Protocols

Preferred channels of communication are generally only vaguely defined in professional relationships. However, a formal communication protocol can be very beneficial. Formalizing communications requires that the project manager and the contractor's representative be identified. It also delineates how and when required communications will occur, and the relationship of communications requirements to project milestones, such as development and delivery of interim and final work products.

Communication relates directly to enforceability of the contract, record building, and effective project management. For example, a communications protocol might address major issues such as: Who can authorize a change in the work plan? Must the authorization be written? It may deal with record keeping for project decisions, such as: Are telephone logs required? Are meeting minutes kept? Are the minutes reviewed and corrected? The protocol also outlines project management systems. How will the project manager provide comments on work products? How will the contractor respond to comments? A planned approach to project communication promotes efficient communication between the contractor and the contracting organization's project manager throughout the process. It can also build a permanent record for public review, or for litigation, as the project progresses.

D. Types of Contracts

Because, at least in theory, everything within a contract is open to negotiation, there may seem to be a dizzying array of possibilities when it comes to contracting. There are, however, several standard contract types, each with it's advantages and disadvantages (see Table 1).

Because of the broad scope of services required for most risk assessments, proposals submitted in response to a RFP are often from teams of contractors. When using a comprehensive RFP, the contracting organization's project manager should anticipate proposals from teams of contractors and prepare to deal with issues inherent in administration of multiple contractors.

There are two schools of thought concerning management of multiple contractors. One approach advocates establishing individual contracts with each contractor on the project, without identifying a prime contractor or subcontractors. This is viewed as an efficient approach because there is direct contact between the project manager and each contractor who works on specific tasks. Arguably, it minimizes the layers of communication. The pattern of communication is like the spokes of a wheel with the project manager as the hub. Such an approach is most viable when three or fewer contractors are involved and the tasks are not interdependent. Those advocating an approach that excludes subcontracts perceive the various tasks involved in completing a risk assessment as to be "highly independent" of one another.

An alternate approach calls for identifying a prime contractor, who oversees subcontractors for various project tasks. Advocates of this approach argue that it is a more efficient, effective project management approach. Even though use of a prime contractor adds additional layers of communication within a contractor's team, it minimizes the contracting organization's project manager's responsibility for completion of individual tasks, placing it instead on the prime contractor. Such an approach should be clearly stipulated in the RFP. By doing so, expectations of those teams submitting proposals can be clearly defined prior to contract formation.

E. Interim Work Products

Effective contract formation defines a set of interim work products within the overall risk assessment project. A contracting organization's project manager and supporting team determine the specific tasks to include in the scope of services. In this way, a systematic approach to the project is outlined before the RFP, or the contract, are developed. When an RFP is developed, a set of work products are specified to establish opportunities to review the progress of the project.

Segmenting the risk assessment project into a set of discrete work products provides opportunities to review, comment, and approve interim work products. A

Table 1 Important Features of Various Types of Contracts

Labor Hour
Features: labor-hour contracts pay fixed rate for each hour of direct labor worked by contractor **Applicability:** used for engineering and design services, repair, maintenance or overhaul work, or in emergency situations **Advantages:** Contractor — least preferred type due to contractee surveillance; potential to maximize profits; minimal risk Contractee — greater flexibility **Disadvantages:** Contractor — least preferred type due to contractee surveillance Contractee — potential for high costs due to surveillance

Time-and-Materials
Features: provide for materials at cost; incorporate indirect costs and profit into fixed hourly rate **Applicability:** typically used for engineering and design services, repair, maintenance or overhaul work, or in emergency situations **Advantages:** Contractor — potential to maximize profits; minimal risk Contractee — greater flexibility **Disadvantages:** Contractor — least preferred type due to contractee surveillance Contractee — potential for high costs due to surveillance

Lump-sum Fee/Firm Fixed Price
Features: pays fixed rate (established before award) which is not subject to any adjustment regardless of contractor's cost experience **Applicability:** used when there are reasonably definite design or performance specifications and a fair and reasonable price can be established at the outset **Advantages:** Contractor — potential for higher profit; minimum contractee control; fewer administrative costs Contractee — risk fixed and limited; contractor bears risk of performance **Disadvantages:** Contractor — greater financial and technical risks; vigilance to initiate and substantiate change claims Contractee — no right to issue technical direction

Cost Plus Fixed Fee
Features: pays allowable cost plus negotiated fixed fee (profit); fixed fee adjusted for changes in work to be performed; either completion or term form **Applicability:** used where performance is uncertain and accurate costs estimates are impossible **Advantages:** Contractor — low risk Contractee — greater flexibility; greater control **Disadvantages:** Contractor — control by contractee; lower fees due to lower risks Contractee — greater risk; demands more resources to monitor costs and performance

Cost Plus Award Fee
Features: pays allowable cost plus base fee (does not vary) and award fee (based on evaluation of contractor's performance) Evaluation and payments of award fee made periodically during performance **Applicability:** cost reimbursement contract; motivates excellence in quality, management, timeliness, ingenuity, and cost effectiveness; used for larger contracts **Advantages:** Contractor — possibility of reward for good performance; limited risk Contractee — able to reward good performance **Disadvantages:** Contractor — increased burden to "prove" itself; fee usually limited to 10%; negotiations complex; performance affected by monitoring and technical direction Contractee — time consuming evaluation process

contracting organization's project manager and supporting team can require contractors to complete each work product to their satisfaction before approving work on subsequent tasks. This does not require every task to be independent of the others. It does, however, require the project manager to grasp which tasks within the risk assessment are interdependent, and address them accordingly. Early, periodic feedback from the contracting organization's project manager to the contractor helps ensure that no significant errors or omissions occur that will undermine subsequent project tasks.

In our opinion, effective contract formation requires awareness of the broader context of the risk assessment project. It also requires the contracting organization's project manager to communicate the specifics of the project; its objectives, related assumptions, and expectations must be conveyed to the contract negotiator. The general circumstances and specific details of the project should be articulated as contract parameters. This systematic approach to the project forms the basis for the RFP and sets the tone for formal contract negotiations. This approach also aids in establishing a series of specific interim work products and deliverables which will ultimately become the final report. Taking this sort of thoughtful approach to project development and management benefits both the contractor and the organization that depends on the contractor's services. These are discussed in the next section.

III. CONTRACT COMPONENTS

Once the circumstances surrounding the contract have been effectively communicated to the contract negotiator, the specific terms can be negotiated and the actual contract can be drafted. An actual contract for services does not have to be a verbose or complex document filled with legalese. Generally, there are four basic components that compose a contract for services: scope of services; schedule; compensation; and standard commercial terms and conditions. This section addresses each component.

A. Scope of Services

A scope of services identifies activities or products a contractor will provide. It may also provide a summary of actions and products that a contracting organization's project manager and support team will perform to support the contractor's efforts. In most circumstances, where an RFP is distributed, a detailed scope of services must be submitted to respond to the RFP. Potential contractors should draft a detailed scope of services with the intent to incorporate it into a formal contract. In this way, a portion of the contract will already be planned, formulated, and drafted prior to reaching this stage in the contract management process.

A scope of services should include a contract preamble identifying the project objectives, as well as: specific tasks; proposed approaches to achieve each task; task outcomes and deliverables; and proposed client involvement in the process. A scope of services should outline individual tasks. For example, a typical risk assessment might be divided into: the kick-off meeting; site characterization; source characterization; toxicity assessment; HHRA; ERA; and final report generation.

In turn, each activity can be subdivided into additional tasks that must occur for each major task. Then, for each subtask, the contractor describes the approach, related outcomes or deliverables, and the client involvement for each subtask. Written descriptions of the approach to tasks should state the type of data to be used. For example, will data be primary or secondary?

Written descriptions should succinctly state each task's relationship to other tasks, outcome and format, and the client's involvement in the task. Involvement can be limited to review and comment of each outcome or deliverable, or it may also require the client to provide information on a specified schedule. Whether for review, or for information-sharing, risk assessment projects generally require a contractor and client to meet. Meetings may be formal or informal, or both. Each formal meeting should be identified and its purpose and length should be incorporated into the scope of services.

Organizing information by subtasks provides the contracting organization's project manager and support team a structured way to identify and review the many different pieces of the risk assessment project. It provides contractors a systematic approach for completing each task and specifies the interrelationships between tasks, and identifies interim deliverables. Finally, it clarifies the items to be included in the scope of services, which generally constitutes the most significant segment of the risk assessment contract.

B. Schedule

RFPs generally state a completion date for the contract, but it is unusual for an RFP to define a schedule for completion for interim deliverables. This is unfortunate, because incorporating a schedule of deadlines for tasks and subtasks in the RFP can inform the contractor of the timing of the project and prevent scheduling conflicts at later dates. If the client incorporates a schedule of tasks into the RFP, the contractor can judge the level of effort that the client expects on each task. If the contractor responds to this schedule in the proposal, the client can assess how a given contractor views the project and can use the information to compare contractor proposals. To create a detailed project schedule for the proposal, a contractor must assess staff availability. A client should review the staff committed to each project task, and draft the contract terms to ensure that staff proposed for a task actually perform that work. Finally, if the schedule for certain deliverables is unrealistic, or conflicts with other project tasks or outside commitments, a schedule allows scheduling conflicts to be addressed in the process of negotiating the contract.

C. Compensation

Completing the detailed scope of services and project schedule, described above, will assist the contractor and the client in projecting realistic cost estimates for the project. Understanding outcomes and interim deliverables, number and purposes of meetings, degree of client involvement, timing of project deadlines, and qualifications of consultant staff to be involved in each phase of the project, greatly simplifies project costs estimation.

A separate issue from the cost of services is the type of compensation. There are several basic compensation types to consider when contracting for risk assessment services including: hourly or "time and materials"; maximum not-to-exceed fee; lump sum fee or fixed price; cost reimbursement; task-by-task fee; and hybrid.

Contractors generally prefer compensation on an hourly or time-and-materials basis. This approach poses the least risk for the contractor and the greatest risk for the client. The most commonly used type of compensation, however, is the maximum not-to-exceed fee. This approach generally requires the proposer to set a maximum price for the entire project that cannot be exceeded. The maximum not-to-exceed price is usually based on the estimated level of effort (i.e., labor hours) needed to complete the project. These hours are then multiplied by salary costs and summed with additional out-of-pocket expenses to determine project costs. Under a maximum not-to-exceed fee approach, the client is only obligated to pay the agreed to costs of completing the project. The contractor bases the price on the scope of services, described in the RFP, and on the schedule for project deliverables, by assessing all cost determinants. Thus, the above approach to drafting the proposal provides the contractor with an efficient means to set a maximum not-to-exceed price.

Compensation based on a lump-sum fee or fixed price provides opportunities for both contractor and client. However, if the client is a government agency, lump-sum contracts are less likely. Lump-sum costs are determined using the same approach as with the maximum not-to-exceed fee approach. If the project requires less labor or fewer expenses than projected, a contractor is awarded the difference as profit. This approach requires minimal accounting by both the contractor and client. Monthly invoices detailing labor and expenses may not be required. Also, actual labor expenses may not need to be tracked to justify compensation. Generally, a lump-sum contract identifies specific milestones to complete to receive lump-sum payments and provides interim payments to the contractor upon completion of these tasks.

If the project involves highly independent tasks, a contract structured with payments on a task-by-task basis may be optimal for both the contractor and the client. The contractor's risk is minimized because the project is actually a series of discrete tasks, with compensation for each on delivery. The client's risk is not greatly increased, but the client must negotiate with a contractor to create a scope of services that explicitly defines each task, and requires formal review and approval by the contracting organization's project manager and supporting team. This approach is consistent with the previously described project management and contract formation approach.

Some projects present a mix of activities, some easy to define and others more ambiguous. If tasks are unclear, the client and the contractor must devise an alternative compensation term. For difficult to define tasks, an established level of effort may be agreed to, coupled with a mechanism for expedited approval for additional compensation if effort and expenses exceed projections. Easily defined tasks can be addressed using a maximum not-to-exceed or lump-sum approach as discussed above.

Compensation incentives or bonuses may be appropriate on certain projects. Their use may be dictated by the client's flexibility, ability to define the scope of services, and project needs. A contractor could earn incentives and bonuses by providing an interim deliverable at a level above the client's expectations. Difficulty may arise in creating a measure that objectively assesses when bonuses are warranted. A review team supporting the project manager is generally required for such an arrangement.

Contract schedules provide the most objective measure of whether a bonus has been earned. However, incentives must be significant to actually influence contractor behavior. Minimal financial incentives are unlikely to impact the behavior of a contractor who is likely to be "juggling" several projects simultaneously.

Selecting the right type of compensation for funding a risk assessment report depends on many factors. Each type of compensation approach has certain advantages and disadvantages for the client and the contractor.

D. Standard Commercial Terms and Conditions

The fourth component of contract formation, standard commercial terms and conditions, should minimally address: contract termination; contractor/client insurance; contractor liability for negligence; reuse of work products; consequences for lack of payment; and dispute resolution.

This primer focuses on the practical aspects of contract formation, therefore, specific terms and conditions will not be presented. Most organizations have standard language for contracts which addresses the above issues and other technical requirements. The issue of dispute resolution, however, varies from contract to contract.

Inevitably, disagreements arise between the contractor and a contracting organization's project manager during a risk assessment project. They often center on expectations of work products. A concise scope of services can serve as a basis for resolving disputes surrounding the breadth or content of interim deliverables. If a dispute escalates, a contractor and contracting organization's project manager may choose to seek some form of dispute resolution. Therefore, a contract should state when the parties will enter into a formal dispute resolution process and the type of process to be used. Alternative dispute resolution techniques (ADR), including formal mediation and arbitration should be considered. Incorporating this process into a contract can benefit both contractor and client by avoiding formal litigation to resolve disputes, and by addressing conflicts efficiently and then moving ahead with a project.

Disputes generally arise because the contractor and client may have fundamentally different interests and expectations. This should be recognized by the contractor and the client. However, both parties should also recognize they can benefit by seeking a mutually acceptable resolution to the conflict, so they can move forward with their business relationship. Recognizing common interests and seeking a win/win solution helps promote efficient resolutions of contract disputes that may result from differing contract interests and expectations.

IV. COMMON CONTRACTING "PITFALLS"

The goal of contract formation is to develop a contract to adequately compensate the contractor for services, and to assure that the client receives a work product that meets all their expectations. After selecting the contractor, the project manager is generally interested in quickly completing the formal contracting process in order to begin project work. If a third party handles contract negotiations, a project manager and a contractor's counterpart may not be involved in the negotiations. If so, the contract negotiators must attempt to develop a contract that minimizes the risk to their organization. The focus usually strays from technical aspects of the contract and focuses, instead, on the terms and conditions associated with contracting for these services.

To assure that both technical and legal aspects of the project are addressed in the contract, the technical staff should work with the contract negotiator. In some circumstances, legal issues related to terms and conditions will not be resolved. There must be a recognition of this possibility in the contract negotiation process and in the subsequent business decision to go forward with the project. If legal terms and conditions overwhelm the contract, the process may be significantly delayed and the contract may not be focusing on its technical objectives. Common "pitfalls" associated with losing the balance between technical and legal issues in contract formation include: lack of a clearly defined scope of work; misapplication of compensation terms to the scope and schedule terms; and failure to modify/amend the contract when necessary.

A. Lack of a Clearly Defined Scope of Work

As described above, the scope of services must be clearly defined to include the tasks, outcome, client involvement, and meeting schedules. Obviously, a lack of clarity in these areas can increase the chance of misinterpretation by the contractor and client, and delay project completion.

For example, lack of clarity in the deliverables can lead to project delays as the deliverables undergo redrafts and reviews. In addition, failure to include the client involvement section, or to specify the form or timelines for client review, will slow down the process, when the client insists on ad hoc review and correction of deliverables. Similarly, failure to specify the length and number of meetings in the contract can result in failure to meet client expectations. Moreover, if a client insists on unplanned meetings, these costs and staff obligations may not have been accounted for in the Scope of Work.

B. Misapplication of the Compensation Terms

Determining the appropriate compensation terms can be a difficult aspect of contract formation. Naturally, the contractor hopes to receive ample compensation. Yet, compensation levels are market driven. The friction between offering adequate compensation without paying more than the market rate makes the compensation terms very important and potentially contentious.

In selecting an appropriate compensation term for a risk assessment project, the request for proposals should specify the preferred compensation term or terms. This provides parties with an opportunity to discuss the issue during the proposal stage. When contract negotiations are later initiated, the previous discussions will have narrowed the range of alternatives and clarified the parties' expectations.

If specific compensation terms are not discussed prior to contract formation, the selection of appropriate compensation terms should be dictated by the scope and schedule. Even so, these instances can complicate the compensation arrangement. For example, the client may want one type of compensation term, such as a maximum not-to-exceed approach, that are inappropriate for the type of work requested. On the other hand, the contractor may want another set of compensation terms, perhaps a lump-sum contract, which may be inappropriate for the client (e.g., a government body with extensive internal accounting requirements). Failing to discuss compensation terms before the contract is formed creates a potential for contract delays. Perhaps the most common pitfall related to compensation terms is failing to require contractors to address specific compensation approaches in their proposals.

C. Contract Amendments

Another common pitfall is failure to amend the contract when necessary during the project. When circumstances change, the need may arise for contract amendments. For example, there may be a need for additional services. If this is the case, these services should be explicitly defined and agreed to by the contractor and client and then should be incorporated into the contract as an amendment. The original scope of services should be consulted to determine whether these new services fall outside the original Scope of Work.

Another related pitfall is the contractor's failure to make timely requests to amend the contract to address the issue of additional services. Delaying such a request may result in conflict and possibly a formal dispute between the contractor and client. The contractor must communicate effectively to the client the services included and those services not included in the scope of services. If the client identifies activities not included in the scope of services prior to beginning the project then the client is more likely to negotiate a change to the existing agreement because their expectations have been addressed early in the process.

V. CONCLUSION

A well-planned contract management process will result in contract formation becoming the process of formalizing the key contract components that have already been defined by a systematic RFP process. As a result, contract formation will be perceived as a viable component of contract management, rather than a burdensome activity that must be completed by individuals with little technical interest in the actual risk assessment project. Basic components discussed above should be integrated into the contract. The compensation term should be carefully chosen to be compatible with the detailed scope of services.

CHAPTER 9

Ecological Risk Assessment Review

Clifford S. Duke and Jan W. Briede

CONTENTS

I. Introduction 257
II. Reviewing an Ecological Risk Assessment 258
    A. Problem Formulation 258
    B. Exposure Analysis 260
    C. Effects Analysis 261
    D. Risk Characterization and Uncertainty Analysis 261
III. Conclusion 262
    References 263

I. INTRODUCTION

ERAs evaluate the likelihood that adverse ecological effects may occur or are occurring as a result of exposure to one or more stressors caused by human activities (U.S. EPA, 1992). The ERA process is described in detail in Chapter 3. Review of contractor produced ERA deliverables is necessary to ensure that the science is consistent with current standards, calculations are verifiable, and all product and performance standards have been met. This chapter offers tools for critically reviewing contractor produced deliverables during the production of interim drafts or following production of a draft final report. No matter when critical review occurs, its purpose is to ensure production of scientifically credible products.

Contracting organization project managers are responsible for ensuring that contractors fully and appropriately respond to all critical reviewer comments. Responsiveness summaries can help verify that the contractor makes all necessary changes in text, tables, figures, and appendices, and that those changes appear in approved interim deliverables and final reports. Contractors should review all comments, make all responses available to the contracting organization in writing, and incorporate designated responses in the report as appropriate.

Regulated entities have a great interest in conducting the critical review before submitting a risk assessment to regulatory agencies. ERAs that follow current guidance and practices and that are critically reviewed are likely to have higher credibility and fare better in regulatory agency reviews. This in turn can reduce costs, shorten the time agencies need to reach risk management decisions, and increase the effectiveness of the risk assessment in negotiating such issues as discharge limits and site remediation goals.

Thorough reviews are essential to high quality reports. Poor reports can result in permit delays or denials and lost opportunities to modify remediation goals or discharge limits. Reports that ignore or downplay ecological risks can contribute to public opposition to projects, increasing the likelihood of regulatory delays and costly lawsuits. Ignoring potential impacts on threatened and endangered species or wetlands can lead to criminal prosecution.

ERA reviewers should have a thorough grounding in ecology, toxicology, and chemistry, as well as a working knowledge of environmental laws and regulations. Contracting organizations that lack personnel with such expertise are advised to hire appropriately trained individuals or contract out the reviewing task to another consultant. The latter strategy can be the preferred one, saving substantial overhead, unless there is a continuing need for ERA staff.

II. REVIEWING AN ECOLOGICAL RISK ASSESSMENT

The major phases of an ERA have been formalized by EPA (1992) as problem formulation, analysis of exposure and effects, and risk characterization (see Table 1). Careful study of the EPA framework and its successor documents (for example, U.S. EPA, 1993a, 1994, 1995) can help reviewers ensure that an ERA uses an up-to-date structural approach and terminology familiar to regulators. At a minimum, the final product should be formally peer reviewed before submittal to a regulator. However, as the most recent draft EPA guidance advocates, each phase of the risk assessment should be discussed with the risk manager and reviewers as it proceeds (U.S. EPA, 1995). This decreases the likelihood that issues of importance to the manager and/or regulators will be overlooked and ensures that the assessment design focuses on the decision to be made. The steps for reviewing an ERA outlined below are based on the EPA framework (U.S. EPA, 1992, 1995).

A. Problem Formulation

Problem formulation includes preliminary characterization of exposure and effects; examination of scientific data and data needs, policy and regulatory issues, and site-specific factors; and determination of the level of detail and information needed. The emphasis on data needs and policy issues is critical, because the purpose of the assessment is to assist efficient and timely decision making. Research in environmental

Table 1 A Checklist for ERA Review

Problem Formulation
___ States purpose of the assessment
___ Defines role of assessment in the project
___ Cites and follows appropriate federal and state agency guidance
___ Identifies ecosystem at risk and sensitive environments (e.g., wetlands) and organisms, especially threatened and endangered species
___ Identifies and justifies assessment endpoints
___ Identifies and justifies measures of effect
___ Describes relationship of measures of effect to the assessment endpoints
___ Describes how stressors of concern may exert their effects
___ Identifies all likely complete pathways
___ Justifies the omission or selection of pathways for analysis

Exposure Analysis
___ Describes stressor characteristics in appropriate detail
___ Describes the basis for selecting stressors for evaluation
___ Describes temporal and spatial distributions of the stressors relative to the measures of effect
___ Provides references for any variables cited
___ Matches tools to the problem
___ Explains selection of biomarkers and models

Effects Analysis
___ Summarizes relevant field data concerning stressor effects
___ Describes the kinds of effects stressors have on measures of effect
___ Describes the shape and extent of the stressor-response relationship, if known

Risk Characterization and Uncertainty Analysis
___ Identifies key sources of uncertainty
___ Describes sensitivity of the conclusions to changes in the values of key parameters
___ Identifies key assumptions and sources of uncertainty
___ States the source and method of calculation benchmark toxicity values used for estimating hazard quotients
___ Provides dates for values obtained from databases that are periodically updated
___ Addresses the weight of evidence supporting the conclusions of the analysis
___ Discusses sufficiency and quality of the data
___ Discusses supplementary information from the literature and other sources
___ Provides evidence that the stressor is causing or can cause the effects of concern
___ Describes additional analyses or field sampling that would strengthen the analysis or answer questions
___ Identifies parameter distributions, ranges, and other inputs to any quantitative uncertainty analysis should be identified
___ Justifies the choices of inputs

General Issues
___ Describes all variables for equations used in the exposure analysis
___ Units on the right side of equations balance those on the left (dimensional analysis)
___ Describes and justifies basis of extrapolation for parameters requiring extrapolation
___ Provides sufficient information to reproduce key calculations
___ States assumptions, potential shortcomings of data, and areas of uncertainty throughout the report

science is requisite to ERA, but is not ordinarily part of its purpose. Exceptions may include cases when no data are available, extrapolation from literature sources is impossible, or sensitive ecosystems or species are investigated. Two key products of the problem formulation phase are a conceptual model and the selection of assessment endpoints and measures of effect. The latter are also called measurement endpoints. These terms are described in detail in Chapter 3.

In this section of an ERA, the reviewer should check that the purpose of the assessment and its overall role in the project are clearly defined. Appropriate federal and state agency guidance must be cited and followed. Guidance in ERA is evolving rapidly, and ERA formats that are acceptable at one point in time may not be later. For example, EPA has recently circulated a draft ERA guidance analogous to those currently used for HHRA (U.S. EPA, 1995). Regulated entities should ensure that their contractors are constantly aware of such efforts. This section should identify the ecosystem at risk and sensitive environments (e.g., wetlands) and organisms, especially threatened and endangered species. This step is critical to designing the conceptual model and choosing appropriate assessment endpoints and measures of effect. Assessment endpoints must be identified and justified. Inappropriate choices may lead the ERA preparer to focus on the wrong issues and provide either insufficient or unnecessary detail. Measures of effect should be identified and justified, and their relationships to the assessment endpoints described. The analyses in ERAs are based on effects measures, which must have a clear relationship to the assessment endpoints that are the ultimate concern of the document. Finally, the conceptual model must clearly describe how stressors of concern may exert their effects, identify all likely complete pathways, and justify the omission or selection of pathways for analysis.

The steps that follow problem formulation depend on the conceptual model. Errors or inappropriate detail (too little or too much) in the conceptual model will result in an ERA of low quality that may be unduly expensive.

B. Exposure Analysis

Exposure characterization may include field measurements of the distribution of a stressor in organisms and environmental media; analysis of biomarkers, which can provide biological evidence of contaminant exposure (McCarthy and Shugart 1990); and computer modeling to estimate exposures in the future or at locations not sampled. The reviewer should check the following items.

Stressor characteristics must be described in appropriate detail. Examples include stressor type (e.g., chemical, physical), exposure intensity, duration, frequency, timing, and scale (U.S. EPA, 1992). The conceptual model, as well as knowledge of the site's characteristics, should help the reviewer to evaluate this factor. Weaknesses in the stressor description may result in either insufficient or unnecessary detail, a less defensible risk characterization, and/or unnecessary expense in the ERA.

The basis for selecting stressors for evaluation should be described either in the ERA itself or in a cited companion document. Not all stressors will necessarily receive detailed attention. For example, if only a few chemicals, out of hundreds, at a site dominate the risk, the others may not need to be addressed in detail. However, it is important to state explicitly the reasons for eliminating any potential stressors from consideration.

Temporal and spatial distributions of the stressors are described in the exposure analysis relative to the measures of effect. The risk characterization depends upon comparing these distributions with a dose-response relationship. If this information is unclear or not provided, the risk characterization cannot be adequately reviewed.

References should be provided for any variables cited, e.g., body weights, feeding rates, etc. The reviewer may wish to do spot checks of values taken from references for quality control purposes. Errors in variables will cause proportional errors in the risk characterization.

Tools, such as biomarkers and computer models, should match the problem. For example, analysis of metallothioneins, which can indicate heavy metal exposure, would have little relevance at a site where heavy metals are known to not be of concern. Fate and transport models designed for use in arid environments may lead to erroneous conclusions when applied to a location with high rainfall. In general, the selection of biomarkers and models should be clearly explained.

C. Effects Analysis

Effects analysis uses literature information and/or laboratory tests to examine both the kinds of effects caused by the stressor and the relationship between exposure and effect. This section of an ERA summarizes relevant field data concerning stressor effects. If such data are not available, this should be explicitly stated. Chemical stressors often have different effects or different magnitudes of effects in the field than in the laboratory, because complex factors in the field alter the availability of chemicals to organisms. For example, metal ions may bind to soils, organic chemicals may degrade, or organisms may be able to avoid the exposure. Conclusions based on field data may therefore differ from conclusions based on laboratory data. The reviewer should be confident that any site-specific studies have been noted and that important related studies have not been overlooked.

The kinds of effects that stressors have on the measures of effect should be described, as well as the shape and extent of the stressor-response relationship, if they are known. The risk characterization depends upon comparing these distributions with the exposure assessment. If this information is unclear or not provided, the risk characterization cannot be adequately reviewed. Errors in the choice of a dose-response relationship may result in underestimates or overestimates of the risk.

D. Risk Characterization and Uncertainty Analysis

The third phase of ERA, risk characterization, uses the data and conceptual tools provided by the first two phases to estimate the likelihood and degree of adverse effects of the stressor(s) on the organism or other ecological components of concern. For screening level assessments, a typical measurement endpoint is the HQ, the ratio of the estimated exposure to the no adverse effect level (or some other toxicity-based benchmark value.) A quotient greater than 1.0 indicates potential adverse effects. More detailed characterizations may combine modeling with site-specific data, toxicity tests, biomarkers, and other information in a "weight-of-evidence" approach. This approach is the preferred one, because it incorporates field and laboratory data, avoiding potential limitations of the quotient method used alone, such as over conservatism or overlooked exposure pathways. The advantages include greater credibility for the analysis, increased confidence that potential risks have been adequately characterized, and potential cost savings on site remediation in cases where field data show that effects predicted by the quotient method are not occurring.

Uncertainty analyses, which typically follow the risk characterization, vary in detail, depending on the needs and constraints of the project, and may be qualitative or quantitative. Whatever the level of detail, the analysis should at least identify the key sources of uncertainty and the sensitivity of the conclusions to changes in the values of key parameters. The basics of uncertainty analysis are outlined elsewhere in this book.

In these sections of an ERA, the reviewer should check to make sure that the source and method of calculation of any benchmark toxicity values used for estimating HQs are clearly stated. For a given intake estimate, the HQ is inversely proportional to the benchmark chosen. Defensible benchmark values are therefore critical in an ERA using the quotient method. Although there is no formal guidance on how to choose "correct" benchmarks, a number of sources of values are available, including current journals, books (e.g., Opresko et al., 1994), and databases such as EPA's IRIS. The dates should be provided for values obtained from databases that are periodically updated, for example, IRIS.

The risk characterization must address the weight of evidence supporting the conclusions of the analysis. It should include a discussion of the sufficiency and quality of the data, supplementary information from the literature and other sources, and evidence that the stressor is causing or can cause the effects of concern (U.S. EPA, 1992). Overlooked site-specific or relevant literature data may result in overly optimistic or conservative conclusions, with consequent impacts on the credibility of the analysis.

Where appropriate, the ERA should describe additional analyses or field sampling that would strengthen the analysis or answer questions that it raises. This will help the contracting organization respond proactively to any regulatory concerns based on the analysis.

Key assumptions and sources of uncertainty should be identified, and the sensitivity of the conclusions to changes in the values of key parameters should be discussed. The parameter distributions, ranges, and other inputs to any quantitative uncertainty analysis should be identified and the choices of inputs, (e.g., distribution type), justified. Inappropriate or unclear choices of parameters may affect the credibility of the uncertainty analysis, and consequently the entire ERA. Clear explanations and justifications, backed up by appropriate literature citations, can help avoid such problems.

III. CONCLUSION

In addition to the section-specific requirements discussed above, the following requirements apply to all sections of an ERA. First, equations should be checked to ensure all required information is provided. Specifically, variables should be described for every equation used in the exposure analysis. Otherwise, an adequate review cannot be conducted. The units on the right side of any equation must balance those on the left (dimensional analysis). If they do not, there are errors in either the equation or the variable descriptions, with potentially catastrophic effects on the ERA. The text or appendices should supply sufficient information to reproduce key calculations. For complex analyses, data may need to be obtained on computer diskettes from the risk assessor, but should be readily available. Second, the basis for any parameters requiring extrapolation must be described and justified (e.g., extrapolation from values measured in one species and applied to another). Although there is no comprehensive guidance on how to do this, it is important for quality assurance purposes and for the credibility of the ERA that the derivations be clear. EPA (1993b) has provided guidance for deriving a number of variables used in wildlife exposure analysis. Finally, assumptions, potential shortcomings of the data, and areas of uncertainty should be clearly stated throughout the ERA. In this light, there is nothing wrong with intuition when reviewing an ERA or related documents. The reviewer, whether an expert in the field or not, should use intuition as a guide in determining if the appropriate steps have been taken and if they make sense. Clear writing often reflects careful analysis; obfuscation nearly always accompanies the opposite.

REFERENCES

McCarthy, J.F. and Shugart, L.R., *Biomarkers of Environmental Contamination*, Lewis Publishers, Boca Raton, FL, 1990.

Opresko, D.M., Sample, B.E., and Suter, G.W., *Toxicological Benchmarks for Wildlife: 1994 Revision*, Oak Ridge National Laboratory, Oak Ridge, TN, 1994.

U.S. Environmental Protection Agency, *Framework for Ecological Risk Assessment*, Risk Assessment Forum, Washington, 1992.

U.S. Environmental Protection Agency, *A Review of Ecological Case Studies from a Risk Assessment Perspective*, Risk Assessment Forum, Washington, 1993a.

U.S. Environmental Protection Agency, *Wildlife Exposure Factors Handbook, Vols. 1 and 2*, Office of Research and Development, Washington, 1993b.

U.S. Environmental Protection Agency, *A Review of Ecological Case Studies from a Risk Assessment Perspective, Vol. II*, Risk Assessment Forum, Washington, 1994.

U.S. Environmental Protection Agency, *Draft Proposed Guidelines for Ecological Risk Assessment, Review Draft*, Risk Assessment Forum, Washington, 1995.

CHAPTER 10

Environmental Chemistry

John P. Cummings and Sally L. Benjamin

CONTENTS

I. Introduction 265
II. Practical Environmental Chemistry 266
    A. Chemists' Shorthand 267
    B. Types of Chemical Reactions 267
    C. Chemical Measurements 269
    D. Physical States 269
III. Major Chemical Disciplines 271
    A. Inorganic Chemistry 271
    B. Organic Chemistry 272
IV. Conclusion 275
References 275

I. INTRODUCTION

Risk assessments are used to determine whether there is (or may be) a threat to the public health or the environment. An assessment of the risks that chemicals pose to living organisms — humans, other animals, or plants — must be based on substantial knowledge of chemicals and their interaction with the life form which may be exposed to these chemicals. A chemical investigation of the past, current, or potential activities on the site is conducted to evaluate the chemical environment and determine whether a release of hazardous or toxic materials has occurred, or is likely to occur.

Basic chemical data are required to determine health effects and risk. Such data help to identify carcinogens, teratogens, or toxic compounds that can cause organ damage if they are ingested, inhaled, or absorbed. Understanding chemistry and the toxic effects of chemicals on life forms in terms of dosage, exposure, and concentrations in soil, water, and air, is key to making an endangerment assessment.

II. PRACTICAL ENVIRONMENTAL CHEMISTRY

An insight into some common chemical properties, at least a rudimentary overview, is necessary. Understanding the basic characteristics of chemical compounds is fundamental to understanding how these materials react with the environment. A conservative starting point to effectively understand chemistry requires an understanding of chemical terms, measures, and properties.

The building blocks for all matter are the elements. They exist, sometimes as separate particles called atoms, but usually in groups of atoms called molecules. Subatomic particles make up atoms; protons and neutrons form the nucleus, or center of the atom. Electrons move around the nucleus. Electrons generally exist in pairs and the pairs form spherical shells which surround the nucleus. Electrons in the outer shell of electrons are termed "valence electrons." Protons are positively charged, electrons are negatively charged, and neutrons have no charge. When an atom has an equal number of protons and electrons, it is electrically neutral. An example of an element which exists naturally as a single atom is helium (He). An example of a molecule composed of the same element is oxygen (O), which usually has two atoms for each molecule, identified as O_2. Molecules composed of different types of elements are termed compounds. Water, composed of two hydrogen (H) atoms and one oxygen atom, is an example of a compound. The chemical formula for water is written as "H_2O."

Elements are organized in the "Periodic Table" according to their atomic number. The Periodic Table uses the chemical abbreviation for each element and specifies the atomic make up (the number of protons, neutrons, and electrons in the atom) of the element. A chemist uses the atomic makeup of an element to indicate its chemical reactivity. A copy of this table can be found in any good chemistry text or handbook. This compilation is helpful because elements with similar characteristics are arranged into families, such as the metal, nonmetal, and metalloid elements.

Atoms bond to other atoms to form molecules. There are three types of chemical bonds: electrovalent or ionic, covalent, and coordinate covalent. The electrovalent, or ionic bond is formed by the transfer of electrons from the outer shell of electrons (called valence electrons) of one atom to the outer shell of another atom. In general, atoms having 1, 2, or 3 valence electrons tend to lose their valence to become positively charged ions, called cations. Metals tend to behave as cations. Atoms having 5, 6, or 7 valence electrons tend to gain electrons to become negatively charged ions, termed anions. Anions include nonmetals.

The second type of bonding, the covalent bond, forms when valence electrons are shared between atoms. The hydrogen molecule, written H_2, is a good example of this type of bonding. In H_2, each hydrogen atom is in a more stable state because it is sharing valence electrons with the other hydrogen.

The third type of bonding, the coordinate covalent bond, forms when one atom supplies both electrons in the electron bond pair. An example of this type of bond is found in the compound ammonium boron trifluoride (BF_3NH_3). The nitrogen-boron bond is formed by the valence electron pair associated originally with the nitrogen atom.

A. Chemists' Shorthand

The chemist's interest is in chemical changes, called "reactions." Reactions are expressed by a shorthand method using symbols and formulas, termed a "chemical equation." $2H_2O$ is a chemical equation. The chemist balances a chemical equation by calculating the types of molecules that form when there is the same number* of atoms of each element on both sides of the equation. By finding this balance, the chemist can predict likely chemical reactions and the resulting products.

In balancing a chemical equation, the reactants are usually on the left side of the equation and the products are on the right side of the equation. Reactants are separated from the products by a symbol, such as →, ↔, or =. A plus sign, +, separates each reactant or each product. When heat is required for the reaction to start or go to completion a symbol called delta, Δ, is placed above or below → or =, like so Δ. If a catalyst is used to speed up or cause the reaction to go to completion, the symbol of the catalyst is often written above the →. For example if the reaction required a Platinum catalyst to cause the reactants to totally become products then "Pt" would be written above the →.

B. Types of Chemical Reactions

There are six general types of chemical reactions:

- Combination reactions
- Decomposition reactions
- Single-replacement reactions
- Double-replacement reactions
- Neutralization reactions
- Oxidation-reduction reactions

1. Combination and Decomposition Reactions

In combination reactions, two or more substances react to produce a single substance. For example, a metal (magnesium [Mg]) plus oxygen react to form a metal oxide ($2 Mg + O_2 \overset{\Delta}{\rightarrow} 2MgO$). In decomposition reactions, one substance reacts to form two substances. For example, when heated red mercury oxide forms mercury (Hg) and oxygen ($2HgO \overset{\Delta}{\rightarrow} 2 Hg + O_2$).

---

* A chemist may also balance the amount of an element on each side of the chemical equation. When amounts are used the chemist works in terms of "moles."

2. Replacement Reactions

In single-replacement reactions, one element reacts by replacing another element in a compound. In fact, in aqueous solution, metals are conceptualized in a series, termed the "electromotive" or "activity" series that denotes their relative activity to each other. Hydrogen is not a metal; however, it is included in this series: Li, K, Ba, Ca, Na, Mg, Al, Zn, Fe, Cd, Ni, Sn, Pb, H, Cu, Hg, Ag, and Au.

In the electromotive series, iron (Fe) replaces copper (Cu) in the copper salt. This reaction is written as, $Fe + CuSO_4 \rightarrow FeSO_4 + Cu$. A series similar to the electromotive series, discussed above, exists for the halogen nonmetals, florine, chlorine, bromine, and iodine (F, Cl, Br, and I), in an aqueous solution. In aqueous solution, F will replace Cl, Br, or I in solution. This series is termed the halogen reduction potential series.

Double-replacement reactions involve two compounds. A positive ion (cation) of one compound exchanges places with the positive ion (cation) of a second compound. The two cations simply change partners by exchanging negative ions (anions). This type of reaction is often referred to as metathesis, meaning "form change." Many double-replacement reactions involve a precipitate, identified by underscoring, AgCl, for example, or by two symbols: $_{(s)}$ or $\downarrow$. When another change of state occurs, such as formation of a gas, it is identified as $_{(g)}$ or $\uparrow$. This reaction would be written as a formula as:

$$AgNO_3 + HCl \rightarrow AgCl_{(s)} + HNO_3$$

3. Neutralizing and Oxidation-Reduction Reactions

In neutralization reactions, an acid (or an acid oxide) reacts with a base (or basic oxide) to form a salt and water. This reaction is written as:

$$HCl + NaOH \times NaCl + H_2O$$

In an aqueous solution the reaction is reversible, depending on the concentrations of the ions and the salt.

Oxidation-reduction reactions occur when one substance loses electrons to another; the first substance said to be "oxidized" and the other substance is "reduced." Two examples of oxidation-reduction reactions follow. Example 1:

$$Ca + S \triangle CaS$$

In this reaction, Ca loses two electrons to S, i.e., it was oxidized. S gained electrons, i.e., it was reduced. Example 2:

$$C + 2 H_2SO_4 \triangle CO_2 + 2 SO_2 + 2 H_2O$$

In this reaction, C loses four electrons and the two S atoms each gain 2 electrons. Notice that the equation is balanced. C is oxidized by S, termed the "oxidizing agent." S is reduced by C, termed the "reducing agent."

C. Chemical Measurements

Chemical units are measured using the metric system. A familiarity with metric prefixes and suffixes relating to multiples of ten is helpful. Tables to convert from the metric system to the English system are readily available. Length is measured in meters [m] and centimeters (1/100 meters [cm]). Weights are measured in grams (gm), kilograms (1000 grams [kg]), milligrams (1/1000 gram [mg]), or micrograms (1/1,000,000 gram [µg]). Liquid volume is measured in liters (l) and milliliters (1/1,000 liter [ml]). Gaseous volume is measured in cubic meters [m^3] and cubic centimeters (cc), or liters and milliliters. Temperature is measured in degrees centigrade ($C°$).

Concentrations of toxic or hazardous materials generally are expressed in parts per million (ppm) or parts per billion (ppb). For solid materials, ppm is generally expressed in milligrams/kilogram (mg/kg) and ppb in micrograms/kilogram (µg/kg). For liquids, ppm is generally expressed in milligrams/Liter (mg/l) and ppb in micrograms/Liter (µg/L). This convention of mixing weight with volume, is derived from the weight of water, i.e., one gram per milliliter. For gases, ppm is generally expressed in cubic centimeters/cubic meters (cc/m^3). These calculations are usually included in the Analytical Laboratory Data Sheets. Analytical Laboratory Data Sheets present the types of chemicals and their concentrations in the medium (air, soil, or water) which were analyzed. The quality control, detection limits (the lowest concentration that the chemist can reliably test for a chemical), and any interferences found in the analysis are also indicated on the Analytical Laboratory Data Sheets. The information found in the Analytical Laboratory Data Sheets is an important factor used in determining risk.

D. Physical States

All matter exists in one of three physical states: solid, liquid, or gaseous. The physical properties of matter affect chemical reactions. These properties include: density (weight per unit volume), solubility (ability to dissolve in a given solvent), volatility (ability to become a gas at ambient temperature and pressure), diffusion (ability to move through and mix with matrix material), vapor pressure (pressure exerted by a substance in equilibrium with ;t~ own pressure), freezing point (temperature where a liquid turns into a solid), boiling point (temperature at which a liquid evolves into a gas), and flash point (the temperature at which a material ignites) of materials.

Some molecules ionize when placed in an aqueous medium. Ions have positive and negative charges.

Materials are considered hazardous, or toxic, if they exhibit at least one of four characteristics: corrosivity, ignitability, reactivity, or toxicity.

1. Corrosivity

Corrosivity is the ability of a substance, in contact with living tissue, to destroy the tissue by chemical reaction. It does not refer to an action by material on an inorganic surface. Materials generally identified as corrosive include acids, bases, and salts of strong acids and bases. Acids are chemical compounds composed of a hydrogen ion and a nonmetal element ion. Bases are chemical compounds composed of a hydroxyl ion (OH$^-$) and a metal element ion. Acids and bases are quantified by the concentration of hydrogen ion (H$^+$) in solution. Water at 25 C° has an H$^+$ concentration of 10^{-7} moles per liter (mol/l) and an OH$^-$ concentration of 10^{-7} mol/l.

In almost every area of chemistry, the acid-base properties of water are extremely important. The fate of chemical pollutants in a water body (or in the presence of water in air or soil), the ability of a metal object to corrode, or the suitability of an aquatic environment to support animal and plant life are all examples of the critical dependence of the acidity or alkalinity (basicity) of water. The acidity or alkalinity of a substance is measured on the pH scale. The pH scale is the negative logarithm of the hydrogen ion activity. A strong acid has a pH of 4 or less and a strong base is one with a pH of 10 or more. The pH of pure water at 25 C° is 7, therefore it is neither strongly acidic or alkaline, and may be termed neutral.

2. Ignitability

Ignitability is a term referring to the readiness with which a substance burns. A liquid is considered ignitable if it has a flash-point at or below 140 F° (60 C°), according to 40 CFR Part 261.21. A gas is considered ignitable when it forms a flammable mixture at 13% or less, by volume, when mixed with air. A solid is considered ignitable when it is likely to cause fire by friction or by heat retained from processing, or if it burns so readily that it poses a serious threat to public health and safety. This term is also used for a liquid, gas, or solid which ignites spontaneously in dry or moist air at or below 130 F° (54.3 C°), upon exposure to water, at any temperature, or any strong oxidizer.

3. Reactivity

Reactivity relates to the ability of a material to detonate, react, or decompose explosively at normal temperatures and pressures. Such material detonate or undergo explosive reactions, but require a strong initiating source (such as dynamite) or require heat under confinement before initiation (such as gunpowder) or may react explosively with water (such as metal sodium). Reactivity also pertains to materials, such as peroxide, that are normally unstable and readily undergo violent chemical change, but do not detonate.

4. Toxicity

Toxicity is the capacity of a material to produce harm to living things. The toxicity of a substance is often defined in the state and federal regulatory codes, or in data published by Federal and State Occupational Safety and Health Acts. This characteristic is often defined in a concentration per unit weight or volume. Toxicity of a material is often defined as the median lethal dose (LD_{50}).

III. MAJOR CHEMICAL DISCIPLINES

Chemists also identify broad classes of chemical substances and reactions in terms of organic chemistry, biochemistry, and inorganic chemistry. Organic chemistry is the chemistry of carbon compounds, specifically, bonding between carbon and hydrogen compounds (C-H) and between carbon compounds (C-C). Biochemistry is the science dealing with the chemistry of living matter. It is a subset of organic chemistry. Whereas organic chemistry is the chemistry of carbon compounds, inorganic chemistry is the chemistry of the other 114 elements.

A. Inorganic Chemistry

Inorganic chemistry may be defined as the chemistry of all compounds or substances, except those containing the carbon-carbon (C-C) or carbon-hydrogen (C-H) bond. This area has also been perceived as encompassing the chemistry of metals and nonmetals. Water, chemically depicted as H_2O, is an inorganic compound.

Rust (iron oxide), sand, (silicon dioxide), carbon dioxide, and carbon monoxide, chemically depicted as Fe_2O_3, SiO_2, CO_2, and CO, respectively, are inorganic compounds. Inorganic compounds number in the hundred thousands. The type of chemical bond generally associated with inorganic compounds is electrovalent, also termed ionic.

Metals include such well known materials as H, Fe, aluminum (Al), calcium (Ca), and cobalt (Co), as well as such relatively obscure materials as cerium (Ce), niobium (Nb), and hafnium (Hf). Metals make up more than 80% of the elements on the Periodic Table. Heavy metals and their compounds, especially their salts, are important environmental pollutants. Nonmetals, such as O, F, Cl, Br, and I are well known materials.

The chemical combination (bonding) of metal and nonmetal elements form compounds called salts. Sodium (Na) and Cl react to form sodium chloride (NaCl), common table salt. The usual method for forming the salts is the chemical reaction of an acid, such as hydrochloric acid (HCl), and a base, such as sodium hydroxide (NaOH), to form the salt NaCl and H_2O.

Some metals are essential to life, such as iron and zinc (Zn). Some metals have no known biological function, but pose no serious toxic hazards. Some metals are extremely toxic. For example, iron is an essential nutrient in small doses, but at excessive dosages it is poisonous to humans. Metals which are most deleterious to humans accumulate in human tissues. The heavy metals — antimony (Sb), arsenic (As), barium (Ba), beryllium (Be), cadmium (Cd), copper (Cu), cobalt (Co), chromium (Cr), lead (Pb), mercury (Hg), nickel (Ni), selenium (Se), and zinc (Zn) — accumulate in human body tissues. However, these metals seldom interface with biological systems in elemental form. Rather, they occur in discrete forms which can pass through biological membranes. Soluble salts in an aqueous solution are easily transported as metal ions. In contrast, insoluble salts are not as easily absorbed. Humans and other organisms may be exposed to toxic metals by inhalation (e.g., mercury vapors), ingestion (eating lead-based paint), and adsorption (such as beryllium dust through skin lesions).

Many metals have, because of anthropogenic activities, accumulated in the environment.

Lead is the most studied and hence best known for its method of entry and toxic effects. Lead interferes with the entry of ferrochelatase, the iron containing enzyme, by bonding with the enzyme, thus stopping "heme" production essential for the creation of hemoglobin. This causes anemia. Arsenic, another heavy metal, has been used as a pesticide for many years. Inorganic copper compounds are also used as agricultural poisons and algicides. Both metallic mercury and organic mercurial compounds react chemically with various body tissue and fluids, generally through ionic bonding. Long-term exposure to either inorganic or organic mercury can permanently damage the brain, kidneys, and developing fetuses. The most sensitive target of low-level exposure to metallic and organic mercury appears to be the nervous system. The most sensitive target of low-level exposure to inorganic mercury appears to be the kidneys. The chemistry behind these health effects is complex and a study in itself. Acceptable levels of metals in drinking water are published by the U.S. Public Health Services and state health agencies.

Two well-known, nonmetal inorganic compounds are carbon monoxide, CO, and cyanide, CN. Both compounds react with and selectively bond with hemoglobin, thus blocking the oxygen supply to the body and causing death, if a lethal amount of either, or both, is present.

B. Organic Chemistry

Organic chemistry is the chemistry of the carbon-carbon and carbon-hydrogen bond, also referred to as covalent bonding. Carbon exists in four forms — amorphous, graphite, diamond, and "white" carbon. It has a strange and diverse range of properties. Whereas graphite is one of the softest known materials, for example, diamond is one of the hardest.

Organic compounds, compounds based on carbon, exist in the hundreds of millions. Compounds with a carbon-hydrogen bond are commonly referred to as "hydrocarbons." Most hydrocarbons are derived from petroleum. The simplest hydrocarbon is methane, composed of one carbon atom bonded to four hydrogen atoms (CH_4). Hydrocarbons composed of carbon-to-carbon atoms in a straight line are termed "aliphatic" compounds, or "paraffins." The names of those carbon compounds with a single carbon-to-carbon bond end in "-ane," hence methane, ethane (C_2H_6), and propane (C_3H_8). The names of carbon-to-carbon compounds with a double bond between carbon atoms end in "-ene," e.g., ethene (C_2H_4) and propene (C_3H_6). Compounds with a triple bond between carbon atoms end in "-yne," e.g., butyne (C_4H_6). Many organic compounds have more than one name. For instance, ethene is also called ethylene.

A major portion of organic chemistry is based on the carbon ring structure and its unique qualities of bonding, stability, and reactivity. The carbon ring structure is composed of six carbons and six hydrogen atoms. The carbon atoms in this ring share electrons and form a bond of incredible strength and stability. These ring compounds are also referred to as "aromatic hydrocarbons."

The simplest compound with a carbon ring structure is benzene, written C_6H_6. Benzene, geometrically speaking, is planar and forms a hexagon. When one hydrogen atom from a benzene ring is replaced by a methyl group ($-CH_3$), methylbenzene, commonly called toluene, forms. When two hydrogen atoms are replaced by two methyl groups, dimethylbenzene, commonly called xylene, is formed. If a hydrogen atom from a benzene ring is replaced by a chlorine atom chlorobenzene is created; if two hydrogen atoms are replaced by chlorine, dichlorobenzene forms; and so on, until all six hydrogens have been replaced by six chlorines, forming hexachlorobenzene. Heavily halogenated benzene ring compounds, such as the highly publicized Polychlorinated Biphenyls (PCBs) are extremely persistent in the environment due to this chemical quality. Hydrogen atoms on a benzene ring can be substituted by other groups and are named accordingly. So, if the hydrogen is substituted with was an ethyl group ($-C_2H_5$), the compound formed is named ethylbenzene. Additional information on this system of chemical nomenclature can be readily found in the references listed in the bibliography.

As a general rule, the greater the bonding between carbon atoms in a compound, the less stable the compound. Thus, butane is more stable than butene, which is more stable than butyne. Carbon atoms can be bonded to other atoms besides carbon and hydrogen. Oxygen, nitrogen, sulfur, the halogens (F, Cl, Br, and I), as well as the metal atoms, bond to carbon atoms.

1. Petroleum Derivatives

Petroleum-based solvents, fuels, and petrochemicals probably contribute the greatest volume and variety of toxic compounds and hazardous wastes to the environment. Petroleum derivatives number in the millions. Fuel and fuel products can and have contaminated the air, water, and soil. The burning of petroleum has caused sulfur and vanadium to foul the air. The incomplete combustion of gasoline, oxides of nitrogen, oxides of sulfur and lead (from additives to gasoline), have caused smog and other air pollution. Leaking underground storage tanks have contaminated the soil and groundwater.

Hydrocarbons exhibit various physical properties and cause various health effects, in relation to their chemical structure. Straight-chain hydrocarbons (aliphatic hydrocarbons), methane, ethane, propane, and butane, are composed of less than five carbon atoms per molecule, and typically exist as gases. They are asphyxiants, but they do not produce systemic effects. Aliphatic hydrocarbons with five to eight carbons generally exist in liquid form. They affect the human nervous system.

Many petroleum-based materials are actually mixtures of aliphatic and aromatic hydrocarbons (e.g., fuels, gasoline, kerosene, diesel, and jet fuels). Solvents used in paints and cleaners are also mixtures of straight-line and ringed hydrocarbons. Mixtures can pose a wide array of health effects.

The toxicity of gasoline and other hydrocarbon solvents, that contain benzene and benzene ring compounds, depends on dosage and routes of exposure. Vapors inhaled in low concentrations are not significantly toxic. If even small amounts are ingested and aspirated into the lungs, however, these compounds can react with the moist lung tissues causing irreparable damage or death. Benzene, because of its ring structure, reacts with biological compounds, generally forming complex covalent bonded compounds. These compounds are very stable and interfere with normal body functions, causing cancer, hemorrhaging, and degenerative changes to the biological systems.

Another significant group of petroleum derivatives which have a unique chemistry are halogenated hydrocarbons. These hydrocarbons contain F, Cl, Br, or I. Halogenated hydrocarbons are widely used for cleaning machine parts, dry cleaning, and paint stripping. These materials react with biological compounds through both covalent and ionic bonds. They cause detrimental effects through complex chemical reactions in the liver, blood, and nervous system. The effects may be rapid or chronic depending on many factors, including length of exposure, routes of entry, concentration of the halogenated hydrocarbon, types of halogenated hydrocarbon, and general health of the exposed subject.

2. Pesticides

Pesticides* are unique environmental contaminants, in that they are hazardous materials that are often deliberately released to the environment to kill or injure other life forms. Ideally, a pesticide is highly specific. However, many pesticides are broad toxins and affect nontarget species, including humans. Because they are manufactured poisons, pesticides as a group are in fact a wide array of materials that do not share a similar chemistry or mechanism of action, and that do not have the same effect on different species.

A pesticide's toxic effect depends on its concentration and the manner in which it penetrates the organism. Some pesticides are acutely toxic in small dosages by inhalation. Some must be ingested to have an effect, while others can be absorbed through the skin. The extent of pesticide contamination and persistence in the environment depends upon soil type, moisture, temperature, ultraviolet ray exposure (sunlight), pH, and soil microbe levels, as well as on the pesticide degradability and original level of usage. In addition, the "parent" pesticide, the material actually applied to the field or crop, may be converted into new compounds of different toxicity as a result of the environmentally or biologically catalyzed reactions. Generally, the nature of these environmental and biological interactions with pesticides is poorly understood. However, the biological and physical degradation of parent pesticides in the environment can give rise to a variety of new compounds. These pesticide "metabolites" and "degradation products" may be of greater, lesser, or equal toxicity, compared to the parent pesticide.

To illustrate, consider the chemical effects on humans of DDT, a highly persistent organochlorine insecticide. The primary site of DDT's toxic effects is believed to

---

* Both insecticides and herbicides are subsets of pesticides.

be the nerve fibers. DDT, and its metabolites DDD and DDE, tend to remain in the fatty tissue of animals. When these tissues are ingested by humans and released during the digestive process, DDT and its metabolites alter the transport of sodium and potassium ions at the nerve endings. Short-term exposure to high doses of DDT has resulted in tremors, excitability, and seizures in humans. People exposed over longer periods to small amounts of DDT, experienced changes in their liver enzymes. DDT and its metabolites have caused cancer in laboratory animals and are, therefore, probable human carcinogens. Pesticide behavior and toxicity is an area where a qualified consultant can be of great help. There are also many texts that discuss pesticide dosage and its relative harm.

IV. CONCLUSION

Understanding and applying chemical data allows risk decision-makers to make informed decisions regarding the existence of risk and whether it exceeds acceptable levels. Ineffective use or misunderstanding of chemical analysis data can cause serious and unacceptable risk exposure, subjecting humans or other organisms to increased risk of irreversible health damage, or even death.

REFERENCES

Amdur, M.O., et al., Eds., *Cassarett and Doull's Toxicology: The Basic Science of Poisons,* 4th ed., McGraw-Hill, Inc., New York, 1991.

Budavari, S., *The Merck Index: An Encyclopedia of Chemicals and Drugs,* Merck and Co., Inc., Whitehouse Station, NJ, 1996.

Hawley, G.G., *Condensed Chemical Dictionary,* Van Nostrand Reinhold Co., New York, 1981.

Lide, D.R., *Crc Handbook of Chemistry and Physics: A Ready-reference Book of Chemical and Physical Data,* CRC Press, Boca Raton, FL, 1999.

Rodricks, J.V., *Calculated Risks: The Toxicity and Human Health Risks of Chemicals in our Environment,* Cambridge University Press, Cambridge, 1992.

Ware, G.W., *Fundamentals of Pesticides,* 2nd ed., Thomas Publications, Fresno, CA, 1986.

CHAPTER 11

Analytical Quality Assurance/Quality Control for Environmental Samples Used in Risk Assessment

Wayne Mattsfield and David A. Belluck

CONTENTS

I. Introduction 277
II. Effective Use of Analytical QA/QC for Risk Assessment 279
III. The Role of Analytical QA/QC in Risk Assessment Preparation,
  Review, and Management..279
  A. Project Description 281
  B. From Sampling to Data Analysis 282
  C. Blanks 287
  D. Choosing Laboratory Analytical Methods 296
  E. Where Analytical QA/QC is Used
   in Risk Assessment Reports 296
  F. Quality Assurance Project Plans (QAPPS) 297
IV. Effect of Data Quality on Data Usability in Risk Assessment 297
V. Conclusion 298
  References 299

I. INTRODUCTION

Risk assessments are designed to calculate site, activity, or facility risks for individual chemicals and chemical mixtures. When environmental releases of chemicals or exposures are known or suspected to have occurred, environmental samples can be collected and chemically analyzed to identify and quantitate sample contaminant residue levels. Regardless of where or how an environmental sample is taken and its chemical composition analyzed, it must meet defined quality parameters or its usefulness is questionable. Sufficient data of known quality must be used in a risk assessment to ensure that risk assessments properly reflect site, activity, or facility risks. Environmental sample quality assurance and quality control is a major focus of chemists and risk assessors during the planning and early phases of the risk assessment process. U.S. EPA recognized the importance of data quality for risk assessment by noting that its quality assurance program goal is to ensure that all data be scientifically valid, defensible, and of known precision and accuracy to withstand scientific and legal challenge relative to the use for which the data are obtained.

Environmental sampling and analytical chemistry work should proceed after a risk assessment team has thoroughly considered why the data is needed, how much data is needed, what kinds of data are needed, how good the data need to be, and who will use and review the data. Sampling and analytical procedures should be matched to the level of risk assessment rigor that is needed to sufficiently understand the nature and extent of contamination and its potential human health or ecological risks.

Several mechanisms have been devised to provide step by step procedures to walk project managers, scientists, risk assessors, and others through the process of designing sampling and analytical plans which provide data of known quantity and quality. Several of these processes have been formalized by the EPA and are recognized by their acronyms: Data Quality Objectives (DQOs), Quality Assurance Project Plans (QAPPs), and Sampling and Analysis Plans (SAPs).

These processes are used to ensure integration of risk assessment data generation activities. This includes design of the work plan or sampling plan, communication with all parties involved in the process, utilization of appropriate sample collection, sample preparation and analytical methods, and validation and assessment of analytical data. This primer provides the basic concepts of QA and QC in field sample collection and laboratory analysis.

Anyone who is about to review environmental data for the purpose of risk assessment is faced with some fundamental questions about its application to the process, such as, how do you differentiate "good" analytical results from "poor" results? Risk assessors are often faced with using data collected prior to their involvement in a case that may not have been produced for their use, and which was obtained and analyzed over time using different sampling, analytical chemistry, and QA/QC protocols. How can this data be appropriately evaluated and combined with other data sets, and can it be combined with new data specifically produced for a risk assessment? As this primer will show, when data is properly collected, analyzed, and reported, data of known quality can be properly considered for use alone or in combination with other data sets of known quality.

Data collected and analyzed for a risk assessment should be collected after several important planning steps have been completed. Before environmental sampling and analysis occurs to supplement historical data, or prior to the first thorough investigation of a site, data quality goals should be clearly defined for collection of analytical data in terms of precision, accuracy, representativeness, comparability and completeness (or PARCC), and DQOs. Failure to use these planning tools may result in collection of data that fails to meet all the needs of risk assessors.

Consultants performing a QA/QC function should be technically trained in physical and chemical sciences and experienced in the design, collection, and interpretation of environmental data. Useful experience includes participation in scoping different environmental investigations, and preparation of SAPs and QAPPs, as well as in data review and validation for these activities. Consultants should thoroughly understand applicable federal and state regulations for risk assessment QA/QC and be able to provide previous work products and reporting formats; a list of laboratories the consultant uses for risk assessment projects (include laboratory audits and relevant certifications); and a summary of the qualifications and experience of the firm and persons proposed to work on the project. If the consultant has their own analytical laboratory, they should provide a prospective client with relevant certifications, approvals, and records of laboratory audits.

II. EFFECTIVE USE OF ANALYTICAL QA/QC FOR RISK ASSESSMENT

Effective use of QA/QC tools results in efficient data collection and chemical analysis of environmental samples and allows for smooth integration of sampling data into the risk assessment. Precious time and money are saved when a properly constituted sampling and analysis plan is followed, because there will then be little need to return to the field to collect and analyze additional samples for the same or supplemental chemical substances not previously sought or analyzed. Effective risk assessment sampling and analysis programs can engender a public perception of those involved as competent, cooperative, and accountable professionals.

III. THE ROLE OF ANALYTICAL QA/QC IN RISK ASSESSMENT PREPARATION, REVIEW, AND MANAGEMENT

Planning the risk assessment must include environmental sampling and analytical QA/QC plans. Obtaining the right type and amount of analytical data begins in the planning or scoping process. During this process, participants should review any previously obtained data and determine the number, location, and media types of samples to be collected. Sample collection techniques; data quality needs; appropriate analytical methods and quantitation limits; QC acceptance criteria for project samples; and the extent and format of the data review/validation report, performed on the analytical data, should also be determined at this time. The planning or scoping meetings can include many parties, but at a minimum should include the project manager, risk assessor, hydrologist or geochemist, and chemist/QA manager (see Tables 1 and 2).

The role of the chemist/QA manager in the planning process is to recommend the sampling techniques; numbers of investigative samples, analytical methods, and quantitation limits; and numbers of QC samples and data reports (deliverables) which are necessary to meet the data quality/quantity needs of the risk assessor.

Table 1   Key Individuals in Risk Assessment Project QA/QC

Individual	Responsibilities
Project Manager	Organizes scoping meeting
	Coordinates actions of all individuals in project
	Oversees preparation of Work Plan, Sampling Analysis Plan
	Coordinates field sampling activities
	Manages subcontractors
Risk Assessor	Reviews historical data
	Determines chemicals of concern for risk assessment
	Assists in preparation of Work Plan, Sampling Analysis Plan
	Reviews validated data for use in risk assessment
	Prepares risk assessment
Chemist/Quality Assurance Manager	Assists in preparation of Work Plan, Sampling Analysis Plan; recommends field and analytical methods to achieve project goals
	Determines quality control samples needed to achieve data QC goals
	Assists project manager in managing field sampling activities; audits field sampling activities
	Provides limited oversight of sample analysis by the laboratory
	Reviews preliminary data
	Validates data
	Provides risk assessor and project manager with report
Geologist /Hydrogeologist	Assists in preparation of Work Plan, Sampling Analysis Plan
	Reviews preliminary data with respect to representativeness to site

During planning, members of a risk assessment team must evaluate:

- relevant historic data to determine the COPC
- the number and types of samples to obtain
- the analytical methods to use
- project-specific QC requirements
- what laboratory will conduct the chemical analyses
- sampling design, data review, and validation protocols and reviewers, balancing good sample collection and analytical procedures with health concerns
- product, process and performance standards
- deliverables
- program constraints.

ANALYTICAL QUALITY ASSURANCE/QUALITY CONTROL

Table 2 Project Scoping Checklist — Sampling/Analytical

What types of media will be sampled and analyzed? _____ Air   _____ Soil   _____ Surface water   _____ Groundwater _____ Other:
What are the chemicals of concern?
Are the methods appropriate for risk assessment?
Will special quality control limits be necessary?
What laboratory will conduct the analyses?
Should analyses be performed by a mobile laboratory, fixed-base laboratory, or both? _____ mobile laboratory   _____ fixed-base laboratory   _____ both
What sampling design is appropriate?
What type of data review is required? Who will perform data review?
How does the data need to be reported? (Data deliverables)
How many background samples are needed? _____ What constraints (budgetary, political) may affect data collection?

A. Project Description

Project descriptions are the summaries of the project location; history of activities; responsible party and/or regulatory agency investigations and monitoring activities; and documents produced from these activities. Project descriptions are used to provide the reader with an understanding of the physical layout of the site; extent of contamination and media affected (if known); the written record of past investigations; and the field and laboratory data acquired from these endeavors.

Project descriptions should be concise and contain several elements. Project descriptions begin with a statement of the decision to be made or questions to be answered. Following this statement of purpose, a description of the site, activity, facility, operating parameters to be studied, and anticipated uses of sampling and analysis results, should be provided. Additional elements include: anticipated uses of sampling and analysis results; a list of all measurements to be performed; a project schedule, indicating when samples are expected to be submitted to the laboratory; and a summary table covering the following for each sampling location — total number of samples (including primary, quality control, and reserve); type of samples (air, water, soil, etc.); analytical techniques employed for each sample; and a list of all measurements to be performed, differentiating, where applicable, the critical measurements (those necessary to achieve project objectives) from the noncritical measurements.

B. From Sampling to Data Analysis

Adhering to proper sample collection procedures is arguably the most important factor in the process leading to the generation of acceptable data. Collection of environmental samples should be carried out after a SAP or Work Plan and QAPP have been developed. Typical contents of a SAP include: a project description (e.g., project purpose, site description and site history, media to be sampled, COC), DQOs (e.g., precision, accuracy, representativeness, comparability, and completeness); sample collection procedures (e.g., standard operating procedures for collecting, handling, and shipping samples); sample shipment and chain of custody; field and laboratory instrument calibration; field and laboratory analytical methods; data reduction, validation, and reporting; and internal quality control checks.

The correct number of samples (e.g., single grab samples, duplicate samples, time sequence samples, or several grab samples to make up a composite sample); depth intervals (soil samples); matrix type and other relevant factors can dictate the type of sampling devices and techniques which will result in the most representative sample for laboratory analysis. Sample collection procedures can range from site specific to those mandated by a given regulatory program. Regardless of the origin of the sampling procedures, they must take into account the type of environmental matrix and substances to be measured. For example, when collecting soil samples containing volatile or quickly degraded substances, special care must be taken to ensure that the chemical will still be in the sample when it reaches an analytical laboratory.

Once a sample is collected, it must be properly labeled, inventoried, and shipped to an appropriate laboratory for analysis. Samples must be stored in a way that minimal loss or change in chemical composition will occur. Proper documentation must be maintained from sample point to laboratory bench to ensure that a sample will not be misidentified. These factors are very important in cases where government enforcement actions or litigation is a possibility.

1. Extraction Methods

Assuming that all sampling, shipping, recipient sample tracking, and storage procedures are adequately followed, the sample can now be analyzed for chemical content. Numerous kinds of methods are used to remove chemicals that are in solution, absorbed, or adsorbed to an environmental matrix. Some of the most common methods used to extract organic chemicals from environmental matrices are discussed below.

a. Purge and Trap

In purge and trap an inert gas is bubbled though an aqueous sample, transferring purgable compounds (organic compounds with boiling points less than 200°C) from the aqueous phase to a vapor phase. Purgeables are trapped on a sorbent material which is heated and back-flushed with a gas to carry the purgables into a chromatographic column for separation.

b. Solvent Extraction

Organic compounds are separated from the aqueous or solid phase of the sample by mixing the sample and organic solvent together, or passing the organic solvent through the sample; in general the solvent has more affinity for the organic compounds in the sample than does the sample matrix. An aliquot of this solvent phase (now containing the organic compounds) is injected directly into the instrument for analysis.

c. Solid Phase Extraction (SPE)

In SPE, an aqueous sample is filtered through or mixed with a solid absorbant that separates the organic chemicals from the sample matrix. After extraction, the organics are eluted or flushed off the solid phase, concentrated, and directly injected into the analytical instrument.

d. Supercritical Fluid Extraction (SFE)

SFE is a low temperature extraction using a gaseous solvent to separate organic compounds from sample matrices, over a short extraction period, with reduced destruction of heat labile compounds.

Metals can be found in aqueous solutions as dissolved ions precipitated out of solution in the form of hydroxides or salts, or bound in organometallic complexes. Water samples that contain relatively few solids (such as drinking waters) may not require sample preparation prior to analysis; water samples with significant solids content typically are digested with an inorganic acid and heat, to free metal ions from precipitates and organometallic complexes. Especially oily samples or media, with significant organic content, may interfere with acid digestion and analysis of samples for metals; under these circumstances the sample may require that the organic interferant be extracted out of the sample prior to digestion.

2. Measurement

Once environmental chemicals are removed from an environmental sample, they can be identified and quantified by laboratory methods, including elaborate and expensive instruments.

Laboratory instruments routinely used for measuring organic and inorganic constituents in environmental samples are discussed below.

a. Gas Chromatography

In gas chromatography organic compounds are separated into individual components based on their boiling point and relative affinity between the gas carrier phase and the solid sorbant phase of the chromatographic column. Compounds are separated by increasing the temperature of the column during sample analysis; compounds of larger molecular weight are eluted from the column last at these high temperatures. After separation, the individual components generate a quantifiable response registered by a detector selected for the specific organic compounds of interest.

b. High Pressure Liquid Chromatography (HPLC)

Organic compounds which are not appropriate for gas chromatography (heat sensitive, high molecular weight) may be analyzed using a liquid carrier and increasing pressure during analysis.

c. Atomic Absorption Spectrophotometry

Both graphite furnace atomic absorption (GFAA) and flame atomic absorption (FLAA) detect metals by the absorption of a light (at a wavelength specific to the metal of interest) passing through an atomized aliquot of the sample injected into the instrument. FLAA is generally less costly and faster than GFAA, but detection limits are lower for GFAA.

d. Inductively Coupled Argon Plasma Spectrophotometry (ICP)

In ICP, atomized samples are heated in a high temperature plasma where metals emit light at one or more wavelengths characteristic of that metal.

3. Data Analysis

a. Data Reduction

Environmental investigations can produce massive amounts of raw data that must be evaluated and reduced into summary tables if it is to be successfully used in a risk assessment report. Data reduction is accomplished by hand entry of analytical data into computer spreadsheets, word processing tables or databases; however, direct electronic data transfer (using computer diskettes, tape, or via modem) is automating the process of the production of tabulated data. There are an ever increasing number of information management systems software that can extract information from electronic databases or spreadsheets and produce graphic displays of chemical concentrations superimposed over site plans. Data reduction procedures produce chemical concentrations at given locations that are used as initial inputs into the risk assessment and are ultimately reflected as calculated risks. However data reduction is accomplished, mathematical methods and logic behind them must be transparent and verifiable by reviewers.

b. Data Validation

Data validation is the process of verification and evaluation which (1) confirms that investigative and QC samples have been properly handled, under appropriate custody, and submitted to the analytical laboratory for the correct analysis, (2) verifies that the laboratory analytical system was in control and capable of generating analytical results of expected quality, (3) verifies that the analytical results reported were accurate as reported, and (4) allows the data validator to qualify or reject reported data based on sample contamination, method deficiencies, or analytical analysis which is out of control. Data validation is accomplished by reviewing field logs and notes, chain of custody forms, laboratory internal QC and external field QC results, instrument raw data and chromatograms, laboratory reports, laboratory standard operating procedures, and the site QA project plan or SAP.

Persons performing data validation work must possess sufficient experience to interpret the analytical data in terms of the project data quality objectives, PARCC, quantitation limits, method performance and risk assessment needs. Validation personnel should have standard protocols (based on U.S. EPA's Contract Laboratory Program [CLP] guidance documents or other method-specific criteria) or contractor specific standard operating procedures to validate project data. Remember that this is the major yardstick by which acceptability of the data will be measured.

c. Data Reporting

Data reporting presents the analytical data to the project manager and risk assessor, along with a description of the limits of usefulness or data qualifiers, for results or analyses that may not have met the designed needs of the investigation. Data reporting is accomplished by providing data summary tables annotated with any appropriate data qualifiers, and a data validation narrative that describes any sampling or analytical difficulties, reporting or detection limit deficiencies, laboratory and validator qualified data, and the data validator's overall assessment of the data. It is important to know who will prepare the project data report, in what time frame, and in what format.

4. QA/QC Measures

Since scientists cannot hold or see individual atoms of single elements or the several atoms comprising compounds, they must rely on the information provided by their laboratory methods and instruments. QC samples are taken to ensure that the analytical methods are performing properly. Any QC method should clearly describe step by step procedures for preparation of standards and reagents, sample preparation, sample analysis, and data reporting, as well as the concentration range of the method, the reporting limits and method detection limits of the method (if different), and potential interferences and limitations of the method (which can be matrix dependent or affected by other substances in the sample medium). Method acceptance criteria for standards, surrogate compounds, spikes, duplicates, and other

Table 3 PARCC Data Quality Indicators

Data Quality Indicator	Importance	Suggested Action
Precision	Reduce uncertainty of data through assessment of the variability in sample measurements; determine confidence in distinguishing site concentrations of compounds of concern from background or upgradient concentrations	Collect and analyze sufficient numbers of field replicate samples; increase frequency of field duplicate samples for heterogeneous matrices (soils and waste)
Accuracy	Increase confidence in distinguishing site concentrations of compounds of concern from background or upgradient concentrations; inaccurate data can result in false positives or errors in the quantitation of compounds of concern	Follow well written, proven sample collection and analytical SOPs that meet accuracy needs for data at key quantitation limits
Representativeness	Avoidance of false negatives and false positives due to field sampling contamination	Use an unbiased sample collection design and mixing of samples to adequately represent the sample conditions; include blanks and QC sample collection/analysis to monitor false positives (blank contamination), false negatives, and biased results (spike sample recoveries)
Completeness	May decrease sample representativeness for identification of false negatives and estimation of average concentrations	Stipulate completeness goals for sampling and sample analysis; require SOPs for sample collection, handling, and analysis to provide for complete and valid sample collection and analysis
Comparability	Ability to combine analytical results across sampling episodes and time periods	Use the same sampling techniques, sampling design, and analytical methods across episodes and time periods

Note: SOPs = standard operating procedures.

internal method performance and quality control checks, should be clearly stated in the method.

There are numerous ways to assure that laboratory methods, instrumentation, and findings are accurate and precise. DQOs are qualitative and quantitative statements that specify the quality of the data required to support decisions. DQOs are determined based on the end use of the data to be collected. PARCC data quality indicators evaluate analytical data precision (measurement of agreement of a set of replicate results, among themselves, without assumption of any prior information as to the true result, and assessed by means of duplicate/replicate sample analysis);

accuracy (nearness of a result, or the mean [X] of a set of results, to the true value and assessed by means of reference samples and percent recoveries); representativeness (extent to which data measure the objectives of the data collection); completeness (measure of the amount of useable data resulting from a data collection activity, given the sample design and analysis); and comparability (measure of the equivalence of the data to other data sets or historical data) (see Table 3).

Achievement of DQOs is measured through attainment of project data quality indicator goals for PARCC. Development of DQOs is detailed in the September 1994 *Guidance for the Data Quality Objective Process,* and the *Data Quality Objectives Decision Error Feasibility Trials Guide and Software.*

Analysis of calibration standards are used to determine that the analytical instrument is correctly identifying and quantifying the chemicals in the environmental samples. This is done by injecting known concentrations of a chemical into a piece of equipment and evaluating the instrument's response. Analysis of calibration standards verify the linearity of the response of the instrument to the concentration(s) of the analyte(s) of interest in the calibration standard.

C. Blanks

Blanks are used to determine if analytical methods, materials, or instruments are reporting chemicals in an environmental sample that are really not there. Blanks are artificial samples designed to monitor the introduction of artifacts into the process. For aqueous samples, reagent water is used as a blank matrix; however, a universal blank matrix does not exist for solid samples, and, therefore, no matrix is used. The blank is taken through the appropriate steps of the process. Several types of laboratory blanks are described below (see Table 4).

1. Trip Blank

A Trip Blank (also known as a Travel Blank) accompanies VOC containers from shipment from the laboratory, to sampling in the field, and receipt by the laboratory. Analysis of the trip blank measures potential contamination of VOC containers and samples by volatile vapors.

2. Field Blank

A Field Blank (also known as a Rinsate Blank) is used to monitor cleanliness of equipment after field cleaning/decontamination of equipment. Laboratory-grade water is dispensed into a clean container for use in the field.

a. Method Blank

Method Blank (also known as a Laboratory Blank) measures contamination introduced by sample preparation solutions; absorption of contaminant vapors or particulates; contaminated sample standards or surrogates; and glassware; and contamination attributable to laboratory instrumentation, equipment, or glassware.

Table 4 Types of Blanks

Blank Sample	Characteristic	Purpose
Trip Blank (Travel Blank)	Laboratory-grade water free of organic compounds; prepared in the analytical laboratory and placed into VOC sample vials prior to shipment of clean vials for sample collection	Accompanies VOC containers from shipment from the laboratory to sampling in the field and receipt by the laboratory; analysis of the trip blank measures potential contamination of VOC containers and samples by volatile vapors
Field Blank (Rinsate Blank if used to monitor cleanliness of equipment after field cleaning/decontamination of equipment)	Laboratory-grade water dispensed into clean container for use in the field	Water is poured into water sampling equipment (bailer) or over soil or waste sampling equipment (augers, split-spoons, hand trowels) and poured or captured in the appropriate sample containers matching the investigative samples of interest; analysis of the field blank measures contamination introduced during sampling or decontamination and cleaning procedures
Method Blank (Laboratory Blank)	Laboratory-grade water	The analytical laboratory prepares the method blank in the same manner as the investigative samples (adds the same digestion or extraction solutions and spikes the sample with standards and surrogate compounds where appropriate); analysis of the method blank measures contamination introduced by sample preparation solutions, absorption of contaminant vapors or particulates, contaminated sample standards or surrogates and glassware, and contamination attributable to laboratory instrumentation, equipment, or glassware
Instrument Blank	Laboratory-grade water	The analytical laboratory analyzes the instrument blank without adding digestion or extraction solutions, spikes, or standards; analysis of the instrument blank measures contamination attributable to laboratory instrumentation, equipment, or glassware

ANALYTICAL QUALITY ASSURANCE/QUALITY CONTROL

Blank Sample	Characteristic	Purpose
Practical Quantitation Limit (PQL) or Estimated Quantitation Limit (EQL)	The PQL has been operationally defined as 5 or 10 times the MDL, or the concentration at which 75% of the laboratories in an interlaboratory study (of the method) report concentrations at $\pm 20\%$ OR $\pm 40\%$ of the true value. The EQL is defined in Solid Waste Methods SW-846 as the lowest concentration that can be reliably achieved within specified limits of precision and accuracy during routine laboratory operating conditions; the EQL is generally 5 to 10 times the MDL	Many methods in Solid Waste SW846 Methods have listed PQLs for each analyte, or provide a conversion factor to multiply MDLs by conversion factor to arrive at EQLs
Laboratory Reporting Limit	No accepted definition; may be statistically derived (a PQL or LOQ), or may be arbitrarily set (CRDL or CRQL)	Laboratories may choose to use reporting limits as contractual targets for compliance with work plans or sampling plans. Reporting limits must not be confused with statistical limits.
Sample Quantitation Limit (SQL)	The SQL is the MDL corrected for sample parameter situations, such as sample dilution, or use of smaller sample sizes for increased sensitivity; reported detection limits are adjusted upwards or downwards to reflect sample-specific action	Reported SQLs account for sample specific conditions and laboratory preparation and analysis steps; where multi-analyte methods (such as a VOC analysis) require dilution to bring one or more compounds into the range of the method, both the diluted and undiluted result should be reported; adjustment of MDLs to SQLs benefits the risk assessor and provides some increase in comparability of samples with varying characteristics
Contract Required Detection Limit (CRDL) and Contract Required Quantitation Limit (CRQL)	The EPA Contract Laboratory Program CRDL (inorganics) and CRQL (organics) are contractual reporting limits required of laboratories participating in the CLP; while these limits are similar to LOQ limits for comparative SW-846 methods, the CRDL and CRQL are by definition not derived statistically by each laboratory	CRDLs and CRQLs are generally achievable by all laboratories following the CLP methods (Statements of Work); these limits have a potential to be used widely, given the frequency that regulatory agencies specify CLP or CLP-like analyses

Note: MDL - Method Detection Limit.
LOQ - Limit of Quantitation.
CLP - Contract Laboratory Program.

Table 5 Types of Quality Control Samples

Sample Type	Characteristic	Purpose
Field Duplicate (aka Field Replicate if more than two samples)	Duplicate sample collected at same time and manner as investigative sample	Measurement of field duplicates or replicates provides data to estimate the sum of sampling and analytical variance; typically measured as the relative percent difference (RPD) between duplicate pairs
Blind Field Duplicate (aka Masked Duplicate)	Duplicate sample collected at same time and manner as investigative sample. The duplicate is given a fictitious or masked sample number so that the laboratory is not aware of the identity of the duplicate pairs	Measurement of the blind field duplicate provides data to estimate the sum of sampling and analytical variance; typically measured as the RPD between duplicate pairs
Performance Evaluation (PE) Sample	Water or soil matrix containing compounds or elements of interest at known concentrations, submitted to the laboratory for analysis with investigative samples	Measurement of performance evaluation samples provides an estimation of overall laboratory accuracy in analyzing for the compounds or elements in the sample; measured as percent recovery
Matrix Spike (MS) and Matrix Spike Duplicate (MSD)	Two extra volumes of the sample matrix (water, soil, or waste) collected with investigative samples for spiking with compounds or elements of interest by the laboratory	Measurement of matrix spike and matrix spike duplicate spiked compound percent recoveries and relative percent differences are generated to determine long term precision and accuracy of the method when used on the sample matrix

ANALYTICAL QUALITY ASSURANCE/QUALITY CONTROL 291

Table 6 Sampling Issues, Impact on Data Usability and Preventative Action

Issue	Situation Causing Data Impact	Impact On Data	How to Detect Effect on Data Usability	How to Prevent Situation
Background Samples	None collected	Cannot compare background concentrations to site concentrations	Review COC, field, and sampling logs and compare to site map	Plan for and collect sufficient background samples
	Contaminated background samples	Background sample results may be elevated or false positives	Review background sample results and all field and lab blanks	Provide for proper sample collection, field decontamination, and sample shipment to lab
	Background and investigative sample not from same media, strata, or representative of each other	Comparison of background and site concentrations not meaningful	Review COC, field, and sampling logs and compare to site map	Sampling locations must include representative background samples
Sample Matrix	Deterioration of sample	May result in unrepresentative, inaccurate data or false negative data	Review sample temperature, preservation, holding time information	Require proper sample preservation, container and temperature conditions during sample transportation to lab
	Incorrect sample (location, depth) collected	Comparison of background and site concentrations not meaningful	Review COC, field, and sampling logs and compare to site map	Sampling locations must include representative location and/or depths

Table 6   continued

Issue	Situation Causing Data Impact	Impact On Data	How to Detect Effect on Data Usability	How to Prevent Situation
	Wrong tissue type collected (biological samples)	Cannot determine concentrations in target organs of animal receptors	Review sampling and lab preparation logs	Collect correct tissue types for lab analysis
Documentation	Sample location poorly or not identified	Cannot compare background and site samples; sample may not be representative	Review COC, field, and sampling logs and compare to site map	Provide instruction and examples of proper COC and log completion
	Sample misidentified	Sample results may be meaningless with respect to representativeness; may affect all samples collected	Compare sample data to historical (if any) and expected concentrations	Provide instruction and examples of proper log completion, and prepare site sampling SOPs
	Break in COC	Sample results may not be valid if challenged	Review COC forms	Provide instruction and examples of proper COC completion; stress need for COC if situation requires
Design of Sampling Plan	Composite Sampling	May lower concentrations of compounds of concern from "hot spots"; could result in volatilization of some VOCs	Review COC, field, and sampling logs	Do not use composite samples unless it fulfills the data quality objectives for the investigation

Table 6 continued

Issue	Situation Causing Data Impact	Impact On Data	How to Detect Effect on Data Usability	How to Prevent Situation
	Wrong area, media or strata sampled	Sample results may be meaningless with respect to representative-ness	Review COC, field, and sampling logs and compare to site map	Provide clear site sampling location maps and descriptions
Sample Handling	Sample collected in inappropriate container	May result in unrepresentative, inaccurate or false positive or negative data, or no data	Review COC, field, and sampling logs	Obtain proper sample containers and preservatives from lab before sampling; provide sampling SOPs or sample container information to samplers
Sample collection equipment contaminated	May result in unrepresentative, inaccurate or false positive data; contaminants may mask low level concentrations of other compounds of interest	Collect field or equipment decontamination blanks for each type of sampling equipment cleaned in field; do not collect samples without cleaning equipment between sampling locations	Use dedicated, clean sampling equipment or disposable sample equipment where possible. Provide SOPs for sample equipment cleaning and decontamination, and collect suitable field or equipment sample receipt information from lab decontamination blanks	

Note: COC - Chain of Custody
SOP - Standard Operating Procedure
From *Guidance for Data Usability in Risk Assessment, Interim Final*, U.S. EPA, Office of Emergency and Remedial Response

b. Instrument Blank

Instrument Blank measures contamination attributable to laboratory instrumentation, equipment, or glassware.

3. Matrix Spikes

In contrast, matrix spikes introduce chemicals into a matrix to determine how well chemical extraction methods are working. Measurement of matrix spike and matrix spike duplicate (spiked compound percent recoveries and relative percent differences) are generated to determine long term precision and accuracy of the method when used on the sample matrix.

4. Duplicate Analyses

Duplicate analyses are used to determine the comparability of sample results. Predetermined quantities of stock solutions of certain analytes are added to a sample matrix prior to sample extraction/digestion and analysis. Samples are split into duplicates, spiked, and analyzed. Percent recoveries are calculated for each of the analytes detected. The relative percent difference between the samples is calculated and used to assess analytical precision. The concentration of the spike should be at the regulatory standard level or the estimated or actual method quantification limit. Types of duplicates are discussed below (see Table 5).

a. Field Duplicate

A Field Duplicate (aka Field Replicate if more than two samples) sample is collected at the same time and in the same manner as investigative sample. Measurement of field duplicates or replicates provides data to estimate the sum of sampling and analytical variance — typically measured as the relative percent difference (RPD) between duplicate pairs.

b. Blind Field Duplicate

A Blind Field Duplicate (aka Masked Duplicate) sample is collected at the same time and in the same manner as the investigative sample. The duplicate is given a fictitious or masked sample number so that the laboratory is not aware of the identity of the duplicate pairs. Measurement of the blind field duplicate provides data to estimate the sum of sampling and analytical variance — typically measured as the RPD between duplicate pairs.

c. Performance Evaluation

In Performance Evaluation (PE), samples of water or soil matrix, containing compounds or elements of interest at known concentrations, are submitted to the laboratory for analysis with investigative samples. Measurement of PE samples provides an estimation of overall laboratory accuracy in analyzing for the compounds or elements in the sample — measured as percent recovery.

5. Detection and Quantitation Limits

Each analytical chemistry method and instrument has limitations. Laboratory methods or recording instruments provide some type of visible and recordable response in the presence of a given substance. Sometimes as simple as a line or curve on a piece of paper, these responses provide chemical identity and concentration information. When a chemical is detected by a method or instrument, it may not be quantifiable because the response is not sufficiently great to make a scientifically defensible identification and quantification. Types of detection and quantitation limits used in risk assessment reports are discussed below.

a. Instrument Detection Limit (IDL)

The limit of detection attributable solely to the instrument (sample preparation, concentration/dilution factors, or other laboratory effects are not assessed).

b. Method Detection Limit (MDL)

The limit of detection attributable to the entire measurement process of a particular method and instrument.

c. Limit of Detection (LOD)

The LOD is the lowest concentration level that can be determined to be statistically different from a blank.

d. Limit of Quantitation (LOQ) or Quantitation Limit

The concentration above which quantitative results may be specified with a specified degree of confidence.

e. Practical Quantitation Limit (PQL) or Estimated Quantitation Limit (EQL)

The PQL has been operationally defined as 5 or 10 times the MDL, or the concentration at which 75% of the laboratories in an interlaboratory study (of the method) report concentrations at $\pm$ 20% or 40% of the true value. The EQL is defined in Solid Waste Methods SW-846 as the lowest concentration that can be reliably achieved within specified limits of precision and accuracy during routine lab conditions. The EQL is generally 5 to 20 times the MDL.

f. Laboratory Reporting Limit

No accepted definition exists. May be statistically derived (a PQL or LOQ), or may be arbitrarily set (Contract Required Detection Limit [CRDL] or Contract Required Quantitation Limit [CRQL], see below).

g. Sample Quantitation Limit (SQL)

The SQL is the MDL corrected for sample parameter situations, such as sample dilution, or use of smaller sample sizes for increased sensitivity.

h. Contract Required Detection Limit (CRDL) and Contract Required Quantitation Limit (CRQL)

The EPA Contract Laboratory Program CRDL (inorganics) and CRQL (organics) are contractual reporting limits required of laboratories participating in the CLP.

D. Choosing Laboratory Analytical Methods

Selecting analytical methods that meet both scientific and regulatory needs and requirements is one of the most critical choices in a risk assessment project. In the past, the most common systematic approach to sampling and data analysis was the EPA's CLP. It provided a standardized format to assess analytical method performance and compliance by supplying the reviewer appropriate documentation. QA/QC methods outside the CLP offer similar information with the same, or tighter, performance or QC acceptance limits than those of the CLP. Therefore, a project is not limited to reliance on only CLP methods.

E. Where Analytical QA/QC is Used in Risk Assessment Reports

For qualitative risk assessments, properly validated data, with defined confidence factors (such as precision and accuracy) associated with the data, should be used. The data validation, or assessment, report submitted with the data should contain a narrative which discusses the effect of associated field and laboratory QC samples, holding time violations, or instrument performance failings on the quality of the sample data. Individual compounds or elements, or entire sample fractions (e.g., all volatile analytes from a multianalyte method) may be qualified as:

- potential false positives or negatives
- estimated
- biased low/high
- usable after completion of validation.

Validated and qualified data is then incorporated into the risk assessment report to address decisions of the identity and concentration of compounds/elements present at the site; the difference between site and nonsite background concentrations; characterization of the spatial and media distribution of compounds/elements; the bioavailability or potential human/animal exposure routes for the compounds/elements; and the need for additional sample collection/analysis at the site.

F. Quality Assurance Project Plans (QAPPS)

QAPPs are used as a systematic method to provide a document that would ensure the quality of project analytical data through written sampling, analysis, and data assessment procedures, including project goals for precision, accuracy, representativeness, comparability, and completeness. In 1980, the U.S. EPA Office of Monitoring Systems and Quality Assurance released the *Interim Guidelines and Specifications for Preparing Quality Assurance Project Plans*, which contained the current QAPP format of sixteen sections or elements which are: title page; table of contents; project description; project organization and responsibility; quality assurance; sampling procedures; sample custody; calibration procedures and frequency; analytical procedures; data reduction, validation, and reporting; internal quality control checks; performance and system audits; preventive maintenance; specific routine procedures used to assess data precision, accuracy, and completeness; corrective action; and quality assurance reports to management. These elements respond to the need to effectively organize, monitor, and evaluate analytical chemistry activities, maintain and repair analytical equipment, routinely evaluate method and equipment performance, and provide quality reports. Subsequent guidance documents on QAPP production include *Preparation Aids for the Development of Category (I, II, III and IV) Quality Assurance Project Plans*, U.S. EPA Office of Research and Development, Risk Reduction and Engineering Laboratory; and *Data Quality Objectives Process for Superfund*, U.S. EPA Office of Solid Waste and Emergency Response. Many U.S. EPA Regional Offices have model QAPPs or region-specific guidance on QAPP writing.

While writing a QAPP would seem relatively straightforward, many elements of these documents seem to become contentious between regional offices of EPA, state regulatory agencies, and consultants. In the past, much of the information in QAPPs were devoted to boilerplate language that did not address the key issues in project data quality — design of the sampling network (through statistically derived sampling strategies), development of PARCC and internal QC goals (through use of DQO procedures), the means to measure the success in meeting the PARCC and internal QC goals (formulas and acceptance criteria), and the final "grading" of the data as to its usability for the project. Frequent comments on field or laboratory procedural language would hold up approval of QAPPs and projects, even if these items did not have a foreseeable impact meeting the project goals.

IV. EFFECT OF DATA QUALITY ON DATA USABILITY IN RISK ASSESSMENT

Contrary to popular opinion, all data are not created equal nor are they equally valid for use in a risk assessment. As individual data points or grouped data decreases in quality, so does its usability in risk assessment. U.S. EPA provides an outstanding review of this topic (U.S. EPA, 1992). In essence, data quality must match data use.

Table 7 Content of Sampling Analysis Plan

____ Project Description
____Description of the purpose of the investigation
____Description of the site and site history
____Description of the media that will be sampled
____Number of samples required
____Chemicals of concern
____Analytical methods
____Required detection or quantitation limits
____ Data Quality Objectives
____Precision
____Accuracy
____Representativeness
____Comparability
____Completeness
____ Description of the project goals for precision, accuracy, representativeness, completeness, and comparability
____ Rationale for the project goals for precision, accuracy, representativeness, completeness, and comparability
____ Sample Collection Procedures
____ Standard Operating Procedures or description of sample collection techniques (including any sample handling techniques such as compositing, placing samples into containers, etc.)
____ Sample Shipment and Chain of Custody
____ Field and Laboratory Instrument Calibration
____ Field and Laboratory Analytical Methods
____ Data Reduction, Validation, and Reporting
____ Internal Quality Control Checks

Note: These elements are Sections of the 16 element Quality Assurance Project Plan developed by U.S. EPA for the CERCLA (Superfund) program.

You cannot use low quality data to produce a scientifically rigorous risk analysis that will have a high level of credibility. To obtain a risk analysis that will have a high level of credibility and withstand piercing peer review, very high quality data must be generated and shown to be so.

The key to successful risk assessment production is to match risk management needs (e.g., screening level to baseline risk assessment levels) to risk assessment expectations and available resources. When a screening level analysis is needed for a gross understanding of site, activity, or facility risks, then a limited sampling and analysis plan could suffice. Thus, make the risk assessment level of rigor match risk managers goals, expectations, and resources, and there will be no need to try and torture the risk assessment team to generate risk conclusions at levels of certainty which the analysis does not deserve, nor can support (see Table 6).

V. CONCLUSION

Project managers need to be aware that obtaining the appropriate quantity of useable data begins with project scoping and planning for the numbers and types of samples required; the compounds of concern and required level of detection and reporting; the degree of precision and accuracy required from the method; and the format and content of the data report and validation summary required to document the integrity of the results produced for the investigation (see Table 7). The project manager must rely on the project team to provide the products required to complete the task of risk assessment. To do this, however, also requires a basic understanding of rigors, limitations, and pitfalls that can be encountered in the process of generating these products, and communication to the team of expectations or goals relating to data quality and quantity.

REFERENCES

Keith, L.H., *Environmental Sampling and Analysis*, American Chemical Society, Washington, 1990.

Simes, G.F., *Preparation Aids for the Development of Category I Quality Assurance Project Plans*, Office of Research and Development, U.S. Environmental Protection Agency, Washington, 1991.

Simes, G.F., *Preparation Aids for the Development of Category II Quality Assurance Project Plans*, Office of Research and Development, U.S. Environmental Protection Agency, Washington, 1991.

Simes, G.F., *Preparation Aids for the Development of Category III Quality Assurance Project Plans*, Office of Research and Development, U.S. Environmental Protection Agency, Washington, 1991.

Simes, G.F., *Preparation Aids for the Development of Category IV Quality Assurance Project Plans*, Office of Research and Development, U.S. Environmental Protection Agency, Washington, 1991.

Stanley, T.W., *Interim Guidelines and Specifications for Preparing Quality Assurance Project Plans*, Office of Monitoring Systems and Quality Assurance, U.S. Environmental Protection Agency, Washington, 1991.

Taylor, J.H., *Quality Assurance of Chemical Measurements*, Lewis Publishers, Ann Arbor, MI, 1987.

U.S. Environmental Protection Agency, *Contract Laboratory Program, National Functional Guidelines for Organic Data Review*, Washington, 1994.

U.S. Environmental Protection Agency, *Contract Laboratory Program, National Functional Guidelines for Inorganic Data Review*, U.S. Environmental Protection Agency, Washington, 1994.

U.S. Environmental Protection Agency, *Data Quality Objectives Decision Error Feasibility Trials (DQO/DEFT): User's Guide*, Washington, 1994.

U.S. Environmental Protection Agency, *Guidance for the Data Quality Objectives Process*, Washington, 1994.

U.S. Environmental Protection Agency, *Guidance for Data Usability in Risk Assessment*, Washington, 1992.

CHAPTER 12

Environmental Sampling Design

Rick D. Cardwell

CONTENTS

I. Introduction 301
II. Sampling Design Team 302
III. Conventional Statistical Approaches 304
    A. Issue Statement 304
    B. Purpose and Goals 304
    C. Statistical Hypotheses to Address Key Questions 305
    D. Defining the Statistical Tests Needed..307
    E. Sampling Design...307
IV. Sampling and Analysis Plan 308
    A. Sampling (Data Collection) 309
    B. Data Analysis and Verification/Rejection of Hypotheses 309
    C. Quality Control and Quality Assurance 309
V. Risk-Based Approach to Sampling Design..310
VI. Conclusion 315
    References..316

I. INTRODUCTION

Where and when to sample? How to sample? How much to sample? These are among the first questions to ask, and answer, when designing a study. Sampling provides the means to answering questions about potential environmental risks; such as, are there any highly contaminated sites and are the risks they pose significant? These questions are closely related to the investigation's goals and objectives.

The reliability of a risk assessment is based on the adequacy of the sampling design. Due to the GIGO principle ("garbage in, garbage out"), a risk assessment cannot be more accurate or credible than the data it employs. Therefore, an adequate sampling design is crucial to ensure that data collected are not only valid, but also capable of answering the questions posed in the investigation. A proper sampling design will be beneficial because it ensures quality control, promotes acceptance by the regulatory authorities, provides useful data for the technical specialists, and promotes cost-effective decision-making. An improper sampling design can undermine the entire risk assessment, requiring re-sampling, and causing delay and added cost.

The sampling design is usually part of the sampling analysis plan and remedial investigation work plan. It is usually discussed in the methodology of sections devoted to collection of environmental data (e.g., water quality data) and risk assessment.

This primer discusses sampling design in two major sections. Section II covers the technical elements of sampling design in general terms and identifies technical articles and books that can be consulted for more information. It describes a "risk-based approach" to sampling design, which focuses sampling and risk assessment on those sites, environmental media (e.g., water, soil), and chemicals that are likely to pose the greatest risk. Section III consists of checklists of items to consider in developing a sampling design.

II. SAMPLING DESIGN TEAM

Designing an effective sampling program should involve a statistician, the technical staff who will use the data (e.g., chemists, biologists, hydrogeologists, toxicologists), and the project manager. It is also wise to involve a representative from the responsible agency, as well, since regulators' perceptions of sampling design deficiencies can lead them to reject work products.

The sampling design team works under the direction of the project manager. The project manager is usually responsible for identifying issues and overall goals, and for directing the technical staff and statistician, accordingly. The technical staff then design the basic sampling approach by identifying questions that must be answered, hypotheses to test, parameters to measure, sampling objectives, study protocols (methods), and standard operating procedures.

The statistician works with the technical staff to ensure that the sampling design proposed by the technical team will provide statistically meaningful tests of hypotheses. The project manager reviews the proposed sampling design, in terms of costs and resulting ability to answer key questions. Usually the technical staff, statistician, and project manager interact iteratively. The sampling design evolves in phases with the team reviewing and refining the design at each of several phases.

Frequently sampling design is addressed solely by the project manager and the technical staff without input from a competent statistician. Such input is essential to ensuring that the study responds to the project objectives and that it is executed in a scientifically defensible manner. A statistician is essential to the process because, typically, data collected by sampling a limited area or population is used to make inferences about risks to larger populations occupying broader areas. Figure 1

ENVIRONMENTAL SAMPLING DESIGN

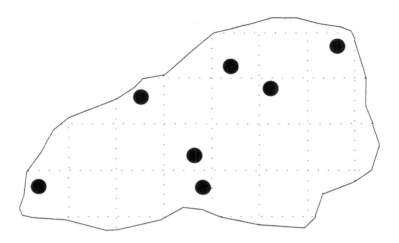

Figure 1 In most cases, limited data (dots) or subsamples are used to make inferences about a larger area of population (cross-hatched).

illustrates how a few samples from a relatively small area can be used to make inferences about larger areas.

Several statistical rules must be followed in order to make valid, scientifically defensible judgments concerning sampling results. They relate to ensuring that the samples collected are truly independent of each other and representative of the population being considered. Representativeness is the key consideration, and the information required to verify the representativeness of the sampling and samples encompasses representativeness over time, over space, repeatability of measurements (replication), uniformity of variability, etc. These requirements are discussed in virtually all statistical text books, and are covered in more detail in texts emphasizing sampling design, such as Green (1979) and Gilbert (1987).

Persons responsible for developing the statistical design must be experienced in sampling design and in analyzing data for the particular type of investigation. For example, if the design involves sampling fish for chemical residues, the statistician should be familiar with the methodologies and intricacies of sampling aquatic life. All team members should have experience designing and conducting investigations on other similar projects. In addition to project experience, familiarity with the statistical principles inherent in all sampling designs is also essential. Consequently, highly experienced staff must be involved in sampling design.

The basic issue in sampling design is how much to sample, given the high costs of sampling and analysis. Technical staff typically perceive abundant information as facilitating interpretation of the data and enhancing work product reliability, and seek to gather as much data as possible. Cost concerns, however, tend to limit the number of samples that can be taken and the areas that can be sampled. Thus, there are typically conflicts between the amount of information that technical staff desire and the financial resources available for sampling and analysis. The challenge is to collect an appropriate amount of data of sufficient quality to create a reliable risk assessment.

This chapter provides a general overview of items that need consideration in developing a scientifically defensible sampling design. It begins with a discussion of conventional statistical approaches used in risk assessment, then presents a risk-based approach that focuses the sampling on the environmental properties: (1) posing the most risks and (2) those most at risk.

III. CONVENTIONAL STATISTICAL APPROACHES

Risk assessors traditionally require specific statistical information to complete their assessments. Figure 2 illustrates how "data" are statistically distributed to define such statistical properties as the mean and confidence limits. The data in Figure 2 follow a "cumulative probability distribution." Risk assessors obtain their information by following a problem-analysis process that is part of the traditional scientific approach, discussed below. The elements and problem-analysis process, illustrated in Figure 3, belong in every sampling plan. Note that the process contains a feedback loop to allow study results to be used to pose additional questions.

A. Issue Statement

The problem-analysis process begins with a statement of the issues. Issues are those concerns that drive the initiative to remediate a site. They usually involve risks perceived to be high enough to warrant proposing remedial action. To illustrate, the example below involves remediation of contaminated sediments in an urban stream. In this case, the issues include:

- Are the fish safe to eat?
- Are the fish and wildlife at risk?
- Is it safe to swim in the water?
- Is it safe to come in contact with the sediments?

B. Purpose and Goals

The purpose and goals of the project are to address the issues. The purpose is a broader term identifying the reason for undertaking the project. Traditionally, it is a statement of the overall values to be protected or gained. The goals are usually the specific values.

Generally, the overall purpose of the project will be to remediate the site to a level that limits risk to an acceptable level. A series of specific goals may be set to address particular aspects of the site remediation. From each of these goals, a number of objectives can be established. Each statement of objectives — e.g., define risks to fishermen — will generate questions that must be answered and data that must be collected.

Consider, for example, the issue "are the fish safe to eat?" The goal is to decide whether fish caught from local waters are safe to eat. Specific objectives responding to this goal could include sampling the most important recreational and commercial

ENVIRONMENTAL SAMPLING DESIGN

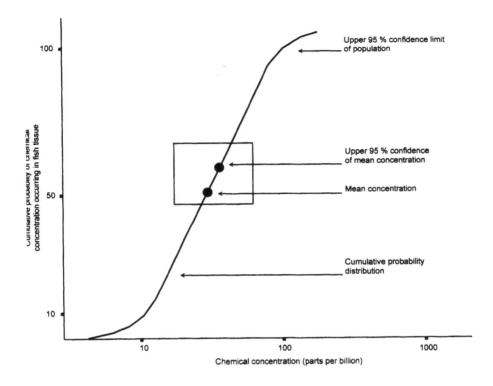

Figure 2 Example of key statistical information used in risk assessment.

species, and conducting a survey of angler and consumer behavior. A survey might include collecting responses to questions such as:

- How many fish are caught?
- How many are eaten?
- What parts are consumed?

C. Statistical Hypotheses to Address Key Questions

Collecting valid, scientifically defensible data generally requires that the data collection respond to explicit statistical hypotheses, since questions will be answered by drawing inferences from a limited set of data. For example, although a stream may be 5 miles long, not every foot section of stream will be sampled for fish. Perhaps only 500 feet of stream can be sampled cumulatively for fish, a mere 1.9% of the total stream miles. This is typically the case, due to the high cost of chemical analysis — in excess of $1000 per sample for a priority pollutant scan of fish tissues.

The challenge is to collect samples in a manner that allows characterization of the entire site, and generates data required for developing the risk assessment. The latter will include such statistical properties as means, confidence limits, and

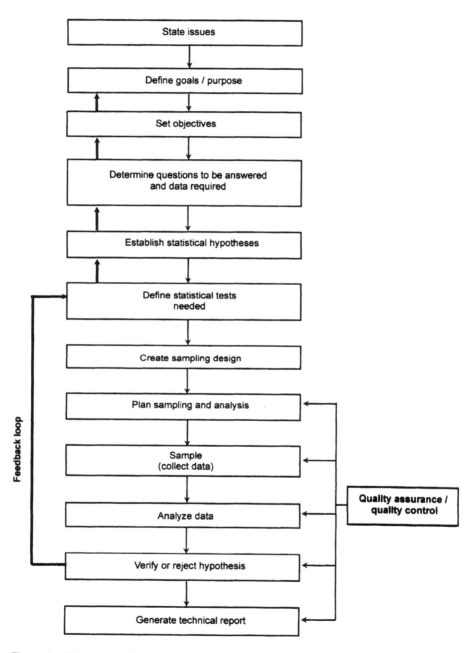

Figure 3 Elements and interrelationships in sampling design.

probability distributions (see Figure 2) for chemical concentrations in sediments, surface waters, and the tissues of fish and shellfish.

Statistical hypotheses help ensure that key questions are tested properly. For example, consider the following issue, posed as a question and, then, as several hypotheses:

- Issue: Are the fish safe to eat?
- Question: Do chemical concentrations in fish from the contaminated stream differ from those in fish from other sites or uncontaminated streams?
- Hypothesis 1: There is no difference in chemical concentrations in fish species 1 from site 1 of the contaminated stream compared to the same species in the reference stream.
- Hypothesis 2: There is no difference in chemical concentrations in fish species 1 from site 1 of the stream of concern compared to sites 2,3, etc.
- Hypothesis 3: Chemical concentrations do not exceed those known to pose significant risks to fishermen or fish-eating birds and mammals.
- Basic Data Needs for Hypotheses 1, 2, &3: Data describing chemical concentrations in the edible tissues of the fish/shellfish (i.e., the means, variances, upper 95% confidence limits, and cumulative probability distributions (see Figure 2). These data are needed for each of the species, streams and sites sampled.

D. Defining the Statistical Tests Needed

Selecting effective hypotheses in an unbiased manner is best accomplished with statistical tests. Experienced statisticians work in concert with other project staff (see Section II) to identify the appropriate statistical tests and the supporting information needed to clarify and interpret the sampling results. Examples of supporting information include the size of the fish, their age (since residues of some chemicals increase with fish age), lipid content, and number of fish constituting a sample.* This information needs are addressed in the sampling design.

E. Sampling Design

The sampling design should provide a "road map" for project data collection. It uses the information and judgements made in all the previous steps and provides specific information on how to conduct sampling, how to conduct the specified laboratory tests and analyses, and how to statistically analyze and report results. The sampling design specifies "how" data will be collected, "where," "when." and "why." The why of data collection was addressed by explicitly defining the questions, objectives, statistical hypotheses, and statistical tests featured in Figure 3. However, the where, when, and how also need description.

Determining where and when to sample and how many samples to collect is essential to a proper statistical design (Gilbert, 1987). It is rarely practical to sample a site completely. Instead, subsamples are taken and, from these, inferences are drawn about the entire population (see Figure 1). For example, limited information

* Fish may be analyzed individually by species or as groups of one or more species (composites), depending on the nature of the questions being posed. Gilbert (1987) discusses compositing.

gathered at a few sampling times and locations is extrapolated to larger time periods and areas, respectively. The extrapolations must deal with the inherent variability of the sampled populations. For example, variation from one location to another (spatial variability) and from one time to the next (temporal variability) are two factors influencing the number of samples required for a statistically valid analysis. It is important to consider variation over time (temporal variability) and over distance or depth (spatial variability). Often variability due to these factors is larger, and more important, than variability from, for example, duplicate laboratory chemical analyses of the same sample, or duplicate samples from the same or similar locations.

Both the sampling design and risk assessment embrace many assumptions, which have the potential to invalidate a study unless they are identified and critiqued at the outset. Accordingly, all assumptions must be explicitly stated in a section describing the sampling design, along with an evaluation of the validity of each assumption, its strengths and weaknesses, and its potential to bias the results.

In general, sampling requirements increase as the size of the area increases, the number of components increase, and variability increases. Statistical texts (e.g., Gilbert, 1987; Sokal and Rohlf, 1981) provide formulas for calculating the number of samples to collect. The key to deciding how many samples to collect depends on the parameter's variability in the population and on a judgment concerning the amount of variability that is acceptable between one sample location and the next. Various measures are available to reduce variability; however, they frequently increase investigational costs. Therefore, the sampling design reflects a balancing of increased costs against reduced variability.

Most statisticians recommend preliminary surveys (sampling), to define how much variability exists over time and space in the parameters being studied, before specifying sample size, including the number of replicates to collect per sample. Replicates are repeat measurements; they can be reanalyses of the same sample or different samples considered representative of one unit (e.g., a fish species, location). Defining what constitutes a replicate is an important consideration. Hurlbert (1984) discusses how study results can be extrapolated incorrectly through designation of inappropriate replicates, termed pseudoreplicates. Some environmental media are more variable than others. For example, chemical concentrations in aquatic sediments collected in urban areas typically tend to be highly variable spatially (Dutka, et al. 1991), as are assemblages of aquatic life (e.g., Boyle, 1985; Elliott, 1978; Hornig, 1983; Schlosser ,1990). Surface waters tend to vary greatly over time (Hensel and Hirsch, 1992), but may be less variable from one location to another unless there are pollutant or riverain inputs (e.g., Weber and Juanico, 1990). Parkhurst et al., (1994) and DeGraeve et al., (1991) specifically evaluate variability in aquatic toxicity tests (bioassays) of some species and media (e.g., effluents).

IV. SAMPLING AND ANALYSIS PLAN

The sampling and analysis plan is the document that contains the sampling design and the specific details on how to collect and analyze the data. It contains maps, for example, that pinpoint all sample locations and provides technical details on how the samples will be collected (e.g., randomly, systematically), processed, and preserved; how they will be analyzed chemically, toxicologically, or biologically; and, how the data will be summarized and evaluated. Equally important, the plan defines the quality control techniques and quality assurance procedures that will be employed to certify the data's reliability and scientific defensibility. An outline for a typical sampling and analysis plan appears in Appendix A.

A. Sampling (Data Collection)

The next step in the process (see Figure 3) is to collect and analyze the samples in accordance with the sampling and analysis plan. The sampling, laboratory tests, and analyses are investigation-specific. The U.S. EPA and American Society of Testing and Materials have developed extensive compendia of field and laboratory methods, protocols, and standard operating procedures (e.g., ASTM, 1993; U.S. EPA, 1983, 1986a).

B. Data Analysis and Verification/Rejection of Hypotheses

Data are analyzed using statistical tests specified by the statistician. The purpose is to generate the statistics needed by the risk assessors to compute risks and test hypotheses. All standard statistical texts (e.g., Sokal and Rohlf, 1981), as well as those that focus on environmental pollution (Gilbert, 1987), contain details on what tests are available and how to run them. Examples of statistical tests include t-tests, analysis of variance, regression, and correlation. As Figure 3 shows, the statistical results feed directly into providing answers to the original questions upon which the investigation was based. Investigations often yield unexpected findings, therefore, the information may also be used to modify the original questions posed, yielding another phase of more focused investigation.

C. Quality Control and Quality Assurance

Quality control and quality assurance are applied extensively to the last five steps shown in Figure 3. These objectives are defined in the sampling and analysis plan and implemented in sampling (data collection), data analysis, and reporting. QC is the process of ensuring the data's accuracy and precision, and QA is an independent verification that the data were collected exactly in the manner described and that the resulting data are accurate.

Examples of QC include analyzing "blanks" and standard reference materials * in the analytical chemistry laboratory, calibrating standards and meters, and testing control organisms and "reference toxicants"** in the ecotoxicology laboratory. Quality control also includes proofing data tables, spell-checking reports, and indepen-

---

* Blanks are samples, containing no test material, that are subjected to the entire analytical process to check on contamination and accuracy. Standard reference materials are test materials known to contain a specific amount of the chemical; they are typically formulated by governmental agencies.
** Control organisms are those subjected to every facet of the toxicity testing procedure, except they are not exposed to the chemical or test material. Reference toxicants are materials possessing a known toxicity.

dently checking all computations, including statistical tests. QC is practiced by technical personnel assigned to the investigation.

QA includes auditing personnel training and competence, auditing the investigations and testing while they are being conducted to ensure the tests and study protocols are followed, and auditing the data and reports to confirm the accuracy and reliability of the reported data. QA uses independent personnel, i.e., personnel not conducting any of the investigations or tests, to verify that all the data were collected and analyzed exactly as specified. They should report to an independent QA unit, including outside consultants, or an officer of the performing organization.

V. RISK-BASED APPROACH TO SAMPLING DESIGN

The risk-based approach is a focusing exercise that may precede the conventional sampling design and statistical testing discussed in the preceding section. It is presented here as an optional technique because not all environmental investigations will be risk-based, i.e., driven by environmental protection concerns. The approach seeks to reduce both the time and costs of sampling and analysis by limiting sampling to the "risky" chemicals,* locations, and media (e.g., sediments, groundwater) and resources at risk. These resources, in risk assessment terminology, are called receptors because they experience chemical exposure along specified environmental pathways (e.g., drinking water). Reducing study scope at the front-end of the project may also reduce costs of analyzing and reporting the data.

The goals of the risk-based approach are illustrated by Figure 4. When presented with environmental contamination, society may be concerned about risks to one or more receptors due to mixtures of chemicals occurring in one or more media at a contaminated site. The site, in turn, may have many locations that vary in contamination potential. The initial goal is to determine which of the media, locations, and chemicals are posing significant potential risks and which of the receptors are potentially at risk. Those determined to be associated with negligible risk can be eliminated from further investigation, which reduces the number of elements studied, in the manner shown in Figure 4. Usually, it is possible to significantly reduce the number of study elements by applying conventional risk assessment methods; the screening-level risk assessment (SLRA) approach described below is a method designed to accomplish this initial screening expeditiously and cost-effectively.

The SLRA process is shown in Figure 5. The screening-level risk assessments are based on abbreviated, conservative calculations, usually accomplished using computer spreadsheets. Compared to detailed risk assessments, they are intended to be performed quickly and economically, allowing project resources to be focused on the sites, chemicals, and media posing potentially significant risks and the receptors potentially at significant risk. They are designed to be performed with limited data. Conservative assumptions, embodying appropriate safety factors, are used to ensure that risks, if any, will tend to be overestimated. In other words, the SLRAs

---

* For example, "risky" chemicals refer to those chemicals shown, using screening-level risk assessments, to pose potentially significant risks to the receptors or resources being assessed. Locations and media posing potential risks are defined similarly.

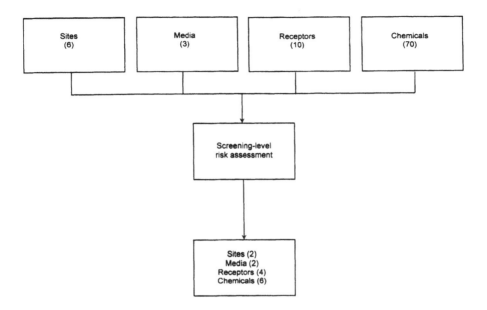

Figure 4 Goals of the risk-based approach to sampling design: focusing on the sampling, analysis and reporting of data.

are intended mainly to screen out the obviously nonrisky* elements. Where an element's risks are either borderline or questionable, due to limited data or other reasons, it should be retained for more detailed examination in sequel, detailed risk assessments.

The SLRAs should be conducted using the most updated methods available from the scientific community and governmental agencies. The human health SLRA should be consistent with EPA risk assessment guidance (1993a). Methodologies for conducting screening-level ERAs with aquatic life are presented by Cardwell et al. (1993), Parkhurst et al. (1994), and Suter et al. (1992). U.S. EPA (1993a) published one of the first complete methodologies for conducting risk assessments with wildlife. Though emphasizing fish-eating birds and mammals, and chemicals with high potential for environmental bioaccumulation, its concepts can be applied to other species. The SLRA methodologies are based on various EPA guidance documents that deal with several different risk assessment aspects. These include control of toxic substances in water (U.S. EPA, 1990), control of bioconcentratable substances in water (U.S. EPA, 1993a), derivation of water quality criteria (Stephan et al., 1985), wildlife risk assessment (U.S. EPA, 1993a), and interpretation of water quality criteria for metals (U.S. EPA, 1993c).

The SLRAs focus on the exposure pathways often associated with the greatest risks, and they are expected to produce quantitative risk estimates that can be used

---

* Nonrisky is used here as being equivalent to negligible risk. Numerically, some risks can always be computed, so there may never truly be zero risk.

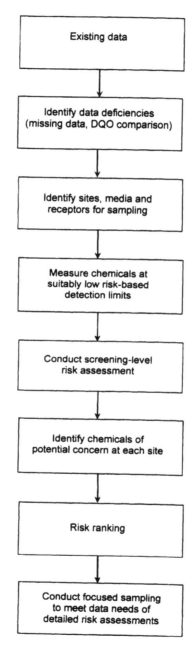

Figure 5   Screening-level risk assessment process.

to identify and rank the risks by receptor, site, media, and chemical. These rankings are consolidated into an overall risk ranking so that higher risk elements can be separated from those posing potentially significant risks, moderate risks, or negligible risks. For example, higher risk sites can be distinguished for more detailed evaluation. Detailed risk assessments may be undertaken to verify the screening level estimates and provide more accurate and reliable estimates of risk based on fewer assumptions.

The process shown in Figure 5 begins with a review of existing data relative to its adequacy for assessing risks to the receptors, etc. that are at issue. Data judged deficient for risk assessment purposes are then identified. Deficiencies stem from missing data, use of analytical detection limits exceeding risk-based detection limits, and questions about data reliability due to inadequate QC or QA.

The next step is to obtain the data needed to support the SLRAs. The latter should use existing data, if available, on chemical concentrations in the environments being studied. Otherwise, new data need collection. If there is any question concerning what chemicals have been released to the environment, it is desirable to measure the list of EPA priority pollutants (U.S. EPA, 1993b) to minimize criticism that chemicals were missed. Unlisted chemicals need measurement if they are known or suspected of being released.

If new data are collected, MDL should be 10-times below the concentration associated with the negligible risk threshold. The latter threshold is called a "risk-based detection limit." The threshold should be 10-times higher than the analytical MDL to account for additive or synergistic toxicological interactions between the chemicals. If desired, chemicals with a frequency of detection less than a specified percentage need not be considered further in the SLRAs. The reasoning is that chemicals not occurring or occurring very infrequently need not be assessed further. However, it must be recognized that chemicals occurring at low frequencies at concentrations near the analytical detection limits may be artifacts, i.e., reflecting contamination or normal variation in the analytical method. It is advisable to only use data where the risk-based detection limit is higher than the PQL or the MDL of the analytical or test method. The PQL or MDL should be used instead of the detection limit because they define the upper limit of variation about the true detection limit. Use of PQLs or MDLs will help minimize the effect of detection limit noise on reported values.

U.S. EPA (1994a) provides definitions distinguishing these detection limits. Maddalone et al. (1993) also provide a thoughtful discussion of which detection limits provide the most reliable data. Chemicals occurring at or below the detection limit (i.e., nondetects or NDs) may be assessed at a concentration equivalent to one half of the detection limit at each sample location. Gleit (1985) discusses other methods for replacing nondetected values.

The screening-level risk assessments estimate risks by calculating HQs for each chemical, medium, site, and receptor. This step is called the risk characterization, which compares the results of the exposure assessment and the ecological effects characterization or toxicity assessment. For the SLRA, the potential for risk is quantified using the quotient method (Barnthouse et al., 1986) which evaluates the ratio of a chemical's expected environmental concentration to its toxic concentration (see Equation 1). The quotient is termed an HQ because it only signifies the potential risk magnitude rather than specifies a specific probability of risk. Risks should be expressed in terms of probabilities of adverse effects. For cancer-causing chemicals (carcinogens) in the human health SLRAs, cancer risks are computed and chemicals flagged according to exceedances of the established EPA allowable risk level (e.g., 10^{-6}). If carcinogens are detected, it automatically qualifies them for detailed risk assessment.

(1) Hazard Quotient = $\dfrac{\text{Expected Environmental Concentration (EEC)}}{\text{Minimum Concentration Causing Adverse Effect}}$

HQs may be computed for acute or chronic toxicity or both. Usually, SLRAs are based on chronic toxicity because this is conservative and limits the number of calculations. Because data are generally limited, the "maximum expected environmental concentration" may be used. If data are sufficient to calculate confidence limits, then the upper 95% confidence limit of the mean can be used for chronic exposure and the upper 95% confidence limit of the population can be used for acute exposure (see Figure 2). Generally, the entire expected environmental concentration is assumed to be bioavailable; corrections for bioavailability are best reserved for the detailed risk assessment.

The minimum concentration causing adverse effect is equivalent to the highest concentration known to cause no adverse effect. Generally, concentrations associated with chronic toxicity are used. For aquatic life inhabiting surface waters and sediments, U.S. EPA water quality criteria are an excellent source of these data. Sediments should be evaluated using equilibrium partitioning for organic chemicals (Di Toro et al., 1991) or using simultaneously extractible metals-acid volatile sulfides for heavy metals (Di Toro et al., 1990, 1992). If there are no aquatic life water quality criteria for a chemical, there are alternate procedures available. For example, Parkhurst et al. (1994) discuss data sources; U.S. EPA has proposed a method for estimating aquatic life water quality criteria for chemicals for which there are limited data; and the OECD (1992) has developed a set of safety factors to apply to situations where data are limited. For toxicological data concerning human health, U.S. EPA's (1994b) IRIS is the standard source. Wildlife toxicological data are available in reports from the Agency for Toxic Substances and Disease Registry (e.g., ATSDR, 1991) and a series of reports edited by Eisler (1985) and the National Research Council (1980).

As Figure 5 shows, the SLRAs will identify the COPC for each site, medium, and receptor. COPCs are those judged to pose potentially significant risk, as a result of the SLRAs.

COPC are those with HQs exceeding specified magnitudes. If a chemical's HQ is greater than 1.0, then it automatically qualifies as a COPC. If it has an HQ less than 1.0, it may pose a negligible risk. However, to account for the possibility of synergistic and additive interactions affecting toxicity, chemicals will be considered to be posing negligible risk only if their HQs are less than or equal to 0.1. Thus, all chemicals with HQs less than or equal to 0.1 are COPCs in the SLRA. The results of the SLRAs can be expressed as risk rankings, which allow prioritization of the chemicals, sites, media, and receptors in terms of risks. Table 1 is an example of a

Table 1 Human Health Risk Ranking for Several Exposure Pathways

Alternative No.	Marine Life Score	Human Health Seafood Score	Swimming Score	Total Risk
18	17.5	18.3	83.3	119.1
5	16.4	0	83.4	99.8
9	11.6	0.62	83.3	95.5
12	10.7	0	83.3	94.0
13	71.9	0	1.82	73.7
14	1.31	0	8.34	9.7
11	4.44	0.25	1.68	6.4
15	2.08	0.21	1.67	4.0
4	2.11	0	0.02	2.1
17	0.49	0.70	0.01	1.2
10	0.74	0.02	0.01	0.77
1	0.55	0.07	0.01	0.63
16	0.45	0.23	0.01	0.69
2	0.53	0.07	0.01	0.61
Total Risk by Receptor Pathway	140.8	20.5	346.9	

human health risk ranking that ranks risks at a site according to several exposure pathways. These cumulative risk scores assume additive chemical toxicity. Additive toxicity is a reasonable assumption for aquatic life (Könemann, 1980, 1981; Alabaster and Lloyd, 1980) and HHRA (U.S. EPA, 1986b, 1989).

Although other interactions may be assumed, they are not amenable to prediction and must be based on chemical-specific toxicological tests. Because so few toxicity tests have been conducted with toxicant mixtures, and the mixtures that will be encountered at each site will be unique, the only method for confirming the interaction is through use of toxicity tests.

The cumulative HQs from the SLRAs are normalized* and summed for each site, medium, and receptor. The ranking's purpose is to rank the sites, receptors, and chemicals in terms of risk. Each ranked variable (e.g., site, receptor, etc.) is sorted from high to low. The rankings may be weighted by receptor, such that risks to one receptor could be weighted higher or lower than another receptor. Weighting reflects public policy judgments.

VI. CONCLUSION

The reliability of a risk assessment depends on the adequacy of the sampling design. No risk assessment can be more accurate or credible than its data. An adequate sampling design ensures that data collected are valid and capable of answering the

---

* Because the risk quotients will vary from one receptor to the next, due to use of different risk assessment methodologies, the risk quotients must be normalized so that the risk rankings are comparable. The quotients in Table 1 have been normalized.

questions posed in the investigation. Proper sampling design ensures QC, promotes acceptance by the regulatory authorities, provides useful data for the technical specialists, and promotes cost effective decision-making. Improper sampling design undermines the entire risk assessment and may result in the need to resample correctly with all the attendant delays and costs.

REFERENCES

Alabaster, J.W. and Lloyd, R., Eds. Mixtures of toxicants, in *Water Quality Criteria for Freshwater Fish*, Butterworths, London, 253, 1980.

Anderson. D.R., Proceedings: risk assessment in aquatic ecology, EPRI Report EPRI EA-4 t 38, Project,1826-12, Electric Power Research Institute, Palo Alto, 1986.

ASTM, *1993 Annual Book of ASTM Standards, Section II*, Water and Environmental Technology, ASTM, Philadelphia, 1993.

ATSDR, *Toxicological Profile for Cadmium*, Center for Disease Control, Atlanta, 1991.

Boyle, T.P., *New Approaches to Monitoring Aquatic Ecosystems*, ASTM Special Technical Publication 940, American Society for Testing and Materials, Philadelphia, a symposium sponsored by ASTM Committee E-47 on Biological Effects and Environmental Fate and by the Ecological Society of America, Minneapolis, MN, 17, 1985.

Cardwell. R.D. et al., Aquatic ecological risk, *Water Environ. & Technol.*, 5, 47, 1993.

Casas, A.I. and Crecelius, E.A., Relationship between acid volatile sulfide and the toxicity of zinc, lead and copper in marine sediments, *Environ. Toxicol. & Chem.*, 13, 529, 1994.

DeGraeve, G.M. et al., Variability in the performance of the seven-day fathead minnow (Pimephales Promelas) larval survival and growth test: an intra- and interlaboratory study, *Environ. Toxicol. & Chem.*, 10, 1189, 1991.

Di Toro, D.M. et al., Acid volatile sulfide predicts the acute toxicity of cadmium and nickel in sediments, *Environ. Toxicol. & Chem.*, 26, 96, 1992.

Di Toro, D.M. et al., Technical basis for establishing sediment quality criteria for nonionic organic chemicals using equilibrium partitioning, *Environ. Toxicol. & Chem.*, 10, 1541, 1991.

Di Toro, D.M. et al., Toxicity of cadmium in sediments: the role of acid volatile sulfide, *Environ. Toxicol. & Chem.*, 9, 1487, 1990.

Dutka, B.J. et al., Use of Bioassays to Evaluate river water and sediment quality, *Environ. Toxicol. & Water Qual.*, 6(3), 309, 1991.

Eberhardt, L.L. and Gilbert, R.O., Statistics and sampling in transuranic studies, in *Transuranic Elements in the Environment*, National Technical Information Service, Springfield, IL, 1980, 173.

Eisler, R. *Cadmium Hazards to Fish, Wildlife, and Invertebrates: a Synoptic Review*, U.S. Fish and Wildlife Service Biological Report No. 85(1), 2, 1985.

Elliott, J.M., Some methods for the statistical analysis of samples of benthic invertebrates, 2nd ed., Sci. Publ. Freshwat. Biol. Assoc. V.25, Freshwater Biological Association, Ambleside, England, 1977.

Gilbert, R.O., *Statistical Methods for Environmental Pollution Monitoring*, Van Nostrand Reinhold, New York, 1987, 320.

Gleit, A., Estimation for small normal data sets with detection limits, *Environ. Sci. & Technol.*, 19, 1201, 1985.

Green, R.H., *Sampling Design and Statistical Methods for Environmental Biologists*, John Wiley & Sons, New York, 1979, 257.

Helsel, D.R. and Hirsch, R.M., *Statistical Methods in Water Resources*, Elsevier, New York.
Herzog, B., Pennino, J., and Nielsen, G., Ground-water sampling in *Practical Handbook of Ground-Water Monitoring*, pp. 449-499, Lewis Publishers, Chelsea, MI, 1991.
Hornig, C.E., Macroinvertebrate inventories of the White River, Colorado and Utah, significance of annual, seasonal, and spatial variation in the design of biomonitoring networks for pollution detection, Environmental Protection Agency, Environmental Monitoring and Support Laboratory, Las Vegas, 1983.
Hurlbert, S.H., Pseudoreplication and the design of ecological field experiments, *Ecological Monographs*, 54, 187, 1984.
Hnemann, H., Fish toxicity tests with mixtures of more than two chemicals: a proposal for a quantitative approach and experimental results, *Toxicology*, 19:29-238, 1981.
Lettenmaier, D.P., Conquest, L.L., and Hughes, J.P., Routine Streams and Rivers Water Quality Trend Monitoring Review, Technical Report No. 75, C.W. Harris Hydraulics Laboratory, Dept. of Civil Engineering, University of Washington, Seattle, 1982.
Maddalone, R.F. et al., Defining detection and quantitation levels, *Water, Environ. & Technol.*, 5, 41, 1993.
McBratney, A.B., Webster, R., and Burgess, T.M., The design of optimal sampling schemes for local estimation and mapping of regionalized variables, Part 1, theory and method, *Computers & Geosciences*, 7, 331, 1981.
National Research Council, *Mineral Tolerance of Domestic Animals*, National Academy of Sciences, Washington, 1980.
Nelson, J.D. and Ward, R.C., Statistical considerations and sampling techniques for groundwater quality monitoring, *Ground Water*, 19, 617, 1981.
Noll, K.E. and Miller, T.L., *Air Monitoring Survey Design*, Ann Arbor Science, Ann Arbor, MI, 1971.
O'Connor, T.P., *Coastal Environmental Quality in the United States*, Chemical contamination in sediment and tissues, NOAA, Rockville, MD, 34, 1990.
OECD, Report of the OECD workshop on the extrapolation of laboratory aquatic toxicity data to the real environment, OECD Environment Monographs No. 59, Pans: Organization for Economic Co-Operation and Development, 43, 1992.
Parkhurst, B.R. and Mount, D.I., Water-quality based approach to toxics control, *Water, Environ. & Technol.*, 45, 1991.
Parkhurst, B.R. et al., *Methodology for Aquatic Ecological Risk Assessment*, Draft Final Report, Prepared for Water Environment Research Foundation, Alexandria, VA, 1994.
Provost, L.P., Statistical methods in environmental sampling, Chapter 9, in *Environmental Sampling of Hazardous Wastes*, Schweitzer, G.E. and Santolucito, J.A., Eds., pp. 79-96. American Chemical Society ACS Symposium Series, Washington, D.C.,V. 267, 1984.
Schlosser, I.J., Environmental variation, life history attributes, and community structure in stream fishes: implications for environmental management and assessment, *Environ. Manag.*, 14, 621, 1990.
Sokal, R.R. and Rohlf, F.J., *Biometry*, W.H. Freeman and Company, New York, 1981.
Stephen, C. E., *Ambient Water Quality Criteria for 2, 3, 7, 8 — Tetrachlorodibenzo-p-dioxin*, U.S. Environmental Protection Agency, Washington, 1985.
Suter, G.W., II., Futrell, M.A., and Kerchner, G.A., *Toxicological Benchmarks for Screening of Potential Contamination of Concern for Effects on Aquatic Biota on the Oak Ridge Reservation, Oak Ridge, Tennessee*, Oak Ridge National Laboratory, Oak Ridge, 1992.
U.S. Environmental Protection Agency, *Methods for Chemical Analysis of Water and Wastes*, Environmental Monitoring and Support Laboratory, Cincinnati, 1983.
U.S. Environmental Protection Agency, *Guidelines for the Health Risk Assessment of Chemicals Mixtures*, Federal Register, 51(185), 34014, 1986a.

U.S. Environmental Protection Agency, *Test Methods for Evaluating Solid Waste*, 3rd ed., Office of Solid Waste and Emergency Response, Washington, 1986b.

U.S. Environmental Protection Agency, *Risk Assessment Guidance for Superfund, Vol. 1: Human Health Evaluation Manual (Part A), Interim Final*, Office of Emergency and Remedial Response, Washington, 1989.

U.S. Environmental Protection Agency, *Risk Assessment Guidance for Superfund, Vol. 1: Human Health Evaluation Manual (Part B, Development of Risk-Based Preliminary Remediation Goals), Interim Final*, Office of Emergency and Remedial Response, Washington, 1991.

U.S. Environmental Protection Agency, *Risk Assessment Guidance for Superfund, Vol. 1: Human Health Evaluation Manual (Part C, Risk Evaluation of Remedial Alternatives), Interim Final*, Office of Emergency and Remedial Response, Washington, 1992.

U.S. Environmental Protection Agency, *Risk Assessment Guidance for Superfund: Vol. 1. Human Health Evaluation Manual (Part D, Standardized Planning, Reporting and Review of Superfund Risk Assesments) Interim*, Office of Emergency and Remedial Response, Washington, 1998.

U.S. Environmental Protection Agency, *Water Quality Guidance for the Great Lakes system and Correction; Proposed Rules*, Federal Register, 20802, 1993a.

U.S. Environmental Protection Agency, List of 126 CWA Section 307(a) priority toxic pollutants. Appendix P, in *Water Quality Standards Handbook*, 2nd ed., Office of Water, Washington 1993b.

U.S. Environmental Protection Agency, Office of Water Policy And Technical Guidance on Interpretation and Implementation of Aquatic Life Metals Criteria, Office of Water, Washington, 1993c.

U.S. Environmental Protection Agency, *National Guidance for the Permitting, Monitoring, and Enforcement of Water Quality-Based Effluent Limitations Set Below Analytical Detection/Quantitation Levels, Draft*, Washington, DC, 1994a.

U.S. Environmental Protection Agency, *Integrated Risk Information System (IRIS) On-line Computer Database*, Information updated regularly by EPA, Washington, 1994b.

Weber, B. and Juanico, M., Variability of effluent quality in a multi-step complex for wastewater treatment and storage, *Water Research*, 24, 765, 1990.

CHAPTER 13

Sampling for Ecological Risk Assessments

Jan W. Briede and Clifford S. Duke

CONTENTS

I. Introduction 319
  A. Sampling Design 320
  B. Vegetation Sampling 321
  C. Animal Sampling 321
  D. Fish Sampling 322
  E. Insect Sampling 322
  F. Bird Sampling 322
  G. Habitat Surveys 322
  H. Diet Determination 323
II. Conclusion 323

I. INTRODUCTION

For purposes of this chapter, sampling for ERAs encompasses environmental sampling and ecological sampling. Chapter 12 describes environmental sampling design, or the collection of samples mainly for chemical analysis. Many items discussed there apply to ecological samples, in particular the discussion of statistics. While environmental sampling deals with concerns such as chemical concentrations in soils, animals, or plants, ecological sampling is concerned with shifts in species composition or alleged disappearance of species caused by chemical or physical impacts.

This chapter focuses on indirect effects. For example, in a recent ERA of potential effects of pesticides on bats, the authors concluded that bats may be affected directly or indirectly. Direct effects result from ingestion, absorption, or inhalation of pesticides, while indirect effects could be caused by disappearance of insects, a major food source for these bats. Sampling schemes to determine direct effects differ from those aimed at indirect effects. Direct effects can be studied by capturing bats for chemical analysis. Indirect effects, on the other hand, require a detailed analysis of the number of insects in the area, analysis of bat diets through the collection of guano, and a comparison with background information.

Sampling design is a crucial item and may often be regulated, in particular when sampling for federally and state threatened and endangered (T&E) species. In that case, regulatory agencies often specify particular methods and periods during which sampling may take place.

A. Sampling Design

Ecological sampling for ERAs includes the determination of the species present in an area and the detection of population shifts caused by the stressor(s) under consideration. Therefore, it is usually essential to include background samples in the sampling design, in particular when examining the ecological effect of a chemical. Typically, background samples are collected in an area near the impact site (the site where the chemical release took place). A background sampling site (reference site) should ideally be located upstream or upwind, when sampling an aquatic or terrestrial habitat, respectively. If such an area is not available near the site, a distant site may be selected as long as the conditions at that site are relatively similar to those at the impact site. The sampling designs at the reference site and impact site should be similar, although the number of samples collected at a reference site may be lower. A proper statistical design should guide researchers in the decision making process. One problem that may be encountered involves organisms that are mobile. For instance, disappearance of a species or changes in ecosystem composition in an area may be difficult to prove when species can easily migrate into an area from adjacent sites.

Care should be taken to collect samples at the appropriate time. For example, sampling for a spring ephemeral bird in autumn or for a winter resident bird in summer will not produce any useful information and may result in the loss of credibility with regulatory agencies. When sampling for T&E species, agencies such as the U.S. Fish and Wildlife Service will specify the appropriate sampling time and method.

Generally, a good consultant can explain the methods that are proposed and/or required, and should be able to supply a contracting officer with information concerning statistical analysis, use of the data, and its shortcomings. Furthermore, the consultant should be able to supply a list of manuals that describe the various techniques. When animals are collected for analysis, commonly, a collector's permit is required. A contracting officer should ensure that the consultant has the needed permits for collection. Reports should clearly discuss methods and assumptions for the investigation and include a list of manuals and field guides used in the study.

B. Vegetation Sampling

Vegetation parameters usually investigated include species composition, vegetation production, and diversity. Vegetation sampling schemes abound. Most of them work; however, some are considered antiquated and should be avoided. Acceptable vegetation sampling methods include the line intercept method; the quadrate method; the nested quadrate method; and point sampling method. The line intercept method involves the measurement of the vegetation along a transect, while the point intercept method involves measurements at a point on the transect. The quadrate methods involve placing a frame (circular or rectangular) over the vegetation and identifying and counting the plants within the quadrate. Nested quadrates use a set of quadrates nested within each other to describe the different vegetation components (e.g., trees, shrubs, and herbaceous plants).

Method selection may depend on vegetation type, as do the size and number of the transects or quadrates. Rules and statistical formulas are available that aid in determining sample number and sample size. For example, in the deserts of the Southwest, many quadrates or transects will contain no vegetation. In that case, sample size needs to be increased to ensure that sufficient vegetation is sampled and results represent the actual vegetation. In a forest in the East, a few square meter quadrates will not result in a good representation of all the trees, shrubs, and herbaceous plants in the vegetation. An increased sample size may be required.

When used properly, most methods will allow for a comparison of diversity and species composition between reference and impact areas. The preferred way to estimate vegetation production includes both quadrate methods. Consultants should be able to explain why a method was chosen and supply background material for review.

C. Animal Sampling

There are as many methods for sampling animals as there are animals. Techniques range from observations along a transect to capture methods. Capture methods attempt either to capture the animal without inflicting harm (thus allowing release), or to kill the animal as humanely as possible. Traps for small animals include pit traps, Sherman traps, and snap traps. Snap traps are like mouse traps and will kill an animal; other methods may allow an animal to survive. A combination of various methods may be used to get accurate information on the animal population in an area.

Sometimes capture and recapture techniques are used. Using these techniques, animals are marked (e.g., banded) for possible recapture. Furthermore, telemetry studies may be warranted. Animals are tagged with a radio transmitter and the animals are followed for a predetermined period. Using radio telemetry, researchers can determine if a disturbance affects the natural movements of an animal. Special consideration is given to bats. These animals require capture at night using mistnets or harp traps (a harp trap looks like its musical counterpart).

D. Fish Sampling

Fish can be captured by using nets or electroshocking equipment. When used properly, both methods are capture-release techniques; however, fish are usually taken to the laboratory for identification and measurement. Sampling at various locations will ensure that a cross section of the total fish population is determined.

E. Insect Sampling

The various methods of insect sampling depend on the type of insect. Popular methods include bait traps, sticky traps, pit traps, sweeps, whitelight traps, and blacklight traps. The sampling methods employed depend on the objective of the study. One of the authors used black and whitelight traps to determine the species composition of night-flying insects. The different lights attracted different insects. Bait traps include fermenting bananas (sometimes impregnated with beer) which are particularly attractive to butterflies. Sticky traps resemble flypaper and are nonselective in capturing animals. Pit traps capture nonflying insects. Sweeping is a method commonly used by entomologists, where a person sweeps an area (the vegetation) with a net.

F. Bird Sampling

Birds are enumerated using a number of methods, including mistnetting, calling, and census. Specific protocols are in place for particular species and the consultant can probably determine the method after agency consultation. Mistnetting includes the placement of nets in certain flight corridors. Birds captured by the nets can be identified and released, or collected for analysis. Released birds may receive identification rings. Some birds will react to bird calls from human researchers or played on tape decks (e.g., the spotted owl and loon). Bird censusing involves traveling a predetermined transect at regular intervals to determine the presence of bird species along this path. Observations should be made concerning vocalization, display, nesting, and feeding behavior. While mistnetting and calling will capture or identify specific species, travel along a transect will give a cross section of birds present in a region.

G. Habitat Surveys

Sometimes it cannot be determined if a species is present. One cause for this may be that a survey needs to be conducted at a time when the species is not present in the area. One way around this predicament is to conduct a habitat survey. Most species have specific habitat requirements and a specific distribution. If a census needs to be conducted for a particular species, a habitat survey can usually suggest whether a species is likely to be present in the area.

H. Diet Determination

Occasionally, it may be useful to determine the diet of a species. A diet analysis may be needed in risk assessment to determine if exposure to a stressor is affecting the diet, and to test transport models. For instance, stressors can alter the behavior of an animal or change prey (food) availability. Three major methods are generally used to determine diet: bite count, stomach analysis, and analysis of feces.

Using the bite count method, researchers observe an animal and write down what the animal is doing/eating at set intervals (e.g., every minute). This method is mostly used with herbivores such as cows, sheep, deer, etc., and researchers generally take notes on specific species that are being eaten at that time. This method can also be used to determine the general behavior of an animal. Analysis of stomach content is usually done after an animal is killed, although some livestock researchers use fistulas to gain access to the stomach of live animals. Analysis of feces (guano, dung, or scat) uses microscopes. For instance, by examining bat guano, fox scat, or deer or mouse pellets, researchers can determine the diet of these animals.

Information on the appropriateness of a sampling method can usually be obtained from regional U.S. Fish and Wildlife Service (USFWS) offices, state departments of natural resources, state heritage departments, state foresters, state agricultural extension service staff, or from the biology, forestry, or agricultural departments of local colleges or universities.

II. CONCLUSION

Many different methods are available to detect the presence or estimate the population of a species. Just a few accepted methods are described in this chapter. Use personal judgement in deciding whether a method described by a consultant applies to the project at hand. A rule of thumb is that, if a consultant can satisfactorily describe a method and discuss its applicability and shortcomings, the method may generally be suitable. Methodologies developed after agency contacts, such as with the USFWS or state heritage departments, can usually be satisfactorily employed. When in doubt, get a second opinion from the agencies mentioned above or from another consultant.

It is difficult to prove that a species is not present. Not finding a species in an area does not mean it is not there. Only presence can be positively proven. However, the presence of a species in an area may be considered unlikely when a species is not found during the appropriate time and in cases when the proper habitat is not available.

CHAPTER 14

Ecotoxicity Testing in Risk Assessment

Guy L. Gilron and Ruth N. Hull

CONTENTS

I. Introduction 325
II. Ecotoxicity Testing: A Technical Review 326
    A. Basic Concepts 326
    B. Important Tools for Implementation 331
    C. Current Issues and Uncertainties 331
III. Consultant Selection 333
    A. Qualifications of Consultant 333
    B. Quality System 333
    C. Accreditation and Certification 335
IV. Conclusion 335
    References 337

I. INTRODUCTION*

Toxicity tests are controlled laboratory experiments in which organisms are exposed to a contaminant (or contaminant mixture) for a specified duration, in order to evaluate potential toxic effects. The type of toxic effect measured depends upon the test organism exposed, the contaminant concentration, and the mechanism(s) of action. Acute lethality in test organisms is a commonly-measured response, since it is relatively straightforward to measure and is biologically meaningful. Examples

---

* The authors wish to acknowledge Ms. E. Jonczyk (BEAK International, Inc.), and Mr. R. Scroggins (Method Development and Applications Section, Environment Canada) for their constructive and useful comments and suggestions on an earlier draft of this chapter.

of acute tests are fathead minnow acute lethality (aquatic) and earthworm survival (terrestrial). Measurement of sublethal test responses are becoming more widely used and generally provide a more sensitive response. Sublethal test responses include changes in growth, reproduction, and behavior. Examples of sublethal tests are daphnid reproduction (aquatic) and radish seedling germination (terrestrial).

Toxicity tests are used in both human health and ecological risk assessment, but in very different ways. In HHRA, data from mammalian toxicity tests are typically used to develop the RfDs and SFs that are used in the risk characterization phase to calculate risk. In ERA, ecotoxicity tests (nonmammalian) are conducted during the hazard/effects assessment phase, and become one of the lines of evidence for the risk characterization. Therefore, ecotoxicity tests are conducted during an ERA, but not during an HHRA. This primer will describe the use of toxicity tests for ERA only, since the HHRA RfDs and SFs are generally developed by the U.S. EPA; they are rarely generated on a site-specific basis.

Toxicity test results are used in ERA to provide an indication of whether the contaminated media are toxic. The other lines of evidence (i.e., biological field surveys and chemical measurements of ambient media) provide information regarding the actual state of the environment (e.g., whether a fish community is typical of unimpacted conditions) and which contaminants are likely to be responsible for observed toxic responses. However, only toxicity tests can directly evaluate whether the contaminated media are toxic to biota. An impact observed in the field may be the result of natural conditions, rather than contaminants. Similarly, chemical concentrations in environmental media often provide little information regarding the bioavailability of contaminants to ecological receptors. Exposing test organisms to environmental media, under controlled laboratory conditions, provides this information.

II. ECOTOXICITY TESTING: A TECHNICAL REVIEW

Figure 1 illustrates a conceptual diagram which represents the framework of an ERA. The shaded box indicates where toxicity testing provides information in a risk assessment.

A. Basic Concepts

1. Purpose of Conducting Ecotoxicity Tests

Ecotoxicity tests are used to measure the combined biological effects of substances present in environmental samples on terrestrial and aquatic plants, animals, and microorganisms. Test organisms used in these tests have become standard because they are:

- Generally representative of biota in soils, sediments, and water bodies
- Are indicators of specific trophic levels in the ecosystem food web
- Are easily maintained or cultured under laboratory conditions
- Are generally sensitive to environmental contaminants

ECOTOXICITY TESTING IN RISK ASSESSMENT

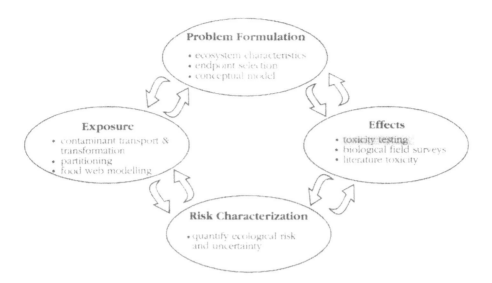

Figure 1 A conceptual diagram of the framework of an ecological risk assessment.

Since different organisms (or trophic levels) vary in their sensitivities to toxicants, it is common practice, especially in the hazard assessment phases of an ERA, to apply a battery (i.e., three or more) of tests to provide an estimation of potential toxic effect(s) for contaminants or substances being assessed.

2. Procedures for Implementing Ecotoxicity Tests

For a given test method, a predetermined number of organisms (e.g., 10) are exposed to each concentration in a dilution series (e.g., for effluent: 100%, 50%, 25%, 12.5%, 6.25%, etc.) of an environmental sample. The sample dilutions are prepared using "clean" dilution material (i.e., for terrestrial — silica sand; for aquatic — control/dilution water). Alternately, test organisms can be exposed to both 100% of an "impacted" site medium and 100% of a "reference" site medium; in this case, no dilutions would be required. At the end of the exposure period, the measured biological response of the test organisms is determined and usually expressed quantitatively as a statistically-derived toxicity endpoint (e.g., lethality, decreased reproduction, growth inhibition).

3. Exposure Duration

There are generally two major categories of toxicity tests with regard to exposure duration. Tests are either acute or chronic. Acute tests cause an effect within a short period in relation to the life span of the test organisms, while chronic tests cause effects which occur during a relatively long-term period of exposure, usually a significant portion of the life span of the organism (e.g., > 10% of its life cycle).

Acute tests (usually measuring lethality) are the most common type of toxicity test, and are used predominantly in regulatory testing. Chronic tests (those measuring sublethal responses, such as growth and reproduction inhibition) are becoming more common in environmental impact and risk assessments, and yield more information regarding the variety of biological effects of contaminant materials.

4. Test Endpoints Determined from Ecotoxicity Tests

The response of test organisms exposed to each concentration of sample is graphically plotted or statistically analyzed in order to estimate the concentration of the sample that produces a level or degree of response. For example, the most widely-used endpoint (in acute tests) is the LC_{50}, which is the concentration of the sample that would cause lethality in 50% of the test organisms during the exposure duration. Other endpoints, such as the no-observed-effect-concentration (NOEC), the lowest-observed-effect-concentration (LOEC), and the inhibition concentration (e.g., IC_{50}) are typical sublethal/chronic endpoints. The endpoints appropriate to the test method utilized are usually calculated using various computer programs (e.g., TOXSTAT).

5. Quality Assurance/Quality Control Program

In consideration of the high priority of ecotoxicity information in environmental risk decision-making, it is crucial that ecotoxicity data be of the highest possible quality. Therefore, a rigorous and comprehensive QA/QC program should be established in the laboratory conducting the tests. This program should comprise a number of important components, which include the following:

a. Standard Operating Procedures (SOPs)

Recognizing the necessity of standardized procedures for all aspects of laboratory operations, a full range of SOPs should be established; these SOPs are dated and an ongoing schedule of review (at least semiannually) is implemented with these SOPs. SOPs should be written for all laboratory-related procedures including: test methods; equipment calibration and maintenance; test organism care and culturing; procedures for handling, treatment, and storage of samples and reagents; and cleaning procedures for test chambers. All SOPs are updated whenever a significant deviation from conventional practice has been implemented to improve the performance or efficiency of the methods.

b. Testing Procedures

Specified methods for each ecotoxicity test should be on hand for reference in the testing lab. Recognized, published international, federal, provincial, state, and other agency test methods are used as appropriate (e.g., ASTM, U.S. EPA, Environment Canada, ISO, OECD). A list of examples of ecotoxicity test methods used in ERA is provided in Table 1.

Table 1 Examples of Ecotoxicity Test Methods Used in Ecological Risk Assessment

Taxon	Test Species	Test Methods*
Freshwater		
Fish	Rainbow trout	OECD, ASTM, US EPA, EC
	Fathead minnow	OECD, ASTM, US EPA, EC
Invertebrates	Daphnia magna	OECD, ASTM, US EPA, EC
	Ceriodaphnia dubia	OECD, ASTM, US EPA, EC
Algae	Selenastrum capricornutum	ASTM, US EPA, EC
	Duckweed	APHA
Marine		
Fish	Inland silverside	US EPA
	Sheepshead minnow	US EPA
	Threespine stickleback	EC
Invertebrates	Sea urchin/Sand dollar	US EPA, EC
	Amphipod	US EPA, EC
	Mussel	ASTM
Algae	Skeletonema costatum	US EPA
	Champia parvula	US EPA
Freshwater Sediment		
Invertebrates	Hyalella sp.	ASTM, US EPA, EC
	_Chironomus sp.	ASTM, US EPA, EC
Soil	Earthworm	OECD, US EPA
	Eisenia sp.	OECD, US EPA

Note: APHA = American Public Health Association
ASTM = American Society for Testing Materials
EC = Environment Canada
OECD = Organization for Economic Cooperation and Development
US EPA= U.S. Environmental Protection Agency

Individual bench sheets for recording chemical and biological data should be provided for each test, and originals should be stored in a central file. Bench sheets document date of sample receipt, date of test initiation, chemical and physical exposure conditions, name of test technician(s) monitoring tests, test observations, and comments identifying unusual observations or deviations from the SOP.

c. Test Organisms

Records on the history of test organisms should be maintained on file and include the quantity and source of each shipment, date of arrival, nature and date of health checks performed, any health certificates, and details of the acclimation history. New stocks of organisms, obtained from off-site suppliers, should be acclimated to the laboratory's holding conditions for at least the minimum period of time specified by the corresponding test method. Laboratory facilities and test stocks should be inspected by the local animal care regulatory body. Unhealthy organisms should never be used in toxicity tests, and survivors of testing are never to be re-used.

d. Quality of Dilution Media

Water/soil/sediment quality and other conditions necessary to the survival of the organisms should be maintained and documented. For example, dilution water for many tests is reconstituted water (water prepared with specific characteristics), while dilution medium for sediment and soil tests is usually silica sand. Laboratory dilution water and/or artificial soil/sediment should also be analyzed for contaminants according to routine and frequent schedules. Also, physical and chemical characteristics (e.g., temperature, pH, dissolved oxygen, conductivity, and light intensity) should be monitored according to laboratory SOPs, especially in stock test-species cultures. Dilution medium quality data should be recorded in a central logbook, and when necessary, on bench sheets. The data are used to report current water/soil/sediment chemistry characteristics at the time of testing on an as needed basis for single tests, or as a matter of routine.

e. Control Response

The control response in an ecotoxicity test is analogous to the blank criterion in chemical analysis. All testing should be conducted using the negative (clean) control vessel consisting of organisms, handled and treated in an identical manner as test sample-exposed organisms, but exposed only to dilution water/artificial soil/sediment. The complete test is usually repeated if more than 10% of control organisms die (or show evidence of sublethal effects), in the case of acute tests, or if more than 20% control mortality occurs in the case of chronic tests. Sediment and soil tests are conducted similarly using clean sediments/soils or artificial soil as controls.

f. Reference Toxicant Testing

Reference toxicant testing should be performed on a regular basis to demonstrate consistency in test performance (e.g., within a defined and limited range of variability) that might be affected by such influences as: changes in test organism sensitivity over time as a result of size, reproductive status, etc.; genetic differences in sensitivity between stocks of organisms obtained from different sources; and, performance of technical staff during training. Control charts should be established and regularly updated to demonstrate that test reproducibility is within established limits. Test-specific, standard reference toxicants should be used and reference tests should be conducted at regular intervals. Stocks of test organismsm which are not cultured in-house, should be tested shortly after organism acclimation to laboratory conditions, and towards the end of stock utilization (as well as monthly) as long as the organism supply lasts.

g. Interlaboratory Test Performance

The ecotoxicity laboratory should, whenever possible and practical, participate in interlaboratory split-sample testing of reference chemicals and toxicant mixtures. Interlaboratory test rounds should demonstrate reasonable agreement with other laboratories using standard test organisms.

B. Important Tools for Implementation

Various pieces of equipment may be necessary for the smooth implementation of ecotoxicity tests. Apparatus commonly used include: environmental chambers (ranging in size from diurnal chambers to full-size rooms), water baths, temperature and light control devices, aeration systems, water and soil quality monitoring equipment, and other specialized devices, such as continuous-flow apparatus.

Although most test methods provide flexibility with regard to the level of equipment sophistication, it is highly desirable that apparatus used in an ecotoxicity laboratory have, as a minimum, the following characteristics:

- Available from a reputable supplier (i.e., replacement parts and customer service readily available)
- Supplied with detailed instructions on operation and maintenance
- Amenable frequent and precise calibration and QC checks
- Cost-effective

C. Current Issues and Uncertainties

There is a wide range of emerging issues in the rapidly growing field of ecotoxicity testing; however, many of these are beyond the scope of the present chapter. Some critical issues are highly relevant in the consideration of ecotoxicity test data and their use in ERA. These are discussed briefly below.

1. "Battery of Tests" Approach

Traditionally, toxicity evaluations have relied upon single-species testing. More recently, in consideration of the complexity of ecosystems and their response to environmental toxicants, a "battery" or "suite" of tests approach has been adopted by many scientists in this field. These test batteries yield greater information, since they encompass ecosystem component effects, which would not be detected through single-species evaluations. Therefore, wherever practicable and feasible, an ecologically appropriate battery of tests should be selected for a site, which will provide a range of dose-response relationships in a range of ecologically-diverse biota. For example, a freshwater toxicity assessment would likely include a battery of three tests comprising: an aquatic vertebrate test (fish acute lethality), an aquatic invertebrate test (a crustacean reproduction test), and an aquatic plant test (algal growth inhibition).

2. Test Sensitivity

In selecting ecotoxicity tests, the risk assessor should evaluate the sensitivity of potential tests to be used. In particular, is the test sensitive to a wide range of toxicants, or is it "toxicant-specific" (with a very specific mode of action)? Some test systems are highly sensitive to one toxicant group (e.g., heavy metals), while relatively insensitive to another toxicant group (e.g., organic compounds). This test sensitivity can have a significant impact on the resultant assessment of risk. For example, it has been demonstrated that aquatic plants are "sensitive" to heavy metals; in other words, plants can be useful species for evaluating heavy metal contamination. If the risk assessor has this information prior to conducting a risk assessment with, e.g., a mine decommissioning project, a green algal growth test can be used to evaluate potential heavy-metal contamination.

3. Ecological Relevance

Although there is a temptation to use standardized, commonly-used test procedures, it is highly desirable to select ecotoxicity tests that yield the most ecologically-relevant data. Tests used to evaluate a site should, whenever possible, utilize test species found in (or near) the receiving environment that is being evaluated. If one is not available, it is advisable to choose a test species similar, either on the basis of ecological niche or taxonomic group, to those in the receiving environment. The concern is that it is sometimes difficult to replace a standard test species with an indigenous native species and still maintain acceptable control sample response.

4. Test Reproducibility

It is crucial that ecotoxicity data be reproducible (i.e., that any laboratory conducting the test in question is able to obtain comparable results). Therefore, it is recommended that tests used in a hazard assessment be highly reproducible. Due to the rapid advances in this field, our knowledge with regard to test reproducibility is not uniform (i.e., based on test validation, large database of test data). Therefore, it is advisable to consult with an experienced ecotoxicologist to determine which tests have proven reproducibility.

5. Logistics

In addition to the above-mentioned issues, the practicability of ecotoxicity testing procedures should also be considered. The timing and cost-effectiveness of the proposed testing should be a major consideration in selecting test procedures. For example, is a test requiring sophisticated and expensive equipment going to yield more valuable information than a simpler, less expensive test?

6. Laboratory-to-field Extrapolations

Uncertainties surround the validity of extrapolating an artificial, laboratory-controlled experiment (i.e., an ecotoxicity test) to actual impacts in the natural environment, where other biotic (e.g., predation) and abiotic (e.g., humidity) parameters cannot be controlled. In this regard, *in situ* experiments (e.g., *in situ* toxicity tests, mesocosms, and artificial streams) would be highly preferable; however, these studies are often very expensive.

III. CONSULTANT SELECTION

In the case of ecotoxicity testing, the "consultant" or "supplier of services" is usually an ecotoxicological laboratory, which may or may not be a part of a larger environmental laboratory or consulting company. A careful evaluation of the laboratory, its operating capability, and the qualifications of its staff are paramount for ensuring high quality ecotoxicity data (see Table 2).

A. Qualifications of Consultant

All laboratory personnel should have education, training, and experience commensurate with their assigned functions in the laboratory. Resumes, job descriptions, diplomas, and other special certification of all individuals working for the laboratory, should be maintained in a personnel file and updated regularly.

B. Quality System

The ecotoxicity laboratory should have a comprehensive, fully-documented, Quality System in place, which includes: a Quality Manual, outlining in detail all of the components of the laboratory's Quality System; a management policy statement, indicating full support for the Quality System; and, a Quality Manager (Unit), responsible for implementation of the Quality System.

Under Good Laboratory Practice (GLP) principles, projects must have a Study Director (or project manager) who ensures that the study is executed according to the procedures and test methods established with the sponsor (or client). The Study Director is responsible for the technical quality of the project and ensures that all project objectives are met. The Study Director also ensures full compliance with QA/QC requirements (see Table 3). In the event of any unforeseen circumstances or responses, records are kept and appropriate actions taken. Project records are regularly updated by the Study Director with respect to findings, schedule, and budget. The Quality System should also specifically identify a laboratory QA Officer (or Unit) who does not directly supervise laboratory staff or deal with laboratory matters on a routine basis, and therefore, provides objective evaluation.

Table 2 Consultant Selection Checklist

PERSONNEL	Organizational and Management Structure:	Is there a clear and well-defined organization structure for the laboratory? Is this structure reflected in an organizational chart?
	Staff Qualifications:	Do staff have qualifications commensurate with their roles in the laboratory (see Table 1 above)?
		Is there an ongoing training program?
	Quality Assurance Officer/Unit:	Does the laboratory have a Quality Assurance Officer or Unit? Is the Officer/Unit independent of laboratory work?
		Are there accurate records kept for all laboratory equipment?
METHODS	Standard Operating Procedures (SOPs):	Does the laboratory have written, comprehensive SOPs?
		Are SOPs established for all procedures implemented in the laboratory?
		Are SOPs routinely and frequently updated?
		Are SOPs reviewed and signed by the QA Officer/Unit?
	Organism Health Criteria:	Are test organisms obtained from reputable and registered suppliers?
		Are test organisms acclimated to lab conditions prior to testing?
		Are accurate records kept for organism acclimation?
		Are there stringent criteria for establishing organism/culture health?
	Dilution Medium Quality:	Does the laboratory have established dilution medium quality criteria?
		Is the quality of dilution medium monitored routinely and frequently?
	Statistical Methods/Software:	Are standard statistical methods used in the calculation of ecotoxicity test results?
		Are calculations and statistical outputs cross-checked for data entry and/or other potential errors?
		Are the methods/software updated regularly?
	Archiving:	Are all bench sheets, study reports, QA/QC data, and other documentation archived?
		Is there a security system in place to address access to archives (both hard copy and electronic format)?
QA/QC PROGRAM	Quality Manual:	Does the laboratory have a Quality Manual outlining (in detail) the Quality System?
		Is the Quality Manual routinely and frequently updated to complement changes in laboratory procedures?
		Is the Quality Manual available for sponsor/client review?

Table 2   continued

	Accreditation/ Certification:	Does the laboratory maintain "second- or third-party" accreditations/certifications?
		Are certifications based on site audits? performance evaluation samples? management review?
	Interlaboratory Testing:	Does the laboratory participate in interlaboratory ("round-robin") testing?
		Do the results obtained compare favorably with other laboratories?
	Internal/External Auditing:	Does the laboratory operation conduct internal audits as part of its QA/QC program?
		Are the results of these audits (including follow-up actions) available for sponsor/client review?
		Does the laboratory permit/encourage external audits from regulatory personnel and/or clients?

Responsibilities and authorities among staff should be clearly identified in an organizational chart and as part of a Quality System. Assignment of authority should involve a signature system of data certification. Each signature verifies that the responsible individual has performed his or her assigned quality assurance function, and is satisfied with the quality of the data as stated, with interpretation provided.

C. Accreditation and Certification

In order to establish and maintain a proper Quality System, it is often useful for the laboratory to participate in "second- or third-party" accreditations and/or certification programs. Accreditation programs recognize the competence of an ecotoxicity laboratory to carry out specified tests. The accreditation is based on an evaluation of laboratory capability and performance evaluations. Certifications (specifically, quality certifications) are also a recognition of the laboratory's proficiency, but focus more on management systems practices. Accreditations and certifications keep the Quality System of the laboratory up to date, and ensure that the laboratory has an established QC Program, and follows standardized QA guidelines. Table 2 presents a checklist that can be used during the review of a proposal or statement of qualifications, or a precontract laboratory audit, for selecting a consultant to conduct ecotoxicity testing in support of an ERA.

IV. CONCLUSION

When toxicity tests are conducted using appropriate species, standardized test methods, and in accordance with quality assurance/quality control (QA/QC) measures, the risk assessor is better able to evaluate the risks to ecological receptors at a site. The combination of chemical and biological response data results in a more credible and scientifically-defensible risk assessment. Risk managers will then have confidence in the decisions they make based on the conclusions of the risk assessment.

Table 3 Discipline Checklist for QA/QC Work Product

Client and sample information
_ Client name _ Sample name and description _ Collection method _ Collection date and time _ Sample collector

Documentation for chain-of-custody of environmental samples
_ Time and date of receipt _ Indication of testing to be implemented _ Condition of sample _ Signature of receiver

Test method reference
_ Complete citation of test method used with an indication of level of compliance with method and laboratory SOP used (if applicable)

Summary of test conditions
_ Complete summary of laboratory-specific (not generic) test conditions, including: duration, test organisms, physico-chemical conditions (and monitoring) during the test, reference toxicant used, test validity criteria

Reference toxicant data
_ Name of reference toxicant _ Most recent reference toxicant results with an indication of agreement with laboratory control charts

Data verification
_ Statement that data entered into statistical programs and data reports have been cross-checked to screen out errors _ Signature of data analyst or laboratory manager

Results including water/soil/sediment quality monitoring
_ Raw data upon which endpoints are based (e.g., # dead, # neonates produced, weight of individuals) _ Raw data of physico-chemical parameters monitored _ Calculated endpoints with full citations of statistical procedures employed _ Relevant comments and observations

Signature of verification (study director, QA officer, quality manager)
_ Final sign off by laboratory representative (as above), guaranteeing that all QC checks have been implemented

REFERENCES

American Society for Testing and Materials, *Standard Guide for Conducting Sediment Toxicity Tests with Freshwater Invertebrates*, American Society for Testing and Materials, Philadelphia, 1992.

Environment Canada, *Biological Test Method: Test of Reproduction and Survival Using the Cladoceran Ceriodaphnia Dubia*, Environmental Protection Service, Ottawa, Ontario, 1992.

Greene, J.C., et al., *Protocol for Short Term Toxicity Screening of Hazardous Waste Sites*, Office of Research and Development, Environmental Research Laboratory, Corvallis, OR, 1989.

Keddy, C., Greene, J.C., and Bonnell, M.A., *A Review of Whole Organism Bioassays for Assessing the Quality of Soil, Freshwater Sediment and Freshwater in Canada*, Ecosystem Conservation Directorate, Evaluation and Interpretation Branch, Ottawa, Ontario, 1994.

Klemm, D.J., et al., *Short-term Methods for Estimating the Chronic Toxicity of Effluents and Receiving Waters to Freshwater and Marine Organisms*, 2nd ed., Environmental Monitoring Systems Laboratory, U.S. Environmental Protection Agency, Cincinnati, 1994. .

Organisation for Economic Cooperation and Development (OECD), *OECD Guidelines for Testing of Chemicals*, Paris, 1987.

Weber, C.I., *Methods for Measuring the Acute Toxicity of Effluents and Receiving Waters to Freshwater and Marine Organisms*, 4th ed., Environmental Monitoring Systems Laboratory, U.S. Environmental Protection Agency, Cincinnati, 1993.

CHAPTER 15

Epidemiology and Health Risk Assessment

Robert A. Kreiger

CONTENTS

I. Introduction 339
II. The Relationship of Epidemiology and Risk Assessment 340
    A. Using Epidemiology in Health Risk Assessments 340
    B. Working with Consultants 342
    C. Data Collection and Evaluation 343
    D. Exposure Assessment 344
    E. Toxicity Assessment 345
    F. Risk Characterization 346
III. Conclusion 348
    References 348

I. INTRODUCTION

The discipline of environmental epidemiology connects risk assessment practice with pure scientific research. Epidemiology has little direct relationship to conducting a risk assessment. Very few projects require a full scale epidemiological study. Even so, project managers should understand how epidemiological studies affect the risk assessment process. Technical comparisons of epidemiological data and animal bioassay results may play an important part in certain projects. Epidemiological studies can be used to set toxicity values (i.e., cancer potency slopes or reference doses) or to classify a carcinogen. Also an epidemiologist's perspective may also be required in certain risk assessment projects. A key project management decision is whether to include an epidemiologist in a project team.

As currently practiced in the United States, risk assessment does not treat proven human carcinogens differently from suspected human carcinogens. The choice to use risk assessment or epidemiological methods may present a project manager with significantly different options on a project. This choice typically arises in complex projects with high visibility and overt liability. At other times, there is no choice. For example, risk assessment is useful for setting preliminary site remediation goals for soil or groundwater, but it may be an inappropriate response to community concern about a local cluster of cancer cases.

In scientific literature of the 1980s, health scientists proposed the integration of epidemiology and health risk assessment methods. Unfortunately, epidemiological methods have still not been integrated into health risk assessment processes. Risk assessment and epidemiology remain discrete approaches with certain intersecting components. Risk assessment affects the practice of epidemiology far more than epidemiology affects risk assessment methodology.

II. THE RELATIONSHIP OF EPIDEMIOLOGY AND RISK ASSESSMENT

Risk assessment and epidemiology are alternate methods for evaluating risks and impacts from known or suspected chemical exposures. Both methods can relate chemical exposures to health effects, and both influence regulatory policies related to chemical exposures. However, epidemiology and health risk assessment have different objectives and use different strategies to link chemical exposures to resulting health consequences. A summary diagram of the relationship between epidemiology and risk assessment appears in Figure 1.

Epidemiology is the study of the distribution of disease in human populations. It is a science that attempts to prove the causes of disease by measuring the consequences of actual chemical exposures. The objective of epidemiology is to create and formally test hypotheses about disease distributions. Epidemiology is a descriptive science. It counts physical events such as deaths, cancer cases, lost days at work, or other recorded data. Epidemiological studies also quantify the factors affecting disease development. These factors can include exposure to chemicals, but may also include genetic, nutritional, and other lifestyle parameters. Epidemiological studies are notoriously slow and expensive.

Risk assessment is not a science. It blends numerous disciplines. Risk assessments cannot be proved or disproved. The objective of risk assessment is to prevent disease from occurring. Risk assessment is predictive, not descriptive. It attempts to estimate the probability of future harm resulting from hypothetical exposures to a particular chemical or source of exposure. The context for risk assessment is rapid decision making in situations involving considerable uncertainty. Risk assessment does not reduce uncertainty in practice, but does provide a tool for dealing with uncertainty.

A. Using Epidemiology in Health Risk Assessments

The main benefit of including epidemiologic data in risk assessment is that epidemiologic data relates directly to human experience. Most risk assessment data is

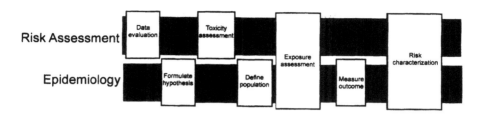

Figure 1 Relationship between epidemiology and risk assessment.

from animal studies. Various mathematical manipulations of animal data, such as species-to-species extrapolations, are used to apply the data to people. The scientific validity of these manipulations cannot be determined. Since epidemiological studies measure observed human responses to chemical exposures, no manipulations are required. This reduces a potentially significant source of uncertainty in risk assessment. Epidemiologic data should be used, if possible, to confirm exposure effect profiles and dose responses obtained from animal studies.

Epidemiologic data has several other benefits. It is used to identify sensitive groups within exposed populations. Animal studies cannot identify human attributes that make some individuals more sensitive than others to chemical exposures. Epidemiology can also indicate the relative importance of lifestyle, genetic, and behavioral factors on responses to chemical exposures. It is the definitive source of information for human dose response and hazard evaluation in risk assessment, when the exposure assessment is sufficient. Epidemiological data can also provide a context for risk assessment processes and results.

Epidemiologic data, however, presents serious pitfalls for the inexperienced user. Few risk assessors have sufficient training in epidemiologic science to avoid these problems. For example, scientific journal articles using epidemiological data to prove or refute risk assessments based on animal studies are now fairly common. Unfortunately, the authors seldom address the tendency of environmental epidemiological studies to underestimate exposure risk. Underestimation of risk usually results from small population sizes, crude exposure-estimation methods, and confounding of small dose response effects by other factors. These problems produce a high level of uncertainty in the results of most environmental epidemiology studies and make it difficult to demonstrate a statistically significant risk increase from exposure. A finding of "no significant increase in risk" due to an environmental exposure is common. The epidemiologist must be cognizant of the fact that negative results can arise from study design limitations.

Reconciling epidemiological studies also requires skill. Integrating disparate results and results from different study types presents particular difficulties. Results from occupational exposures to high chemical concentrations producing large effects must also be interpreted with caution. Valid extrapolation from high to low dose effects in epidemiological studies requires a high degree of training and specific skill.

B. Working with Consultants

An epidemiological consultant has two primary roles related to applied project work. The most common role is reviewing and evaluating epidemiological data used to set site-compliance criteria or establish regulatory or legal strategies. In some cases, this consists of reviewing the use of epidemiological data by a regulatory agency. Conversely, if regulatory and legal actions are based only on animal test results, and fail to fully evaluate existing human data, the consultant may have a role in preparing alternative criteria proposals based on current epidemiological information.

A second, more demanding role is for the consultant to determine the appropriate response level if public health impacts from site releases are likely. This most typically occurs when a community expresses concerns over perceived disease clusters. Appropriate responses can range from establishing a dialogue with affected parties to collecting data for a formal feasibility study of the epidemiological issues. The need to carry a site release investigation forward to a full epidemiologic study is a rare event, but it does happen. A competent epidemiological consultant will guide you to selecting the best response to the situation. This is usually a step-wise process satisfying the concerns of the affected parties and regulatory community.

Five situations trigger involvement of an epidemiologist in the risk assessment process:

- When known past or current human exposure to site releases occur
- When class-action litigation claiming health effect damages is pending
- If the Agency for Toxic Substances and Disease Control (ATSDR) is planning, or completed, a public health assessment of the site
- If a public agency or university is, or has been, investigating disease rates in a community near the project site
- If site chemicals of concern produce short-term exposure effects

Also, consider using an epidemiologist when the community or media are actively scrutinizing site related activities, when local residents report subjective symptoms they attribute to site releases, when public complaints have been filed with regulators related to site emissions or conditions, and when the site COC have ARARS or other key criteria based on epidemiological studies.

An epidemiologist skilled in the completion of biomarker studies may also be needed when past or current population exposures are known to occur, but where analytical exposure data are weak or missing. For example, mass balance facility data and site demographics may indicate a probable air exposure, but no air samples were taken of either stack emissions or ambient air. Instead of trying to construct an exposure scenario in absence of this data, the epidemiologist can conduct an exposure assessment by measuring blood enzymes, urine chemicals, effects on blood cells, or some other index of biological change in the exposed group resulting from exposure.

A project epidemiologist has two types of work product. As a consultant to the risk assessment team, the epidemiologist advises on the necessity for intervention and health studies as part of a larger project context, serving as a link between community health concerns and successful completion of the risk assessment project. The epidemiologist may actually conduct a public health evaluation in parallel with the project risk assessment effort. The second area of responsibility is more familiar. Epidemiology can be a component part of all the standard risk assessment tasks. A description of this role is described in the following sections.

C. Data Collection and Evaluation

Epidemiologists assist a risk assessment project team in the identification of potential COCs. A comprehensive data search is usually conducted using computer databases to obtain the most current information possible. Regulatory agencies are slow to modify potency slopes and reference doses. The epidemiologist will look for studies conducted after the last regulatory review. The epidemiologist will also scan the data base for epidemiological data for chemical mixtures that may be relevant for the project, and population factors (age, gender, race, lifestyle) that affect expression of toxic effects. In most cases, chemicals will be identified that do not have regulatory reference doses, potency slopes, or ARARS, but do have indications of human toxicity potential. The epidemiologist will work with the project toxicologist to derive a process for including or excluding a chemical from further evaluation. The epidemiologist also has a role in characterizing the expected hazards resulting from chemical exposure at the concentrations relating to site conditions. The work product for this step in the risk assessment is a list of COCs cross-referenced to documented human exposure consequences, and correlated to exposure levels. It has become customary to produce a small encyclopedia of toxicology for the COCs as a product of this risk assessment step. Most of these have included horrific descriptions of human exposure consequences without relating effects to dose. The epidemiologist has a primary function in tempering these lurid lists of adverse effects with common sense discussions of the exposures producing the effects. In addition, the epidemiologist may relate toxic concentrations to other factors like odor detection thresholds, and analytical detection limits. Figure 2 shows an example of such an evaluation.

An epidemiologist may produce a list of COCs based on epidemiologic studies, or a list of chemicals with documented human toxicity, but no quantitative data on exposure or dose which may be candidates for qualitative risk assessment. At times, an epidemiologist may also:

- Identify data sources linking effects and symptoms to specific exposure levels
- Provide insight into the relationship between exposure intensity, exposure duration, and exposure pattern (these types of insights may be critical if regulatory criteria equate intensity with duration, and epidemiological studies show that these factors are interdependent)
- Identify populations of sensitive subgroups
- Identify interactions among components of chemical mixtures
- Address variability of human responses in large studies (this affects the legitimacy of using a 95% confidence interval of the human dose response extrapolation for potency slope or reference dose calculations)
- Provide a basis for modifying established regulatory criteria due to newer epidemiology data that were not considered during the promulgation of a reference dose or cancer potency slope

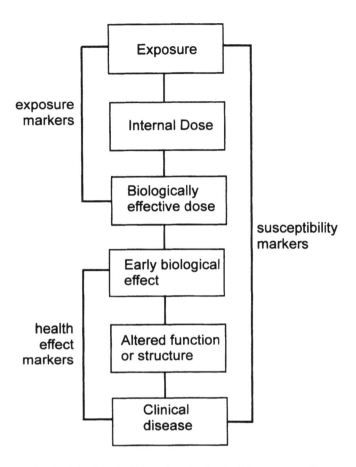

Figure 2 Example of epidemiological hazard evaluation in risk assessment.

D. Exposure Assessment

Other disciplines will adequately characterize the physical exposure setting. Epidemiology has a particularly important role in identifying the exposed populations in a risk assessment, and describing behaviors that affect exposure potential. These tasks are particularly important when the target population is not expected to correspond to standard default exposure assumptions.

The epidemiologist may evaluate the physical parameter match between the potential target population and the EPA exposure assessment defaults. Specifically, EPA risk assessment practice frequently ignores epidemiological data on body weight, tap water consumption, and age-specific respiration rates. The standard use of default values of 70 kg body weights, 2 l/day tap water consumption, and 20 m^3/day respiration volume still predominates in agency risk assessments, despite the existence of better validated data. If the target population is not well represented by an 18-year-old, white, male model, the epidemiologist has a vital role in selecting relevant physical parameters for the exposure assessment. The project epidemiologist can evaluate site demographic data to select appropriate physical parameters for the exposure assessment.

The project epidemiologist can review local demographic trends to construct a relevant exposure duration estimate. Exposure frequency and duration are largely a function of human activity. The current risk assessment paradigm expends tremendous resources on collecting site specific chemistry data, and almost no effort on characterizing specific exposure patterns. Target populations containing commuter groups, occupational cohorts, or transients are not well represented by the current risk assessment paradigm. Similarly, the average time at one residence is location specific. Similarly, if food chain exposures are likely, local food consumption patterns can be determined by the epidemiologist.

In many cases, the project manager must attempt to estimate exposures without adequate field measurements. Release sources may be poorly characterized, or highly variable with time. Data at the release site may be of poor quality or totally absent. An estimate of the mass of release is not even available. When site-release data are missing, or the steady-state assumptions do not make sense, the quantitative modeling of chemical transport from the release point to the target population is not usually practical. The project epidemiologist can provide an alternative approach for exposure assessment. This alternative involves testing exposure directly in the affected receptor population. For this option, urine, blood, hair, or other human media can be tested for the presence of site chemicals, their byproducts, or unique effects. The project epidemiologist can determine if these alternative approaches are practical for a particular project, and design a work plan to accomplish this type of exposure assessment.

The project epidemiologist can also determine how the exposure factors interact. For example, most risk assessments poorly define what a reasonable maximum exposure is for a particular target population. The definition of a reasonable maximum exposure is usually derived in absence of data and is based on unvalidated assumptions. The epidemiologist can determine how factors like age, body weight, and gender interact with behavioral patterns to influence exposure. A customized population exposure estimate can then be constructed using site demography, and reasonable maximum exposures can be calculated for well- defined population subgroups.

E. Toxicity Assessment

This stage of risk assessment usually involves hazard identification and dose response calculation. The primary contribution the project epidemiologist can make to this stage of the risk assessment process is to update the existing database used for deriving regulatory criteria. Another contribution is to help establish no observed effect levels (NOEL) and lowest observed effect levels (LOEL) for common symptoms related to exposure to site chemicals. This information would be useful in responding to symptoms reported in the potentially exposed populations. It would also be important to compare these results with animal test results. Inconsistencies in the response between human and animal tests of exposure consequence would have impacts on the risk characterization discussion. The epidemiologist can have a significant role to play in describing the certainty of the risk assessment results, and the interpretation of the risk assessment product.

If the chemical in question has no EPA reference dose or cancer potency value, but there is sufficient human dose response data, dose response data must be developed. This will be a particularly important function for sites with chemicals producing acute irritation effects. Similarly, many toxic chemicals are in regulatory limbo, with no established formal dose response criteria. Site responses may have to address these chemicals before definitive reviews have been completed. The epidemiologist may have to generate dose response data in these cases.

The project epidemiologist may need to revise uncertainty factors used in reference dose calculations, based on newer data. Similarly, the use of HEAST data requires review and interpretation. The project epidemiologist reviews any HEAST determinations involving human exposure effects and dose responses. Human dose response data may have to be adapted in cases where exposure is for less than a lifetime. Since most human exposure data is derived from occupational studies, the project epidemiologist needs to interpret the importance of exposure duration.

F. Risk Characterization

Typically, this is usually the most complex and poorly executed task in the health risk assessment process. The project epidemiologist can help to convert risk characterization into the most meaningful stage of the risk assessment process. Work tasks would include dealing with acute effects and making sure exposure pathway combinations make sense to the interested stakeholders. Compounded nonsensical exposure scenarios have been common in the past. For example, some risk assessments have been based on a hypothetical individual living on an island next to a source area, subsisting on fish caught from an impacted river, drinking polluted groundwater, eating only homegrown vegetables affected by site releases, from birth to death, without leaving the island. Public reaction to this scenario ranged from amusement, to confusion, to outrage. It failed as a risk characterization tool because no one could take it seriously. By looking at the site demographics, the project epidemiologist can make sure that the driving exposure scenario is one that makes sense.

It is important that the project epidemiologist provides interpretation guidelines for affected parties not modeled in the risk assessment. For example, most readers of the assessment are at a loss if their characteristics do not match the risk assessment model. The project epidemiologist can help to explain issues like less-than-lifetime exposure consequences, effects of background exposures, group sensitivities based on genetic attributes, and lifestyle complications (smoking, drinking, diet), based on the human effects data in hand.

The project epidemiologist should evaluate the logic behind combining noncarcinogenic exposure effects into a hazard index. The list of chemicals producing common effects in animals should be checked against the human exposure data to assure consistency. The creation of a hazard index that has value to the risk assessment stakeholders is a major concern. The common practice of slopping unrelated effects into a communal hazard index confuses the risk communication process, and damages the credibility of the risk assessment. The epidemiologist has a critical role in how the uncertainty for the risk assessment is explained. The human perspective on exposure effects from multiple chemicals, the contrasts between animal and human carcinogenesis, and the magnitude of site exposures compared to other sources are all important issues.

Epidemiology can sometimes help determine the most likely risk estimate, instead of just the upper bound 95% confidence level risk estimate. This is particularly important for chemicals with a sufficient human exposure data base.

Risk could be characterized based on local epidemiology data. This is not the common practice, but may be more meaningful for the average reader. The product of a risk assessment usually provides an absolute risk estimate number, or range. The official EPA explanation of a risk estimate is that the modeled exposure produces a risk no higher than the estimated number, and that actual risk could be much less or even zero. For anyone but a risk analyst, this definition of a risk estimate is not very satisfying. An alternative is to present a risk estimate based on local disease data. For carcinogens, this type of characterization would pose the question: If a new cancer case was observed in your community, what is the probability that it is related to a particular site exposure? Intuitively, this may be a more relevant question for the affected population. It may not be possible to apply this approach to every situation, but does show a potential for characterizing risk as attributable to a particular source, or looking at relative risk as a characterization tool.

The project epidemiologist can help eliminate the confusion of terminology in the risk characterization terms "individual risk" and "population risk." Conventional health risk assessment uses these terms in a faulty manner. The product of a conventional health risk assessment does not truly make estimates of individual risk. The only individual risk estimated by a typical risk assessment would be for a receptor that matches the exposure model exactly. Usually, the odds of any individual in the receptor population matching standard default assumptions is very low, and mostly unknown. For example, conventional risk assessment results won't tell a person what their risk is if they weigh 80 kg, only have exposure during working hours, or use chemicals on their jobs. The project epidemiologist can make the default exposure scenario match the majority exposure pattern of the receptor population using site demographics. Furthermore, it may be possible to include qualitative modifiers for certain population attributes. After conducting a sensitivity analysis, the epidemiologist can review the factors affecting risk, and produce individual modifiers that personalize the risk estimate accordingly. For example, epidemiological data may show that heavy smokers have twice the risk of cancer from site emissions compared to non-smokers.

The project epidemiologist should decide which exposure factors are integrated into the general model, and which factors should be treated as separate issues. Effective execution of this task keeps the baseline risk scenario simple. The focus of the risk assessment relates to the receptor population's attributes, and reduces confusion in risk characterization.

Population risk, as described by the EPA model, is strictly a numbers game. The erroneous practice of multiplying the risk estimate by the number of individuals in a receptor population does a poor job of estimating population risk. Again, the epidemiologist can use specific site demographics for age, gender, race, residence time, and background disease rates to construct a relevant estimate of population risk. Most risk assessments will look at children and adult receptors for some pathways. Few look at age or gender distributions. However, the risk assessment product can hardly claim to characterize population risk if there is no weighting of risk based on membership numbers for these various receptor types.

The project epidemiologist can turn a generic risk characterization describing no actual exposed group in the receptor population into a practical tool for discussing predicted risk in an affected community.

III. CONCLUSION

Is epidemiological intervention needed? If the answer is "yes" to any of the questions posed in Table 1 consider implementing an intervention program. These actions parallel the risk assessment effort. However, if the situation arises before or during the risk assessment, the intervention must be successfully accomplished before risk assessment activities can proceed. Most of these activities will be conducted by public health agencies for high-profile sites. However, if agency response lags, or a proactive stance is desired for non-Superfund sites, the recommended responses should be considered.

REFERENCES

Agency for Toxic Substances and Disease Registry, *Public Health Assessment Guidance Manual*, U.S. Department of Health and Human Services, Atlanta, 1992.

Aldrich, T. and Griffith, J., Eds., *Environmental Epidemiology and Risk Assessment*, Van Nostrand Reinhold, Florence, KY, 1993.

Bernier, R.H. and Mason, V.M., Eds., *Episource: a Guide to Resources in Epidemiology*, 1st ed., Epidemiology Monitor, Roswell, GA, 1991.

Gordis, Leon, Ed., *Epidemiology and Health Risk Assessment*, Oxford University Press, New York, 1998.

Hulka, B., Wilcosky, T., and Griffith, J., Eds., *Biological Markers in Epidemiology*, Oxford University Press, New York, 1990.

Lilienfeld, A.M. and Lilienfeld, D.E., *Foundations of Epidemiology*, Oxford University Press, New York, 1980.

McMahon, B. and Pugh, T.F., *Epidemiology: Principles and Methods*, Little, Brown Publishers, Boston, 1970.

National Research Council, *Environmental Epidemiology Volume 1, Public Health and Hazardous Waste*, National Academy Press, Washington, 1991.

Table 1 Determining the Need for Epidemiological Intervention in Risk Assessment

Situation	Response: If Yes
Is the potentially exposed population worried about exposure? Are they having a problem understanding their risk?	Consider a community education program if concern in the affected population is high, if the population lives along potential exposure pathways, or if exposure is possible.
	Implement health professional education if there are site concerns and a lack of information available from local health professionals. This may also come as a request from local physicians, interested groups, academia, or governmental agencies.
	Consider helping the residents set up a voluntary tracking system if exposure has been documented or is reasonable to assume, if concern is high, and if formal studies look likely.
Is exposure now occurring, or has it occurred in the past?	Consider the possibility of an exposure study using biomarkers, when practical. This is a viable option when exposure is known to have occurred, and when specific tests are available for the chemicals of concern. This strategy is most useful in situations where current knowledge is insufficient to predict if actual chemical uptake is occurring or illnesses are likely.
	Evaluate biomedical testing as an option. This may be useful when exposure is certain, the exposed population can be identified, and laboratory tests are available to measure the effects in question.
	Consider specific population studies (cross-sectional, cohort, case-control) of exposure and effects.
	Consider a site-specific medical surveillance program if the methods exist and the population exposure profile is thought to be fairly constant.
Are there indications or allegations of adverse health conditions in an exposed population?	Same as above, but include cluster investigation as an option. Cluster investigations may be useful when geographical and temporal descriptions of cases are available.

CHAPTER 16

Surface Water Modeling

Bruce T. Rodgers

CONTENTS

I. Introduction 351
II. Model Objective 352
III. Model Classification 352
    A. Degree of Detail 353
    B. Principal Model Components 353
IV. Model Selection 355
    A. Budget 355
    B. Function 355
    C. Options 355
    D. Resources 355
V. Sources of Models 355
VI. Conclusion 356
    References 356

I. INTRODUCTION

Surface water models serve as an integral part of the risk assessment process when the contaminant pathway or exposure media includes streams, rivers, estuaries, coastal regions, or open ocean. These models quantify the physical and chemical characteristics of the environmental compartments through which a contaminant moves. They also quantify the spatial and temporal extent of contaminant concentrations in surface water environments. These models are analytical tools, used to support ERA planing and subsequent assessment of remedial options, or pollution prevention strategies.

Surface water models are used when the required information cannot be measured directly. As an example, consider the case of a river containing contaminated sediment. Field measurement alone can identify the quality of the sediment and size of the contaminated area. However, a surface water model together with field measurement is necessary to determine the cause of contamination and the effectiveness of remediation.

In general, surface water models are comprised of a series of mathematical equations representing various processes occurring in an aquatic environment. Each equation, or group of equations, mathematically reproduces the abiotic or biotic interactions of a chemical with its environment. This produces a simulation of what would happen if a chemical were to enter a surface water environment and be transported to some point away from its point of entry.

A wide variety of models are available to surface water modelers. Each model provides a different degree of detail about each environmental component. The art of model usage lies in the selection of an appropriate model to reflect the needed level of detail for use in a specific application.

A project manager and project team must generally understand the various types of surface water models, and the advantages and disadvantages of each. This understanding aids in determining the appropriate level of effort to invest in modeling activities.

II. MODEL OBJECTIVE

Surface water models explain and predict the interaction of a chemical with its environment. Explanation refers to the relationships between source loading, water and contaminant movement, water and sediment chemistry, and biological response. Using models to understand these relationships helps in planning an ERA by identifying critical elements affecting contaminant pathways and exposure.

Prediction involves the simulation of risk. Unlike field measurements where risk assessors must deal with existing environmental conditions, models can use any specified physical or chemical input to quantify change in risk. For example, the model can quantify the change in risk attributed to a reduction in contaminant loading, or a change in the physical or chemical characteristic of the aquatic environment. This information is useful to assess the effectiveness of remedial options or pollution prevention strategies.

III. MODEL CLASSIFICATION

An array of models exist for the prediction and assessment of surface waters. Each model is classified based on the degree of detail provided and the environmental component simulated.

A. Degree of Detail

The degree of detail provided by a surface water model refers to the complexity of the mathematical equations used to simulate the environmental processes, such as the movement of water or contaminants. At one end of the spectrum are screening level models which are simple to use and fairly generic. At the other end of the spectrum are complex models which are difficult to use and highly specific to a particular application. A wide variety of models reside between these extremes.

Screening level models describe environmental processes in general, "coarse" terms based on simple linear relationships. Such simple models require very little site specific information and can be applied by contractors with only a general understanding of modeling and the environment being simulated. Accordingly, screening level models are appropriate in applications which focus on general or relative impacts without regard for high accuracy, as is the case during the initial planning of the ERA. However, these models may oversimplify the critical environmental component required for a sufficiently accurate assessment of imposed risk.

Complex models simulate the physical, chemical, and biological environment using complicated mathematical equations. The degree of resolution and precision of these models depends on the quality of site specific field data, and by the understanding and experience of the contractor. Such models are appropriate where a high degree of confidence is required, but are generally cumbersome for planning level analysis.

B. Principal Model Components

The principal components of the surface water model are source loading, water movements, contaminant movement, water and sediment chemistry, and biological response. Biological response is the most significant component for an ERA because it specifically addresses risk to exposed organisms. The validity of the model component used to determine biological response is limited by the accuracy of each individual component comprising the overall model.

1. Source Loading

Source loading refers to the amount of contaminant released to the aquatic environment by an identified source. Examples of sources include treated waste waters from industries and municipalities, urban stormwater runoff, combined sewer overflows, agricultural runoff, groundwater inflow, and contaminated sediments. In each of these cases, and where possible, source loading is best determined through direct field monitoring. The model is useful to estimate source loading when field information is not available. The model is also useful to establish a relationship between the amount of contaminant and factors contributing to its release. This relationship allows assessment of the probability of source loading, and the effectiveness of source control.

2. Water Movement

Water movement refers to the physical characteristics of the water body which affect the movement of contaminants. Specific physical characteristics of concern include the quantity of water available for mixing with the contaminant, and the pathway and speed of contaminant transport. The appropriate degree of detail depends on the spatial and temporal resolution required for the risk assessment. At one extreme are models which provide no space or time resolution. These models are appropriate for the most simplistic application involving uniform exposure over prolonged durations, such as a small lake with contaminated sediments throughout. At the other extreme are three dimensional dynamic models which resolve the variations in water movement throughout the water body and through time. Such a complex model may be appropriate for assessment of a chemical spill or waste water discharge within a tidal estuary.

3. Contaminant Movement

Contaminant movement refers to transport of chemicals from the identified source to a receptor. Since chemicals are often transported with water, contaminant movement relates to water movement. However, water and contaminant movement are distinguished as separate modeling components since the effort required to provide the same spatial and temporal detail differ greatly. This distinction is important when the project manager compares bids, since the estimated cost depends more on the degree of detail specified for water movement than contaminant movement.

4. Water and Sediment Chemistry

Water and sediment chemistry refers to the physical, chemical, or biochemical change of a contaminant while in transport from source to receptor, and at rest in sediment. Sediment is included since many of the contaminants of concern occur in the organic fraction of suspended or bottom sediment. The proper degree of detail is dictated by the contaminant of concern and by the nature of the issue under consideration. First order kinetics are the simplest application but may over simplify the transformation of complex chemicals. More complex models account for parent-compound/daughter-product sequences, ionic speciation, and sorption to dissolved organic carbon and solids.

5. Biological Response

Biological response refers to the uptake and accumulation of contaminants in the aquatic receptor. This component is often addressed independently of the surface water model, however, it is included to stress the importance of overall integration. The biological response is the focal point of a surface water model. It defines the contaminants of concern, means of exposure, and nature of imposed risk. It is the basis from which the model function is defined.

IV. MODEL SELECTION

In many ways, model selection is similar to purchasing a new automobile — the selection seems infinite, the price seems irrational, and clear answers are difficult to find. As with selecting an automobile, it is difficult to outline a series of steps to ensure the appropriate selection. Personal choice always plays a factor, yet several basic criteria must be recognized. These criteria include budget, function, options, and resources.

A. Budget

A realistic budget for the surface water model must be identified at the beginning of the ERA. If cost is not a factor then this should be stated. The degree of detail provided by the model is limited by effort spent. Budgetary constraints will focus selection towards the appropriate class of model.

B. Function

Function relates to the principal model components and degree of detail required of the model. It is subtle, and requires a strong understanding of modeling, the environment being simulated, and issues to be resolved. The contaminants of concern, means of exposure, and nature of imposed risk must be clearly rationalized prior to defining function. For a complex ERA, a screening level model is often necessary to determine the components and detail required to accurately and realistically resolve the application.

C. Options

Options refer to the "bells-and-whistles" of the model. Surface water models are merely a tool to support the risk manager. As such, they must test ideas and communicate results. Options, such as graphical user interfaces, graphical display, and linkages to databases and information systems, open the model to a much wider audience than otherwise possible.

D. Resources

Resources refer to the quality and quantity of site-specific field data. Although it is easy to be tempted, one cannot use a sophisticated surface water model without supporting data. If data are limited, then the degree of model detail must also be limited.

V. SOURCES OF MODELS

Surface water models can be obtained through various sources. The U.S. EPA and the U.S. Army Corp of Engineers are excellent sources of well-tested and documented models. Universities are also excellent sources of models. Regulatory agencies are generally familiar with models from these sources, and are usually ready to accept them as legitimate.

Contractors may also offer models that they have either developed or modified. These models demonstrate the expertise and experience of the contractor; however, they are generally less familiar to regulatory agencies than those developed in the public sector. Therefore, they should be thoroughly reviewed and tested prior to use.

VI. CONCLUSION

Surface water models are a tool to support the risk assessment. They are used to explain and predict the interaction of a chemical with its environment. The components of the surface water model include: (1) source loading, to estimate the amount of chemical entering the water; (2) water movement, to estimate the transport pathway; (3) contaminant movement, to estimate the region of exposure; (4) water and sediment chemistry, to estimate the change in chemical composition; and (5) biological response, to estimate the risk to exposed organisms. A wide variety of models are available to address each component of the surface water model. Each model provides a different degree of detail ranging from screening level to highly complex. The art of model usage lies in the selection of the appropriate model to reflect the needed level of detail for use in a specific application.

REFERENCES

Ambrose, R.B., WASP4, *A Hydrodynamic and Water Quality Model Theory, User's Manual and Programmer's Guide,* Office of Research and Development, U.S. Environmental Protection Agency, Athens, GA, 1987.

Brown, L.C., and Barnwell, T.O., *Enhanced Stream Water Quality Models QUAL2E and QUAL2E-UNCAS: Documentation and User Manual,* Office of Research and Development, U.S. Environmental Protection Agency, Athens, GA, 1987.

Burns, L.A., Cline, D.M., and Lassiter, R.R., *Exposure Analysis Modeling System (EXAMS): User Manual and System Documentation,* Environmental Research Laboratory, U.S. Environmental Protection Agency, Athens, GA, 1982.

Thomann, R.V., and Mueller, J.A., *Principles of Surface Water Quality Modeling and Control,* Harper & Rowe, New York, 1987.

U.S. Army Corp of Engineers, *Cequal-w2: A Numerical Two-Dimensional, Laterally-Averaged Model of Hydrodynamics and Water Quality,* U.S. Army Corp of Engineers, Washington, 1986.

CHAPTER 17

Groundwater Modeling in Health Risk Assessment

Jeanette H. Leete

CONTENTS

I. Introduction 357
II. Groundwater Modeling Reports 358
III. Technical Aspects of Groundwater Modeling 358
   A. Definition of "Model" 358
IV. Technical Aspects of Contaminant Transport 365
   A. Physical and Chemical Forces Influencing Movement 365
   B. Model Misuse, Limitations, and Sources of Error 366
   C. Groundwater Quality Monitoring 367
V. Conclusion 367
   References 367

I. INTRODUCTION

Groundwater modeling refers to the construction and operation of a model that can mimic the actual behavior of groundwater in an aquifer system. There are several kinds of groundwater models: electrical analog, physical (most physical models look like ant farms packed with layers of sand and clay), and mathematical. For this primer, we use "groundwater model" to mean a mathematical model. A mathematical model is a set of equations and assumptions chosen to represent a groundwater system. Computer programs then solve these sets of equations.

Groundwater modeling is extremely useful for developing credible risk assessments where groundwater is a potential exposure pathway. Groundwater modeling is employed during the risk assessment process in the hazard evaluation and exposure assessment steps. Modeling is used to evaluate the possible contaminant transport pathways (so that exposure potential can be evaluated). Groundwater modeling can provide information about changes in concentration from source to discharge or withdrawal point.

Assessing the risk to humans or to the environment of a constituent of groundwater requires the ability to predict exposure to this groundwater. This requires knowledge of the direction and amount of groundwater movement; the chemical nature of the water; concentration of the undesirable constituent at the point or area of entry into groundwater; possible interactions with the aquifer material and natural groundwater; interconnections to other water sources (discharge to springs, pumping from wells, hydraulic connections between aquifers); and potential for transformations during transport, such as adsorption, dilution, and dispersion.

Direct measurement of this information is generally impossible because of limited access to subsurface information. Movement of water or contaminated water in the subsurface cannot be directly observed, nor can continuous measurements over an area be taken. Available information is always limited to point information at a limited number of locations. If done efficiently and well, groundwater modeling can combine sparse data into a coherent representation of the workings of a hydrogeologic system. That information can then be used to predict the current and future extent of contamination and pathways of exposure.

II. GROUNDWATER MODELING REPORTS

Groundwater modeling reports are typically produced as one large deliverable. This format is acceptable for simple physical and geochemical settings, and for situations where previous work has created a credible understanding of the geology of the area, and has defined the existing hydrogeochemistry and extent of contamination.

A groundwater modeling report should be broken into several interim deliverables for complex or poorly understood settings. Examples of complex settings include multiaquifer problems, flow in fractured formations, and situations where groundwater withdrawals are variable in amount, timing, and location. Possible logical subreports include Site Geology and Conceptual Hydrogeologic Setting; Ground Water Flow Model Calibration and Verification; and Ground Water Transport Modeling. A series of smaller reports allows the project manager to review intermediate results and ensure that the project is on "solid ground" before authorizing subsequent work. If necessary, the project manager may arrange for peer review by a second consultant, selected to review the specific report segment.

III. TECHNICAL ASPECTS OF GROUNDWATER MODELING

A. Definition of "Model"

A model is a characterization of a real system. In hydrogeology, as mentioned in the introduction, there are several classes of models. These classes are discussed below.

GROUNDWATER MODELING IN HEALTH RISK ASSESSMENT

1. Conceptual Models

Conceptual models describe and offer an explanation of "how groundwater works" in a given system. Conceptual models should always precede data collection. For example, the regional geology would be described in a conceptual model along with the locations and nature of the bounding conditions on the aquifer (which might be rivers, discharge areas, recharge areas, faults, and areas where the aquifer is not present). An example of a conceptual model could read:

> The groundwater system in the study area consists of a stack of three regional aquifers, within a vaguely bowl-shaped basin of horizontally layered Paleozoic sedimentary rock, over a crystalline bedrock surface. The uppermost unit consists of varying thicknesses of glacial materials. Where these materials are sandy and of sufficient thickness, they too can serve as local aquifers. Preglacial drainage systems have cut through all but the deepest of the aquifers. Recharge to the system is focused where aquifers subcrop beneath sandy glacial deposits, and where aquifers appear at the surface. A major river system bisects the study area. The valley is incised from 100 to 300 feet below the general surface elevations, and forms the major discharge zone for the regional aquifer system, and thus a major boundary to the system.

From such a model (i.e., the description and accompanying geologic cross-sections and maps), the risk assessment professional can form a mental picture of regional groundwater flow directions and groundwater/surface water interactions. General opinions of cause and effect are given in a conceptual model, but for predictions and analysis of local conditions, dynamic models are necessary.

2. Dynamic Models

A dynamic model can be changed to reflect changing conditions, that is, it can be manipulated. Physical models, scale models of the groundwater system, can be built in aquariums or narrow plexiglass "ant farms" of sand, gravel, and clay or other porous materials. The surface topography, complete with lakes and/or streams, can be represented — wells can be built of acrylic or other clear tubing (so that water levels can be observed); leaky underground storage tanks can be made from empty film canisters with pinholes and an access pipe made from a straw. With some imagination, patience, and visits to the hardware store, most types of groundwater problems can be built into a physical model.

A physical model can show groundwater movement in response to regional flow, and can show response to pumping of the model's wells. Food coloring can be added to the recharge water or to water at a contaminant source in order to reveal the flow paths of the water. Because of the difficulty in deriving quantifiable results from such models, and the amount of time needed to rebuild it every time a change is needed, these models are rarely used today to solve groundwater problems. They have, however, proven to be very useful in the public meeting forum where they can be used to demystify the concepts of groundwater flow and contaminant transport.

The flow of electricity through a conductor is analogous to the flow of groundwater through an aquifer, the realization of which was the breakthrough which allowed the development of mathematical solutions to groundwater flow problems. Accordingly, some of the first groundwater models were built as networks of resistors and capacitors. The aquifer characteristics were scaled into the model by using different resistors to represent the transmission of water and different capacitors to represent the storage of water. When such a model was finished, current represented the flow of water and voltage represented the hydraulic head (which can be understood as the water level in wells which penetrate the aquifer). These models are called electrical analog models. Electrical analog models can take months to build and adjust so that they represent the aquifer system under study, and they tend to take up quite a bit of lab space. Three-dimensional flow can be modeled by connecting two or more horizontal models to each other with the appropriate electronics to represent leakage between the layers. Electrical analog models are rarely built today because other models are easier to work with. Today's uses still include permanent museum and public education displays.

Stochastic models are statistical models. Much recent research, and possibly hundreds of recent papers, have explored the use of stochastics in the modeling of groundwater flow and contaminant transport, but the method has not gained wide acceptance among practitioners. This is almost certainly due to the complexity of the concepts employed and the fact that none of the many modeling approaches presented has become a standard. It is possible that rapid progress toward an accepted standard could be made in the next several years.

Mathematical models are derived from the physical laws that govern the situation (e.g., conservation of mass, conservation of momentum, and Darcy's equation) with simplifying assumptions about the aquifer and about the edges of the modeled area. Analytical models can be used to solve very simple problems (e.g., the aquifer can be assumed to be the same in every direction and only one value for each parameter is needed). Equations are set up which represent the system variables (e.g., hydraulic head) over the domain of the model. The resulting analytical model of groundwater flow will be a set of partial differential equations that can be solved directly using calculus. Analytical models for solute transport can be created in a similar fashion.

The results of more than one analytical model run can be combined to produce a solution to a more complicated situation. One could, for example, set up an analytical model which produced a solution for the hydraulic gradient over the area of concern, then use a different analytical model to predict the movement of contaminants in response to those gradients.

The data requirements for an analytical model are not extensive, because after all, only one number can be used for each system parameter. Analytical models can be solved quickly with an inexpensive programmable calculator or personal computer.

Graphical solution of some of the less complicated flow equations is possible. For example, flow nets combine lines which describe flow paths and lines which represent equal hydraulic head to provide a visualization of the groundwater flow field. Once constructed, a flow net can be used for prediction of flow directions and amounts.

It is clear that many real world problems are not simple enough to be accurately assessed with simple analytical or graphical models. Where enough is known about a hydrogeologic or contaminant transport problem to be able to characterize the system with variable aquifer parameters and detailed boundary conditions, the groundwater flow equations cannot be directly solved with calculus; rather, they must be approximated by systems of algebraic equations. Groundwater models using this technique are termed numerical models. Calculations must be carried out repeatedly over the entire system of equations until a solution is reached. The process is repeated every time a change in any of the model data is made.

Mathematical models have replaced other types of models as the speed of computers has increased and the cost of computers has decreased. As few as 15 years ago, the best high speed computers the major universities had to offer often took hours to complete one run of a numerical model. Because of computing costs, these model runs were often done overnight at lower rates. The results were picked up in the morning (if indeed the program had run without fatal errors), and during the day necessary changes were made in model input for the next night's run. Today's personal computers provide the speed and flexibility needed to handle many model runs, of even very complex models, in one day, and advanced workstations allow the calculation of detailed three-dimensional models, and provide graphical color output of the results in seconds.

3. Model Selection

The particular problem at hand will determine which of the methods is appropriate. Each of the modeling approaches discussed above has its limitations, advantages, and disadvantages. The essential question is: Can this method answer my question most efficiently? There is a tendency in the groundwater profession to turn to the numerical models without consideration of the less elegant methods. To counteract this bias, the following questions should be posed as part of the model selection process:

- What is the model's purpose? The scope of the study may be such that answers can be obtained from analytical models or from graphical solutions.
- What data are available to characterize the aquifer system? If the aquifer system can only be described in general terms, what is the justification for the use of a complex model?
- Is the collection of additional data to be part of the study? If so, then a preliminary model can be constructed to guide data acquisition and eventual construction of a full model.

If the decision is made that a numerical model is indeed necessary, an appropriate numerical method should be selected. There are three basic approaches to numerical modeling: finite difference, finite element, and analytic element.

As mentioned above, the continuous equations that describe conditions in the aquifer at every point can only be directly solved for very simple situations. To accommodate more complex aquifer characteristics, the study area is divided up into smaller pieces. In both the finite difference and finite element approaches, each aquifer segment is described by an algebraic equation or set of equations, all of which must be solved for each time step. In the analytic element method, the aquifer is divided into segments (elements), each of which can be simplified so that analytic solution techniques can be used.

The finite difference method superimposes a grid system over the study area. The method has developed to the point where the grid need not be regular, nor even rectangular. A finer mesh can be located over the area of greatest concern so that more detail can be obtained. Within each cell or aquifer segment there is a node point at which the equations are solved. The nodes can be in the centers of the cells or at the intersections of the grid lines. Any pumping occurring within the cell, or any water added to the cell, will be treated as if it were added at the node. All water levels are calculated at the node and applied over the whole cell. To avoid a "stair-step" effect in predicted water levels, areas of concern should have finer meshes.

A first approximation of the head at each of the nodes is the starting point. The computer recalculates heads at each node (some nodes may have fixed heads as part of the boundary conditions), based on the heads of adjacent nodes, until the changes between successive recalculations is less than the predetermined error limit. The solution of the finite difference model is an iterative process. Sometimes during iterations, errors can start to build on each other and the resulting solution may be nonsense. Only by comparing the computer's answer to reality can you know if a real solution has been reached. This problem is called numerical dispersion.

The finite element method divides the aquifer into polygonal elements (often triangular) by connecting irregularly placed nodes into a mesh where each element has multiple nodes, termed "discretization," this allows more accurate representation of irregular areas than does the finite difference model, even though the finite element model will usually have fewer nodes.

Values of system variables are interpolated over the element by basis functions. The basis functions are specified in terms of the node coordinates and the results are combined into an integral system which is then approximated using finite difference techniques. Be aware that these methods are also subject to numerical dispersion.

If you divide any groundwater system into small enough pieces, you reach a point where the pieces are internally simple enough that analytical solutions can be used. Superposition (the adding together of analytical solutions) is used to deal with complexity. Relatively uniform portions of the aquifer can be turned into model elements, and because there are no restrictions on element size or shape, the model can be built with exactly the level of detail needed to meet the modeling requirements with no excess elements. As this solution is continuous over the model domain, it is independent of scale (this means that detailed local formation and comprehensive regional information can easily be obtained from the same solution).

Use of the model has been limited in the past, because it required more computing power than did equivalent finite difference or finite element models, and because some types of situations could not be simulated as analytical solutions had not yet been developed. The analytic element technique shows great promise and is quickly gaining in popularity as more analytical solutions are added to the code to handle more types of situations, and the typical computer available to the groundwater modeler gains speed and memory.

Most groundwater problems can be addressed using models that have been created and tested by others. The advantage to this, aside from not having to devise a solution method and write the computer program, is that the model will have undergone peer review. This could be very important if the issue might go to court. By the time a model is published, whether in the public domain (e.g, a model produced by the U. S. Geological Survey [U.S.G.S.]) or a commercial code, it should run without internal errors and should produce accurate results.

A recent survey of groundwater modelers revealed that widely used models are those developed by the U.S.G.S. (Geraghty and Miller, 1992). The International Ground Water Modeling Center at the Colorado School of Mines in Golden, Colorado, a clearinghouse for information on models, offers information on types of groundwater models and computer codes.

4. Modeling Process

The conceptual model is the first description of the aquifers' nature, relationships with surrounding water resources, and boundaries of the system. It forms the framework for the mathematical model. The ideas in the conceptual model are then quantified by data collection and located in space through mapping. The first step is to decide where the logical boundaries of the study area should be. The boundaries referred to, in this sense, are typically hydrologic boundaries, and will not necessarily coincide with boundaries used in any other context. Faulty assumptions about the conditions at boundaries, and oversimplified boundary conditions, are among the most common problems in setting up a valid model.

Once the boundaries are set, and the regional extent of the model is determined, the grid or mesh network can be designed, or the domain of the model can be divided into elements. It makes sense that the finer the divisions, the more accurate the solution; it is also logical that finer divisions mean more work for the modeler and for the computer. A typical approach is to use finer spacing or smaller elements near areas of concern to provide both accuracy and efficiency. Finite difference grids should be arranged so that boundaries are represented as accurately as possible. Changes in grid spacing, or finite element sizes, should be gradual. When a well is part of the model, a node should be located close to the actual location of the well(s).

When the model structure is complete, the process of data collection and preparation begins. A listing of data requirements for predictive models is given in Table 1. Data must be formatted and structured for computer input. Much of the information starts out as maps where different parameter values are portrayed as areas of different colors or as areas between contour lines. Such data must be discretized so that parameter values are known for each of the grid cells, or elements, of the models. The discretization process could entail overlaying the scaled grid or mesh network on the map and interpolating values from map data, or it could be done by computer.

The above tasks are part of an iterative process. Information gained during data collection may lead to changes in the conceptual model, and it may be necessary to change how the domain of the model is divided, or to revise the boundaries as the system is better understood.

Table 1 Data Requirements for Predictive Models

Groundwater flow	Topographic base map of the study area (adequate cultural features to identify and understand project location; streams, rivers, wetlands)
	Cross sections showing three-dimensional relationships between aquifers
	Maps of surficial aquifer: water table contours; saturated thickness and transmissivity distribution; boundaries and boundary conditions
	Hydrogeologic maps of all other aquifers: area underlain by aquifer; Hydraulic head distribution; boundaries and boundary conditions; Transmissivity and storage coefficient distribution
	Hydrogeologic maps of confining beds: areal extent; transmissivity and specific storage map
	Hydraulic connections between surface water and aquifers and between aquifers
Solute transport	Background information on natural water quality in the study area
	Effective porosity distribution
	Estimates of the hydrodynamic dispersivity
	Estimates of the variation in and distribution of fluid density
	Boundary conditions for the concentrations of any groundwater quality constituents of concern
	Constituent dispersion, adsorption, desorption, ion exchange, biological or chemical degradation, oxidation, reduction, complexation, dissolution, and precipitation
Groundwater	Precipitation and evapotranspiration
	Natural and cultural recharge areas
	Timing and volumes of stream discharge
	Timing and volumes of stream water withdrawals
	Timing and volumes of groundwater withdrawals
Solute transport	Contaminant sources and concentrations
	Ambient water quality distribution (areal and temporal) in the aquifer
	Quality of streamflow and of any imported recharge water

As soon as an initial set of input data is ready, trial runs of the model can be carried out to see if the model can match the observed conditions for one of the data sets. Comparison of observed and modeled heads gives the modeler an idea about the accuracy of the data that was entered into the model. Adjustments to the model data are made until the comparison is satisfactory. The model is then run with a different set of initial conditions or stressors and adjustments are made to the input data. These model adjustments are called calibration or history matching.

Sensitivity analysis can be carried out to guide the collection of new aquifer data. This analysis involves changing model parameters in a systematic fashion to learn which parameters the calculated heads are most sensitive to. For example, if order-of-magnitude changes in transmissivity in one part of the model have very little effect on changes in head at your area of concern, then that area is not where you want to spend money on an aquifer test.

The ultimate test of the quality of the model is whether it can match real conditions in a data set that the model was not calibrated with. If the model passes this test, it is said to have been verified and can be used with greater confidence for predicting similar conditions. The prediction phase of groundwater modeling is a process of assessing planned changes in water and land use for their possible effects on groundwater. Sensitivity analysis can also be performed in this stage and can quantify the range of possible outcomes.

IV. TECHNICAL ASPECTS OF CONTAMINANT TRANSPORT

A. Physical and Chemical Forces Influencing Movement

Contaminant transport models calculate the movement of constituents of concern in groundwater as a function of their movement with (or in proportion to) the movement of groundwater (advection), and spreading or mixing of the contaminant with the natural groundwater. Movement of contaminants in the subsurface is either under saturated or unsaturated conditions (above the water table). Unsaturated conditions are difficult to model because of the effects of periods of wetting and drying, and because of the addition of a gas phase into the problem. This discussion will cover contaminant transport in saturated media.

Dissolved substances in water will move from areas of high concentration to areas of low concentration by diffusion. The water does not have to be moving for diffusion to occur because the driving force is the concentration difference. The flux of dissolved material is proportional to the concentration gradient. When the water is in pore spaces of an aquifer, the diffusion process cannot work as fast because the dissolved material has to move around the matrix of solid matter. Advection carries dissolved substances along with flowing water. It relates to the average linear velocity and the effective porosity of the matrix. If the water is not moving, no advection occurs.

When a contaminant is moved through an aquifer by advection, some of the water will travel faster than the rest. This could be due to different flow path lengths, to water movement through pores of different sizes, and to friction which slows water movement adjacent to the wall of the pore. These differences will ultimately bring water with the contaminant into contact with other water and create a diluted mixture of the two. This mixing is mechanical dispersion. At the same time, diffusion will be occurring between the contaminated water and the other water due to the concentration gradient. The combined effect is called hydrodynamic dispersion. In addition to the above, the concentrations of solutes in groundwater may also change due to transformation by biological or chemical processes, and due to adsorption of the contaminant to the matrix. Appropriate parameters for many common contaminants have been derived from laboratory and limited field tests.

All of these processes are relatively well understood. Where the hydraulic characteristics of the aquifer are well-known, a model based on the advection-dispersion equations will be useful. Analytical solutions of the advection-dispersion contaminant transport equations are possible for less complex problems, and numerical solutions are available for more complex problems. Solution methods are analogous to those discussed for groundwater flow, with the exception that grid sizes must be kept small to avoid numerical problems. This requirement for small grids or meshes may make these solution methods too inefficient for practical use.

Models for contaminant transport incorporate all of the uncertainties of the groundwater flow models plus all of the uncertainties involved in the movement of contaminants. For this reason, several contaminant transport models rely on statistical approaches to describe both the aquifer materials and movement of solutes. This approach acknowledges the reality that there is, in fact, a range of possible starting conditions and a range of possible outcomes.

Many popular contaminant transport models follow theoretical particles of contaminant along flow paths. These methods are called particle tracking methods. Each particle tracked represents a certain amount of contaminant, and by figuring out how many particles are in a given volume of the aquifer, the concentrations of contaminants are calculated. One simple method, which accounts for both advection and dispersion, is called the Random Walk Model (Prickett et al., 1981). In this model, particles are moved along their flow paths by advection, and a statistical function is used to add an additional movement to each step, which represents dispersion.

B. Model Misuse, Limitations, and Sources of Error

Modeling in general is subject to several types of errors. First, one might have started the process with an erroneous conceptual model, or have used an underlying groundwater flow equation that cannot handle the specific site conditions. Second, round-off error can accumulate during internal calculations, and truncation errors may have happened during the translation of the flow equations into algebraic computer code. Third, your input data may be wrong.

Mathematical modeling of the fate and transport of contaminants in the subsurface is used to simulate the transport and behavior of substances when monitoring data are inadequate. Obviously, this means that many times model results cannot be physically verified. Trust in the results of these models can only come from an understanding of the underlying assumptions and confidence that the best possible data underlie the computer's calculations.

Direct sampling of groundwater at exposure points (ambient monitoring) can and should be used in conjunction with groundwater modeling to determine concentrations of a substance of concern at a particular location. Data from ambient monitoring is then used to refine the predictive value of the groundwater model.

Limitations to any model include uncertainties in input data, uncertainties in the simplifying assumptions, validity of the computer code, ability of the model to handle complexity, and the adequacy of model calibration, sensitivity analysis, and verification. In some cases, the amount of underlying real data does not justify the use of a complex computer model (because most of the input data would have been created or extrapolated from the few real data points). It is thus possible that a relatively simple analytical model may incorporate less uncertainty than a complex numerical model in certain situations.

C. Groundwater Quality Monitoring

Groundwater monitoring results, both for assessment of natural water quality and water quality impacted by the constituents of concern, must be collected and analyzed under very specific, predefined conditions in order for the data to be representative of exposure concentrations. In any sampling study there is variation in concentration over time, there are errors introduced by sampling methodology and analysis, and there is the potential for study design errors. Groundwater quality monitoring study plans must carefully consider the placement, depth, construction method, construction materials, and design of monitoring wells. In addition, samples must be collected and handled following a specified field and laboratory QA/QC protocol.

V. CONCLUSION

When logical and scientifically supportable assumptions, mathematical methods, and professional judgments are used in a groundwater model, the model will add to the strength of the whole report. Also, a model that correctly characterizes the physical setting, and that can be manipulated to assess the effect on groundwater flow and transport, due to a given set of environmental conditions, will facilitate efficient monitoring of contaminated sites and ensure that monitoring is done in the proper locations, at the proper depths, and at reasonable intervals in time and space.

If the hydrogeologic setting of a contamination problem is misunderstood through faulty or hasty analysis, money and time may be wasted and opportunities lost, in an effort to protect people and the environment from nonexistent hazards. In the worst case, real pathways of exposure will remain unrecognized, putting people and the environment at risk. Obviously, a weak groundwater flow and transport model will weaken the entire risk assessment report, and it may be judged unacceptable. Unfortunately, weaknesses in groundwater modeling reports are not always evident. A polished report, with impressive graphics derived from powerful computer programs, may mask an inadequate assessment of the geology and contaminant hydrogeology of the site. Weaknesses may be discovered through peer review, however, or through detailed and diligent tracking by the project manager.

REFERENCES

Anderson, M.P. and Woessner, W.W., *Applied Ground Water Modeling Simulation of Flow and Advective Transport*, Academic Press, Inc., San Diego, 1992.

Bear, J. and Verruijt, A., *Modeling Groundwater Flow and Pollution*, D. Reidel Publishing, Dordrecht, Holland, 1987.

Bredehoeft, J.D. and Pinder, G.F., Mass transport and flowing groundwater, *Water Resources Research*, 9, 194–210, 1973.

Driscoll, F., *Groundwater and Wells*, Johnson Division UOP, St. Paul, Minnesota, 1986.

Fetter, C.W., *Contaminant Hydrogeology*, Macmillan, New York, 1993.

Fetter, C.W., *Applied Hydrogeology*, Macmillan, New York, 1994.
Freeze, R.A. and Cherry, J.A., *Groundwater*, Prentice-Hall, Engelwood Cliffs, NJ, 1979.
Geraghty and Miller, Inc., Survey of groundwater model use, *Geraghty and Miller Software Newsletter*, Summer 1992.
Hem, J.D., *Study and Interpretation of the Chemical Characteristics of Natural Water*, U.S. Geological Survey Water-Supply Paper 2254, 1985.
Konikow, L.F. and Bredehoeft, J.D., Computer model of two-dimensional solute transport and dispersion in groundwater, Chapter 2, in *U.S. Geological Survey, Techniques of Water Resources Investigations, Book 7*, 1978.
McDonald, H.R.M.G. and Harbaugh, A.W., *A Modular Three-dimensional Finite Difference Groundwater Flow Model*, U.S. Geological Survey, 1984.
Moore, J.E., Contribution of groundwater modeling to planning, *J. Hydrology*, 43, 121–128, 1979.
Pinder, G.F. and Gray, W.G., *Finite Element Simulation in Surface and Subsurface Hydrology*, Academic Press, Inc., New York, 1977.
Prickett, T.A. and Lonnquist, C.G., *Selected Digital Computer Techniques for Groundwater Resources Evaluation*, Illinois State Water Survey Bulletin 55, 1971.
Prickett, T.A., Naymik, T.G., and Lonnquist, C.G., *A Random-Walk Solute Transport Model for Selected Ground Water Quality Evaluations*, Illinois State Geological Survey Bulletin 65, 1981.
Strack, O.D.L., *Groundwater Mechanics*, Prentice-Hall, Englewood Cliffs, NJ, 1989.
Trescott, P.C., Pinder, G.F., and Larson, S.P., Finite differences model for aquifer simulation in two dimensions with results of numerical experiments, Chapter 1, in *U.S. Geological Survey Techniques of Water Resources Investigations, Book 7*, 1976.
van der Heide, P. and Beljin, M., *Model Assessment for Delineating Wellhead Protection Areas*, U.S. Environmental Protection Agency, Washington, 1988.

CHAPTER 18

Air Toxics Dispersion and Deposition Modeling

Richard A. Rothstein

CONTENTS

I. Introduction 370
    A.   Regulatory Drivers Affecting Risk Assessment
           Modeling Studies 371
    B.   Consultant Selection 372
II. Overview of Air Modeling Process for Risk Assessment 373
    A.   Reliability of Air Model Predictions..373
III. Practical Air Modeling Considerations, Approaches, and Issues 374
    A. Basic Air Modeling Concepts 374
    B. Dispersion Modeling 376
    C. Deposition Modeling 378
IV. Sources of Air Quality Models 380
V. Sources of Data 380
    A.   Air Quality and Meteorological Data 380
    B.   Sources of Air Emissions Data 381
    C.   Evaluating and Interpreting Air Emissions Data
           for Risk Assessment Modeling..382
VI. "Cutting Edge" Air Modeling Issues for Risk Assessment
    A.   Air Pathway Fate and Transport Issues..383
           for Contentious Multiphase Contaminants.......................................383
    B.   Atmospheric Fate And Deposition Modeling —
           Always Needed? 385
    C.   Limitations of Deposition Modeling...385
    D.   Micrometeorological Effects 386

VII. Collection of Emissions Data Appropriate for Site-Specific,
Multi-Pathway Risk Assessments 386
VIII. Conclusion 387
References 388

I. INTRODUCTION

Regulatory agencies increasingly require use of air dispersion and deposition modeling to evaluate the environmental risk of facility remediation, construction, or operation. Mathematical models calculate air contaminant (plume) dispersion and deposition — the changes in concentration of substances from the source to some location at a given distance from the release point. Typical air emission sources evaluated by regulatory agencies include superfund and hazardous waste sites undergoing groundwater or soil remediation; municipal solid-waste incinerators and landfills; industrial source operations that use various chemicals in the manufacturing process; industrial and municipal wastewater treatment facilities; and microelectronics industries which use specialty gases and chemicals.

Air modeling analyses are used in risk assessment to evaluate three aspects of atmospheric releases:

- The type of activity, including permitted normal or routine facility operations, or unlikely or unavoidable malfunction of operation conditions
- The type of exposure, for example effects from predicted short-term (acute) and long-term (chronic) impacts from different exposure routes
- The exposed population, such as on-site workers and facility operators or on people off-site

Off-site exposures are often characterized as the potential impacts to the "reasonably" maximum exposed individual or as the "average" exposed individual within the modeled site region.

Air emissions are also modeled from sources under consideration for air permits, environmental impact reports; facility engineering design; air monitoring network design; and input to exposure assessment and risk characterization studies, the focus of this textbook. In addition to their use in risk assessments, such air modeling results are also used to help properly site air-monitoring equipment for remediation projects. Air dispersion and deposition modeling results are used to select technically feasible and commercially available state-of-the-art control technology so as to minimize source air emissions and the resulting exposure impacts.

Air modeling for risk assessment can be broadly subdivided into two major categories: (1) those analyses conducted for stationary point sources, e.g., sources whose air emissions to the atmosphere come from a facility vent or stack; and (2) those conducted for near ground-level area type sources, e.g., an open area of emissions, such as a solid or hazardous waste landfill site, or a lagoon. Depending on the source category, the air quality analyst needs to ensure that models are properly selected and applied to provide for reliable exposure assessments and risk characterization predictions.

A. Regulatory Drivers Affecting Risk Assessment Modeling Studies

Over the past two decades, facilities involved with the generation, treatment, storage, and disposal of hazardous waste have been affected by U.S. EPA regulations developed to minimize and maintain air emissions at safe levels. These rules include those developed under the Resource Conservation and Recovery Act (RCRA) and the Comprehensive Environmental Response, Compensation, and Liability Act (CERCLA), also referred to as the Superfund Act. The siting and design of new treatment facilities, or cleanup of existing contaminated waste disposal sites, often triggers a myriad of state and federal environmental permitting and impact assessment requirements to receive necessary approvals. Depending on applicable agency rules, or when planned project actions have the potential to adversely affect human health and the environment, a risk assessment is conventionally performed. Such assessment will evaluate potential multimedia impacts, and where applicable, ensure that appropriate risk management plans and mitigation measures are implemented in the facility design, construction, and operation.

Agencies also frequently require risk assessments for a variety of stationary combustion sources to confirm the necessary air-emission control levels. These include municipal solid waste and medical-waste incinerators, hazardous-waste incinerators, and boilers and industrial furnaces (BIFs) that burn hazardous wastes. On May 18, 1993, the EPA Administrator issued a policy directive that included a draft combustion strategy intended to minimize toxic air emissions from new and existing hazardous-waste incinerators, as well as from BIFs. The policy directive requires: (1) site-specific, comprehensive multipathway risk assessments to quantify potential risks to public health and the environment, and (2) facility-specific permit emission limits for dioxins/furans and particulate matter, to control unacceptable risks from trace organic compounds and hazardous metal emissions, respectively. EPA's Industrial Source Complex (ISC) dispersion and dry/wet deposition model can evaluate explicitly potential risks due to indirect exposures to combustor emissions.

More recently, Title III, of the 1990 Clean Air Act Amendments, addresses control of 188 hazardous air pollutants (HAPs) that were identified initially by Congress. EPA and states will be promulgating new air rules throughout this decade to control HAP emissions from hundreds of major new and existing stationary source categories. These include municipal, industrial, manufacturing, petrochemical, waste processing, and power generating facilities. Hence, major sources of HAPs will need to implement new control-technology measures, mainly during the next ten years, to reduce HAP emissions. EPA may later promulgate more restrictive emission control regulations for the affected HAP source categories based on the outcome of residual risk assessment studies.

Unlike for RCRA, the HAP emission reduction rules that EPA is developing are mainly control-technology based rather than risk-assessment based. State agencies may, nevertheless, require certain source owners and operators to continue to perform site-specific multipathway risk assessments. This requirement may be part of the permit approval process for major or controversial projects to ensure that adequate levels of control will be used.

Notwithstanding the regulatory drivers, air modeling to support the risk assessment process is an important tool for all affected parties to confirm the appropriate facility designs, remedial action cleanup levels, or source emission control technologies to be employed.

B. Consultant Selection

This section summarizes the preferred education, experience, and special qualifications that the air modeling practitioner should possess. The criteria given below are germane to the project or task manager responsible for the air modeling. This individual is responsible for managing and/or providing the model output which drives the exposure assessment and risk characterization studies, whether they be human health related or ecologically related.

The art and science of air modeling is in selecting the proper model for a given situation, and then choosing scientifically credible model inputs. It takes considerable scientific training and experience to ensure that the proper model data input are developed, and that model output and its implications for driving the risk assessment are properly interpreted. Notwithstanding the continued advent of user-friendly computerized air dispersion models being readily available to the technical community via electronic bulletin boards and software vendors, air modeling for risk assessment should be performed by qualified and experienced individuals.

The individual (or firm) selected for the air modeling should have application experience in evaluating air emission impacts from (1) proposed and existing stationary combustion or process emission sources; and (2) releases to the air, soil, ground water, and surface water from existing waste disposal sites, or from proposed waste remediation alternatives. The diverse nature of risk assessment necessitates an individual who is well-versed in technical, regulatory, and public health and environmental issues, with a particular sensitivity to public perception. A basic understanding of both carcinogenic and noncarcinogenic risk assessment methodologies pertaining to hazard identification, dose-response assessment, exposure assessment, and risk characterization is essential, so that the air models can be selected and applied properly.

Technical knowledge and capabilities need to include an understanding of the physical, chemical, and toxicological properties of the contaminants in question, including proper identification and evaluation of the exposure pathways, transport, and fate of contaminants. A basic understanding of both carcinogenic and noncarcinogenic risk assessment methodologies (e.g., multistage linear models for assessing carcinogenic impacts; and hazard indices, quotients, and reference doses for assessing noncarcinogenic impacts) is also important, to ensure that modeling goals and objectives will be satisfied.

The individual should be experienced in technical and regulatory criteria for properly selecting and applying approved EPA computerized air dispersion and deposition models. To properly interpret the air model output, the individual should have an understanding and appreciation of the limitations and uncertainties of applying models. These uncertainties pertain to adequacy of source emission and meteorological databases, and applicability and appropriateness of model algorithms to properly simulate the site and regional setting.

The individual should possess strong project management and people skills as he or she will be dealing with a wide variety of multidisciplinary specialties and interested parties. The individual should possess a B.S. degree in a scientific or engineering discipline (M.S. or Ph.D. preferred) with at least 10 years of direct air modeling experience for risk assessment applications. Certification in an air quality, meteorological, or multidisciplinary environmental science or engineering discipline is also preferred.

II. OVERVIEW OF AIR MODELING PROCESS FOR RISK ASSESSMENT

Air quality analysts are vital members of a risk assessment team whose task is to evaluate the transport and impact of substances released to the environment via the air release pathway. From a list of contaminants of concern, air emission rates are calculated, based on media concentrations (e.g., soil, air) of air contaminants at their source (e.g., fugitive emissions, trans-media movement of chemicals), and the emission flux to the atmosphere. Air quality analysts also use physical source characteristics (e.g., stack height, volumetric flow rate) and emission data to predict what the contaminant concentrations will be at a receptor point some distance from the contaminant source location. Air quality analysts use computer mathematical models designed to simulate environmental processes that are thought to occur in the atmosphere from the source to a receptor location. They use their computer simulation capabilities to evaluate how different environmental conditions will affect a chemical's concentration and environmental distribution over the study area. Receptor-point air concentrations and deposition rates are provided to risk assessors for one or more exposure case scenarios, where these predictions are used as inputs in exposure equations that are designed to calculate chemical intakes and uptakes for risk characterization.

A. Reliability of Air Model Predictions

Two important roles for air modeling for risk assessment include: (1) making reasonably accurate and reliable predictions about the transport and fate of air emissions and (2) satisfying technical, regulatory, and public perception concerns about potential source air impacts.

Reliable air quality modeling provides for more reliable exposure assessments and risk characterization predictions. Model predictions are only as good as the model itself, and the quality of data input. As such, an air quality model can only be as good as the databases and assumptions that are incorporated into its application. Hence, proper quantification of site and regional characteristics, source operation parameters, emission rates, and meteorological data is essential in any risk assessment.

Regardless of how carefully one selects and applies air quality models, a number of unknowns, data gaps, and technical uncertainties still remain about the myriad of chemical reactions and physical processes actually taking place in the atmosphere that affect the transport and fate of air contaminants. Many computerized air models have been developed over the years for risk assessment applications. While models continue to be developed and refined, they are predictive tools. They should not be perceived as yielding "absolute" accurate numerical estimates for all air contaminants of concern, and for all conceivable environmental circumstances encountered. Modeling uncertainties normally are addressed by making simplifying or conservative assumptions to avoid underestimating the potential risk.

III. PRACTICAL AIR MODELING CONSIDERATIONS, APPROACHES, AND ISSUES

Air dispersion and deposition models are used to estimate the atmospheric transport, the ambient air concentrations, and the surface deposition flux of specific air contaminants. An overview of dispersion and deposition models, including model application concepts, is given in terms of "what," "where," and "how" to model.

A. Basic Air Modeling Concepts

Physical source parameters and emission characteristics of contaminants of concern describe the nature of the discharges to the atmosphere. Contaminant emission rates can be calculated for point and area (nonpoint) sources. These rates are input to air models whose outputs are used to predict ambient air concentrations or deposition rates to various surfaces such as vegetation, soils, and water bodies. Receptor-point concentrations are used in exposure models to calculate exposure levels.

Calculation of point source emissions, from stack and vent emissions data, are generally straightforward in that source test data, emission factors, or mass balance calculations can be used. Point-source emission rates based on testing are normally derived from the flue gas concentration of the contaminant and the volumetric flue-gas flow rate. Emission rates for continuous point source operations are normally expressed as mass per unit time (typically in g/sec for air modeling).

Point-source physical parameters include stack height, internal stack top diameter, flue-gas stack exit velocity or volumetric flow rate, and flue-gas stack temperature. It is also important to specify dimensions of building in the vicinity of the stack. For relatively short stack to building height ratios, the stack plume dispersion in the near field can be dramatically affected by turbulent building-wake effects caused by winds blowing over and around the structure(s). Such effects can cause the magnitude of the concentration impact to be higher, and the location of maximum impact to be closer to the stack, than would otherwise be the case in the absence of such building wake effects.

Other point-source configurations to be modeled may include exhaust fans and louver vents that discharge air contaminants to the atmosphere. In these cases, the physical height of the emission point above ground is normally modeled, along with the specified building dimensions, to account for turbulent building-wake effects.

Area sources result from underground or aboveground sources, typically referred to as "fugitive emissions," since they do not emanate from a stack or vent. Contaminants in the subsurface can exist as a free product (pure compound), absorbed to soil or other deposited substances, as vapor, or as solutes in groundwater. Air emissions from the subsurface can be quantified from flux chamber type measurements;

gas emission models; or "back-calculation" air modeling analyses that use site perimeter ambient-air monitoring and meteorological data to quantify the source term in the model.

Aboveground area sources are typically associated with storage piles, landfills, ponds, and lagoons. Fugitive dust or vapor emission rates are quantified from air emissions modeling or monitoring that relies on chemical and physical properties of the contaminant, the type of medium hosting the contaminant, and associated meteorological influences (temperature, wind speed). Area-source emission rates are normally expressed in mass/area/unit time (typically in $g/m^2/sec$ for air modeling).

Area source parameters to specify in the modeling include the area-source dimensions and the effective emission height above local grade. If the distance separating the area source and nearby receptors is too small, particularly for large area sources with nearby fence-line receptors, the model may require that the area source be divided into smaller "squares" to predict impacts at the close-in receptors.

Contaminants of concern selected for the risk assessment modeling usually satisfy the following general criteria — they are known to be routinely emitted, or have been detected in the air emissions from the source category in question, and they are irritants or potentially toxic to humans and/or have a propensity to bioaccumulate or bioconcentrate in the environment. Quantifiable air emission data from representative source tests, or from other data sources exist for inclusion in air modeling analyses. The actual number of contaminants of concern that are quantitatively evaluated throughout the risk assessment is a function of factors including report rigor, economics, and availability of actual or surrogate data sets for a particular emissions source. In many cases, relatively few air contaminants are routinely monitored at certain facility source categories. As a result, chemical identity/source emission data gaps can limit the robustness of air modeling for risk assessment.

Air model selection and application depends on addressing several source and site-specific questions. For example, is the release to the atmosphere (1) <u>quasi-instantaneous</u>, such as from gas cylinder or chemical tank ruptures, or sudden soil venting during remedial excavation or construction work; (2) <u>intermittent,</u> such as from fugitive dust emissions from remedial equipment operations or windborne effects, or vapor emissions from contaminated soils; or (3) <u>continuous,</u> such as from combustion or process vents and stacks? Is it a (1) <u>point source,</u> such as fuel combustion stacks, solid and liquid waste incinerators, storage tanks, soil and landfill venting operations, and air stripper columns; (2) an <u>area source,</u> such as aggregate storage piles, landfills and hazardous waste storage sites, ponds, and lagoons; or (3) a <u>line-type source,</u> such as trenches from remedial excavation and cleanup, perimeter venting at landfills, and vehicular traffic operating on, or egressing from, contaminated property? Moreover, are released substances reactive, non-reactive, vapors, particles, buoyant, neutrally buoyant/passive, or denser than air? Other considerations include defining the location and nature of land use at receptor locations (e.g., on-site, at the fenceline, on complex terrain, in a high rise building); the type of meteorological data available (e.g., collected on-site data, representative off-site data, worst-case screening meteorological data); the appropriate modeling time frame (e.g., short or long-term impacts); and the type of exposure pathways to be considered (e.g., concentration predictions for inhalation exposure; deposition predictions for dermal and ingestion exposures).

Air modeling requirements and approaches for risk assessment applications may differ between political jurisdictions and governmental agencies. An air modeling protocol prepared at the onset of a project for approval by the regulatory permitting entity serves to establish the "bench mark" for the conduct of the air modeling study. If certain modeling assumptions or considerations later need to be revised or updated during the course of the study, it is easier for the analyst to justify such changes, to the state or EPA, via comparison to the previously approved modeling protocol. Considerable project time and expense can be saved if an approved modeling protocol is used.

Air models are used to calculate air concentrations or deposition rates for specific receptor locations, to evaluate risks to human health and the environment. Modeled receptors can be: (1) onsite to predict exposure to workers; (2) fenceline and offsite to predict exposure to the general public and environment; (3) over land to predict (concentration) inhalation impacts, and (deposition) dermal and ingestion impacts; (4) over water to predict (deposition) ingestion impacts); and (5) over elevated terrain to predict stack plume impaction concentration impacts. Model outputs can cover broad areas or can focus on particularly sensitive locations such as hospitals. The study area varies based on regulatory agency requirements and case-specific determinations.

B. Dispersion Modeling

Air dispersion models are mathematical representations that approximate the physical and chemical processes in the atmosphere governing the transport and dilution of gaseous and particulate air contaminants between the source and receptor. They serve by using the source emission rate to the atmosphere to calculate the resultant ambient-air concentration at specified downwind receptor locations (usually at ground-level). The basic model algorithms which treat the source emission releases, plume rise, transport, and atmospheric dilution have not changed significantly over the past several decades. However, the computational features of models have advanced to the point of providing a significant amount of model input and output data being available to sift through. This allows the model user a greater degree of resolution to conform with applicable modeling regulations, guidelines, and study objectives.

Gaussian air dispersion models are often used in support of risk assessments. When Gaussian models are applied, the atmosphere is assumed to be homogeneous, with the source and meteorological parameters being steady-state for the interval of time that the air concentrations are predicted (e.g., one-hour average). This model assumes that maximum chemical concentration occurs at the center of the cloud or along the plume centerline axis, and that the concentration drops off with increasing vertical or crosswind distance from the plume centerline, thus appearing like the familiar bell-shaped "normal distribution" statistical curve in the vertical and horizontal.

Not all Gaussian models are the same, and their dissimilarities can generate quite different answers from the same input data. Dispersion coefficients define the rate of plume spread with distance in models, and depend on whether the study region is considered urban or rural. Selecting urban or rural scenarios results in changes in dispersion coefficients, wind profiles (e.g., rate of change in wind speed with increasing height above ground), and atmospheric mixing height (depth of atmosphere that the plume readily disperses within).

Gaussian model outputs vary but are generally a concentration or deposition rate for a unit time interval (e.g., hour, day, annual average, etc.) at a given receptor point. Standard model averaging times used for exposure assessment purposes range from 1-hour to 24-hours to evaluate acute impacts (irritants, systemic toxicants), and up to annual average to assess long-term chronic noncarcinogenic and carcinogenic impacts.

Regulatory agencies typically require either one year of on-site, or five years of representative off-site meteorological data to be used in refined modeling analyses. When more than one year of meteorological data is used for risk assessment modeling, the year producing the highest annual average impact within the five year data block is commonly used in the exposure assessment. However, it is not unreasonable to average the multiyear impacts, predicted at each modeled receptor, to derive a five-year average impact when performing long-term (e.g., 70 year lifetime) average carcinogenic and noncarcinogenic impact assessments.

Gaussian dispersion models are relatively straight forward and easy to apply compared to other statistical and physical models. They produce results that agree with experimental data as well as any model. Hence, most of the air modeling formulations for risk assessment applications are Gaussian models. The most popular and versatile Gaussian dispersion model, to develop air contaminant concentration and deposition predictions for use in risk assessments, is EPA's ISC model. The ISC model, originally developed in 1979, remains the "work horse" model for a wide variety of model applications in relatively "simple" terrain settings. ISC can be used to simulate dispersion from point, area, and line-type sources. ISC is also the only EPA-approved dispersion model capable of estimating the effects, from building-induced downwash, on the distribution of downwind ground-level concentration impacts. ISC can be used to calculate maximum 1, 3, 8, and 24-hour, monthly, calendar quarter, and annual average concentration impacts at each receptor location with a full year (8,760 hours), or for multiple years, of hourly meteorology data. This model, along with numerous other Gaussian dispersion models, are described in EPA's *Guideline on Air Quality Models* (1993). Other EPA dispersion models are available to evaluate impacts in complex terrain settings where terrain height exceeds stack top height.

Dispersion models can be used in either a refined or screening fashion depending on the application. Screening modeling produces "worst case" concentration estimates. Screening modeling can be relatively quick to apply, less computer intensive, and more conservative. The standard approach for screening modeling is to assume a set of hourly meteorological data that represents a wide range of possible meteorological conditions (about two dozen combinations of hourly wind direction, wind speed, and atmospheric stability class). Screening modeling can help to: (1) initially confirm which sources in a multisource region or complex may cause the greatest concentration impacts at key receptor locations, (2) confirm whether complex terrain models also need to be applied, and (3) confirm the receptor grid configuration for the refined dispersion modeling. Screening modeling usually yields overly conservative results which are typically inappropriate for detailed risk assessment analysis purposes. As discussed before, refined modeling uses at least a full year of hourly meteorological data.

C. Deposition Modeling

Deposition modeling is a method of accounting for the transfer of air contaminants from the ambient air to environmental surfaces. Deposition modeling accounts for the concentration of the contaminant in ambient air that is subsequently deposited onto the surface feature at ground-level (e.g., vegetation, soil, lakes). This transfer, or deposition, affects the availability of air contaminants for human (or ecological) exposure via indirect pathways (e.g., dermal and ingestion exposure routes) rather than from direct inhalation. The removal of pollutants from the atmosphere can be represented by two processes — dry and wet deposition. Dry deposition modeling accounts for both gravitational settling and deposition due to other atmospheric processes, and hence, can be used for all particle sizes. Dry deposition of particles is modeled as the result of several processes including gravitational settling, eddy motion (atmospheric turbulence), Brownian motion, and electrostatic attraction. Wet deposition of particles can account for precipitation washout from a dispersing stack plume.

The approach used in the ISC model is especially well-suited for predicting deposition of submicron particles for which deposition rate increases with decreasing particle diameter. It is these finer particles in which certain trace organic compounds, such as PAH, PCB, and dioxins/furans, and heavy metals, such as lead, cadmium, and mercury, are assumed to be primarily associated with, such as from waste combustion sources. Due to the greater ratio of the particle surface area to volume, these trace contaminants will preferentially adsorb or condense onto the finest-sized particulates. Dry deposition model algorithms handle different particle sizes, in the analysis of the surface deposition of air contaminants, that are either bound to the particle surface or included as part of the matrix of the particle. The particle surface-area fraction distribution is used in the analysis of air contaminants that are bound to the particle surface, while the mass fraction distribution is used if the contaminants are part of the matrix of the particle. The dry deposition rate is proportional to the ambient air contaminant concentration immediately above the ground surface.

Dry deposition modeling is generally based on applying a calculated particle deposition velocity which is based on particle size, particle density, wind speed, atmospheric stability, air temperature, and surface roughness parameters. The particle deposition velocity is multiplied by the predicted ambient air concentration at each modeled receptor, which results in a deposition rate to the ground or water body surface. Compared to water surfaces, the calculated dry deposition rate is normally greater over land surfaces, due to the greater associated surface roughness, which increases the particle deposition velocity.

Dry deposition of an air contaminant, associated with each particle size category (as a function of particle mass or surface area fraction), is calculated as the product of the hourly predicted ground-level concentration, at each receptor location, and the calculated hourly deposition velocity. Thus, obtaining an hourly flux or deposition rate at each receptor. The hourly deposition rates calculated at each receptor are then summed to compute the annual average deposition rate at each receptor (in units of g or $\mu g/m^2/yr$.

An alternative screening methodology to estimate conservatively the dry deposition flux is as follows: (1) assume an "upper bound" average particle deposition velocity of 2 cm/sec (0.02 m/sec); (2) multiply the deposition velocity times the predicted ambient air contaminant concentration at the given receptor for the time period in question, e.g, annual average; (3) determine the deposition rate in units of g or $\mu g/m^2/yr$.

In recent years, the emphasis on multimedia impacts of waste combustion sources on water quality, coupled with the realization that potentially hazardous levels of air contaminants attached to particulate matter may be washed out of stack plumes, has prompted an examination of wet deposition on a case-by-case basis. While gaseous wet deposition can also be simulated by adaptation of precipitation scavenging coefficients in these models, the primary focus for air permitting and risk assessment has been with particulate deposition. To simplify the analysis, it is conventionally assumed that below-cloud scavenging (particle washout) is the primary source of wet deposition. This assumption is applicable for particulate deposition within several kilometers of a source, where the maximum impact is expected, and reasonable for risk assessment applications that focus on exposure assessments in the near-field region. Once air contaminants in a stack plume become incorporated into the cloud/precipitation forming process, i.e., in-cloud rainout scavenging, the fate and transport mechanisms become much more complex to address in standard models.

The principal approach used to calculate the wet deposition of particulates is that the total mass deposited at a given receptor for each particle size category, in g or $\mu g/m^2/yr$, depends on: (1) the precipitation scavenging coefficients (a function of particle size category and precipitation intensity), and (2) the fraction of time precipitation occurs during a given hour. The atmospheric scavenging process consists of repeated exposures of particles and soluble gases to precipitation or cloud elements, with some chance of collection onto the elements for each time exposure interval. Two basic wet deposition modeling assumptions are that the intensity of precipitation is constant over the entire path between the source and the receptor, and that precipitation originates at a level above the top of the stack plume that precipitation passes through. A number of simplifying assumptions commonly used in wet deposition models may lead to unrealistic model predictions due, in part, to limitations in available precipitation meteorological bases, and the empirical precipitation scavenging coefficients that are used. Wet deposition models, by virtue of their assumptions and limitations, tend to maximize the predicted impacts in the immediate vicinity of the stack; as a result, maximum predicted wet deposition impacts, and hence, calculated risks due to wet deposition, will be highest near the source.

EPA recommends the use of its ISC model for performing both dispersion, and dry and/or wet deposition modeling for stationary combustion sources located in flat or complex terrain regions. Direct inhalation exposures based on ambient air concentrations of vapors (and fine particulate matter), and indirect exposures based on dry and/or wet deposited particulates (e.g, dermal and ingestion pathways) can be determined with the ISC model. It is beyond the current model capabilities to reliably account for dry deposition of gaseous pollutants, or in-cloud rainout scavenging of gases or particulates (only plume washout is accounted for in ISC).

IV. SOURCES OF AIR QUALITY MODELS

State and federal agencies involved with the risk assessment process generally require contractors to use approved EPA models such as those listed in EPA's *Guideline on Air Quality Models* (1993). This EPA's guideline on Air Quality Models identifies numerous air dispersion and deposition models that may be applied to the analysis of source emissions. EPA's computerized air quality models and users guides are also maintained on EPA's Office of Air Quality Planning and Standards (OAQPS) Technology Transfer Network (TTN) Electronic Bulletin Board. This bulletin board system historically allowed remote users, with either terminals or microcomputers, to dial up via a phone modem connection and exchange information without an operator at the other end. Those with microcomputers had the additional ability to download computer programs, as well as text files. Internet access is now commonly used to access the TTN.

EPA's Source Receptor Analysis Branch of the TTN maintains its air quality models on the Support Center for Regulatory Air Models (SCRAM) bulletin board system. The SCRAM bulletin board system provides a forum for technical interchange at the working level among EPA, state and local agencies, and the private sector. The system offers computer model code, test data, utility programs, bulletins, news, messages, and E-mail service. The system is open to all persons involved in air quality modeling.

The same EPA air quality models and users guides are also available from the National Technical Information Service. Several private sector consulting firms in the United States also develop and sell enhanced or more "user-friendly" software versions of the same EPA models, and offer hands-on, air modeling short courses at various locations in the United States.

V. SOURCES OF DATA

A. Air Quality and Meteorological Data

For noncarcinogenic impact analyses of trace organic and metal contaminants emitted to the atmosphere, regulatory agencies may require the inclusion of representative background ambient air quality data to provide for the cumulative impact of the source emissions, plus background levels. Other regional source emissions may also need to be modeled in cumulative impact analyses. Criteria pollutant emissions from combustion sources may also need to be evaluated for compliance with applicable state and federal ambient air quality standards. Most ambient air quality data for modeling analyses are available from state agencies that routinely monitor for at least the criteria air pollutants. With the exception of the criteria air pollutant, lead, air toxics monitoring data are not normally available from state agencies. Therefore, it may be up to the source owner or permit applicant to conduct such monitoring programs as part of the permit application and approval process.

For refined dispersion and deposition modeling analyses, hourly average meteorological data files need to be developed for wind speed, wind direction, atmospheric stability class, mixing height (i.e., height above ground at which vertical dispersion becomes blocked or suppressed), and ambient air temperature. For wet deposition modeling, hourly precipitation data records (intensity and precipitation type) are also required. In lieu of conducting on-site meteorological data monitoring programs, most risk assessment modeling studies rely on using representative off-site meteorological data, available from governmental agencies and private sources, such as utilities.

Regardless of the data source, it is important to ensure that the format of the acquired meteorological data is compatible with the model input data requirements. Hourly meteorological data used in risk assessment modeling are commonly collected from National Weather Service stations located throughout the United States at hundreds of major airports. These raw hourly observations are compiled and archived by the National Climatic Data Center (NCDC), located in Asheville, NC, and are also available for a large number of airport locations from EPA's SCRAM electronic bulletin board system, discussed previously. Meteorological preprocessor computer programs such as RAMMET and MPRM from the SCRAM electronic bulletin board are used to convert the raw hourly meteorological data into a suitable format for use in refined EPA dispersion models such as ISC.

B. Sources of Air Emissions Data

When acute exposures are of concern in the risk assessment, the source emission release rate should be reflective of maximum short-term emissions, during normal or routine operation conditions, with the source emitting at full-load design. It may also be necessary to address maximum short-term emissions during sporadic or nonroutine operation conditions, (e.g., as a result of equipment malfunctions, facility start-up and shutdown, possible accidental releases, and intermittent releases from site remedial cleanup). If long-term chronic exposures of carcinogenic and noncarcinogenic air contaminants are of concern, then the direct and indirect exposure assessments should generally be reflective of the expected average emissions from the source over the long-term (e.g., over the engineered life of the facility). Standard air emission data sources include:

- Field monitoring emission measurements for area type sources such as flux chambers, stack test data, and soil vapor (ground) probe techniques

- Theoretical and empirical emissions modeling predictions, including equipment vendor design data, EPA compilations of air pollution emission factors such as those contained on EPA's air CHIEF TTN electronic bulletin board system
- Literature reports and studies
- Upwind-downwind ambient air monitoring using conventional sampling (e.g., sorbent tubes, particulate and semi-volatile filter traps and resins, Summa canisters, and release of gaseous tracers) and open-path monitoring using optical remote sensing methods techniques

C. Evaluating and Interpreting Air Emissions Data for Risk Assessment Modeling

For stationary combustion sources in operation, EPA prefers direct stack measurements using EPA recommended chemical-specific (and wherever possible, species or congener-specific) stack sampling, analytical and quality control, quality assurance protocols and procedures. An arithmetic mean emission rate for each substance, derived from a series of representative, source-specific stack test data, will properly characterize the potential modeled exposure levels at the impacted receptors.

For constructed facilities not yet operating, or those in the planning stages, EPA prefers the use of stack test data from surrogate or "representative" facilities. Such facilities include those with similar technology, design, operation, capacity, auxiliary fuels, waste feed types and composition, and air pollution control systems. Stack test data should satisfy sampling and laboratory protocols recommended by EPA.

When combining data from several representative facilities, stack concentrations and flue-gas parameters must be converted to a common basis and consistent units of measurement that are appropriate for the facility under consideration. Ranges and average emission values should be developed for exposure assessment and risk characterization purposes. Should source test emissions data for a given contaminant be skewed or log-normally distributed, then the geometric mean is a better representation of the characteristic emission rate, rather than the arithmetic mean. If no data exist relevant to a specific facility, then EPA's compilations of air pollution emission factors from the CHIEF TTN electronic bulletin board system should be used. In the absence of suitable EPA emission factors, engineering evaluations should be used to derive the emission estimates.

Air modeling analysts must evaluate numerous other site and chemical-specific factors when using models and interpreting model outputs. Modelers must account for temporary increases, i.e., "upsets" in emissions that may occur as a result of start-up and shutdown in operations, malfunctions or perturbations in combustion process and/or air pollution control technology systems. For areas source emissions, the analyst must consider numerous physical and chemical processes (e.g., partitioning of chemicals into the air from soil, water, or other materials). Fugitive dust emissions can be a principal mechanism for transporting semivolatile organic compounds from hazardous waste sites. Both remedial construction activities and wind erosion contribute to fugitive dust emissions.

VI. "CUTTING EDGE" AIR MODELING ISSUES FOR RISK ASSESSMENT

There currently exist a number of challenging air modeling issues associated with the risk assessment process. These include:

- Assessing the validity and accuracy of models using facility and field monitoring data
- Determining proper use of worst case vs. typical/average emission rates to characterize air concentration and deposition impacts
- Evaluating partitioning between vapor-phase and solid-phase substances for input into the dispersion and deposition models
- Determining appropriate number of years to model vs. method of averaging impacts at each receptor, and over the entire modeled region, to characterize potential exposures and risk
- Developing methods to estimate emission rates of trace organic compounds which may be emitted, but not yet adequately quantified to properly characterize a source emission term
- Determining how changes made in certain model-input parameters and assumptions affect the resultant calculated modeled impact and estimated risk
- Evaluating risks based on more frequent compliance stack testing (e.g., quarterly) for chemicals of potential concern rather than overly conservative bounding or worst case risk analyses
- Developing more comprehensive and representative lists of contaminants of concern to estimate risks from both direct and indirect exposure routes from specified activities, sites or, facilities
- Designing air emission data collection programs specifically for risk assessment purposes and not just for facility design acceptance testing and/or compliance testing demonstrations

A. Air Pathway Fate and Transport Issues for Contentious Multiphase Contaminants

For air contaminants such as mercury, which can exist in both the vapor and solid phases in the stack and atmosphere, one of the most important factors determining the fate and transport of stack emissions is the forms or species that occur during the combustion process, and the relative amounts of each form that is emitted to the atmosphere. The speciation of mercury plays a significant role in determining whether mercury will be deposited locally, or be further dispersed and transported over longer distances in the atmosphere, before being deposited on the ground surface and water bodies. A major impediment to the permitting of new solid and hazardous waste incinerators in certain states, regardless of how well emissions can be controlled, has been the issue of modeled mercury stack emission impacts, as compared to surface water quality standards and fish ingestion guidelines, which were originally developed to control industrial wastewater point source discharges.

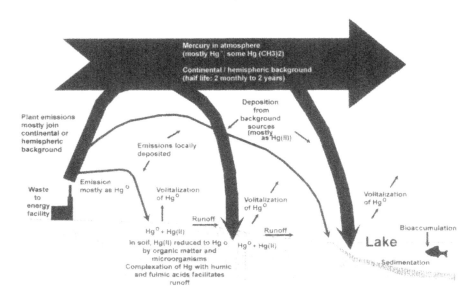

Figure 1 Schematic of the fate and transport of mercury emissions from a stack source.

The main form of mercury in the atmosphere is gaseous elemental mercury, which is relatively insoluble, and, therefore, can remain in the atmosphere for long periods of time (months to years). Oxidized forms such as mercuric chloride have a much shorter residence time in the atmosphere (days to weeks) since they are soluble in rain or snow, and can be deposited by dry and wet deposition processes. Figure 1 portrays the fate and transport of mercury emissions in the environment which initially emanate from a stack source. Some pollutants, such as mercury, can cycle between various media in the environment. This cycling can significantly complicate the fate and transport evaluations that comprise air modeling studies for risk assessments. During the combustion process, mercury experiences several different temperature and chemical regimes within the combustion chamber, the air pollution control equipment, the stack, and then the atmosphere. The specific forms of mercury emitted from waste combustion stacks will vary, depending on the nature and composition of the waste stream, facility operating conditions, flue gas characteristics, and air pollution control technology used.

Data suggest that the only forms likely to occur for municipal solid waste combustion are elemental mercury and oxidized mercury, predominantly mercuric chloride. However, the data base for mercury speciation is quite limited, and sometimes inconsistent, and there are some significant differences of opinion regarding the interpretation of the data. Sampling and analytical methods that can accurately identify different forms of mercury in the stack and atmosphere are still being developed and tested. The speciation of mercury in stack emissions between oxidized and elemental mercury is a very complex issue, and more research is needed to confirm the various amounts of each potential form.

For air contaminants such as mercury, which can exist in both the vapor and solid phases, the standard (conservative) air modeling approach for risk assessment purposes is to model twice by first assuming the emission behaves as a gas, or fine particulate, for inhalation exposure, and then as a particulate which can deposit for dermal or ingestion exposure.

B. Atmospheric Fate and Deposition Modeling — Always Needed?

For certain air contaminants, deposition modeling may not be necessary, or appropriate, if they are either emitted to the atmosphere predominantly in the vapor phase, and if phase changes in the atmosphere are unlikely to take place from the source emission points to the modeled receptor locations. For example, the chemical EGBE (ethylene glycol monobutyl ether), in the glycol ether chemical family, which is listed as one of the HAPs in the 1990 Clean Air Act Amendments, is commonly used as an inside spray coating during the manufacturing of beverage and food cans. In terms of atmospheric fate, glycol ethers do not absorb ultraviolet light in the environmentally significant range (> 290 nm), and, therefore, should not undergo direct photolysis in the atmosphere. Based on a vapor pressure of 0.88 mm Hg at 25°C, EGBE is expected to exist almost entirely in the vapor phase in the atmosphere. Vapor phase atmospheric reactions with other photochemically produced hydroxyl radicals may be important, with an associated atmospheric half-life of about less than a day. The complete miscibility of EGBE in water suggests that physical removal, via wet deposition processes, or dissolution in clouds may occur. However, EGBE's relatively short residence time in the atmosphere suggests that wet deposition is of limited importance.

C. Limitations of Deposition Modeling

Notwithstanding the previous uncertainties raised, about developing reliable wet deposition modeling estimates, due to inherent limitations in the model assumptions and available databases, localized wet deposition can be an important removal mechanism, but not necessarily more important than dry deposition. Unlike dry deposition which occurs continuously, wet deposition due to the precipitation scavenging process is an occasional event. Wet deposition may be quite variable, both spatially and temporally, over a typical 10 kilometer radius study area around a combustor stack. Temporal and spatial variability of precipitation events over a modeled region can potentially lead to unreliable predicted wet deposition modeling results. For example, wet deposition could actually be greater at more distant receptors, than what is predicted, if the precipitation is more showery in nature than uniform over the modeled region. On the other hand, uniform precipitation could scavenge out air contaminants near an emission source, so that actual wet deposition might be inconsequential at more distant receptors. The standard wet deposition model assumption of homogeneity, that reported hourly precipitation events occur uniformly over the study area, means that whenever precipitation occurs, it also occurs at the stack emission point. Because standard wet deposition models account for plume mass depletion, it is likely that they overpredict wet deposition at receptor locations near the stack, and underpredict impacts at more distant receptors.

As a result, for moderate to tall stack heights, the locations of maximum predicted dry and wet deposition may not necessarily coincide. It is also possible that the total annual dry deposition impact may be overpredicted at a given receptor when wet deposition effects are excluded in the modeling. Dry deposition can also be overpredicted at a given receptor if the model does not explicitly, or implicitly, account for any possible effects of plume depletion of the air contaminant by the ground surface upwind of the receptor.

D. Micrometeorological Effects

The highest inhalation exposures are associated with periods of highest air concentrations of the air contaminants of concern. Temperature inversions, or other unusual meteorological conditions that cause the atmospheric stability to be more stable, can minimize atmospheric turbulence, and hence dispersion. High ambient air concentrations may then result for facilities or sources which either have near ground-level releases (e.g., routine or accidental releases of fugitive dust or vapors), or for very short stacks. Stable atmospheric dispersion conditions also may be important if complex terrain is present in the immediate site region. However, stable atmospheric dispersion conditions generally do not result in the maximum ground-level concentrations for taller, nondownwashing stacks that have large thermal plume buoyancy, (i.e., large plume rise). There will be a critical combination of atmospheric stability and wind speed which produces the maximum ground-level concentration during any given hour. The critical wind speed is that condition which minimizes both stack plume rise and dilution of the stack plume in the atmosphere.

Sources located near large water bodies or in deep valleys may experience meteorological conditions unique to their setting (e.g., seabreeze effects, or mountain-valley wind flows) that are not routinely addressed in standard EPA dispersion models.

It may be necessary, on a case-by-case basis, for the contractor to acquire site-specific meteorological data, and/or adapt current EPA, models to adequately address unusual flow regimes that exist in the site region (unless screening modeling or other conservative refined modeling assumptions that are made eliminates such a need).

VII. COLLECTION OF EMISSIONS DATA APPROPRIATE FOR SITE-SPECIFIC, MULTI-PATHWAY RISK ASSESSMENTS

Currently, air emission data has been used mainly for facility design acceptance testing and/or compliance testing demonstrations, and not specifically for assessment of risks. As such, a limited amount of emissions data may be available for performing air pathway risk assessment modeling for all of the potential contaminants of concern. Additional waste stream evaluations should be conducted, along with additional testing for trace organic compound and trace metal pollutants, to aid in a more reasonable and accurate risk assessment. Emissions data for routine and nonroutine facility operations should be collected or estimated, along with the frequency of occurrence and duration of nonroutine operations over the annual period.

However, in lieu of conducting extensive stack testing programs, the following approach could be applied to noncommonly tested organic compounds to ensure that "enough" toxic air contaminants are being evaluated in the risk assessment. To evaluate if certain organic compounds that are not an inherent part of the waste stream might pose any potential health risk concern, there is an alternative screening approach to starting with a "shopping list of chemicals" and attempting to address the question, "Are they emitted and in what concentrations?"

1. Determine the expected total nonmethane hydrocarbon emissions from the waste combustor from routinely available stack test data or vendor design data
2. Calculate the maximum annual average ground level concentration using dispersion modeling
3. Resolve the question of "Are there any compounds which could conceivably be present, as a constituent of the total nonmethane hydrocarbons, that could be significant on a health-related basis at the calculated exposure concentrations?"

One would first assume (conservatively) that no more than one percent of the total non-methane hydrocarbon emissions could represent any single hypothetical toxic organic compound of concern. Using the hypothetical organic compound emission rate in a dispersion model, the maximum annual average ambient air concentration of the organic compound would be determined for comparison with an applicable exposure guideline level. Conversely, the acceptable ambient criteria for the organic compound in question could be used to back-calculate the acceptable stack concentration, in the event EPA needed to set permit emission limits for the organic compound. This approach assumes that one is simply attempting to ascertain the potential importance of potential products of incomplete combustion (PICs) in the stack flue gases, rather than addressing a prime organic component that may be included as part of the wastestream to be incinerated.

In addition, EPA could also use direct stack test measurements of dioxins/furans, carbon monoxide, and particulate matter to determine the effectiveness of controlling trace metal emissions, and other organic compounds of concern, at a waste combustion source.

VIII. CONCLUSION

Air quality impacts can be one of the most sensitive and controversial issues to be encountered in the siting, permitting, design, construction, and operation of stationary combustion and process emission sources, or remediating existing sources. Dispersion and deposition modeling for risk assessments identify, evaluate, and resolve air pathway analysis issues to satisfy associated regulatory and project design issues. Such issues affect project decisions rendered in terms of facility siting, source operations, or degree of control technology or remediation required. The goal is to ensure that facilities are constructed and operated, or remediated in a safe and reliable manner, and within established permit limits, applicable agency rules, and guidelines.

A properly conducted air modeling/risk assessment study, coupled with a good understanding of the modeling limitations and uncertainties, promotes "good science" being used to render opinions about proposed environmental actions that have an air quality component.

REFERENCES

Eklund, B., *Procedures for Conducting Air Pathway Analyses for Superfund Activities, Interim Final Document: Vol. 1 — Overview of Air Pathway Assessments for Superfund Sites (Revised)*, Office of Air Quality Planning and Standards, U.S. Environmental Protection Agency, Research Triangle Park, 1993.

Randerson, D., *Atmospheric Science and Power Reduction*, Technical Information Center of the U.S. Department of Energy, Washington, 1984.

Turner, D. B., *Workbook of Atmospheric Dispersion Estimates: An Introduction to Dispersion Modeling*, 2nd ed., Lewis Publishers, Boca Raton, FL, 1994.

U.S. Environmental Protection Agency, *Methodology for Assessing Health Risks Associated with Indirect Exposure To Combustor Emissions, Interim Final*, Environmental Criteria and Assessment Office, Office of Health and Environmental Assessment, Cincinnati, 1990.

U.S. Environmental Protection Agency, *Addendum to the Methodology for Assessing Health Risks Associated with Indirect Exposure to Combustor Emissions*, Office of Research and Development, Washington, 1993.

U.S. Environmental Protection Agency, *Guideline on Air Quality Models, Revised, (40 CFR Part 51 Appendix W)*, Office of Air Quality Planning and Standards, Research Triangle Park, 1993.

CHAPTER 19

Using Statistics in Health and Environmental Risk Assessments

Michael E. Ginevan

CONTENTS

I.	Introduction	390
II.	Statistical Thinking and Regulatory Guidance	391
	A. Risk Assessment	391
	1. The Hazard Index	391
	2. Assessment of Chemical Cancer Risk	392
	B. Risk Assessment of Radionuclides	393
	C. Evaluation of Exposure	394
	D. Data Quality Objectives (DQOs)	394
	1. The Data Quality Assessment Process	395
III.	Evaluation of the Utility of Environmental Sampling for Health and Environmental Risk Assessments	395
	A. Graphical Methods	
	B. Distributional Fitting and Other Hypothesis Testing	400
	C. Nondetects	402
	D. Sample Support	402
	E. Does Contamination Exceed Background?	403
IV.	Estimation of Relevant Exposure: Data Use and Mental Models	404
V.	Finding Out What is Important: A Checklist	408
VI.	Tools	409
	References	411

I. INTRODUCTION

This chapter reviews the role that statistical thinking and methodology should play in the conduct of health and environmental risk assessments. What do I mean when I refer to "statistics?" *The Random House Unabridged Dictionary* defines statistics as "the science that deals with the collection, classification, analysis, and interpretation of numerical facts or data, and that, by use of mathematical theories of probability, imposes order and regularity on aggregates of more or less disparate elements." This has a simple translation: statistics finds ways of coping with uncertain, incomplete, and otherwise not wholly satisfactory data. Therefore, if you know the answer exactly, you don't need statistics . . . .

How might this methodology apply to planning, generating, and evaluating risk assessment reports? This question can best be approached by considering the four components of the risk assessment process, described in Chapters 2 and 3. The first step of the assessment is "hazard identification," which reviews the inventory or materials present in the environment and uses information from toxicology or epidemiology studies to determine which of these might pose a risk to human health and/or the environment. Statistical principles play important roles in epidemiology and both environmental and laboratory toxicology studies, but the form of these studies, and the role of statistics in them is so diverse that a meaningful discussion is beyond the scope of this chapter. Many hazards are quite well characterized (e.g., there is little debate that high levels of environmental lead are hazardous in a variety of contexts), so the identification can be taken as a given. However, in some cases the hazard identification of a material may rest on one or two studies that are of dubious validity. If the risk assessment is driven by such materials (we will discuss how to determine the factors that are of greatest importance to the estimation of risk) it is often worthwhile to reconsider the underlying literature to determine how valid the studies underlying the hazard identification actually are.

The next step, toxicity assessment, requires development of a dose-response function. A dose-response function provides the risk coefficients used to translate exposure into risk. In essence it answers the question, "Given that substance X is bad, how rapidly do its effects increase with increasing dose?" Many such coefficients are specified by regulatory agencies and will not be readily open to reevaluation. However, in our discussion we will consider how a dose-response function is developed. We will also treat the problem that arises because many "approved" dose-response coefficients are either 95% statistical upper bounds, or incorporate "safety factors" of between 100 and 10,000. That is, if one is assessing the risk of one material, an upper bound or safety factor estimate is arguably appropriate because such assessments should err on the side of safety. However, when, as is the case for hazardous waste sites, many risk coefficients are used, many materials are relevant to determining overall site risk. It has been observed that if one sums 95% upper bounds for 10 dose-response coefficients, the probability of all of the coefficients being at or above their 95% upper bound is 0.05^{10}, or about 1×10^{-13}. As it turns out, this calculation, though correct, is not entirely relevant to the question of the conservatism inherent in a sum of upper bounds. We will discuss some approaches to getting a better answer to this problem.

The "exposure assessment" step frames the question of what actual exposures are likely to be. Estimation of exposure is what drives (or should drive) environmental sampling efforts and subsequent exposure assessment modeling. Both areas have substantial statistical content and will be treated in some detail. Important topics include the pattern of environmental sampling, and why many "engineering judgement" or "compliance monitoring" samples may be nearly useless in terms of assessing actual exposures; the importance of having a model of human (or animal) behavior as the basis for estimating actual exposure; and the necessity of understanding the origin of your environmental contamination numbers.

The final step, risk characterization, is the product of the estimated exposure and the risk coefficients adopted. In practice both quantities may have substantial uncertainties. We will examine the source of such uncertainties, and the use of analytic and Monte Carlo methods for obtaining an overview of the uncertainties in the final risk estimates.

II. STATISTICAL THINKING AND REGULATORY GUIDANCE

There is a lot of good (and some not so good) statistical advice to be found in regulatory guidance documents. This section will review three pertinent areas: risk assessment (U.S. EPA, 1989), data quality objectives (U.S. EPA, 1993), and data quality assessment (U.S. EPA 1996).

A. Risk Assessment

The Risk Assessment Guidance for Superfund (RAGS) document codifies many of the standard procedures used in HHRA. This describes three distinct subprocesses: risk assessment of nonradioactive, noncarcinogenic, chemical toxicants using a quantity referred to as the Hazard Index (HI); risk assessment of chemical carcinogens using q_1^* values (also termed "slope factors" or "cancer potency factors"); and, risk assessment of radioactive materials (radionuclides).

1. The Hazard Index

The HI is given by:

$$HI = \sum_{i=1}^{N} D_i \div RfD_i \qquad (1)$$

where D_i = dose received from the i_{th} toxicant; RfD_i = reference dose from the i_{th} toxicant.

The origin of the RfD deserves some consideration. It is generally taken from a single animal or, rarely, human study. The starting point is the dose at which no biological response was observed (the no observed effect level or NOEL), the lowest dose level at which an effect was observed (the lowest observed effect level or

LOEL), or either the dose predicted to yield a response in 10% of the individuals (the ED10) or a 95% lower bound on this dose (the LED10). Once a starting dose has been determined, various safety factors of 10 are applied. That is, the value is usually divided by 10 to reflect uncertainties in animal to human extrapolation, and a second factor of 10 to reflect interindividual human variability. Additional factors of 10 may be invoked if the starting dose is an LOEL, rather than an NOEL, if the study from which the dose number was derived was a subchronic, as opposed to a chronic, bioassay, and if the person developing the RfD had reservations about the quality of the study from which data originated. Thus most RfDs are 100 to 1000fold below a dose which caused no or minimal effect, and reflect substantial regulatory conservatism.

The site may be considered safe if the HI is less than 1. Actually, following the approach in RAGS, many HIs must usually be defined for the same site. For example, there may be HIs of chronic (long-term or lifetime) exposure and subchronic exposure (shorter term than chronic; usually weeks or months); inhalation HIs, ingestion HIs, and HIs for developmental toxicants; or HIs broken out by mode of action of the toxicants involved (e.g., all liver toxicants). It should be stressed that, despite this variety, the HI is not a quantitative measure of risk. A quantitative measure of risk is the RfD, which may be loosely defined as a dose at which we are quite sure nothing bad will happen. Three elements are lacking from the HI: a quantitative description of the degree of conservatism inherent in a given RfD, a definition of what bad is, and some notion how rapidly things get worse as the RfD is exceeded (a slope factor). For example an HI of 5 might mean that an exposed individual would suffer a small chance of a small depression in cholinesterase activity (an event of dubious clinical significance), or it might mean that an exposed individual could experience acute liver toxicity and possibly death. Likewise, while HI values less than 1 may be taken as safe, it does not follow that a site with an HI of 0.3 is safer than a site with an HI of 0.5.

From a statistical perspective there is not much to say. The HI is intended as a screening index, not a quantitative statement of risk. Moreover, the diversity of the origin of the RfDs, and the arbitrary degrees of conservatism inherent in their derivation, makes it futile to discuss "distributional" properties of the HI. One can, however, make some quantitative statements. First, if one has a report with a single HI for all toxicants at a site, it is almost certainly too large, and its derivation contrary to regulatory guidance. That is, as noted above, RAGS clearly states that HIs should be calculated separately for toxicants with different modes of action and differing exposure scenarios. A second area of concern, which also applies to cancer risk assessment, is the accuracy of the exposure numbers used to derive the HI. These statistical issues will be discussed in detail in subsequent sections.

2. Assessment of Chemical Cancer Risk

At first look, the determination of cancer risk for chemical carcinogens, CRC, looks much like the HI calculation:

$$\text{CRC} = \sum_{i=1}^{N} D_i \times q_{1\,i}^{*} \qquad (2)$$

where D_i = the dose or exposure from the ith carcinogen of interest; q_1*_i = the cancer potency for that carcinogen.

However, this is an actual quantitative expression of risk, with units given in lifetime cancers per exposed individual. Thus, any calculation of this type has a common endpoint. An important feature of this calculation is that each q_1*_i i is an upper bound on the risk calculated on the basis of some model (usually the linearized multistage model of carcinogenesis).

The derivation of these upper bounds deserves discussion. The starting point is usually an animal study, where 3 to 4 groups of animals are exposed to different doses of a carcinogen, and a separate control group of animals is left unexposed. The cancer response in these groups is fit with a dose-response model and the resulting dose-response model is used to develop a linear 95% upper bound on dose-response, referred to as the cancer potency factor, or q_1* value.

Thus, one statistical issue is that Equation (2) involves the summing of possibly many upper bounds, which seems to many to be excessively conservative. One approach to determining the conservatism inherent in Equation (2) involves Monte Carlo simulation methods. These methods first assume that the estimate of cancer potency, q_1* follows a log-normal probability density (Putzrath and Ginevan, 1991). The logarithmic mean (μ) is calculated as:

$$\mu = \ln(q_{mle}) \quad (3)$$

where q_{mle} = the maximum likelihood or "best" estimate of q_1*.

The logarithmic standard deviation (σ) can also be estimated as:

$$\sigma = [\ln(q_1*) - \mu] \div Z_{0.95} \quad (4)$$

where $Z_{0.95} = 1.645$ (the normal score associated with an upper 95% bound on q_1).

After μ and σ have been determined for each carcinogen of interest, a large number (500 – 1000) of realizations are generated of Equation (2) using randomly generated q_1s, and the 95th percentile of this empirical distribution can be determined. Use of this approach can show that the supposed conservatism is less than one might think, in that the result of Equation (2) using q_1*s is rarely more than twice as large as the 95th percentile of the Monte Carlo empirical distribution. Still, Monte Carlo calculations like those described may be worthwhile when the number of carcinogens considered in Equation (2) is large. Differences of a factor of 5 or more are possible when the number of carcinogens is greater than 20.

A more important aspect of Equation (2) is that D_i is the lifetime average daily dose for the carcinogen in question. Thus D_i must be a dose estimate derived from very long-term average exposure. This brings us again to the importance of exposure estimation to the entire risk assessment process.

B. Risk Assessment of Radionuclides

The situation for radiation is somewhat different from the situation for chemical carcinogens. First, there is an extensive literature on the epidemiology of humans exposed to radiation. Thus cancer risk coefficients are well known and relatively precise. Second, the physical means by which radiation damages cells are well known, and precise dosimetric calculations are nearly always possible. Finally, radiation is relatively easy to measure in the environment, and actual concentrations can be determined unambiguously. It should also be mentioned that, because of the superior database, radiation cancer risk coefficients are usually based on best estimates rather than upper bounds.

C. Evaluation of Exposure

A general theme running through the RAGS document is that exposure assessments and, thus, doses should be based on values which are conservative, but not too conservative. Yet the question of uncertainty is treated in a way which would be surprising to most statisticians: "Highly quantitative statistical uncertainty analysis is usually not practical or necessary for Superfund site risk assessments for a number of reasons, not the least of which are the resource requirements to collect and analyze site data in such a way that results can be presented as valid probability distributions." It seems clear that this is not so, and given that cleanup costs are often in the tens of millions dollars it is hard to see why resources to do the job right would not be forthcoming.

Moreover, the two U.S. EPA documents which outline the DQO and Data Quality Assessment process, give careful guidance and recommend many good statistical tools which can be used in assessing data needs and data quality, and which are directly relevant to the issue of assessing environmental contamination, and hence the potential for exposure to human beings or other biota. This contrast is interesting because the DQO and Data Quality Assessment documents are much more recent than the RAGS document, and reflect the evolving position of U.S. EPA in the area of desirable levels of statistical sophistication. In terms of regulatory risk assessment, we are moving from a qualitative to a quantitative world and from simple deterministic models to more sophisticated probabilistic ones.

D. Data Quality Objectives (DQOs)

The DQO process as defined by U.S. EPA is useful for any data collection, not just the collection of data for Superfund sites (see Chapter 11). This process has seven steps:

1. State the problem:
   What sort of environmental contamination is it that you want to characterize? One might be interested in gas phase contaminants (e.g., radon, volatile organics), particulates (e.g., asbestos), or soil contamination. One might be concerned with exposure from inhalation (e.g., radon, asbestos), dermal contact (e.g., pesticides), or soil ingestion (metals). Likewise the exact exposure scenario will affect data needs.
2. Identify the decision:
   What sort of question needs to be answered? Do you want to know about long-term average exposures (carcinogens), short-term maxima (neurotoxicants), or

episodic exposure resulting from particular human activities? What is the exact form of the question you want answered?

3. Identify the inputs to the decision:
How will you use the data? A hypothesis testing exercise might have different data requirements from a modeling study.

4. Define the study boundaries:
Where and when should the data apply? Are you interested in current risks, or a particular site, or risks that may evolve over time (e.g., groundwater)?

5. Develop a decision rule:
You want to be able to say that given these data the exposure of interest is: a quantity, acceptable, unacceptable; or to precisely define the extent of remediation required.

6. Specify limits on decision errors:
How precise do exposure estimates need to be? What is the "loss" of calling an acceptable exposure unacceptable or vice versa. If you are trying to infer dose-response, will your study lose an unacceptable amount of power because of imprecise exposure data?

7. Optimize the design for obtaining data:
Define the most resource-effective sampling and analysis design for generating the data needed to satisfy the DQOs of the project.

The purpose of this seven-step process is to identify the characteristics of the data required, and to arrive at a strategy for collection. It should be noted that the interaction between Step 7 and Steps 1 – 6 is iterative. That is, if one defines DQOs that exceed one's resources, one must rethink the question to identify DQOs with more reasonable resource requirements. In the extreme case, one might be forced to abandon a particular study because meaningful data cannot be collected for reasonable cost.

1. The Data Quality Assessment Process

This process assumes that you already have environmental contamination data and want to determine whether or not this data is adequate to the task at hand, i.e., assessing exposures and thus risks of a particular site or activity. It is nearly the same as for defining a data collection effort, except here one must identify DQOs that can be met by the data at hand. That is, the whole process of data quality assessment is aimed at defining whether or not a set of data meets a particular set of DQOs, or alternatively defining what set of DQOs a given data set will support.

III. EVALUATION OF THE UTILITY OF ENVIRONMENTAL SAMPLING FOR HEALTH AND ENVIRONMENTAL RISK ASSESSMENTS

As noted above, exposure assessment is the factor which most often drives the uncertainty in a risk assessment, and environmental monitoring data are the factors which most commonly drive the exposure assessment. Following the DQO process, we need to state the problem, which is to characterize the risk a given site or activity might pose to human health and the environment. The ultimate decision (DQO, Step 2) we want to make is whether environmental contamination poses an unacceptable risk. We must also specify the model we will use to determine whether or not unacceptable risks are present, because the form of the model will determine the inputs required (DQO, Step 3). We must also specify where and when the decision applies. That is, what is the extent of the area of interest, and what time frame applies the decision of interest (DQO, Step 4)? Having defined the parameters of our decision, we must then determine what overall scale will determine whether risks are unacceptable (DQO, Step 5), and how sure we want to be about our decision (DQO, Step 6). Finally, armed with a clear description of what we want to accomplish, we can either plan our environmental sampling efforts, or evaluate the data at hand (DQO/DQA Step 7). This, of course, is not how things usually happen, but it is good to have an ideal as a yardstick.

Perhaps the most frequent flaw in environmental sampling efforts is the substitution of "compliance sampling" for "characterization sampling." Compliance sampling is embodied by the "engineering-judgement sample" also described as the "sample-the-dirty-spots strategy." This approach focuses sampling efforts on those areas assumed (*a priori*) most likely to be contaminated. This approach evolved from disciplines like industrial hygiene where the goal is worker protection. Here, if one samples all high exposure areas and these are found to be in compliance, exposures from the process may be assumed to be acceptably low. For a well-defined process, this strategy is excellent, but for most environmental contamination problems the purpose of the sampling effort is to determine the nature and extent of contamination. Thus, it is a characterization problem, not a compliance monitoring problem.

A. Graphical Methods

Figure 1 shows a pseudo 3-D ball and stick plot of contamination for "bad stuff" at a hypothetical hazardous waste site. There are four quadrants, each with 150 potential samples. Quadrant 1 is uncontaminated, quadrant 2 is lightly contaminated, quadrant 3 is moderately contaminated, and quadrant 4 is heavily contaminated. What would happen if we followed a compliance monitoring approach and sampled almost exclusively in quadrant 4? Clearly the site is heavily contaminated, but is this the correct answer? An evenly distributed sample would give a better overview of the extent of contamination and would allow a more reasonable risk assessment. Plots like Figure 1 can give a very good idea of the distributions of the samples taken at a site, and can indicate whether a given sample is unbiased and representative with respect to defining environmental contamination.

One should also be interested in the distribution of contamination in the sample used to characterize the site. The box and whisker plot is a graphical aid that is useful in this context (see Figure 2). The line in the center in Figure 2 marks the 50th percentile or median of the data. The upper and lower "hinges" appear at the 25th and 75th percentiles of the data. The "whiskers" connect the upper and lower hinges to the largest and smallest data point within 1.5 times the distance between the hinges, termed the interquartile range (IQ) from its respective hinge. Outside points are between the hinge plus (upper) or minus (lower) 1.5 times the IQ and 3

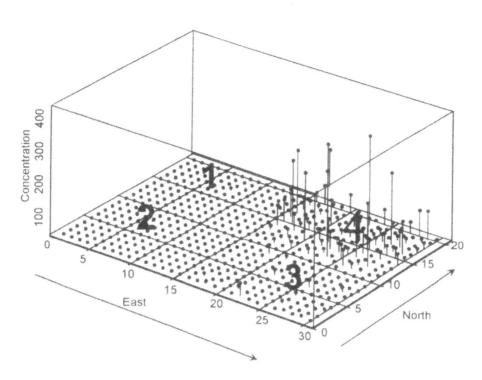

Figure 1 This hypothetical hazardous waste site has four areas. Areas 1 and 2 have little contamination, while 3 has moderate contamination and 4 is heavily contaminated.

times the IQ. Far outside points are beyond the hinges plus or minus 3 times the IQ. Outside points are atypical of the data and may represent statistical outliers.

Figure 3 shows box plots for the log-transformed data from 4 quadrants of our hypothetical site. It is easy to see that, as one goes from quadrant 1 to quadrant 4, each data for each quadrant has a reasonably symmetrical distribution and a median about 10fold greater than the median for the preceding quadrant.

While box plots are a simple way to convey the central tendency and form of a set of data, one can use even simpler graphics. See, for example, the dot plot in Figure 4. This plot was generated by sorting the data into "bins" of specified width (here about 0.2) and plotting the points in a bin as a stack of dots (hence the name dot plot). Dot plots give a general idea of the shape and spread of a set of data and they are very simple to interpret.

Aside from the spatial structure of the data and its general shape and central tendency, we are often interested in the temporal structure of a data set. Pesticide risk studies, for example, frequently involve 5 or 6 sets of data collected on the day of pesticide application and at several time periods postapplication. Figure 5 illustrates a temporal set of environmental measurements. In Figure 5 we see a set of log-transformed pesticide residue measurements plotted against the time since application. The plot shows clearly that residue measurements diminish over time and

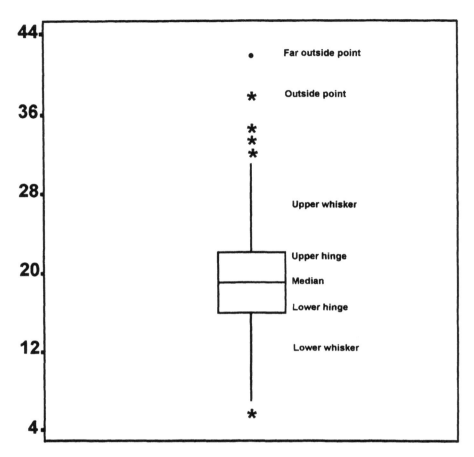

Figure 2 A sample box plot. The median is the 50% point of the data; the upper hinge (UH) is the 75% point of the data. The lower hinge (LH) is the 25% point of the data; the upper whisker extends from the UH to the largest data value less than the UH plus 1.5 times the difference between the UH and the LH hinges [the interquartile range (IQ)]; the lower whisker extends from the LH to the smallest data value greater than the LH minus 1.5 times the LQ; outside points are either between the UH plus 1.5 IQ and UH plus 3 IQ or LH minus 1.5 IQ and LH minus 3 IQ. Far outside points are beyond UH plus 3 IQ or LH minus 3 IQ.

that the rate of decline in the logarithm of residue levels is well-approximated by a straight line. If we fit a linear regression to such data, the equation is of the form:

$$\mathrm{Log}(C_t) = A - B \cdot t \tag{5}$$

where C_t = the concentration at time t; A and B = fitted regression coefficients.

If we rearrange (5) we get

$$C_t = \exp(A - B \cdot t) \tag{6}$$

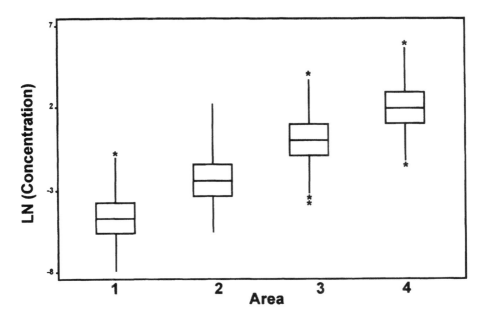

Figure 3 A box plot of contamination levels in areas 1-4. Note that median contamination level increases about an order of magnitude as one moves from area to area.

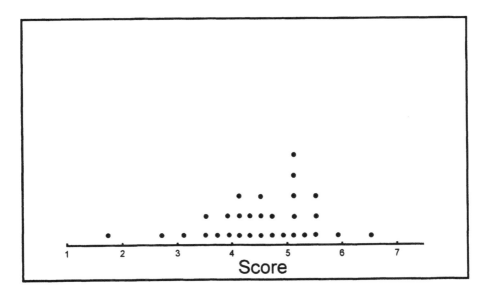

Figure 4 An example dot-plot.

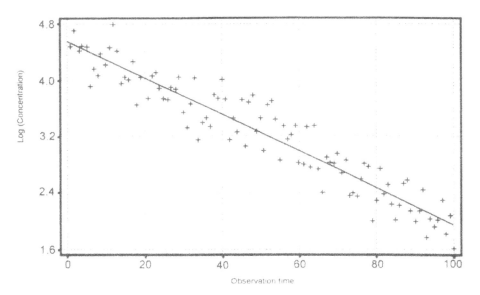

Figure 5 A scatter plot showing exponential decay.

That is, the pesticide in question is following an exponential decay process in measured residue values, which in turn suggests that a constant fraction of the material per unit time is being broken down. Thus, such plots can give an idea of the magnitude of concentration change over time and can also suggest functional forms and even general processes responsible for concentration changes.

We have focused, up to this point, on the use of graphical methods to determine the general form of the distribution of environmental samples. Graphical methods are also useful to gain insight into what "statistical" distribution, such as the normal or log-normal distribution, approximates the observed data distribution. Figure 6 shows a rankit, or normal-scores, plot of 100 random numbers from a standard normal (mean zero, variance one) distribution. A plot like this is constructed by plotting the values of the data against their expected normal scores or "rankits." The expected normal scores are the Z scores predicted from an observation's rank and the total number of observations. For example, the largest value in a sample of 50 has an expected normal score of about 2.2. If, as is the case here, the plot tends to fall on a straight line, it provides evidence that the data fit a normal distribution. Note also that rankit plots can be constructed using the logarithms of the data. If such transformed data produce a linear rankit plot, it suggests that the data fit a log-normal distribution.

B. Distributional Fitting and Other Hypothesis Testing

Figure 6 illustrates a graphical method of evaluating fit to a normal of log-normal distribution. The Wilk-Shapiro statistic is a goodness-of-fit statistic often included with a normal quantile plot. It represents the correlation between the observed data

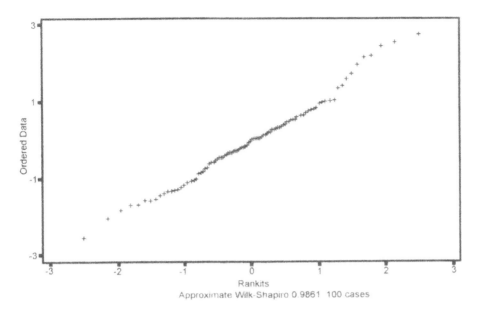

Figure 6 A normal scores plot of 100 random normal numbers. Note that the points lie approximately on a straight line.

and their expected normal scores. Note that, since the expected value of this statistic is much greater than zero, the usual tests cannot be used to assess its significance. Use of this statistic for assessing normality is discussed in Gilbert (1987).

There are a variety of alternative ways to test for fit to a normal distribution (remember that for a log-normal the same tests apply to the log-transformed data). One good method is to standardize the data by subtracting the sample mean and dividing by the sample standard deviation, and applying Lillifors test for normality. This test is a modification of the Kolmogrov-Smirnov goodness-of-fit test which takes into account the fact that the data have been standardized (Wilkinson et al., 1994).

A larger question is whether or not you should care what distribution a set of measurements follow. It is often helpful to know whether or not a set of measurements appears consistent with a log-normal or normal distribution and, under certain circumstances, it may be helpful to know if measurements fit some other distribution suggested by an *a priori* hypothesis. In the last case, one of the best goodness-of-fit tests is the Kolmogrov-Smirnov (KS) test discussed in Conover (1980). However, checking goodness-of-fit to a variety of esoteric distributions to say that the data are most consistent with, e.g., a Lapace distribution, seems pointless. If you have a limited amount of data, a statistical "fit" only says that one cannot reject the candidate distribution. If you have a large amount of data, the fit" may offer fairly strong evidence that the data are in fact from a particular candidate distribution, but is this useful information? That is, you can use normal theory statistics, or you can use nonparametric statistics, but in general other distributions are not immediately helpful for the development of statistical tests or confidence intervals. You might argue that a known distribution might be an advantage for a Monte Carlo modeling exercise. However, if data are sufficient to strongly suggest a particular distribution, they can also be used to construct a nonparametric density function (Silverman, 1986). Such nonparametric density estimates have the added advantage that, given adequate data, they are always appropriate.

C. Nondetects

Another distributional problem concerns nondetects. These occur when the analytical method used for a particular substance cannot distinguish the measured concentration from zero. Data sets which contain nondetects are said to be "left-censored" because all one knows about the low values (on the left side of the axis) are that they are less than the detection limit.

If the number of nondetects is low, the easiest approach is to simply assume that nondetects are worth one-half the detection limit. This assumes that the distribution of nondetects is uniform between the detection limit and zero. To be more conservative, you could also assume that nondetects follow a triangular distribution between zero and the detection limit and, thus, assign a value of two-thirds of the detection limit to each nondetect (see Figure 7).

In practice, these approaches work acceptably well if the number of nondetect values is less than 10%. Where larger numbers of nondetects occur, you can make use of the fact that observations from a normal distribution tend to fall on a straight line when plotted against their expected normal scores (see Figure 6). This is true even if some of the data are below the limit of detection. For example, if only 50 of the values in Figure 1 were above the detection limit, they would still tend to fall on a straight line when plotted against the Z-scores derived from their ranks. Calculating a linear regression of the form

$$C = A + B \cdot Z\text{-Score} \quad (7)$$

where C = the measured concentration; A = an estimate of the mean; B = an estimate of the standard deviation; Z-Score = expected normal score based on the rank order of the data (Gilbert, 1987; Helsel, 1990).

Finally, what if nondetects are numerous and detects do appear to follow a log-normal or normal distribution? There is no really good answer. Nonparametric techniques might be used (Cleveland, 1993; Silverman, 1986) to try to get an idea of the likely shape of the nondetect part of distribution. However, this is an area for further research.

D. Sample Support

A related point on environmental contamination measurements concerns the analytical "support" for a particular measurement. Support refers to the actual volume of material that a particular concentration measurement represents. For example, in soil sampling it is not uncommon for a grab sample to contain about 500 grams of

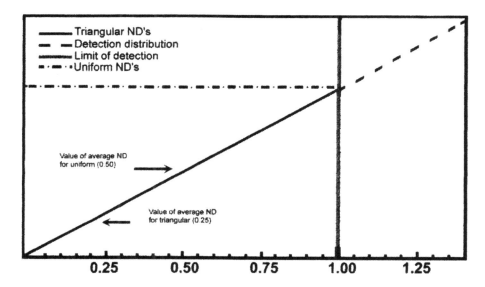

Figure 7 An illustration of two options for assigning a value to nondetects.

material. In the laboratory, the chemist will select a 10 gram subsample, which will be extracted with a solvent. The actual analysis may involve a milliliter of the extract being put through the analyzer. Why is this important to risk assessment? Because "outliers" sometimes occur which are a result of this process. In one data set we encountered a sample that was listed as being 30,000 ppm (3%) lead. The lab was requested to reanalyze the sample, and they reported that the reading was accurate. Further investigation revealed that "replicate" analysis consisted of injecting a second aliquot of extract into the massspectrometer. The lab was then asked to replicate the measurement, starting with the original 800 gm sample. The result was somewhat lower: 0.8 ppm. Thus, a "heavily contaminated" area was found to be essentially clean. The moral of this story is that if a few really high numbers drive the risk assessment, concerns about sample support may be in order.

E. Does Contamination Exceed Background?

It is often assumed that health and environmental risk assessments should not concern themselves with background contamination levels. In practice, however, an intelligent approach to risk assessment requires two things: first, a definition of background contamination; second, a definition of exceedance.

Definitions of background can range from that contamination found in a pristine environment to the contamination that would be present if the activity of interest had not occurred. In quantitative terms any definition of background should specify some distribution and include a measure of central tendency and a measure of variability. For example, background measurements might have a geometric mean of 10 units and a geometric standard deviation of 5 units.

Having defined background, we must then define exceedance. A sample might be deemed to be background, for example, if it is not significantly different from background on the basis of a t-test. The definition might also require a sample size above some specific N (usually 10 – 30) to guarantee that findings of no difference were based on comparisons of sufficient statistical power. Definition of an exceedence could also specify that any sample exceeds background if it is more than the background mean, plus 2 background standard deviations. This definition, however, will result in the finding that 2% of background samples are above background. Discussions of useful statistical tools for approaching the background question are provided in the U.S. EPA DQA document and in Gilbert (1987).

IV. ESTIMATION OF RELEVANT EXPOSURE: DATA USE AND MENTAL MODELS

A reasonable exposure assessment proceeds from a careful characterization of the spatial and temporal aspects of contamination and a model of how this contamination reaches the receptor. Otherwise, the risk assessment assumptions may give rise to highly improbable or impossible events. We have encountered risk assessments for carcinogens, for example, with a D_i based on an assumed long-term exposure to the maximum concentration encountered at the site for every toxicant sampled at the site. Similarly, maximum concentrations may be combined for pairs of toxicants which do not occur together anywhere on the site; soil ingestion by children at a site may be based on toxicant concentrations from soil core samples taken from a depth of 20 feet. In one pesticide risk assessment we encountered, it was alleged that the pesticide in question posed unacceptable risks to nesting birds. The pesticide concentration present in the crop on the day of application was used to support this argument, despite the facts that pesticide application occurred about 4 months before bird-nesting season, and that the pesticide in question has an environmental half-life of about 16 days.

These sorts of assumptions may result in alarmingly high risk numbers, although they do have value as a screening exercise. That is, if a given situation is found to be safe under extreme assumptions, it is clearly not a problem. Unfortunately, if a site, facility, or activity fails to pass such a screen, it may prove difficult to later dislodge the assumptions with better science. For example, if a screening exercise suggests cancer risks as high as 10^{-3} (assuming a family that lives in the sludge disposal pit, eats lots of soil, and grows all of their food in their backyard), it may be very hard to convince stakeholders that more careful analysis shows little or no remedial action is necessary.

Assuming that we have a defensible exposure model, the next question is how to turn environmental contamination measurements into exposure estimates. Such a determination depends on the toxic endpoint of concern. For acute toxic endpoints, such as neurotoxicity, an upper 95% bound or even the maximum of the distribution of sample measurements might be appropriate. For cancer risk, at the other end of the spectrum, U.S. EPA guidance suggests that an upper 95% bound on the arithmetic mean of the sample measurements is an appropriate "conservative" exposure. This is often reasonable guidance. If there is a persistent chemical in the environment, a person or animal living in that environment will receive a dose proportionate to the arithmetic mean of the environmental concentration. Risk assessments of chronic exposures, based on geometric means of environmental contaminants, are always incorrect. The geometric mean is a good measure of central tendency for a log-normal distribution, but the central limit theorem guarantees that the average of a large number of samples from a log-normal distribution (or any distribution for that matter) follows a normal distribution with a mean equal to the arithmetic mean of the environmental contamination (log-normal) distribution. Moreover, an exposure estimate based on the geometric mean is always too low and, thus, understates actual exposures.

This brings up the question of a defensible upper bound on this mean (see Figure 1). Here there are four quadrants, each with 150 potential samples. As Figure 2 shows, going from quadrant 1 to quadrant 4, each quadrant has a geometric mean about 10fold greater than the one preceding. A simple random sample of this site has an expected logarithmic mean (M) of -1.12 and an expected logarithmic standard error (S) of 3.04. Using these quantities in the well-known formula for the arithmetic mean μ of a log-normal distribution:

$$\mu = \exp [M + (S_2 \div 2)] \qquad (8)$$

generates a μ value of about 33.4* (which overstates matters a bit, given that the actual arithmetic mean is about 8.6!). One might assume that this example is contrived in that such a great disparity among areas would show as a lack-of-fit to a log-normal. This is not the case. Figure 8, where a rankit plot of the entire 600 sample universe is shown, suggests a tolerably good fit to a log-normal. Clearly, if the entire sample cannot readily reject a log-normal, it is unrealistic to expect subsamples to do so.

If one goes for a conservative 95% upper bound on the arithmetic mean, $UB_{0.95}$, as a "worst-case" exposure:

$$UB_{0.95} = \exp [M + (S_2 \div 2) + S \times C_{0.95} \times (N - 1)^{-1/2} \qquad (9)$$

where N = the sample size; $C_{0.95}$ = a tabled constant.

$C_{0.95}$ is a tabled constant which depends on both S and N (Gilbert, 1987). Assume we take a sample of 80 from our universe and obtain our expected M and S values (M = -1.12; S = 3.04). The resulting worst-case exposure is 241 or about 30 times the actual expected exposure. Moreover, this result is still not a worst-case because it assumes expected values for M and S when, in fact, the result from a small sample could yield an even higher $UB_{0.95}$.

There are at least two morals in this story. First, in environmental risk assessment you ignore spatial (and temporal) heterogeneity at your peril. Second, if you can avoid making a lot of parametric assumptions, it is best to do so. This raises the question of how to estimate a reasonable upper bound exposure. Actually, Equation (9) may

---

* In this example all concentrations are unitless.

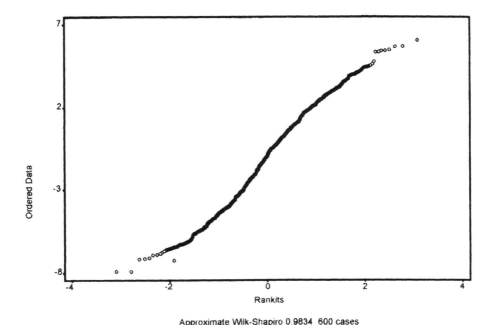

Figure 8 A rankit plot of the data shown in Figure 1. Note that the data appear to fit a log-normal fairly well.

give a pretty good answer if the data are really log-normal. Since this is not always easy to verify, the bootstrap method (Efron and Tibshirani, 1993), a nonparametric method for generating error bounds, is generally preferred. In the bootstrap method, the data is resampled with replacement and the mean of these samples is calculated. In the example above, we would take samples of our 80 values with replacement (which means that a measurement is likely to occur more than once in a given sample) using one of two possible sampling strategies. In the first strategy, about 30 resamples are taken; their mean and variance are calculated, and standard normal statistical theory is used to calculate an upper bound. This generally works well, but the result may be affected by outliers. An even safer strategy is the second; a thousand bootstrap means are generated and the 950th largest is taken as a 95% upper bound. This gives an utterly defensible, totally nonparametric 95% upper bound. The bootstrap method may appear to give "something for nothing" because resampling data does not seem like a valid way of generating new information. This is not so, but an explanation is beyond the scope of our discussion (see Efron and Tibshirani, 1993 for reassurance).

At this point a good question is, "what do I do if I am stuck with a 'dirty-spots' sample?" If there is a great deal of money riding on the decision, redo the sampling. Note also that nothing is ever so bad that it cannot be made worse. In one case, for example, a dirty-spots sample was taken first. This was pointed out to the client, who then went out and took a comparable number of samples from an area known to be clean. At this point the formula given by Equation (9) was applied to the combined data (which were strongly bimodal because of the clean/dirty dichotomy). The resulting upper bound on the mean exceeded the largest observation from the dirty spots sample! These data were beyond salvage by even the bootstrap method. The original sample had been taken to find dirty spots and, thus, was simply not representative of the site. The clean sample had been taken to compensate for the bias of the first sample and, thus, was likewise unrepresentative of the site. The result was a set of about 100 measurements which told us almost nothing about the nature and extent of contamination at the site. The client finally instituted a statistically designed sampling plan.

Problems can also arise with HI calculations. At one waste site, for example, a very large number of toxicants were sampled. A total of 50 samples were taken, using relatively imprecise,* but cheap, analytic methods in the interest of cost-savings. The site was not heavily contaminated, so for most toxicants all samples were below the LOD. Nonetheless, the risk assessment showed 5 HI calculations greater than 1. This was because, in the absence of other information, the risk assessor assumed 1/2 the LOD as the concentration of the toxicants that have not been found. Since many toxicants had not been found and the LODs were high, relative to the RfDs, the HIs exceeded 1.

A risk-oriented approach could have avoided this problem. First, consider the decision to sample for so many toxicants. There was no good reason for this, so fewer toxicants should have been sampled. Second, the usefulness of LODs which are 20%-30% of the RfD, should have been questioned, since for this case, a sample with a number of contaminants at levels below the LOD is not necessarily "safe" from a regulatory viewpoint. LODs should be specified to reflect this concern. Finally, when 50 samples are all LODs, order statistics can be used to show that the mean is much less than 1/2 the LOD, if you have an idea of the variability of higher concentration compounds at the site (Ginevan, 1993).

Monte Carlo methods are an important recent development in exposure assessment methods. To illustrate the use of this tool, assume that we have an estimate of environmental contamination C, an exposure scenario S, and we want to calculate an average daily intake of toxicant in milligrams (mg) per kilogram (kg) of body weight. We can find the total daily intake of toxicant D in mg as a function of environmental contamination level and our exposure scenario. An example:

1. Assume the upper 95% bound on the mean of "bad stuff" (BS) in soil is 10 ppm (or 10mg/kg).
2. Assume our exposure of interest is to a 20 kg child and the child eats 100 mg of soil per day.
3. Intake in mg/kg is given as 10 mg/kg (mgs bad stuff per gram of soil) × 0.1 g (100 mg) soil ÷ 20 kg (child weight) = 0.05 mg/kg.

If we put this model in Monte Carlo terms, we would specify a probability distribution for the concentration of bad stuff in soil, a probability distribution for the amount of soil ingested per day, and a probability distribution for the body weight

* The limit of detection (LOD) was 20-30% of the RfD for most chemicals

of children. A large number of BS intakes are generated using randomly selected values for BS concentration in soil, ingestion amount, and child body weight. The result is a distribution of daily intakes in mg/kg. An extensive discussion of this sort of Monte Carlo modeling can be found in Holland and Sielken (1993).

In some cases, this might be a real advance. For example, if we were concerned about an acute toxicant, we could use a Monte Carlo model to calculate proportion of soil ingestion episodes which would result in an unacceptable intake of toxicant. For this model we would use the distributions discussed in our example. If we were, instead, interested in the distribution of chronic intake we would use the same distribution for child body weight, but would need to define a distribution for the average soil intake across children (e.g., how much soil per day does a specific child eat on the average?), and we would need a distribution for the uncertainty in the arithmetic mean soil concentration (probably from the bootstrap method). Also, we would need to be very careful how the resulting chronic intake distribution was employed because children do not remain children for a lifetime, so the resulting chronic distribution applies to a limited period. Monte Carlo methods are often useful, but like any other tool they carry with them the potential for misuse. A bad model translated into a Monte Carlo simulation is still bad, and even a good model requires correctly specified inputs.

V. FINDING OUT WHAT IS IMPORTANT: A CHECKLIST

The preceding discussion was intended to give the reader an overview of those statistical issues we feel are most important in planning or evaluating risk assessments. Here we summarize those points in the context of planning and evaluating actual risk assessments.

The first requirement is to understand the origin of the environmental measurements used as the basis of the exposure analysis. This includes a good understanding of the distribution of the measurements in space (see Figure 1) and time (see Figure 5) and the rationale of the sampling plan used to collect the data. There should be clear graphical displays of the data and a description of the sampling plan, which clearly states its purpose. Samples that appear to be clustered, rather than evenly distributed across the area of interest, should raise concern. Likewise, if the report casts its discussion in terms of a bounding exercise, there is reason to be concerned that the samples are taken in a manner that tends to overstate contamination levels.

Sample support is another important issue. Does the report say how a particular type of sampling was conducted and exactly how chemical analyses were performed? It is likely that it does not, but if outliers appear to be a problem, or if the data appear odd in other ways, this is an area worth exploring.

Also look for information on the distributional form of contamination data. This should include both goodness-of-fit tests and graphical representations of the data. If the contamination data appear to have been derived in a reasonable way, the next area of concern is how they are used to estimate exposure. For human cancer risk, or other endpoints based on chronic exposure, exposure estimates should be based on either the arithmetic mean of exposure measurements or an upper bound on the arithmetic mean of these measurements. Upper bounds based on bootstrap procedures are generally preferable to those derived from assumptions of log-normality. The latter are acceptable, however, if the fit of the data to a log-normal distribution is good. If cancer risk is determined by several chemicals, be aware that it is rather conservative to assume the mean concentration of each chemical will be at its upper 95% bound. Not all compounds are likely to actually be at their respective upper bounds. A better answer could be derived in a manner similar to the Monte Carlo procedure described above for obtaining an upper bound on the sum of cancer potency factors. Consider the temporal aspects of exposure. For example, many pesticides have rather short environmental half-lives, which need to be taken into account in estimating exposure (e.g., see Figure 5). Similarly, assuming long-term exposure to volatile organic compounds (VOCs) may result in incorrect answers. One assessment we reviewed assumed a constant 30-year emission rate for a VOC. The result of this model was a total emission that was 5 times the total amount of VOC present at the site. A final concern is the choice of exposure scenario. Is it the hypothetical family living in the sludge pit, or is it a reasonable scenario? This is important because one scenario which fails the "laugh test" will cast doubt on the whole risk assessment.

If exposure estimates seem reasonable, then consider the dose response models used to determine risk. As noted earlier, these tend to be numbers that are approved by U.S. EPA or some other regulatory agency. However, if the assessment is driven by one or two compounds, a review of the origin of their dose response data may be in order.

The foregoing assumes a deterministic risk assessment, based on statistical upper bounds. For a Monte Carlo-based risk assessment, examine the derivation of each input distribution, the structure of the model and the assumptions behind it. A good quick check is to run the best estimates of all of the input parameters through the algebraic form of the model (e.g., consider the soil ingestion example). The result should be near the center (median) of the Monte Carlo result. Similarly, run the reasonable upper bounds for each input; the result should be above the 95th percentile of the Monte Carlo result (sometimes quite far above). If Monte Carlo results in a substantially different central value, or an even more extreme upper bound, something is wrong.

VI. TOOLS

So, now that you know how to apply statistical principles to planning and evaluating health and environmental risk assessments, what tools should you use? If a lot is riding on the assessment, the first thing you should get is a good statistical consultant. What are some traits to look for in such an individual? One important point in problems of this sort is that their primary focus is not data analysis. The first step in solving these sorts of problems always involves the client and the statistical consultant working together to define the questions to be asked. Look for someone who asks lots of questions and who wants to work with you to understand the problem. If a statistician's first instinct is to start crunching numbers, you are almost certain to be badly served.

Another point is that much of the preceding discussion probably did not seem very "statistical." Consider the problem of combining cancer potency estimates. The actual problem was to identify a plausible distribution for q_1. Similarly, the point of the example of the family living in the sludge pit is that bad assumptions can ruin a technically correct analysis. At first glance these might both be viewed as qualitative problems; in fact they are vital to sensible quantitative analysis. Sometimes qualitative problems are quantitative problems disguised by uncertainty. A statistician can help you tell the difference.

This discussion also placed a lot of emphasis on graphics. This is because it is very important to present analyses and results as clearly as possible and graphics are a powerful way to clarify things (imagine Figure 1 as a table). A good consultant uses lots of graphics. If your consultant's reports are so "scientific" that few can understand them, they are unlikely to be useful in decision making.

Finally, problems of the sort discussed here often require a large library of specialized statistical, modeling, and graphics software, and considerable computer power. Even large companies may not have good resources for statistical analysis and modeling, and may try to tailor your problem to fit what they can do, rather than do what you need to be done. If you are hiring a statistical consultant, don't be afraid to ask him about his resources.

What tools might be appropriate for do-it-yourself analysis? The short answer is that if you have a recent version of one of the major statistical packages, such as SPSS, SAS, or S-Plus, they will do all of the graphics and tests discussed here, and much more. An environmental statistics module for S-Plus is now available (Millard, 1997). If you are purchasing new software, however, note that all major statistics packages are expensive and require training for effective use. At the other extreme, the release of the final U.S. EPA DQA document contains a mini-statistical package which will do most, but not all, of the analyses and graphics discussed here. For the actual analyses discussed here, I use three packages: Statistix for Windows (Analytical Software, 1996), Axum for Windows (MathSoft, 1996), and Systat for Windows (SPSS, 1996). Of the three, Statistix will do much of the analysis presented here (but not 3-D plots), and is very easy to use. Axum is likewise quite easy to use and will do a very broad range of statistical graphics. Systat is another comprehensive package which will do everything, but it is probably not a good choice for the casual user. For those intrigued by the bootstrap method, I recommend Resampling Stats for Windows (Bruce, Simon, and Oswald, 1996). It is one of the easiest ways to implement this sort of methodology and is inexpensive.

The foregoing discussion should give the reader an idea of the spectrum of tools available, but it is by no means an exhaustive list of tools. There are many other excellent statistical packages on the market. For additional advice, talk to people who are doing the kinds of statistics you need to do. For further reading I particularly recommend the U.S. EPA DQO and DQA documents, and two general references: Holland and Sielken (1993), and Gilbert (1987). I hope the reader finds this discussion useful and that it clarifies the use of statistics in health and environmental risk assessment.

REFERENCES

Analytical Software, *Statistix for Windows*, Tallahassee, FL, 1996.

Bruce, P., Simon, J., and Oswald, T., *Resampling Stats for Windows, Operation Guide*, Resampling Stats, Inc., Arlington, VA, 1996.

Caulcutt, R. and Boddy R., *Statistics for Analytical Chemists*, Chapman & Hall, London, 1983.

Cleveland, W.S., *The Elements of Graphing Data*, Wadsworth and Brooks/Cole, New York, 1985.

Cleveland, W.S., *Visualizing Data*, AT&T Bell Laboratories, Murray Hill, NJ, 1993.

Conover, W.J., *Practical Non-parametric Statistics*, John Wiley & Sons, New York, 1980.

Efron, B. and Tibshirani, R.J., *An Introduction to the Bootstrap*, Chapman & Hall, London, 1993.

Gilbert, R.O., *Statistical Methods for Environmental Pollution Monitoring*, Van Nostrand Reinhold, New York, 1987.

Ginevan, M.E., Bounding the mean concentration for environmental contaminants when all observations are below the limit of detection, *American Statistical Association 1993 Proceedings of the Section on Statistics and the Environment*, Hayward, CA, 123, 1993.

Helsel, D.R., Less than obvious: statistical treatment of data below the detection limit, *Environ. Sci. and Technol.*, 24, 1766, 1990.

Holland, C.D. and Sielken, R.L., *Quantitative Cancer Modeling and Risk Assessment*, PTR Prentice-Hall, Englewood Cliffs, NJ, 1993.

MathSoft, *Axum User's Guide*, Cambridge, MA, 1996.

Millard, S.P., *Environmental Statistics for S-Plus, Users Manual*, Probability, Statistics & Information, Seattle, WA, 1997.

National Academy of Sciences, *Health Effects of Exposures to Low Levels of Ionizing Radiation: BEIR V*, National Academy Press, Washington, 1990.

National Academy of Sciences, *Health Effects of Radon and Other Internally Deposited Alpha Emitters: BEIR IV*, National Academy Press, Washington, 1988.

Putzrath, R.M. And Ginevan, M.E., Meta-analysis: methods for combining data to improve quantitative risk assessment, *Regulatory Toxicol. and Pharmacol.*, 14, 178, 1991.

Scheaffer, R.L., Mendenhall, W., and Ott, L., *Elementary Survey Sampling*, PWS- ent, Boston, 1990.

Silverman, B.W., *Density Estimation for Statistics and Data Analysis*, Chapman & Hall, London, 1986.

SPSS, *Systat 6.0 for Windows, Statistics*, SPSS Inc., Chicago, 1996.

U.S. Environmental Protection Agency, *Data Quality Objectives Process for Superfund: Interim Final Guidance*, Office of Emergency and Remedial Response, Washington, 1993.

U.S. Environmental Protection Agency, *Guidance for Data Quality Assessment: Practical Methods for Data Analysis, EPA QA/G-9, QA96 version*, Washington, 1996.

U.S. Environmental Protection Agency, *Risk Assessment Guidance for Superfund, volume I: Human Health Evaluation Manual (Part A), Interim Final*, Office of Emergency and Remedial Response, Washington, 1989.

Wilkinson, L.M., et al., *SYSTAT for DOS: Using SYSTAT, Version 6*, SYSTAT Inc., Evanston, IL, 1994.

CHAPTER 20

Uncertainty Analysis

Maxine Dakins and Carol Griffin

CONTENTS

I. Introduction 413
II. Uncertainty in Risk Assessment 414
III. Technical Aspects of Uncertainty Analysis 415
    A. Uncertainty and Variability 415
    B. Sources of Uncertainty 415
    C. Describing and Summarizing Data 416
    D. Sensitivity Analysis 418
IV. Uncertainty Analysis 418
    A. Monte Carlo Methods 420
V. Communication of Uncertainty 421
VI. Conclusion 423
    References 424

I. INTRODUCTION

Uncertainty in risk assessment refers to the lack of definiteness that exists about the procedures, quantities, and data used and, therefore, to the lack of sureness about the resulting values and conclusions. Uncertainties exist in risk assessments whether or not they are acknowledged, incorporated into the analysis, or used by the risk manager in decision making. Ignoring or mishandling uncertainty may paralyze decision makers or generate controversy in risk assessment and management. Instead, uncertainty can be explicitly modeled, discussed, and incorporated into decision making through a quantitative uncertainty analysis, resulting in decisions that are more thorough and, hopefully, less contentious.

This chapter discusses technical aspects of analyzing uncertainty, including the difference between uncertainty and variability; sources of uncertainty in risk assessments; describing and summarizing data; sensitivity analysis; and quantitative uncertainty analysis. Communication of uncertainty and the use of uncertainty information are also discussed.

II. UNCERTAINTY IN RISK ASSESSMENT

The processes of exposure and effect are analyzed and modeled in a risk assessment. Uncertainty can enter the risk assessment through both analyses. In an exposure process, subjects are exposed to the possibility of some change, usually negative. The effects process is the change that the subject or process undergoes as a result of an exposure. For example, children are exposed to lead through ingestion (paint chips, dirt, dust, and contaminated food and water), inhalation (dust particles), and dermal contact. Possible effects of lead poisoning in children include decreased intelligence, impaired neurobehavioral development, decreased stature and, in severe cases, death.

Normally, an exposure model is developed, single-value estimates of model coefficients are selected, and calculations are performed to generate base-case (nominal) predictions of exposure levels. Risk assessors input these predictions into an effects model, again using nominal values for model coefficients, to arrive at estimates of effects. Risk managers then base management decisions on these predictions. Some risk assessments include a calibration step which is the adjustment of coefficient values to obtain a good fit between predictions and observations, or a sensitivity analysis which is the determination of the effects of changes in model input values, coefficients, or assumptions on risk predictions. A quantitative uncertainty analysis, the computation of the total uncertainty induced in the output of the risk assessment by quantifying the uncertainty in the inputs, coefficients, or model structure, is less frequently performed. The formal assessment of uncertainty and the determination of its effect on a risk management decision can be useful in assessing the reliability of predicted values, exposing areas of controversy or disagreement, making underlying assumptions explicit, combining information from multiple sources, documenting the details of the analysis, identifying whether additional information should be collected, and determining how a data collection or research program should be structured.

Uncertainties and unknowns pervade situations where quantitative risk assessments are used. They undermine the quality of risk management decisions that rely on single-value risk assessment predictions. Nevertheless, these predictions are rarely accompanied by information about their reliability. Instead, an uncertainty factor is often used to adjust the risk estimate to account for uncertainty. Uncertainty factors, often set at ten, are applied to reflect the uncertainty of extrapolating from animals-to-humans, from acute-to-chronic exposure, and for sensitive subgroups. The problem with using uncertainty factors is that the factors themselves are uncertain; thus, their use makes the degree of conservatism of the final decision unknown and controversial.

Another way to avoid underestimating risks is to use conservative values for some, or all, model inputs and coefficients. Use of reasonable maximum exposure values is an example of this approach. Using conservative values poses some problems, however, including a lack of consistency of results from one analysis to another, controversy about the degree of conservatism in the final result, and unquantified social costs of conservatism. Again, the solution may lie in using uncertainty analysis to quantitatively assess the uncertainty in the model output, and then formally incorporating this information into the decision making process.

III. TECHNICAL ASPECTS OF UNCERTAINTY ANALYSIS

A. Uncertainty and Variability

Uncertainty can arise from a lack of knowledge or from natural variability. Uncertainty arising from ignorance, the first situation, can be reduced through scientific research and information gathering. Examples of variables posing ignorance-based uncertainty are concentrations at a source, average-daily exposure at a particular place, and average-uptake efficiency.

Uncertainty arising from variability, the second situation, has an irreducible component. This inherent variability exists regardless of the amount of information obtained. Variables possessing inherent variability include characteristics of the exposed population (such as age and body weight) or the natural environment (such as temperature, wind speed, and rainfall). Many variables in environmental risk models have both ignorance-based uncertainty and inherent variability.

B. Sources of Uncertainty

In models of exposure and effects processes, uncertainties arise in several areas. Major sources of uncertainty are limited scientific understanding of fundamental biological mechanisms and of environmental fate-and-transport phenomena, as well as inadequate mathematical representations. Since the relationships between variables, which serve as the basis for risk assessments, are often unknown, incorrect, or incomplete, uncertainty can arise from the model structure. For example, model structure uncertainty includes choices about which aspects of a system to include in the model, selection of equations to represent relationships between variables, and choices of appropriate surrogate variables if relevant characteristics cannot be directly measured.

In addition to the uncertainty arising from limited knowledge or inadequate models, the values of the coefficients used in risk assessments are often unknown and must be estimated. Uncertainty can arise due to our limited ability to measure model inputs and coefficients; sampling error due to the need to draw inferences about a population characteristic from sample data; and disagreement between data gathered at different times or in different laboratories because of differences in procedures, personnel, or materials.

Yet another source of uncertainty in risk assessments is extrapolation. Risk assessors extrapolate from high-to-low doses, species-to-species, acute-to-chronic exposures, or from laboratory data to field situations. Uncertainty associated with extrapolations may relate to the model structure uncertainty, discussed above, as in the case of a dose-response model, or may involve extrapolations for which no reliable model exists.

Systematic error or bias in measurements is another source of uncertainty. Systematic errors can arise from incorrectly calibrated equipment, poor laboratory procedures, or inaccurate assumptions used in calculating inferred quantities from observations. Systematic error cannot be reduced by additional observations, but careful design of equipment, procedures, and calculations can prevent it.

Some aspects of the exposure and effects processes are inherently variable. Variability occurs when a quantity that could be modeled as a single value consists, in reality, of multiple values depending on time, space, or other factors. This type of variability may reflect biological differences between individual organisms, differences in activity and behavior patterns, seasonal differences, year-to-year variations, differences due to spatial variations across a geographic area, or random fluctuations.

Finally, value judgements are often treated as if they are constants instead of decision variables. An example is the choice of the population to model in a risk assessment for a household chemical. The most sensitive group (fetuses, infants, immune-suppressed, or elderly) could be targeted even though it represents a small percentage of the population. Alternatively, a typical individual representing the majority of the population could be modeled. These value judgements can influence the outcome of the assessment as well as subsequent risk management decisions.

C. Describing and Summarizing Data

The sources of uncertainty already discussed result in collections of data points that must be analyzed and summarized to facilitate their use in risk assessments and uncertainty analysis. Data points can be envisioned as having been sampled from a probability distribution that has a location, spread, and shape. A probability distribution function describes the way measured values are expected to vary.

A data set can be summarized in several ways to provide information about a variable. Measures of location provide information about the center of the distribution, measures of spread provide information about the different plausible quantities the variable could take on, percentiles provide information about high or low values, and graphs provide an estimate of the likely shape of the probability distribution (see Figure 1).

The simplest summary statistics calculated from data are point estimates of location. Point estimates representing the middle or center of the data include the sample mean and the sample median. The sample mean, or average, is the sum of the data points divided by the number of points. The sample median is the data value which is greater than half of the data points and less than the other half. When the underlying distribution of the data is symmetric, the sample mean and sample median should be similar. Examples of symmetric distributions include the

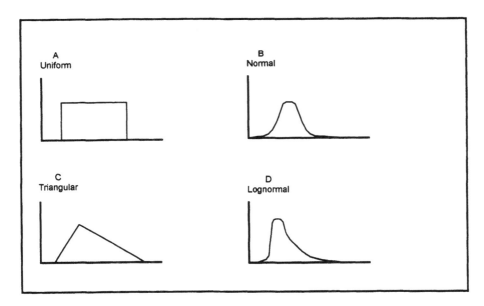

Figure 1   Probability distribution functions.

uniform distribution (see Figure 1A) and the normal distribution (see Figure 1B). When the underlying distribution of the data is not symmetric, the sample mean and sample median will differ. For a distribution with a long tail to the right, for example, the mean will be larger than the median. Examples of distributions that may be asymmetric are the triangular distribution (see Figure 1C) and the lognormal distribution (see Figure 1D).

A second type of summary statistic describes data spread or variability. The simplest measure of variability is the sample range, the difference between the largest and smallest values. A better, more commonly used, measure of spread is the sample standard deviation. The sample standard deviation uses information from all the data points in the set and, as such, is a more powerful measure of variability than the range. Calculations of confidence intervals rely on estimates of the sample standard deviation.

Point estimates are sometimes used to estimate a characteristic of the probability distribution other than the center. For example, a value at the upper end of the distribution may be used to ensure that the risk for the population is kept below an allowable level. If the maximum value of the distribution is known, as might be the case if the underlying distribution is uniform or triangular, it can be used. More commonly, the maximum possible value would be so far from the bulk of the data that it would not be useful. If so, percentiles of the distribution, usually the 90th, 95th, or 99th, are used instead. A percentile is the value that has a specified percent of the probability below it; for example, the 95th percentile is the value with 95% of the probable data values below it. However, estimates of upper percentiles made from small data sets can be highly variable.

Table 1 Sensitivity Analysis

X = (A − B)/C				
where: Variable	−10%	Best Estimate	+10%	Change in X
A	90.00	100.00	110.00	200
B	54.00	60.00	66.00	120
C	0.09	0.10	0.11	81

D. Sensitivity Analysis

A sensitivity analysis can be used to determine the effect of changes in input variables, coefficients, or model form on the model output. It is conducted by holding all uncertain quantities at their nominal or base-case values, except one. As values for a single variable are modified, changes in the risk assessment results are analyzed. For example, the best estimate for each variable in a simple model is increased and decreased by 10% to determine the change in the output X. As Table 1 shows, a 10% change in A has the biggest effect on X.

A sensitivity analysis can be used to screen a large set of candidate variables to identify those that contribute significantly to the output uncertainty. The sensitivity analysis identifies important variables which should be included in the uncertainty analysis. It can also provide insights into resource allocations that will achieve the most cost-effective reduction in uncertainty.

IV. UNCERTAINTY ANALYSIS

The analysis of uncertainty involves the estimation or computation of the total uncertainty in the output of a risk assessment. Different types of uncertainty analyses are appropriate depending on the level of information available, the sophistication of the analyst, and the requirements of the risk manager. Uncertainty analyses can be conducted by simply listing the sources of uncertainty. Also, judgement or analogy to a similar situation can be used to infer an upper limit or range. Alternately, a single-value best-estimate analysis can be performed, followed by a sensitivity analysis; or an order-of-magnitude bounding analysis can be conducted to obtain a range bounding the possible values. More rigorous methods include repeating the assessment for various plausible models or coefficient values and reporting each set of results; estimating probability distributions for important model coefficients and using analytic methods (calculus) to propagate the uncertainty through the analysis; and estimating probability distributions and using Monte Carlo simulation techniques to propagate the uncertainty through the assessment.

Listing sources of uncertainty, using judgement or analogy to a similar situation, or performing an order-of-magnitude bounding analysis have the advantage of being quick to perform and easy to understand. These techniques are appropriate when insufficient information exists to carry out a quantitative uncertainty analysis, when time or resources do not allow a more rigorous analysis, or when the risk manager requires only a general idea of the range of risks that may exist. Some disadvantages of these informal methods are that the resulting information is subjective, may be overly conservative, and may be difficult to defend. In addition, uncertainty analyses of these types may not yield much useful information for the risk manager and may, in fact, lead to less confidence in the analysis. A single-value best-estimate analysis followed by a sensitivity analysis will provide some information about the important sources of uncertainty, but falls short of providing an overall estimate of the uncertainty in the risk prediction.

More rigorous methods exist to quantitatively assess uncertainty. For example, the risk assessment can be repeated using different models or different plausible values for each uncertain variable and the resulting model outcomes can be reported and used to estimate the overall uncertainty in the assessment. This type of analysis is particularly appropriate when experts disagree about the fundamental mechanisms involved. An example of disagreement about fundamental mechanisms would be a case where both threshold and nonthreshold dose-response models are plausible. In such cases, the differing estimates should not be averaged or combined in any way, but should be treated separately, and the outcomes from each assessment should be reported along with information about the theories and assumptions that went into generating them.

A formal quantitative uncertainty analysis involves the quantification of the uncertainty in model inputs and coefficients and the propagation of that uncertainty through the assessment. Formal quantitative uncertainty analysis is appropriate when general agreement exists about the mechanisms involved, when an accepted model exists to incorporate those mechanisms, and when there is sufficient information to allow definition of probability distributions for important model inputs and coefficients. In practice, these criteria are more often met for exposure assessments than for effects assessments. If these criteria are only partially met, a quantitative uncertainty analysis may still be useful and may lead to important insights about uncertainties in the risk assessment.

When quantitative uncertainty analysis is appropriate, analytic techniques such as calculus can be used; however, the equations are sometimes difficult to solve mathematically. Instead, numerical techniques using Monte Carlo simulation may by used to propagate the uncertainties in the model coefficients and input variables through the model. Quantitative uncertainty analysis gives the most complete and rigorous estimate of the uncertainty in the risk prediction by providing a range of risk values along with estimates of the probability of each value.

Conducting a quantitative uncertainty analysis requires the determination of the form of a probability distribution function (pdf) and its associated parameter values for each uncertain variable. Distributions should be selected to reflect the amount of information available and parameter values chosen to scale the distributions to the estimated minimum and maximum values. Where little or no information exists beyond a plausible low and high value, a uniform distribution may be most appropriate (see Figure 1A). If a most likely value exists in addition to a range, a triangular distribution could be used (see Figure 1C). If the uncertainty in the variable is believed to follow a bell-shaped curve, a normal distribution should be used (see Figure 1B). If the uncertainty in the variable is believed to follow a distribution which has a longer tail to the right than the left, as in cases where the values are

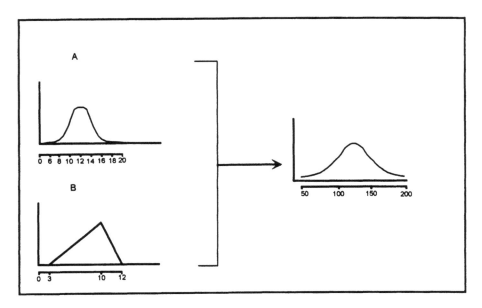

Figure 2 Example of input and output probability distribution functions for a Monte Carlo simulation.

small but have a lower limit of zero, a lognormal distribution may be most appropriate (see Figure 1D).

For ease of analysis, uncertain variables are often assumed to be independent although correlations among them exist. This assumption may lead to an under- or an overestimation of uncertainty in the output of the risk assessment which, in turn, may lead to a poor risk management decision. Although specifying the correlation structure between variables may be difficult, every effort should be made to estimate the correlations and to incorporate this information into the analysis. Algorithms for sampling from correlated input distributions are now available in commercial software packages.

A. Monte Carlo Methods

Monte Carlo techniques are numerical methods for propagating uncertainty through models. Because they are easy to use, Monte Carlo techniques have become widely accepted for analyzing uncertainty in risk assessments. They generate representative samples from probability distribution functions of the model inputs and coefficients and propagate them through the mathematical model, producing corresponding samples from the pdf of the model output.

A scenario is defined by the random selection of values, one from each input pdf, that are used in the model to compute an output value. The procedure is repeated for N iterations, yielding N output values that characterize the uncertainty in the model prediction. Simple Monte Carlo sampling involves random selections of

Table 2 Example of a Monte Carlo Simulation

Iteration$_i$	\multicolumn{2}{c	}{X = A * B}	
	\multicolumn{2}{c	}{Inputs}	Outputs
	A	B	X
1	8	3	24
2	10	11	110
3	11	6	66
4	16	10	160
5	12	9	108
6	9	7	63
7	12	10	120
8	13	12	156
•	•	•	•
•	•	•	•
•	•	•	•
N	11	7	77

values from input pdfs while Latin hypercube sampling takes a stratified approach; the input pdfs are subdivided into N intervals of equal probability and a value is selected at random from each interval. Latin hypercube sampling is often used because it ensures that each input distribution is entirely sampled and thereby characterizes the output distribution with a smaller number of iterations.

Figure 2 and Table 2 illustrate a Monte Carlo simulation for a model where A*B = X. For each iteration, samples are drawn from the pdfs for variables A and B and are multiplied together to yield predicted values for output variable X. The values for X can be combined to form a probability distribution that describes the uncertainty in the output.

V. COMMUNICATION OF UNCERTAINTY

The outcome of an uncertainty analysis will generally be a statement, table, or figure containing information about the possible values of the risk being assessed. Uncertainty estimates should be communicated both within the text of the risk assessment and using graphical formats. Methods for describing and depicting uncertainty vary as to the amount of information expressed, the clarity of the description, and the ease of interpretation. Since simple, easy to understand, descriptions often do not contain the same information as more sophisticated graphical methods, different methods should be combined to maximize understanding.

Within the text of the assessment, expressions of uncertainty can take several forms. The simplest is a probability estimate, such as a statement that there is a 60% chance that X is a human carcinogen. While a statement of this type is easy to understand, it may be an oversimplification of the risk situation.

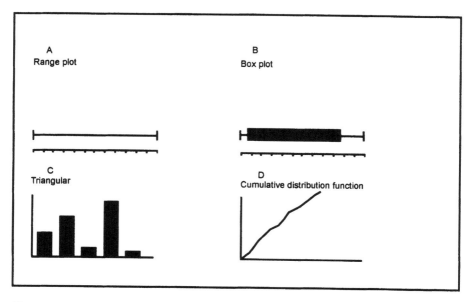

Figure 3 Graphical depictions of uncertainty.

Other ways of expressing uncertainty include giving an upper limit on the risk or defining a range involving both a lower and upper limit. Using only an estimate of the upper limit implies that the best estimate of the lower limit is zero, an assumption which may or may not be true. The advantage of a range is that it provides two values which are believed to bracket the risk. The disadvantage of using a range to describe uncertainty is that it provides no information about the probability that the risk lies within the range (the implication is that it is 100% likely), the relative likelihood of the values within the range, or the best estimate of the risk for the population. Using a confidence interval to express uncertainty in the risk is similar to using a range but a confidence interval also provides an estimate of the probability that the true risk falls within the interval. For example, a 95% confidence interval is a range around the best estimate which has a 95% probability of containing the true value of the risk. A 99% confidence interval will be somewhat wider than the 95% interval; however, its probability of containing the true risk value is higher.

If the uncertainty analysis has yielded a probability distribution function as its output, an estimate of the center of the distribution (mean or median) and of the standard deviation may be given. The advantage of this method is that it gives a best estimate and information from which confidence intervals can easily be calculated; the disadvantage is that, unless the distribution is normal, important information is lost. A better approach is to give a description of the shape of the probability distribution, such as a graph, along with a mean or median value and the values of selected percentiles of the distribution (e.g., the 5th and 95th).

There are several types of graphical expressions which can be used to pictorially express estimates of uncertainty. The simplest, a range plot, is used in conjunction with a range estimate (see Figure 3A). A variation on this type of graph is a boxplot (see Figure 3B) which shows the range, the best estimate, the 25th and 75th percentiles. A histogram is a discrete estimate of the underlying probability distribution function and shows not only the range of plausible values but also the relative likelihoods of various values (see Figure 3C). Also useful, particularly in conjunction with the histogram, is the cumulative probability distribution function (cdf) (see Figure 3D). While the histogram shows the relative probability of different estimates, the cdf shows the cumulative probability. The advantage of the cdf over the pdf is that it allows the probability that the risk lies below a specified value to be directly read from the graph. Some people have suggested that the optimal strategy is to use these two curves together with the mean and median clearly marked on both.

VI. CONCLUSION

An uncertainty analysis should be performed in tandem with the risk assessment. A checklist can be used to evaluate the adequacy of the uncertainty analysis effort or, better yet, to help direct the project and ensure that the proper steps are included (see Table 3). There are many ways to use uncertainty information to make better informed risk management decisions. If the overall uncertainty is considered too large for a good decision, the results of a sensitivity or uncertainty analysis can be used to prioritize information gathering efforts to ensure the most cost-effective use of resources. Perhaps the most common use of uncertainty information occurs when the decision maker examines the estimated range or probability distribution and uses this information informally to develop a risk management strategy. When given only a single value, the risk manager must trust that this value has the appropriate degree of conservatism for the situation, or must adjust the number in some way. Providing the risk manager with a clear and complete picture of the uncertainty that exists, however, allows for a decision that balances over- and underconservatism, creates confidence that the decision is based on sound science, and helps document and communicate the degree of conservatism involved.

The information obtained about uncertainty can also be useful in communicating risk to the public and in gaining credibility for the scientists involved. Whether or not uncertainty is explicitly acknowledged, the public understands that these techniques yield numbers which are not absolutes and, therefore, are subject to debate and controversy. Experts may initially produce different estimates, but if uncertainty is quantified, closer examination may reveal that the ranges, confidence intervals, or pdfs overlap. Explicitly acknowledging what we do know and what we do not know, while on the surface muddying the water, may lead to less debate and more trust in the long term.

REFERENCES

Finkel, A.M., *Confronting Uncertainty in Risk Management: A Guide for Decision-makers,* Center for Risk Management, Resources for the Future, Washington, 1990.

McKone, T.E. and Bogen, K.T., Predicting the uncertainties in risk assessment, *Environ. Sci. Technol.,* 25(10), 1674, 1991.

Morgan, M.G. and Henrion, M., *Uncertainty: A Guide to Dealing with Uncertainty in Quantitative Risk and Policy Analysis,* Cambridge University Press, Cambridge, UK, 1990.

Thompson, K.M., Burmaster, D.E., and Crouch, E.A.C., Monte Carlo techniques for quantitative uncertainty analysis in public health risk assessments, *Risk Analysis,* 12(1), 53, 1992.

Table 3 Checklist to Evaluate Adequacy of Uncertainty Analysis

Required Task	Was Required Task Satisfactorily Performed?	
Exposure Assessment:		
Was an uncertainty analysis performed on the exposure assessment?	Yes	No
Were all potentially important variables included?	Yes	No
If probability distributions were used, were they of the appropriate type?	Yes	No
Was the analysis, either analytic or numeric, carried out correctly?	Yes	No
If Monte Carlo simulation was used, were enough iterations performed to characterize the full uncertainty in the risk value?	Yes	No
Were the results of the analysis communicated both in the text and graphically, and were they presented in a format that is easy to understand and use?	Yes	No
Was there a discussion of the implications of the uncertainty for the risk management decision?	Yes	No
Effects Process:		
Was there an attempt to discuss or characterize uncertainties in the effects process?	Yes	No
If fundamental disagreement exists, such as in model structure, was the analysis repeated using the different theories?	Yes	No
Risk Characterization:		
Was there an attempt to combine the uncertainties from the exposure and effects processes, either formally or informally, to get an overall estimate of uncertainty?	Yes	No

CHAPTER 21

Risk Communication

David Weitz and Sally L. Benjamin

CONTENTS

I. Introduction ... 426
II. Communicating Risk Assessment Findings and Results ... 426
    A. Overcoming the Need to Know Everything ... 426
    B. Creating a Meaningful Dialogue ... 427
    C. Planning for Useful Communication ... 428
    D. Benefits of Early Risk Communication ... 428
III. Understanding Environmental Risk ... 428
IV. Turning on the Porch Light — Building Trust ... 429
V. Managing Risk Communication ... 430
    A. The REACT Loop Method ... 430
        1. Media Needs ... 430
        2. Network Contacts ... 430
VI. What to Look for When Hiring a Consultant in this Field ... 431
    A. Trust ... 431
    B. Training ... 432
VII. Practical Applications ... 432
VIII. An Abrupt Catastrophe — A Train Derailment, Chemical Plant Leak, or Tire Fire ... 433
    A. Preliminaries ... 433
    B. On-Scene Priorities ... 434
        1. The Risk Communicator's Message ... 435
        2. Others On-Scene ... 435
        3. Other Benefits of Risk Communication ... 435
IX. A Problem Discovered as the Result of Testing — Groundwater Contamination ... 436
X. Conclusion ... 436
    References ... 437

I. INTRODUCTION

Risk communication happens on several levels and requires a multifaceted approach. Of course, to communicate about environmental risk requires technical understanding. Risk estimates, behavior and effect of environmental contaminants, and efficacy of remediation strategies are complex subjects and require risk communicators to possess capability in science, as well as communication. It also requires the ability to communicate technical subjects in ordinary terms, but this level is not where risk communication really lives.

Risk communication also requires human understanding. What does it feel like to learn that your well has been contaminated? What is your first reaction? What answers are most important? How does the feeling change? What helps? What doesn't? This understanding is not available through classes or books. It comes from being involved with the people affected by environmental contamination.

This chapter introduces both levels of risk communication. First, technical aspects are addressed in the context of communicating risk assessment findings and results. Next, the human aspects of risk communication are presented in the context of coping with three emergency response scenarios. Finally, the two levels are integrated in a discussion of information as one type of control that can be provided to people suffering from a loss of control as a result of environmental contamination.

II. COMMUNICATING RISK ASSESSMENT FINDINGS AND RESULTS

Often those responsible want to delay risk communication until the end of a project when a "solution" has been decided upon. This is unfortunate. It escalates the emotional reaction of people to the problem and reduces the credibility of the experts. Risk communication, at its best, is an integral part of each step in the process of dealing with an environmental issue — from deciding to perform a risk assessment, through the process of conducting the assessment and making the risk management decision, to the implementation of that decision.

A. Overcoming the Need to Know Everything

One reason people in charge of a project hesitate to address risk communication throughout every step is that they do not have firm answers. Technical experts are plagued with a desire to answer people's questions. They prefer not to ever say, "I don't know," but that's exactly what will happen in the early stages of any project.

In fact, in the early stages of risk communication, the communication tends to be from lay people to the project managers. Managers will be asking questions and listening to answers at this point, allowing the public to educate them about the site and public perception of the issues. The communication can deal with a variety of topics (see Table 1).

Table 1 Early Information Exchange on a Risk Assessment Project

The Site	What activities occur at the site?
	What activities have occurred in the area in the past 50 years?
Potential Exposure	Who uses the site?
	What do people do on the site?
	Do children or elderly people live near the site?
	How long do people usually spend on the site?
	How many days a year?
Local Interest	What groups are interested in the site?
	Who else is interested in the site?
Fear and Concern	Do people feel that they have enough information about the site/project?
	What do people know about the project?
	What worries people about the site/about the project?
	What information do people need immediately?
Logistics of Communication	What are the best times/locations for discussing the project with the people who are concerned?

B. Creating a Meaningful Dialogue

Over time, there will be opportunities for greater public involvement on certain aspects of the site. The nature of the involvement will depend on the situation. A few examples of public involvement, however, include:

- Conducting neighborhood surveys
- Simple data collection (e.g., the lake watch techniques in use with volunteers in Wisconsin, Minnesota, and other states)
- Values-based issues

Values-based issues are really the area where public input can be most valuable. Here lay opinion is as legitimate as the views of technical experts. As other sections of this book have made clear, risk assessment and risk management involve personal and social values, as well as science, math, and technology. The choice of remediation will be a value-ladened choice. Unless a single method is legally mandated, there is probably room for public participation in making a choice about a remediation plan. Similarly, risk levels deemed "of concern" are highly value-laden. Discussions about these choices can be expected to be emotionally charged. The assistance of trained conflict managers and risk communicators will be a great asset to the project staff. It is important, however, for the technical decision makers to participate in the discussions to benefit from the information offered, and to demonstrate a genuine concern and interest in the information and ideas concerned laypeople have to offer on the project.

C. Planning for Useful Communication

The discussions suggested above must occur in tandem with technical phases of the project to make the best use of citizen information and keep a tight interconnection between the project's progress and risk communication activities. To do so, it is only necessary to ask what information would be helpful (to people and from people) at each step in the project; consider who would have or need information and how they can be reached; then, simply structure opportunities for effective information exchange. There are myriad techniques from which to choose. Focusing on the goal of effectively communicating with people who care about the project and have something to share will help project staff select techniques tailored to the project needs.

D. Benefits of Early Risk Communication

Perhaps the most difficult problem in risk communication is cutting through the intense emotions that surround many environmental risks. By informing people completely, early in the project, managers usually can reduce suspicion and conflict. Over a period of time, emotions will usually lessen if people have the opportunity to voice their concerns and feel their views are seriously considered and acted upon by someone in a position to take effective action.

Also, emotions will return to a more even temper if people can regain a sense of control over their lives. Learning that your neighborhood, water supply, property, or air has been contaminated shatters the sense of security most people enjoy in their homes. Being allowed to help solve the problem greatly enhances peoples' recovery from this loss of security because they come to see that there are actions they can take to control the problem. As this happens, fear and anger will subside somewhat and be replaced, in many cases, by fierce interest and a determination to get the job done. Project managers willing to tap into this source of human energy can gain tremendous support. Risk communication is the way to tap into this source of political, financial, and psychological support, by involving the people who care most about the problem.

III. UNDERSTANDING ENVIRONMENTAL RISK

What then about the unseen, untasted, and unsmelled mystery that may be in the air we breathe or the water we drink? It is a horror left in childhood come back to haunt us; the spirit of the older child taunting at Halloween, that a monster truly does exist behind the bushes; an ethereal thing that has neither form nor substance, but will surely envelop and overcome us. Like the brave, small child who, with lip trembling, moved forward praying silently for Mother to turn on the porch light, today we face environmental contamination with uncertainty and fear hoping for help.

Straight talk is the porch light of protection for today's concerned citizens. In a highly technical world, where gas chromatography measures contaminants in billionths and toxic effects are defined as percentages for chronic impact and acute impact, sometimes honest efforts to communicate only cloud issues. Business leaders, elected officials, and the people in state and federal agencies attempting to explain issues may fail to communicate with the people as they dance to a fearful tune of legal liability, future damages, and budget uncertainties. A policy of caution is a prescription for failure because it neglects uncertainties which underlie the problem. It fails to respond to the perception of environmental risk, often far more influential than factual analysis.

It is not uncommon for an elected official to fear carrying bad news about environmental contamination to the electorate. Few welcome the opportunity to carry bad news. Yet, when news about an environmental problem is carried to the people forthrightly, honestly, and with conviction that all that can be done to remedy the problem will be done speedily, there is seldom a significant adverse reaction from citizens.

Problems happen when the porch light is on only fleetingly and people believe they are not getting all the available information.

IV. TURNING ON THE PORCH LIGHT — BUILDING TRUST

There are some simple rules for building trust. First, do not patronize the people affected. Nearly always, those near a contaminated site will resent any inference that others are best able to decide their future. It is essential to realize that risk communication addresses both the scientifically assessed risks and the perceived risks. People act on perceptions of risk. Individuals who feel at risk will act in accord with that belief, and must be given serious attention.

Second, open communication channels and keep them open. Even when there are no new laboratory results, no major discoveries or progress, it is important to continue to report to all those who believe they have a stake in the problem. Whether the problem is being addressed by a business, municipality, or state or federal agency, it is essential that people in the affected community have open avenues both to seek information and to provide it.

Third, recognize that risk communication is one piece of a bigger project; that of building a genuine partnership between the contaminated community, elected officials, the business community, and regulatory agencies. If offered merely to placate people, rather than to establish a real working partnership, risk communication can backfire. Citizens will quickly become frustrated, civil actions will spring up against the business or elected officials, and regulatory agencies will heighten enforcement actions or become embroiled in political battles.

Fourth, risk communication will not instantly banish the uncertainty, fear, and anxiety of people living with the contamination. Providing the facts about environmental contamination and dealing with a sense of helplessness by assisting them in taking control of their destinies are only first steps toward helping them regain their sense of empowerment.

Finally, risk communication experts can provide a framework for assessing communications needs, establishing a program to further communication efforts, and assisting in the ongoing effort to "keep the porch light on" as progress is made in resolving the problem.

V. MANAGING RISK COMMUNICATION

There are no "cookie-cutter" solutions to risk communications. The proper action depends on careful analysis of the situation, including consideration of the needs of the people in the community, businesses, and government. Just as marketing efforts, election campaigns, and citizen outreach efforts are adjusted to meet changing needs, so a risk communication strategy must be fine tuned. A trained consultant can provide objective feedback to guide such an effort.

A. The REACT* Loop Method

A REACT team may be formed in advance of any risk communication emergency (see Figure 1). It can consist of a staff or contract communication team and should include several key staff members committed to working together to serve the best interests of the company by serving the needs of the affected community. A model REACT team can include the following:

- Risk communicator
- Chief executive officer or designee
- Corporate counsel or designee
- Public affairs staff lead worker

In practice the team will be tailored to the needs and personality of the company. It is essential that the team be able to act with authority and speed.

1. Media Needs

Members of the popular press will want information quickly and will use the best available information within the constraints of their deadlines. They will not delay to secure information and will look elsewhere if it is not promptly available from the risk communicator. It is important, when dealing with the electronic media to recognize the need for visuals and statements that are concise and clear. Local reporters will often be more likely to spend the time to more deeply research a story and will benefit from experts who the risk communicator suggests as sources. In all dealings with the media remember that everything is always "on the record" and any assertions are subject to be checked.

2. Network Contacts

It is essential that a risk team develop a network of contacts which can be used before any emergency exists requiring risk communication. Individuals who should be available by telephone, on weekends as well as during work hours, are: toxicologists, health professionals, doctors, veterinary doctors, communication experts, risk communicators, or appropriate experts in academic institutions or government.

---

* REACT = Research, Evaluation, Application, Communication, and Troubleshooting.

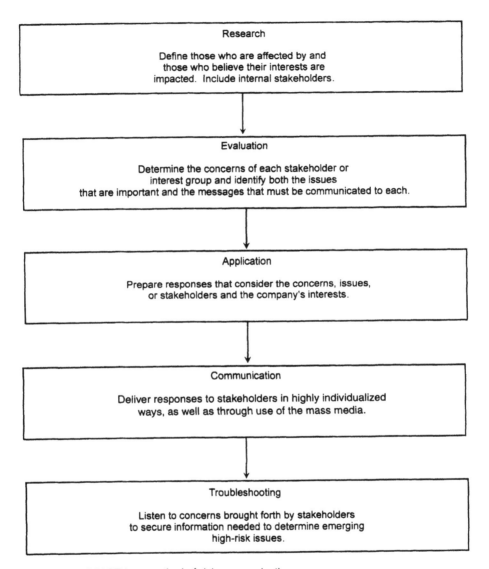

Figure 1 REACT loop method of risk communication.

VI. WHAT TO LOOK FOR WHEN HIRING A CONSULTANT IN THIS FIELD

A. Trust

No individual consultant or firm can serve an employer without a trusting relationship. Risk communication is a process to reduce fear. It is a process of supplanting uncertainty with knowledge, and an ongoing process of sharing both what is known and what is not known. During this process mutual understanding and respect emerges. A consultant must be trusted by the elected officials or business leaders he or she represents, and by the community, or the process will not work.

B. Training

Professional risk communicators may come from a variety of educational backgrounds. Journalism, as a major or minor, provides a solid basis in many skills required in risk communication. It also enforces a sense of ethics that can serve as a valuable touchstone amid the turmoil of dealing with public concern over environmental risks. There are a growing number of college and graduate level programs that include training in stakeholder involvement. This training, whether in college or on-the-job, is very useful.

The credibility and ability to translate between scientists and lay people can be enhanced by a technical degree, such as chemistry, biology, or geology. However, a sound grounding in the humanities and social sciences is also vital.

Experience in the public eye can be a good basis for risk communicators. Staff work in the media, such as newspapers, radio, or television, provides valuable insight into the "news" angle offered by environmental contamination, and may offer a risk communicator the edge in working with the media to respond to the public's need for balanced coverage.

A risk communicator should be able to work with technical staff to enhance their ability to work with the press and public. Other excellent training for risk communicators includes emergency training for crisis handling of environmental or governmental emergencies; familiarity with federal Superfund, Clean Air and Clean Water Acts; familiarity with newsletter and brochure design, video tape production, and broadcast media; and familiarity with working with legislative and congressional contacts.

VII. PRACTICAL APPLICATIONS

To turn on the porch light requires us to provide practical knowledge to people who face unusual environmental contamination cases. It must involve face-to-face communication. Above all, it must be an honest effort to provide candid information to people who believe they are affected by an unusual incident.

The questions you can always expect about environmental contamination include: Is there a hidden threat I should know about? How will the situation impact my life? Is my life threatened? Is my family in danger? If there is no immediate risk, what about tomorrow? How much in dollars will this cost me? Will this harm my pets, livestock, trees, and the vegetation around my home, business, or farm? Will this affect my economic wellbeing? Will my business be worth less? The real issue behind these questions is, "How can I control my fate?"

It is typical for the private citizen to worry first about immediate health, then about the physical well-being of pets and valued livestock, and finally about the value of real estate investments. These are serious uncertainties experienced by people who communicate directly to risk communicators, technical experts, and political representatives. They are also experienced by people "left at home" who receive information secondhand from parents, spouses, neighbors, or media. A major difficulty for emergency risk communicators is the likelihood that information will become garbled when carried to those at home. When that happens, the risk communicator may suddenly be confronted by angry or frightened persons, or news media whom others have misinformed. Written information, distributed widely and quickly to all who believe they are at risk, is the only realistic way to prevent such problems.

Whether the risk communicator works as an agent for a municipality, for industry, or in a regulatory agency, the needs of individuals affected by an incident are largely the same. After the initial concerns about immediate health, safety, and economic wellbeing are addressed, the fundamental question emerges, "How can I control my fate?"

Few of us can consider ourselves truly in control of our fate. Surprises, whether unexpected professional opportunities or freak accidents, happen to us all. Still, because the risks are normal, they are acceptable. When an environmental hazard suddenly "invades" our personal environment, and we lack the ability to "fix" the problem, normal equilibrium is upset. Risk communicators help people to understand the problem, and what is needed to fix it, and, in doing so, regain equilibrium.

The timeline for providing information changes with each risk scenario, but the information must always be conveyed to the people involved with reasonable speed and in credible fashion.

Consider the following scenarios:

1. A major environmental incident stemming from an abrupt catastrophe, such as a railroad train derailment, chemical plant leak, or tire fire.
2. A major environmental problem that has been an ongoing problem and is discovered as the result of routine testing, such as pesticides, volatile organic compounds, nitrates, or other contaminants in groundwater.

In both situations private citizens ask similar questions, but their priorities may differ with the urgency of the problem.

VIII. AN ABRUPT CATASTROPHE — A TRAIN DERAILMENT, CHEMICAL PLANT LEAK, OR TIRE FIRE

A. Preliminaries

The risk communicator working in such a situation must be at the scene of the incident, representing his or her constituents and coordinating work with local authorities. It is important for the risk communicator to be briefed on the major issues (pro and con) that could affect the attitudes of local officials, local residents, and local regulators or business entities. When a catastrophe occurs virtually all the facts in the case will be unearthed and examined within hours or days. By honestly analyzing all the problems facing the responsible organization, the risk communicator will be able to provide background, both positive and negative, establish credibility, and, thus, become a trusted information source.

A fundamental organization includes the risk communicator, chief executive officer, and REACT loop risk team, and should address:

1. The problem
   - Gather information on what happened.
   - Identify any past environmental problems, spills, or accidents at the facility; within 100 miles; or in the state (appropriate scope will reflect the size of the current incident).
2. The people
   - Based on past relationships, project the likely attitude of state and federal regulators toward the firm, in the face of the problem.
   - What are the names of local fire, police, and sheriff's lead staff; or at least the names of the fire chief, police chief, and sheriff?
   - If a command staff has been established, who is in charge? It is essential to become part of the integrated response team if one is established. Often, when a local law enforcement organization is overwhelmed with media it is willing to work with a risk communicator who can establish good relationships with the media and fulfill a valid information-sharing role.
3. The team
   - Establish a network of individuals who can either be called upon to make an appearance on-scene or to provide telephone interviews for indepth background for news reporters or local officials.
   - Ensure that communication links between the risk communicator and home office remains frequent by taking advantage of electronic transmission facilities.
   - Be able to prepare printed information within 4 hours to be distributed to persons throughout the affected area.

B. On-Scene Priorities

Upon arriving on the scene of the emergency the risk communicator must report to the law enforcement officials or emergency response officials in charge and establish a cooperative relationship based on mutual agreement to serve the people of the area. It is likely that the law enforcement/emergency services chief will be extremely busy and not eager to discuss corporate concerns or interests of municipal leaders. It is essential that the risk communicator establish that he or she is on-scene to help work with the emergency and provide service to emergency responders and people of the area. After establishing a working relationship with the law enforcement/emergency services chief the risk communicator must talk with people in the area to begin to analyze their immediate information needs and secondary information concerns likely to develop in the near future.

The risk communicator must then work with existing media relations staff, or create a media information staff if none exists to provide local media with timely information about the incident. A team approach offers the best opportunity to succeed in such operations. One staff person can brief the media from clipboard notes and accompany television or news photographers to vantage points where photographs can be taken safely. Providing escorted tours of an emergency area allows media professionals to shoot film or footage, while allowing law enforcement staff to control access and assure safety. Other team members provide updates on the fast-breaking situation to help keep the media informed.

By working with media reporters, the risk communicator quickly grasps developing lines of questioning. Secondary lines of questioning can be predicted enabling team members to benefit from the presence of technical staff or other credible experts. By communicating with a risk team at staff offices, the risk communicator can arrange for assistance from an appropriate person. Concerns of private citizens become apparent if an on-scene risk communicator listens carefully to local people. Consider, for example, the concerns of local residents when a toxic chemical cloud released from a derailed train in Wisconsin forced evacuation of a 26–square mile area. Area farmers were urgently seeking information and permission to return to their farms to care for livestock. State officials contacted university specialists and were informed that the impacts of the chemical would not last long or harm livestock. The farmers were informed and, by mid-afternoon, most were able to take care of their farm animals.

1. The Risk Communicator's Message

It is vital for a risk communicator to convey the message that the organization, while accepting no liability, is anxious to do everything it can reasonably do to ameliorate adverse impacts on people or their property. It is perfectly consistent with this stance to explain that settlements and formal apportionment of liability must await further consideration. The immediate message is that the organization cares about people and is acting to protect their health and welfare.

2. Others On-Scene

It is absolutely essential that a risk communicator on-scene assist any citizen or official who wants to discuss the situation. The chief executive officer or high-level officer should also appear quickly on-scene to demonstrate that the incident and its impact on people is of serious concern to the company. Elected officials and responsible state and federal agency staff should realize a personal appearance is the best way to show people that they are not forgotten. High level officials need not state solutions to the problem but merely be on-scene to examine what can be done for people or the environment.

3. Other Benefits of Risk Communication

Providing information not only helps people at the time, it provides a framework helpful in determining the level of service and response provided during the emergency and the degree of liability of those responsible if civil suits occur later. An industry, municipality, or state or federal agency with significant responsibilities during emergencies must recognize that its actions will be scrutinized and it will be held accountable when the emergency ends. Such recognition should not prevent frank comments to the press and public during the time of the emergency, but should inspire the organization to ensure that people in the community realize their concerns have been recognized and served to the greatest extent possible.

IX. A PROBLEM DISCOVERED AS THE RESULT OF TESTING — GROUNDWATER CONTAMINATION

Discovering that an environmental hazard has existed unseen for some time can shake a community's sense of security. In such cases the risk communicator faces the problem of conveying health information to people before all analyses have been completed. At no time, in such cases, can the risk communicator guarantee "safety" to anyone. In addition, since in most cases the health risk will be one of chronic exposure, rather than acute exposure, the difference in terms must be carefully explained.

The message that can be conveyed immediately is that state and federal standards exist, what those standards are, and what is currently known about contaminant levels. If nothing is known yet, the process of investigation can be explained and timelines announced for the test results and an information sharing process.

It is important to realize that the relationship with the community may last for years. Groundwater contamination problems do not go away quickly. Therefore, people living near the "problem sites" must deal with the problem for years, and over time will develop an intense relationship with those responsible for the problem. It can be an intensely unpleasant, costly relationship, or, it can be a close working relationship. As time passes, the risk communicator will be faced with the challenge of dealing with long-standing residents knowledgeable about the problem as well as new residents only recently aware of the situation. This will require continually communicating about the problem on several levels of sophistication.

X. CONCLUSION

The key to communicating about environmental health risks is to provide a framework of certainty in an uncertain situation. The risk communicator must:

- Clearly describe what is known and what is unknown.
- Demonstrate how regulators and local industries are working to develop the best possible understanding of the problem and how interim measures are being implemented while studies continue.

Risk communication helps people know where they stand, the first step toward taking control of a situation and regaining a sense of security, despite environmental risks. Organizations responsible for remediating or regulating environmental risk can choose to empower affected communities by involving them in the decision-making process, and gain strong partners. Risk communicators guide organizations in the process of building partnerships.

REFERENCES

Benjamin, S.L. and Belluck, D.A., Risk feedback: an important step in risk communication, *American Water Works Association*, 82(11), 50, 1990.

Bingham, G., *Resolving Environmental Disputes: A Decade of Experience*, The Conservation Foundation, Washington, 1986.

Committee on Risk Perception and Communication, *Improving Risk Communication*, National Academy Press, Washington, 1989.

Fisher, R. and Ury, W., *Getting to Yes: Negotiating Agreement Without Giving In*, Penguin Books, New York, 1983.

Hance, B.J., Chess, C., and Sandman, P., *Improving Dialogue with Communities*, New Jersey Department of Environmental Protection, New Brunswick, NJ, 1988.

Spencer, L.J., *Winning Through Participation: Meeting the Challenge of Corporate Change with the Technology of Participation*, Kendall/Hunt Publishing, Dubuque, IA, 1989.

Suskind, L. and Cruikshank, J., *Breaking the Impasse: Consensual Approaches to Resolving Public Disputes*, Basic Books, New York, 1987.

Talbot, A.R., *Settling Things: Six Case Studies in Environmental Mediation*, The Conservation Foundation, Washington, 1983.

CHAPTER 22

Clear Communication in Risk Assessment Writing

Wendy Reuhl Jacobson

CONTENTS

I. Introduction 439
II. Readability 440
III. Graphics 442
IV. Clear Writing Problems 443
V. Conclusion 444
    References 444

I. INTRODUCTION

Clear communication of ideas is essential to people in all areas of life and all types of work. It is important in the formation of ideas and opinions, and their analysis. Clear communication is essential for changing these ideas and opinions into a form that can be easily understood by others. Clear communication of ideas relies heavily on readability, but typography of text and the proper use of illustrations also play major roles. Together, these factors heavily influence the document user's ability to understand and fully utilize the information presented.

To define these parts of writing, readability is a measure of the text's level of difficulty. It is usually described in terms of the grade level at which an average student can understand the material. Graphic content of writing includes the two nontext elements mentioned earlier: typography and illustrations. "Typography" encompasses a variety of page layout factors including line length, use of highlighting, white space, and type size and design (see Table 1).

Table 1 Ten Typographic Criteria and Their Suggested Use

Criteria	Suggested Use
Type Size	10–12 point type
Type design	Roman typefaces with a serif
Line Length	50–70 characters or 3–5.5 inches
Justification	Unjustified or ragged right
Headings	Large type size, boldface, italics, underlining, or all caps
Highlighting (boldface, all caps, color, underlining, italics)	To emphasize single words or short phrases within text; for use in headings
Spacing: between paragraphs	One blank line
Spacing: between sections	Two blank lines
Margins	One inch on all sides
Reproduction	Paper with dull finish; black ink on white paper; paper thick enough so no ink shows through
Page Size	8.5 × 11 inches
Binding	Along long side with text printed across narrow width of page
Space between lines of text	Single space
Consistency	No changes in typography within document

Note: Currently accepted use of typography to achieve clear communication. Source: Gallagher and Jacobson, The typography of environmental impact statements: criteria, evaluation, and public participation, *Environmental Management*, 17(1): 99–109, 1993. With permission.

"Illustrations" includes proper presentation of maps, tables, graphs, charts, diagrams, and photos (see Table 2). Both readability and use of graphics can affect the reading rate and accuracy of risk assessments. Documents should be designed in a way that is user-friendly: easy to understand and use.

II. READABILITY

The first formulas for measuring the readability of textbooks were developed in the 1920s. Currently there are many formulas in use for all types of documents, from training manuals to consumer information. However, one of the most widely used formulas is the Flesch formula (Flesch, 1949). It calculates readability by two objective measures, sentence length and number of syllables per word, multiplied by constants, resulting in a reading ease score. The score may range from 0 (very difficult) to 100 (very easy). In Table 3 the scores are correlated with school grade levels and to material written at each level.

Although this formula is quite accurate in predicting readability of text, it does not reflect other important components of comprehension, including typography and illustrations. Therefore, it should not be used as the only guide when preparing a legible document.

Table 2 Ten Criteria for Illustration Clarity

Criteria applicable to all illustrations:	
	1. Title or caption
	2. Explanation or legend
	3. Source or reference
	4. Located near text reference
	5. Horizontal labels
	6. Legible
Criteria applicable to specific types of illustrations:	
Maps	7. Key of colors, shades, or symbols
	8. Arrow to show direction
	9. Border to separate from text
	10. Tick or grid marks
Tables	7. Numbers in evenly spaced columns
	8. Numbers rounded off
	9. Table numbered
	10. Numbers in ascending/descending order
Line graph	7. Axes begin at zero; horizontal X-axis depicts time; vertical Y-axis represents quantity
	8. No more than four different lines used
	9. Equally spaced time divisions
	10. Key of colors, shades, or symbols used
Diagrams	7. Labels consistent
	8. Diagram simple (not cluttered)
	9. Parts of diagram labeled
	10. Arrows to indicate direction
Photographs	7. Large enough to show relevant detail
	8. Cropped to remove extraneous detail
	9. Clear photo
	10. Border around photo
Bar graph	7. Equal spacing between bars of same width
	8. Two axes with baseline showing zero
	9. Numerical values on/above bars
	10. No more than 2-12 bars on one graph

Table 2 continued

Pie chart	
	7. Largest segment at 12 o'clock
	8. Slices consistent with values; totals 100%
	9. No more than six slices
	10. Key of colors or shades used

Note: Currently accepted design of illustrations to achieve clear communication. For each type of illustration, all 10 of the criteria associated with that type of illustration should be followed to create a simple and clear graphic. Source: Jacobson, 1990.

Researchers have not yet established a universal reading level, and there exists no definitive guidelines for achieving "plain language" in writing. However, in an overview of recent studies, Gallagher and Patrick-Riley (1989) concluded that to achieve plain language, documents should be written at or below tenth-grade level. They found that "the average person would prefer simpler language, about the eighth-grade level, and will accept more difficult text if motivated." In their book, *Reporting Technical Information* (1988), Houp and Pearsall offer guidelines for technical writers to follow when creating a readable document:

1. Use shorter sentences (those which are too complex in structure or are too dense with content are more difficult to understand).
2. Use simpler words (give all necessary background information and define technical terms). Do not use large words to impress the reader.
3. Use the active voice when appropriate.
4. Use graphics to explain the major points of your message.

Additional suggestions for achieving plain language can be found in style manuals such as the *Chicago Manual of Style*, and other scientific publications like Day's book, *How to Write and Publish a Scientific Paper* (1983).

III. GRAPHICS

Although graphics predate writing, graphic criteria were not established until 1902. Often, typography was designed to be aesthetically pleasing rather than useful to the reader. In 1963, Tinker found that poorly designed typography tired the reader, leading to low comprehension. Recently, researchers have developed typographic criteria for many types of publications including books, scientific papers, and government documents. Through a review of research and phone surveys of federal departments, Gallagher and I (1993) compiled a table of what is currently considered the most accepted use of typography. Typographic criteria are divided into ten categories (see Table 2). Although some criteria may be more important than others, we did not find sufficient information in the literature to weight each criterion.

Illustrations have the potential to clarify ideas not easily communicated with text, but they must be easy to understand (Houp and Pearsall, 1988). As with weak typography, researchers have found that poorly designed illustrations not only tire

Table 3 Description of Flesch Readability Scores

Reading Ease Score	Description of Style	Estimated Grade Level	Typical Magazine
90–100	Very easy	5	Comics
80–90	Easy	6	Pulp fiction
70–80	Fairly easy	7	Slick fiction
60–70	Standard	8/9	Digests, *Time*
50–60	Fairly difficult	10/11	*Harper's, Atlantic*
30–50	Difficult	College	Academic, scholarly
0–30	Very difficult	College graduate	Scientific, professional

Note: Comparison of Flesch Reading Ease scores to grad levels and publications written at each level. Source: Flesch, R.F., *The Art of Readable Writing*, Harper, NY, 1949 (revised 1974), 352 pages. With permission.

the reader, but may also confuse the reader or fail to communicate important ideas. Through a review of relevant studies (Jacobson, 1990), I compiled a list of ten criteria essential in the proper design of illustrations (see Table 3). Previous researchers have most commonly suggested these criteria to achieve a clear use of illustrations. I found no research to determine how often illustrations should be used, or which types are most effective in presenting data. Writer's discretion must guide these decisions.

In recent years, the evolution of desktop publishing has provided writers with the opportunity to modify typography and create illustrations. Some programs even give the user the ability to calculate the readability of their documents by using the Flesch formula. Because of these capabilities, writers are now able to make more choices about how their risk assessment is presented. Accordingly, writers should develop and follow a program of defensible criteria when preparing risk assessments for the purpose of achieving clear communication.

IV. CLEAR WRITING PROBLEMS

Researchers have uncovered many problems in document design. In a study on the readability of land management plans, Gallagher and Patrick-Riley (1989) found that most agency writing is difficult to read because it is written on an average of seven to ten grades above the eighth-grade reading level of the average person. In preparing these and other documents, writers must often struggle to present complex and technical information in a clear manner. Day (1983) suggests, however, that when a subject is difficult, it is more desirable to use a simpler writing style. Although risk assessment writers may find it more challenging and time-consuming to write simply, assessment users will find the reports easier to comprehend.

In a similar study on the graphic content of Environmental Impact Statements, Gallagher and I found several common typographic problems (Gallagher and Jacobson, 1993; Jacobson, 1990). The first of these is the "dense-packed page" that, through a combination of factors, contains too much text and not enough white space. A second problem is the "ghost page," or poor reproduction of the print. A third major problem is the "uncoordinated page" stemming from improper page design. Writers achieve proper page design by using headings and spacing so that the document makes sense to the reader. This study emphasizes the need for writers to create and follow a system of typographic criteria to be used every time a document is prepared.

We also found three major problems with the use of illustrations in Environmental Impact Statements. First, they often lack information important to the reader's clear understanding of the illustration. Second, many illustrations are difficult to locate because they are not placed near their textual reference. And third, there is a substantial variation in the use of illustrations throughout many of the documents. Often, illustrations are presented and used differently from section to section. This variation most often occurs when several people work on different sections of the same document. Such a lack in consistency emphasizes the problems writers face due to the absence of criteria to guide their use of illustrations. The lack of continuity causes additional work for readers and creates the opportunity for misunderstanding of the material.

V. CONCLUSION

Writing clearly has several benefits. First, the information being presented will be easier for the reader to understand. Second, the information will be more useful to them. However, writing clearly often means putting additional time into your document.*

REFERENCES

Council of Environmental Quality, National Environmental Policy Act regulations, *Federal Register*, 34(112), 1978.
Day, R.A., *How to Write and Publish a Scientific Paper*, ISI Press, Philadelphia, 1983.
Fazio, J.R., *Communication with the Wilderness User*, Forest, Wildlife, and Range Experiment Station, University of Idaho, 28, 1979.
Felker, D.B., Ed., *Document Design: A Review of the Relevant Research*, Document Design Center, American Institute for Research, Washington, 1980.
Flesch, R.F., *The Art of Readable Writing*, Harper, New York, 1949 (revised 1974).
Gallagher, T.J. and Patrick-Riley, K., The readability of federal land management plans, *Environmental Management*, 13(1), 95–90, 1989.
Gallagher, T.J. and Jacobson, W.S., The typography of Environmental Impact Statements: criteria, evaluation, and public participation, *Environmental Management*, 17(1), 99–109, 1993.
Houp, K.W. and Pearsall, T.E., *Reporting Technical Information*, Macmillan, New York, 1988.
Jacobson, W.S., *The Appropriate Use of Graphics in Federal Environmental Impact Statements*, unpublished MS research report, University of Alaska, Fairbanks, 1990.
Klare, G.R., *The Measurement of Readability*, Iowa State University Press, Ames, IA, 1963.

---

* As my advisor and friend, Tom Gallagher, is fond of saying,"I would have written you a shorter letter, but I didn't have the time."

Klare, G.R., A second look at the validity of readability formulas, *J. Reading Behavior*, 8(2), 129–152, 1976.
Moen, D.R., *Newspaper Layout and Design*, Iowa State University Press, Ames, IA, 1989.
Tinker, M.A., *Legibility of Print*, The Iowa State University Press, Ames, IA, 1963.
University of Chicago, *Chicago Manual of Style*, The University of Chicago Press, Chicago, 1985.

CHAPTER 23

Scientific Library Research for Risk Assessment

Kathy Malec and David A. Belluck

CONTENTS

I. Introduction 447
II. Library Resources 448
   A. Electronic Media 448
   B. Surfing the Net for Risk Assessment Data 449
   C. Hard Copy 452
III. Selected Environmental Information Sources 455
IV. Conclusion 455

I. INTRODUCTION

Library research is one of the most important factors in the development of a successful risk assessment. Modern environmental research libraries contain journals, reference books, government documents, and CD-ROMs (containing important guidance documents, laws, and databases). They allow access to resident or on-line public and commercial technical databases, and library holdings around the nation. Documents not immediately available on research library shelves can usually be quickly obtained via interlibrary loan requests.

Library staff are indispensable guides through the sometimes bewildering array of hard copy and electronic media resources. They understand the strengths of the different resources that are integral to their mission. For example, public libraries, especially larger libraries, contain basic reference works, directories, and indexes to scientific literature. A number are also depositories for a variety of federal government documents. University/college libraries contain a more substantial amount of detailed chemical, environmental, and legal information because they support teaching programs and research. Federal, state, and local government agencies support numerous technical libraries. Many federal agencies and their field components maintain libraries. For example, EPA has libraries in Washington, D.C., in many of its regional offices, and in various EPA laboratories around the country. This type of arrangement is mirrored by other federal agencies. In addition, small but important library collections may be held in hard copy, microfiche, or electronic form by government agency division, sections, bureaus, or offices. Depending on the organization of the state government, there may be pollution control, natural resources, health department, or other state libraries containing information needed for risk assessment research. Many city or county governments have substantial environmental programs with a library or collection of materials helpful for risk assessment research. The type of library needed varies with the focus and technical rigor of a risk assessment report.

II. LIBRARY RESOURCES

A. Electronic Media

Risk assessors and risk assessment project teams use library resources to define the risks associated with environmental releases and known media contamination. They want the most recent data available in order to ensure the usefulness of their risk report and findings. At the same time, they need to build a large body of information of historical, technical, and policy information that will be used in the risk assessment report. After defining the level of scientific rigor needed to answer their questions, risk researchers head for the technical library to begin their work.

An important first step in any risk assessment project is to confer with technical librarians about a given research problem. These professionals can save the researcher considerable time in finding answers to their problems by acting as a guide to library resources. In many cases, technical librarians are also trained to perform computer database searches. Modern libraries offer many services and data sources that are not obvious to the researcher and can differ significantly among libraries.

Risk assessment researchers need to determine the types of data they require to perform their risk assessment. Will general publications for lay audiences suffice or will highly technical publications targeted at a narrow band of specialists be required? Does the researcher need publications from a geographic region, a particular language, or from a particular time period? Is historical data or cutting edge data needed? Answers to these questions will determine the types of library resources a researcher will need to obtain and will help a technical librarian to focus their suggestions for your research.

One of the most powerful tools currently available to risk assessment researchers is the computer database search. For many researchers, this type of search has replaced handsearching abstracting service hard copies still found in most technical libraries. Whether they are resident on CD-ROMs or via telephone connections to remote computer sites, computer database searches offer the researcher a way to scan the world's literature. Searching languages used by computer databases can range from simple logic to highly stylized syntax that must be precisely followed.

Selection of single or multiple key words to use in a computer database search is a critical initial step in data acquisition. Using dioxin as an example, a researcher might match the key word dioxin (or dioxins) to the media of concern (e.g., groundwater, soil, air), human or environmental health, or a specific organism. Computer databases allow the user to combine words to expand the scope of a search or to limit the number of possible data sources that would contain a specific combination of key words.

Many libraries maintain computer accessible databases at no cost to users. These same libraries may also have access to government or commercial databases that operate on a pay as you go basis. The more complex the search the more it costs to run.

There are a great many databases available to the risk assessment researcher. There are so many that contain environmental information, in fact, that it would be an advantage to the researcher to learn about the variety available. An excellent survey of the breadth available is *Environment Online: The Greening of Databases* (Eight Bit Books, Wilson, CT, 1992). The book was originally published as a series of three articles in *Database* magazine. It includes a number of other columns published in *Database* and *Online* magazines, as well as chapters on environmental information in general interest, scientific and technical, and business and regulatory databases; a list of environmental terms and phrases; search tips; and strategies for locating legislative materials, legal literature, and information from the *Federal Register*. It aids the database searcher in choosing databases to search, and then may also help the risk researcher evaluate information located during a database search. Table 1 presents a summary of available databases, vendors, ease of use, cost, and helpful and explanatory notes.

There are numerous and ever-increasing numbers of private and public databases available commercially as on-line systems or as CD-ROMs. One of the best compilations of these services can be found in *Environment Online*.

It is often necessary to have indepth training to effectively use a given database. Consult with a reference librarian to determine if you should perform a given database search yourself or with the assistance of a librarian trained and experienced in using a particular database.

B. Surfing the Net for Risk Assessment Data

The Internet has become a key source of toxicological and other data used in risk assessment. Risk assessment data on the Internet can come from government and private vendors. While finding risk assessment related sites is not difficult, determining which key terms will access important sites can be difficult. Search engines (e.g., Yahoo, Lycos, Magellan, Excite, and Alta Vista) are used to find risk assessment related sites. These are sites where typing in key terms (e.g., toxicology, risk assessment) and hitting enter will result in a database search and display of sites which match your terms. Each search engine has its own strength and weaknesses and should be evaluated by the user for his or her own purposes. Once a search engine has produced its listing of sites, clicking on their icons or names will result in the

Table 1 U.S. EPA Environmental Information Documents

Document /Source	Contents/Services
Environmental Criteria and Assessment Office, ECAO-Cin, 26 Martin Luther King Drive, Cincinnati, OH 45268.	Prepares human health-based risk assessment documents and conducts toxicology research. Serves as focal point for the collection, summarization, evaluation, and assessment of toxicology data for environmental pollutants. Call 513-569-7531.
Environmental Information Management: A State Resource Guide, Information Sharing Branch, Information Management and Services Division, Office of Information Resources Management (PM-211D), U.S. EPA, Washington, D.C. 20460.	Brief compilation of environmental information sources.
Environmental Law: A Selective, Annotated Bibliography and Guide to Legal Research, May 1993, Library Management Series, EPA 220-B-93-009.	An outstanding reference guide to resources in environmental law.
Environmental Monitoring Assessment Program (EMAP), 401 M Street, SW, Washington, D.C. 20460.	Provides framework for integrating existing and new environmental data. Supplies environmental data to EPA's Center for Environmental Statistics. Call 202-260-7238 for assistance.
EPA Locator.	Call 202-260-2090 for U.S. EPA employee telephone numbers.
EPA Telephone Directory (EPA Headquarters Telephone Directory- WITS Edition).	This indispensable document contains telephone numbers for U.S. EPA regional and field components. Call GPO at 202-260-2118 to order the latest edition.
Ground-Water Research Technical Assistance Directory.	Contact Office of Research and Development, Washington, D.C. for latest edition.
Guide to Key Environmental Statistics in the U.S. Government, Center for Environmental Statistics, Office of Policy, Planning and Evaluation, U.S. EPA, 410 M Street, SW, Washington, D.C. 20460.	Programs generating key environmental statistics. Call 202-260-3726.
Health Effects Summary Tables (HEAST).	Provides summary tables of toxicology data, some of which may be on the IRIS system. Contact NTIS at 703-487-4650 or 800-336-4700.
Information Systems Inventory (ISI).	Computerized inventory of EPA data systems. Updated summaries of more than 500 EPA data systems. Available through NTIS or EPA libraries.
Integrated Risk Information System (IRIS).	Up-to-date health risk and EPA regulatory information for selected chemicals. For many regulatory agencies, IRIS data supersedes all other data sources. Available via computer hookup. IRIS user support at 513-569-7254.
National Computer Center (NCC), Research Triangle Park, NC 27711.	Most of EPA's mission critical data systems reside at this facility. For information concerning access to these databases call 800-334-2405 or 919-541-7862.

computer opening up the home page of the selected site. From this point it is a matter of exploring the site, clicking on each offered subject, or using a site search engine to narrow the list of possible pages to be individually evaluated by the user.

There are several excellent sites offered by U.S. government agencies. They include the ATSDR and U.S. EPA websites. Many of the publications listed by these sites can be downloaded to a personal computer, for example from the U.S. EPA's on-line library service, http://cave.epa.gov. Many of the databases listed in these sites are searchable and the information sources or references they list are readable and can be downloaded. Examples of what these two sites offer for risk assessment projects are listed below.

1. ATSDR (http://atsdr1.atsdr.cdc.gov:8080/atsdrhome.html)
   - HazDat, ATSDR's Hazardous Substance Release/Health Effects Database
   - ToxFAQs, short, easy to read summaries about hazardous substances excerpted from ATSDR Toxicological Profiles
   - Public Health Statements, easy to read summaries of many hazardous substances
   - A Primer on Health Risk Communication Principles and Practices, a practical guide for effectively communicating health risk information to the general public
   - Cluster Version 3.1, PC/DOS software to help researchers determine the statistical significance of a disease cluster
   - Access to the Consortium for International Earth Science Information Network (CIESN) Gateway, a way to obtain datasets from other organizations, containing environmental, earth science, and global change information
   - Case Studies in Environmental Medicine, an excellent series of documents that relate chemical exposures to human disease
   - Information Center Bookmarks to Web Resources, a comprehensive listing of extremely useful computer accessible information sources for risk assessors
   - Electronic links to the Public Health Service, Centers for Disease Control and Prevention, and U.S. EPA
2. U.S. EPA (http://www.epa.gov/epahome/index.html)
   - Rules, regulations, and legislation
   - U.S. EPA publications
   - Environmental test methods and guidelines
   - EPA datasytems and software
   - Finding EPA information libraries, hotlines, and information locators

Each program office has its own home page from which information can be accessed. For example, persons working on pesticide risk analyses can access the Office of Pesticide Programs and obtain the following types of information:

1. Reregistration Eligibility Decisions (REDs) and RED fact sheets
2. The "Rainbow Report" on pesticide reregistration review status of individual pesticides
3. Pesticide (re)registration progress reports
4. Special Review Reports
5. Environmental Federal Register Notices

6. Pesticide Effects on Health and the Environment — At the time of writing this site was under construction. It will offer reports and databases which EPA uses to determine the impact of specific pesticides on health and the environment. This site notes that the following resources are useful for this purpose.
   - Pesticide Information Network (PIN) bulletin board system that provides an on-line collection of files containing current and historic pesticide information. Currently available information includes the Pesticide Monitoring Inventory (PMI) (including the Pesticides in Ground Water Database), the Ecological Incident Information System (EIIS), a Regulatory Status database, and a Biological Pesticides dataset.
7. GOP (http://www.access.gpo.gov/su_docs/aces/aces140.html)
   - Code of Federal Regulations (all titles)
   - Federal Register, 1995 to date
   - Public laws
   - Congressional documents, bills, hearings
   - U.S. government manual

C. Hard Copy

The world of risk assessment and its associated sciences and disciplines are in a constant state of change. Keeping up with these changes means learning effective use of environmental library resources. While computer databases provide an excellent and efficient method to find relevant citations, the risk assessment researcher must still rely on hard copies of texts, government documents, reference materials, and telephone contacts with appropriate persons in the private and public sectors. Although data in these printed works can rapidly become obsolete (e.g., changes in telephone numbers, addresses and key personnel, regulatory concentrations), they offer a wealth of background information vital in the development of a risk assessment. Examples of such documents include:

- Clayton, George D. and Clayton, Florence E., Eds., *Patty's Industrial Hygiene and Toxicology, Vol. 2, Toxicology,* 1991-1994, John Wiley & Sons, New York. Currently published in six parts. Compounds are included in classes of substances, e.g., metals, epoxy compounds, or esters. Each chapter discusses various human and animal studies which have been conducted on the class of compounds.
- *Current Contents,* Institute of Scientific Information, Philadelphia, PA. Weekly. Tables of contents of a large number of journals, published weekly, in several parts. Of particular interest are: agriculture, biology, and environmental sciences; engineering, technology, and applied sciences; physical, chemical, and earth sciences; and life sciences.
- *The Merck Index,* The Merck Co., Rathway, NJ.
- *Sax's Dangerous Properties of Industrial Materials.*
- *Hazardous Substances in our Environment: A Citizen's Guide to Understanding Health Risks and Reducing Exposure,* 1990, U.S. Environmental Protection Agency, Policy, Planning and Evaluation report no. EPA 230/09/90/081, Washington, D.C. Includes general information on how to identify hazardous substances in the environment; how to estimate risk; and government programs to reduce risk and inform the public of possible risks. It also contains a glossary of terms; a

bibliography of EPA publications on hazardous substances; and directories of state and EPA contacts, and private and nonprofit organizations.
- *EPA Publications Bibliography,* National Technical Information Service, Springfield, VA, 1970 - present. Published quarterly. Contains abstracts of EPA publications published by NTIS. The October-December issue contains indices for the entire year. In addition, there are presently, cumulations for 1970-1976, 1977-1983, and 1984-1990. Documents are indexed by title, key word, personal and corporate author, sponsoring office, and report number. The user should keep in mind, however, that not every EPA document is distributed by NTIS and that EPA offices should be contacted directly if the publication cannot be located elsewhere. A complete NTIS database is also available on CD-ROM.
- *Pollution Abstracts,* Cambridge Scientific Abstracts, Bethesda, MD, 1970- . A quarterly publication with annual cumulations, this index contains a section on toxicology and health, including toxicology of pesticides, heavy metals, and agricultural chemicals, and the effects of toxic materials on humans, other animals, and plants. Pollution Abstracts is also available in some libraries on CD-ROMs as part of a database called Poltox.
- *Access EPA,* Information Access Branch, Information Management and Services Division, Office of Information Resources Management, U.S. Environmental Protection Agency, Washington, D.C. Published annually. An extremely valuable tool for obtaining information from the U.S. EPA. The volume contains clearinghouses and hotlines (e.g., Superfund), EPA and state agency libraries, and major EPA dockets.
- *EPA Headquarters Telephone Directory,* Government Institutes, Inc., Rockville, MD, Published periodically. Contains a detailed breakdown of various EPA offices in Washington, in the regions, and at the environmental laboratories located nationwide. Indices by subject and personnel title are included.
- *Environmental Telephone Directory,* Government Institutes, Inc., Rockville, MD, Published annually. Detailed directory to Federal legislative committees and subcommittees, the U.S. EPA headquarters, other Federal agencies dealing with environmental issues (including the Dept. of Agriculture, Dept. of Energy, Health and Human Services, and Dept. of Fish and Wildlife), and state environmental agencies. Also includes a list of clearinghouses and hotlines from the EPA, DOT, and U.S. Coast Guard, and other agencies. Since this directory is published annually, it is probably more reliable than the *EPA Headquarters Directory* for correct telephone numbers. It is also much easier to use.
- *Directory of Environmental Information Sources,* 1995, 5th ed., Government Institutes, Rockville, MD. In addition to the governmental sources included in the *Environmental Telephone Directory,* it lists professional and scientific trade organizations (e.g., Association of State and Territorial Health Officials, the National Environmental Association, and the Sierra Club), publications, and databases.
- *Federal Yellow Book,* Monitor Leadership Directories, Inc., New York. Published quarterly. By far the most up to date directory available. It includes the departments and the independent agencies (e.g., U.S. EPA) of the Federal government. Under each major division, there is a detailed breakdown of offices and staff. Indices are both by personal name and by major office. There are no index entries for the offices listed under departments/agencies (e.g., Office of Solid Waste and Emergency Response), however, so users need to refer to the department/agency entry in order to locate a specific office.

- *Gale Environmental Sourcebook: A Guide to Organizations, Agencies, and Publications*, 1992, Gale Research, Inc., Detroit. A fairly comprehensive directory to government agencies and programs; research facilities and educational programs; clearinghouses and hotlines; publications; databases; and library collections. Entries on risk assessment and toxicology, for example, include the Center for Risk Management, Risk Science Research Center, Syracuse Research Center, Risk Reduction Engineering Laboratory, *Toxicological and Environmental Chemistry*, and the Toxics Use Reduction Institute. It also includes an appendix containing the EPA National Priorities List. Some of this information will become dated (e.g., Federal telephone number changes), so an updated version should be consulted if available. This sourcebook is a good starting place for those unfamiliar with the field.
- *Technical Assistance Directory*, 1993, Office of Research and Development report no. EPA/600JK-93J006, Washington, D.C. Includes various EPA programs and staff with their areas of expertise and telephone numbers. This volume is particularly valuable for contacts in various areas such as Risk Assessment Forum, Office of Health Research, Health Effects Research Laboratory, and Office of Health and Environmental Assessment. The contacts contained in this directory would be particularly valuable in interpreting regulation language and for sources for particular kinds of information.
- Howard, Philip, H., Ed., *Handbook of Environmental Fate and Exposure Data for Organic Chemicals*, Lewis Publishers, Chelsea, MI. Four volumes published in this series to date: *Large Production and Priority Pollutants* (vol. 1), *Solvents* (vol. 2), *Pesticides* (vol. 3), *Solvents* 2 (vol. 4), and *Solvents* 3 (vol. 5). The chemicals in each volume were selected from chemicals included in the National Library of Medicine's (NLM) Hazardous Substances Data Bank (HSDB). Listed for each chemical (if data are available) are substance identification, chemical and physical properties, toxicity and environmental fate, and exposure potential (e.g., natural and artificial sources, terrestrial, aquatic and atmospheric fate, and biodegradation).
- Lewis, R. J., *Dangerous Properties of Industrial Materials*, 1992 and 1993 Update, 8th ed., Van Nostrand Reinhold, New York. Volume 1 is an index by chemical name and includes many synonyms for each chemical. The information included for each chemical varies widely, depending on the information available. The basic record includes synonyms, Chemical Abstracts number, formula and molecular weight, dose information, inclusion in various federal government hazardous chemical lists, and available references.
- *The Merck Index: an Encyclopedia of Chemicals, Drugs, and Biologicals*, 1989, 11th ed., Merck & Co., Rathway, NJ. A new edition of this index is published approximately every 8 or 9 years. It includes physical descriptions, chemical properties, history of research, and indices by chemical name, synonym, formula, and Chemical Abstracts number.
- *Pesticide Fact Handbook: U.S. Environmental Protection Agency*, 1988- , Noyes Data Corp., Park Ridge, NJ. Presently published in two volumes and contains Pesticide Fact Sheets issued by the U.S. EPA, arranged alphabetically, with numerical, common name, generic name, and trade name indexes. They include description of chemicals; use pattern and formulations; science findings (including toxicological characteristics, oncogenicity, mutagenicity, and teratogenicity); summary of regulatory positions and rationales; summary of labeling statements; summary of major data gaps; and the name of the contact person at the EPA.

III. SELECTED ENVIRONMENTAL INFORMATION SOURCES

Information presented in this chapter is designed to help the reader locate information sources and information that could be critical to his or her project. Many of the sources listed in this primer should be used as a first contact for finding information. For example, when calling the Safe Drinking Water Hotline, you may want to obtain very technical documents or information that persons working for the Hotline may not have or be qualified to answer. However, they can find out who in EPA has the documents you need or the technical person you need to contact. This basic method works well if you have patience and don't give up as you get bounced from office to office in your search for a person to help you get the information you need. Since institutions are constantly changing their internal structures and telephone numbers, the reader is advised that the addresses and telephone numbers provided can change at any time. The reader should obtain an organization's general telephone number from commercial telephone directories to locate telephone numbers that have changed since publication of this book (see Tables 1 – 6).

IV. CONCLUSION

Each section of a risk assessment report requires specific types of information. Information can be obtained by mail from private and public organizations or through library research. Modern environmental research libraries and their professional staffs offer the researcher an electronic and paper highway to find appropriate references for use in their risk analysis.

Table 2   Clearinghouses, Hotlines, Bulletin Boards, and Docket

GENERAL	Clearinghouses, Hotlines, Bulletin Boards, and Dockets	Clearinghouses are central access points for technical reports and documents. Hotlines and bulletin boards provide access to information for persons via telephone or computers. Dockets are collections of documents used by EPA to make regulatory decisions.
	Center for Environmental Research (CERI)	Exchange of scientific and technical information. 513-569-7562.
	INFOTERRA	International environmental information. 202-260-5917.
	Pollution Prevention Information Clearinghouse	Reference library, electronic reference, hotline, and outreach efforts. 703-821-4800.
AIR AND RADIATION	Air Docket	Public record information on Clean Air Act matters. 202-260-7548.
	Air Risk Information Support Center (AIR RISC) Hotline	Toxic pollutant health, exposure, and risk assessment. 919-541-0888.
	BACT/LAER Clearinghouse	Best Available Control Technology at Lowest Achievable Emission Rate. Air pollution control technology information related to new source review permitting requirements. 919-541-2376.
	Control Technology Center (CTC) Hotline	Air emissions and air pollution control technology for all pollutants. 919-541-0800.
	EPA Model Clearinghouse	Interpretations of modeling guidance. Electronic bulletin board. 919-541-5683.
	National Air Toxics Information Clearinghouse (NATICH)	Noncriteria air pollutants and air toxics control program development. 919-541-0850.

Table 2 continued

HAZARDOUS AND SOLID WASTE	Hazardous Waste Technology	Hotline, electronic bulletin board, and reference library. 301-670-6294.
	CERCLIS Helpline	Superfund help. 202-260-0056.
	Emergency Planning and Community Right-To-Know Hotline	SARA Title III information. 800-535-0202.
	National Response Center Hotline	Reporting of accidental release of oil and hazardous substances to the environment. 800-424-8802.
	RCRA Docket Information Center	Materials used to make RCRA regulatory decisions. 202-260-3046.
	RCRA/Superfund/OUST Assistance Hotline	Assistance with RCRA, Superfund, underground storage tanks, and pollution prevention/waste minimization questions. 800-424-9346.
	Superfund Docket and Information Center	Superfund inquiries, primarily dockets and documents. 202-260-9760.
	UST Docket	Documents related to underground storage tank regulatory actions. 202-260-9720.
PESTICIDES AND TOXIC SUBSTANCES	Asbestos Ombudsman Clearinghouse/Hotline	Asbestos abatement. 800-368-5888.
	FIFRA (Pesticides) Docket	Documents related to regulatory actions under the Federal Insecticide, Fungicide, and Rodenticide Act. 703-305-5805.
	Toxic Substances Docket	Documents related to regulatory actions of Office of Toxic Substances. 202-260-7099.
	TSCA Assistance Information Service	Regulatory information on Toxic Substances Control Act.

Table 2 continued

WATER	Clean Lakes Clearinghouse	Lake protection management and restoration. 800-726-5253.
	Drinking Water Docket	Documents related to regulatory decision on Safe Drinking Water Act Section 1412. 202-260-3027.
	National Small Flows Clearinghouse	Small community water and wastewater treatment. 800-624-8301.
	Nonpoint Sources Pollution Exchange	Nonpoint water pollution. 202-260-7109.
	Safe Drinking Water Hotline	Information related to Safe Drinking Water Act and Amendments. 800-426-4791.

Table 3 General Non-EPA Sources of Information

Agency for Toxic Substances and Disease Registry (ATSDR) Toxicology Profiles	In depth toxicology profiles for selected chemicals. Contact NTIS at 703-487-4650 or 800-336-4700 for profiles.
California Environmental Protection Agency	The Toxics Directory, Fourth Edition. References and Resources on the Health Effects of Toxic Substances. Berkeley, California.
Code of Federal Regulations (CFR)	Books codifying federal regulations. Available at many libraries and by GPO subscription.
Council on Environmental Quality	Environmental Quality report. Council on Environmental Quality, 722 Jackson Place NW, Washington, D.C. 20503. 202-395-5750.
Directory of Environmental Information Sources	Book providing the name of organizations and contacts for environmental information. Government Institutes, Inc., 4 Research Place, Suite 200, Rockville, MD 20850. 301-921-2323.
Federal Geographic Data Committee	Promotes coordinated development, use, sharing, and dissemination of surveying, mapping, and related spatial data. Executive Secretary, Federal Geographic Data Committee, U.S.G.S., 590 National Center, Reston, Virginia 22092.
Federal IRM Directory	Identifies information resource management contacts throughout the federal government. Information Resources Management Service (IRMS-KAP). U.S. General Services Administration,18th and F Street, NW, Washington, D.C. 20405. 202-501-2426.
Federal Register (FR)	Provides information on proposed and final federal agency rules. Available at many libraries and by GPO subscription.
Fish and Wildlife Data	Contact state fish, wildlife, or natural resources department.
General Accounting Office	Assesses many government programs and issues. Document Handling and Information Services Facility, U.S. GAO, P.O. Box 6015, Gaithersburg, MD 20877. 202-275-6241.
Government Printing Office	Government publications. Superintendent of Documents, Government Printing Office, Washington, DC 20402. 202-783-3238.
Local Health Department	Provides information about health-related problems associated with a given site, activity, or facility. Contact local town government office.
Local Fire Department	Provides records of underground storage tanks, copies of Material Safety Data Sheets for locally stored chemicals, and other hazardous substance information for local businesses. Contact local town government office.
Local Tax Assessor	Provides information related to land ownership and structures. Contact local town government office.
Local Water Authority	Provides public and private water supply information including maps, well locations and depths, and water intake locations. Contact local town government office.
Local Well Drillers	Provide data on public and private wells. Check local government offices and yellow pages for local drillers.

Table 3 continued

Local Zoning Board or Planning Commission	Provides information on local land use and ownership. Contact local town government office.
National Technical Information Service	Primary source for government scientific and technical information. Can also be accessed via hard copy, electronic databases or CD-ROMS at many libraries. U.S. Department of Commerce, 5285 Port Royal Road, Springfield, VA 22161. 800-553-NTIS.
National Cartographic Center	Provides information on national soils geographic databases and their interpretive attribute files, and GIS resource data and maps. National Cartographic Center, Soil Conservation Service, U.S. Department of Agriculture, P.O. Box 6567, Fort Worth, TX 76115. 817-334-5292 or 817-334-5559.
National Wetlands Inventory	Information on wetlands. National Wetlands Inventory, U.S. Fish and Wildlife Service, 9720 Executive Center Drive, Monroe Building, Suite 101, St. Petersburg, FL 33702-2440. 813-893-3624. For National Wetlands Inventory maps call 800-USA-MAPS.
Natural Heritage Program	Provides information on federal and state-designated endangered and threatened plants, animals, and natural communities. Contact state environmental, natural resources, or conservation departments for state specific information on availability of lists, maps, and general information.
State Geological Surveys	Geologic and hydrologic information.
U.S. Army Corps of Engineers	Records and data involving surface waters.
U.S. Fish and Wildlife Service	Provides environmental information including toxicology data. U.S. Fish and Wildlife Service, 18th & C Streets, NW, Washington, D.C. 20240, or regional offices.
U.S. Geological Survey	Geologic, hydrogeologic, and hydraulic information including maps, reports, databases, and studies. U.S. Geological Survey, 12201 Sunrise Valley Drive, Reston, VA 22092.

Table 4 Sources of Maps and Aerial Photographs

Aerial Photographs	Contact state departments of transportation, local zoning and planning offices, county tax assessor's office, college and university libraries, geology or geography departments, EPA's Environmental Monitoring Services Laboratory (EMSL), EPA's Environmental Photographic Interpretation Center (EPIC), U.S. Army Corps of Engineers, U.S. Department of Agriculture, and U.S. Geological Survey.
Geologic and Bedrock Maps	Surficial exposure and outcrop information for interpreting subsurface geology. Contact USGS Regional or Field Offices, State Geological Survey Office, or U.S. Geological Survey, 12201 Sunrise Valley Drive, Reston, VA 22092 to obtain maps.
Flood Insurance Rate Maps (FIRM)	Maps delineating flood hazard boundaries for flood insurance purposes. Contact Federal Emergency Management Agency (FEMA), Federal Insurance Administration, Office of Risk Assessment, 500 C Street, SW, Washington, D.C. 20472 or local zoning and planning offices to obtain maps.
National Wetland Inventory Maps	Provides maps delineating environments and habitats. Contact U.S. Geological Survey, 12201 Sunrise Valley Drive, Reston, VA 22092 or U.S. Fish and Wildlife Service, 18th and C Streets, NW, Washington, D.C. 20240 to obtain maps.
State Department of Transportation Maps	State maps detailing road systems, surface water systems, and other important geographical and political features. Contact state or local government agencies for copies.
U.S. Geological Survey (USGS) Topographic Quadrangles	Maps detailing topographic, political, and cultural features that are available in 7.5 and 15 minute series. Contact USGS Regional or Field Offices or U.S. Geological Survey, 12201 Sunrise Valley Drive, Reston, VA 22092 to obtain maps.

Table 5 Government and Private Databases

CERCLIS (Comprehensive Environmental Response, Compensation, and Liability Information System)	EPA's inventory of potential hazardous waste sites. Contact EPA Regional Offices for access information.
Chemtox.Dialog (file 337)	Includes approximately 10,000 chemicals. For each, includes identification information, properties, regulatory information, toxicity, first aid, and spill, storage, and response information. Cost: $1.00/connect minute; $10.00/full record.
Compliance Monitoring and Enforcement Logs (CMELs)	EPA's summary of compliance monitoring and enforcement logs for facilities. Contact EPA Regional Offices for access information.
Federal Reporting Data System (FRDS)	General information on public water supply utilities using ground or surface waters. Contact EPA for access information.
Geographical Exposure Modeling System (GEMS)	EPA's database of U.S. census data. Contact EPA for access information.
HWDMS (Hazardous Waste Data Management System)	EPA's inventory of hazardous waste producers. Contact EPA Regional Offices for access information.
National Planning Corporation (NPDC)	Commercial database of U.S. census data. Contact National Planning Data Corporation, 20 Terrace Hill, Ithaca, NY 14850.
NPDES (National Pollutant Discharge Elimination System) Database Printouts	EPA's list of sites with current or past wastewater disposal permits. Contact EPA Regional Offices for access information.
PATHSCAN	Identifies surface water drinking water intakes and populations served. Contact EPA for access information.
RCRA (Resource Conservation and Recovery Act) Database Printouts	EPA inventory of hazardous waste generators. Contact EPA Regional Offices for access information.
STORET	EPA's repository of water quality data for U.S. waterways. Contact EPA Regional Offices for access information.
WATSTORE	U.S. Geological Survey's National Water Data Storage and Retrieval System contains the Ground Water Site Inventory file (GWSI). Contact USGS Regional or Field Offices or U.S. Geological Survey, 12201 Sunrise Valley Drive, Reston, VA 22092 for access information.
WellFax	National Well Water Association's inventory of municipal and community water supplies. Contact National Well Water Information (NWWA), 6375 Riverside Drive, Dublin, OH 43017 for access information.

Table 6 Technical Guidance Documents

SOIL SAMPLING AND EVALUATION	U.S. EPA. 1986. Test Methods for Evaluating Solid Waste (SW-846); Physical/Chemical Methods. Office of Solid Waste.
	U.S. EPA. 1986. Field Manual for Grid Sampling of PCB Spill Sites to Verify Cleanups. Office of Toxic Substances. EPA/560/5-86/017.
	U.S. EPA. 1987. A Compendium of Superfund Field Operations Models. Office of Emergency and Remedial Response. EPA/540/P-87/001 (OSWER Directive 9355.0-14).
	U.S. EPA. 1989. Soil Sampling Quality Assurance Guide. Environmental Monitoring Support Laboratory, Las Vegas, NV.
	U.S. EPA. 1990. Rationale for the Assessment of Errors in Sampling of Soils. PB90-242306.
	U.S. EPA. 1991. Description and Sampling of Contaminated Soils. A Field Pocket Guide. EPA/625/12-91/002.
	U.S. EPA. 1991. Characterizing Soils for Hazardous Waste Site Assessment. EPA/540/4-91/003.
	U.S. EPA. 1992. Preparation of Soil Sampling Protocols: Sampling Techniques and Strategies. PB92-220532/AS.
GROUNDWATER SAMPLING AND EVALUATION	U.S. EPA. 1985. Practical Guide to Ground-water Sampling. Environmental Research Laboratory, Ada, OK. EPA 600/2-85/104.
	U.S. EPA. 1987. A Compendium of Superfund Field Operations Models. Office of Emergency and Remedial Response. EPA/540/P-87/001 (OSWER Directive 9355.0-14).
	U.S. EPA. 1987. Handbook: Ground Water. Office of Research and Development. EPA/625/6-87/016.
	U.S. EPA. 1988. Statistical Methods for Evaluating Ground Water from Hazardous Waste Facilities. Office of Solid Waste.
	U.S. EPA. 1988. Guidance on Remedial Actions for Contaminated Ground Water at Superfund Sites, Interim Final. Office of Emergency and Remedial Response (OSWER Directive 9283.1-2).
	U.S. EPA. 1989. Ground-water Sampling for Metal Analyses. Office of Solid Waste and Emergency Response. EPA/540/4-89-001.
	U.S. EPA. 1992. Potential Sources of Error in Groundwater Sampling at Hazardous Waste Sites. EPA/540/S-92/019.
	U.S. EPA. 1993. DNAPL Site Evaluation. PB-93-150217.
	Wilson, N. 1995. *Introduction to Soil Water and Ground Water Sampling*. Lewis Publishers, CRC Press, Boca Raton, FL.

Table 6 continued

SURFACE WATER AND SEDIMENTS SAMPLING AND EVALUATION	U.S. EPA. 1981. Procedures for Handling and Chemical Analysis of Sediment and Water Samples. Great Lakes Laboratory.
	U.S. EPA. 1984. Sediment Sampling Quality Assurance User's Guide. Environmental Monitoring Support Laboratory, Las Vegas, NV. NTIS PB-85-233-542.
	U.S. EPA. 1985. Methods Manual for Bottom Sediment Sample Collection. Great Lakes National Program Office. EPA 905/4-85/004.
	U.S. EPA. 1987. A Compendium of Superfund Field Operations Models. Office of Emergency and Remedial Response. EPA/540/P-87/001 (OSWER Directive 9355.0-14).
	U.S. EPA. 1987. An Overview of Sediment Quality in the United States. Office of Water Regulations and Standards.
AIR SAMPLING AND EVALUATION	U.S. EPA. 1983. Technical Assistance Document for Sampling and Analysis of Toxic Organic Compounds in Ambient Air. Office of Research and Development.
	U.S. EPA. 1987. A Compendium of Superfund Field Operations Models. Office of Emergency and Remedial Response. EPA/540/P-87/001 (OSWER Directive 9355.0-14).
	U.S. EPA. 1988. Procedures for Dispersion Modeling and Air Monitoring for Superfund Air Pathway Analysis.
	U.S. EPA. 1990. Compendium of Methods for the Determination of Air Pollutants in Indoor Air. PB90-200 288/AS.
	U.S. EPA. 1993. Particle Total Exposure Assessment Methodology. PB93-166957.
BIOTA SAMPLING AND EVALUATION	Asante-Duah, D.K. 1993. *Hazardous Waste Site Risk Assessment*. Lewis Publishers, CRC Press, Boca Raton, FL.
	U.S. EPA. 1987. A Compendium of Superfund Field Operations Models. Office of Emergency and Remedial Response. EPA/540/P-87/001 (OSWER Directive 9355.0-14).
	U.S. EPA. 1989. Guidance Manual for Assessing Human Health Risks from Chemically Contaminated Fish and Shellfish. Office of Marine and Estuarine Protection. EPA/503/8-89/002.

CHAPTER 24

Risk Assessment of Airborne Chemicals

Jeanne C. Willson

CONTENTS

I. Introduction 466
II. Conceptual Site Models 466
   A. Indirect Exposure Pathways 469
   B. Project Manager Role in Conceptual Site
      Model Development 469
   C. Developing Data Quality Objectives 470
      1. State the Problem 471
      2. Identify (Define) the Decision 471
      3. Identify Inputs to Decision 471
      4. Define Study Boundaries 471
      5. Develop a Decision Rule 471
      6. Specify Limits on Decision Errors 471
      7. Optimize Design for Obtaining Data 472
   D. DQO Process: Final Check 472
III. Estimating Chemical Concentrations at Exposure Points:
    Transport Models 473
IV. Occupational Exposure and Risk Assessment 473
    A. Describing Toxicity: Reference Concentrations
       and Unit Risk 474
V. Conclusion: Risk Characterization and Informing
   the Risk Manager 476
   References 477

I. INTRODUCTION

Risk assessment professionals argue endlessly about how much soil people eat, if any, or whether certain groundwater sources will be used as sole sources of residential drinking water, and a host of other risk assessment exposure questions. But nobody argues about whether people breathe air. When chemicals are in the air, people are exposed. Discussion of airborne chemical risk assessment centers around modeled predictions, the toxic effects of the chemicals (especially at low doses), probabilities of accidental releases, the hazards of inhaling small particulate matter, and indirect pathways. Project managers have many opportunities to inject rationality into the air toxics risk process, regardless of their level of technical involvement. In this chapter, we will discuss the typical issues that arise in evaluating air toxics, with special emphasis on what managers should watch for, and we will discuss the general approach to risk assessment* as it applies to air toxics, including:

- Developing a conceptual site model
- Applying the DQO process
- Using appropriate exposure and toxicity information to develop a risk characterization

II. CONCEPTUAL SITE MODELS

Evaluating risk from chemicals in air is not quite as simple as detecting its presence somewhere and plugging detected values into a model. While a consulting risk assessor will probably do this evaluation, direct input and oversight from project managers at this point is significant and critical. Project managers know the site (or situation) and know what happens. That knowledge, plus common sense, provides 90% of what is needed to develop a conceptual site model, which describes all of the significant ways in which people may contact site-related chemicals and which will be the foundation of the risk assessment.

Fortunately for all of us, the mere presence of a chemical anywhere is not enough to cause a risk. Enough of it must (1) move to and (2) contact someone (a receptor) before there is a risk. Actually we can be more specific than that about the requirements for significant exposure that might indicate complete exposure pathways from a source to a receptor, via air.

1. A *source* must exist, such as an incinerator or ventilation stack, an evaporation pond, fugitive (nonpoint) emissions from an industrial facility, or any other significant source of chemical that is open to the air. A *secondary source* might be water in a home that releases aerosols when used for showering, cooking, flushing toilets, watering the lawn, watering a vegetable garden, and so forth.
2. A *release mechanism* is required. For air, look for (1) volatilization, (2) wind release of particulates from contaminated soils, (3) emission through ventilation

---

* For additional information on this general approach, the EPA's 1999 risk assessment guidance is still the best single summary available at this writing (see References).

of stacks, etc., and (4) negative pressure that develops inside basements that sucks in volatile chemicals, radon, etc. from soils or groundwater-derived vapors surrounding building foundations.
3. A *transport mechanism* may be required if potentially exposed people are located away from the release point. For air, transport mechanism means "wind."
4. The obvious *exposure medium* is air, but consider also (1) deposition of particulate matter on outdoor soils that may be eaten directly, tracked inside, and eaten as "incidentally ingested" house dust, or absorbed by garden vegetables, which may be eaten; (2) deposition in indoor house dust; (3) attachment to dust particles, which are then readily deposited in the lungs (this is the mechanism for radon exposure). Other potential routes to exposure media are possible.
5. An *exposure point* is required. The amount of chemical that actually reaches a person or an ecological receptor is the amount that is significant, not the amount emitted from the source. The selection of transport models and placement of monitoring stations should account for this distinction. Air measurements should be made in the breathing zone — 3 to 6 feet off the ground — not at the ground from a flux chamber or far above head on a telephone pole. Also note that direct measurements made away from a source are likely to measure other sources as well. We found, for example, that measured cadmium and other metals may originate from domestic wood burning, not from metal mines.
6. Receptors must be present, now or in the future. Is that downwind cabin a year-round residence or just a summer home? Did the transport model predict concentrations at the housing development or in the middle of a fallow field? If the only possible receptors are maintenance people or occasional visitors, what is their expected exposure frequency? At this point in the analysis, risk assessors generally note that there must also be a route of exposure: oral, inhalation, or dermal absorption. Whether there is a route of exposure can be debated for certain contaminated media; for example, not everyone has to pump and drink the groundwater. The debate is a minor issue for air; since everyone breathes, inhalation is an obvious route of exposure. Another exposure route that may be important for airborne chemicals is eye exposure that may result in significant irritation and tearing.

Potentially complete pathways are compiled into descriptive lists and graphic presentations to provide guidance to the risk assessment (see Figure 1). Conceptual site models can be elaborate, with molecules of a chemical being chased all over the countryside. Perhaps this tendency is an ill-guided response to public pressure and concern. The author even heard of a serious proposal to evaluate the risk to humans posed by being bitten by wild animals exposed to windblown (radioactive) particulate matter deposited on the soil. The best way to argue against such foolishness is to identify a limited number of exposure pathways that will cause the greatest potential exposure. If those pathways are managed so that risk is negligible, then other pathways derived from those are also almost guaranteed* to be negligible.

A factor that is not always considered in risk assessments is degradation of chemicals. Chemicals in air may be photodegraded or oxidized, and this may result in greatly reduced risk. On the other hand, it also results in smog formation in cities.

---

* Note that it is part of the job description of a risk assessor to never be virtually 100% certain of anything.

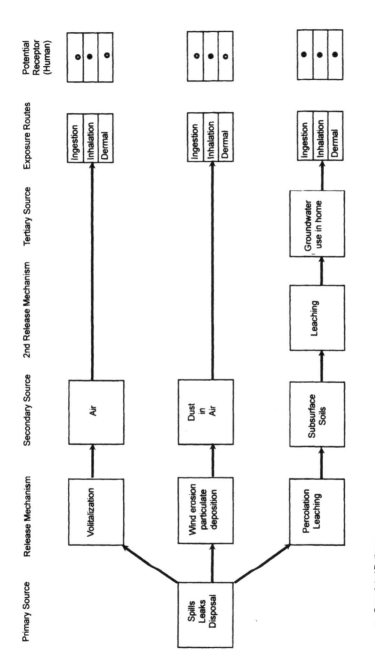

Figure 1 Elements of site conceptual model.

Since this may be overlooked, project managers should suggest it to risk assessors if some exposure pathways are found to pose a potential risk of health effects based on hypothetical modeling.

A. Indirect Exposure Pathways

In the early 1980s, the infancy of environmental risk assessment, it was deemed sufficient to make simple assumptions about daily intake of water, soil, and air by adults living full time with chemicals in those media. Simple calculations of chemical intake were made and risk was calculated. As risk assessment grew up, it became apparent that these simple assumptions were inadequate in some important ways: children are not tiny adults, exposure rates can vary widely, and chemicals move from one environmental medium to another. These things need to be accounted for in risk assessments. In the field of air toxics, additional pathways became known as "Indirect Pathways." Guidance for evaluating indirect pathways has been formalized under some programs (see, e.g., guidance for hazardous waste and other combustion facilities: U.S. EPA; 1990, 1993a, and 1994. Such federal and state guidance documents show how a full and complex set of exposure pathways can be evaluated. Pathways include deposition on plant leaves and deposition on soils, root uptake and translocation, uptake by cattle and accumulation in beef and milk, and so forth. The approach is conservative and protective, but the resulting risk models have not been validated. By combining many conservative decisions together, there is a real danger of producing an unrealistically high estimate of risk. Some people argue that in the absence of data such estimates are appropriate to fully protect people. The models provide a good starting point for understanding the fate of air toxics. In many cases, it will be worthwhile to gather supplementary information to refine the risk estimates. It may also be worth using quantitative uncertainty analysis methods such as probabilistic or Monte Carlo analysis, or fuzzy logic analysis (for further discussion, see Burmaster and Appling, 1995).

B. Project Manager Role in Conceptual Site Model Development

As the risk assessor develops the conceptual site model, the project manager should gather and provide as much information as possible about the site, historic conditions and occurrences, known or potential exposures, worker behavior and job duties, recreational visitation rates, current land use and likely future land development plans, and so forth. It may be important at this stage to collect additional data. While mangers with bottom-line accountability are naturally reluctant to spend project resources for additional data collection, however, it can pay off in lower remedial costs or improved public confidence. A cost-benefit analysis may help a project manager decide whether additional data collection might be cost-effective. For example, assume that the choice is either to collect air monitoring data in a nearby housing development or to use conservative air modeling in the risk assessment. On one hand, data collection can be expensive. If air monitoring data in the nearby housing development is not collected, the project will save a certain sum of money, but conservative default assumptions will be used to compensate for uncertainty about air concentrations. This conservative risk assessment approach may generate risk findings that indicate the potential presence of a significant, but hypothetical risk, and, as a result, a risk management decision to require greatly reduced stack emissions. On the other hand, if air monitoring data is collected, it may show that actual exposure is minimal, and less costly alternatives protect health adequately. The question is whether scrimping on data collection is penny-wise and pound foolish.

Appropriate data needs will become apparent as the conceptual site model is developed. Often, in an enthusiastic rush to solve an environmental problem, both agencies and industries may be guilty of collecting data that does not help the risk assessors generate a better estimate risk. The Data Quality Objectives (DQO) process, described below, helps all parties think through all stages of the complex risk assessment process to avert such errors. It also saves money in the long run. If the formal DQO process isn't being used, the project manager should demand that it be used.

C. Developing Data Quality Objectives

The conceptual site model is really a collection of hypotheses about what could happen to chemicals from a site, facility, or activity. An investigation leading to a risk assessment is an evaluation of these hypotheses. The best (some would say only) way to evaluate hypotheses is to use the scientific method; the first steps are these: ask your questions, design your investigation to answer the questions, and check to see that your investigation will really answer the questions you originally asked. Variations of this process have been formalized in many fields (e.g., economics, psychology) under different names. In environmental investigations, it is called the DQO process. Good guidance from U.S. EPA describes the process (U.S. EPA, 1993b). Often, however, people think that the DQO process is nothing more than getting a high enough count of soil or air samples and a low enough detection limit ("Gee, 5 nanograms per microgram sounds low enough to me!"). The DQO process is much more than that: it is a project manager's most powerful tool to demonstrate to senior management, agency personnel, and the public that the environmental project is doing what it should. The process documents decisions that are made, so that if project personnel change, or the project is so long that at the end no one can remember the beginning, it is less likely that previously settled matters will be reversed or challenged.

EPA has proposed a three-step, and more recently a seven-step, process for developing DQOs. The original three steps can be restated as questions. Exactly what question are you trying to answer? What decision are you trying to make? What do you need to know or learn to answer that question or make that decision? What data collection and study design will provide the needed information? The seven-step process, laid out in detail in U.S. EPA guidance cited above, is outlined below to demonstrate its value and scope.

RISK ASSESSMENT OF AIRBORNE CHEMICALS

1. State the Problem

Summarize the contamination problem that will require new environmental data, and identify the resources available to resolve the problem. For example, an old industrial site is found to have specific chemicals in its soil; the weather is dry and the area is dusty.

2. Identify (Define) the Decision

Identify the decisions to be made and identify those that require new environmental data to address the problem. For example, determine whether chemicals released by wind erosion from a bare site pose a risk to nearby residents or determine whether dust raised by driving trucks on-site poses a risk.

3. Identify Inputs to Decision

Identify the information needed to support the decision. This may use existing information or require new measurements. For the above examples, it may be necessary to take air quality measurements in the residential area or during typical truck usage.

4. Define Study Boundaries

Specify the spatial and temporal aspects of the environmental media or potential exposure that will bear on the decision. For example, weather patterns through the course of a year may result in different emission rates. These differences must be factored in to arrive at realistic health risk estimates.

5. Develop a Decision Rule

Develop a logical statement defining the conditions that would drive the decision maker's choice among alternative actions. In air toxics risk assessment the decision rule often takes the form of, "If measured levels do not exceed calculated levels then a 'No further actions' alternative is appropriate; otherwise conduct additional evaluation." The series of small decisions that comprise the major decisions are also laid out in this step (see Figure 2).

6. Specify Limits on Decision Errors

Specify acceptable limits on decision errors, which are used to establish performance goals for limiting uncertainty in the data. Performance goals are translated into sampling protocols, detection limits, statistical power calculations, laboratory performance requirements, and specific DQOs. However, effective and appropriate DQOs can only be established in the context of the rest of this process.

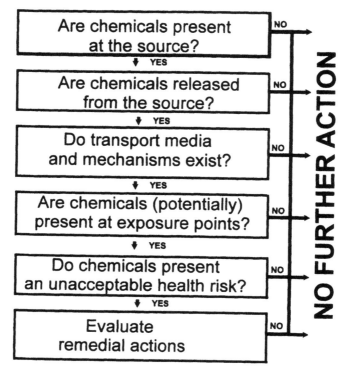

Figure 2 Considerations to incorporate into a decision rule.

7. Optimize Design for Obtaining Data

Identify the most resource-effective sampling and analysis design for generating data that are expected to satisfy the DQOs.

D. DQO Process: Final Check

To review a study that has been planned using the DQO process, a project manager asks: "What is missing?"

- Do all proposed data, tests, analyses, etc. address an identified need or decision? If not, do not collect the data.
- Have all of the significant questions and decisions been identified?
- Will the selected inputs answer the questions? If not, what additional information is required?
- What uncertainties in the risk assessment make the assessment too conservative? Could uncertainties be reduced with additional data?
- What are the weaknesses of the study from a logical standpoint? Can they be strengthened in the design phase?

III. ESTIMATING CHEMICAL CONCENTRATIONS AT EXPOSURE POINTS: TRANSPORT MODELS

A first-pass risk assessment might simply evaluate concentrations in air emissions from a stack, other point source (vents), multiple point sources (motor vehicles), or area sources (landfill). If these concentrations are safe for full-time residential exposure, then a more complex risk assessment may not be necessary. Analysis of multiple chemicals or additional pathways may be required, however, as a matter of U.S. EPA region or state policy. If the concentrations appear to be too high, it's appropriate and reasonable to model potential concentrations to realistic exposure points. Measuring air concentrations at exposure points away from the source should also help to arrive at more realistic air concentrations, although other sources (such as wood-burning stoves, fireplaces, other industrial facilities, home chemicals) often contaminate samples.

In some cases modeling is a cost-effective alternative to data collection. In others, modeling is the only option (see Chapter 24). Modeling may be necessary when emissions are hypothetical and cannot be measured for a proposed facility, for example, or for a facility expansion that is not operational. Models may be complex or simple. In large area dispersion models, airborne chemicals move over hill and dale. The hills and dales and other land forms are important factors in these models, as are large buildings. Complex-terrain models address these factors. In contrast, simple models ignore terrain and buildings. Small-scale models for predicting concentrations of chemicals in buildings can also be quite complex. Such models include:

- Vadose zone models of how vapors move from the groundwater table through soils
- Basement models of how vapor enters houses from soils surrounding the foundation (about one-third of the made-up air inside of a house enters from subsoil cracks)
- Shower models of how aerosols form in houses from water use for flushing toilets, showering, and cooking (groundwater-borne chemicals are significant sources of air toxics in some versions of these models)
- Deposition models of how airborne contaminants accumulate on soils and garden vegetables
- Soil-to-vegetable uptake models of how contaminants move from soil into plants

Some models are so simple that nonmodelers can use them at great cost savings. However, their accuracy may be questionable. Models are available from various sources, including EPA's Center for Exposure Assessment Modeling (CEAM), and the published literature. We recommend using the most current versions available of the very simple models included in U.S. EPA's *Risk Assessment Guidance for Superfund, Volume 1, Part B*, (U.S. EPA, 1991).

IV. OCCUPATIONAL EXPOSURE AND RISK ASSESSMENT

The good news is: we have lots of information from actual inhalation human exposure about the toxicity of many significant chemicals. The bad news is: we can't use much of it for public health risk assessment. This is because most of the data is from worker exposure on the job. Worker exposure information is not directly applicable to residential exposure issues. This is because workers can be exposed to more chemicals in larger quantities and, thus, be at greater risk than the general public. The thinking goes that a worker is (generally) protected by medical monitoring programs, guarded by the watchful eye of industrial hygienists and physicians, and works with chemicals willingly.* Also, people in the work force are often healthier or less naturally susceptible than the general population, which includes the elderly, children, and the infirm. Another reason that worker exposure may not be very useful for public health risk assessment is that the exposure may not be high enough, or the number of people exposed may not be large enough, to determine whether chemicals cause cancer even though hundreds of workers may receive exposures. In any case, information from occupational exposures is not often used in evaluating toxicity and risk of chemicals typically found at hazardous waste sites or emitted from industrial facilities.

A. Describing Toxicity: Reference Concentrations and Unit Risk

Risks are described a bit differently for exposure to chemicals by the inhalation route. In the past decade, the U.S. EPA has developed toxicity values,** which it generally requires to be used in Superfund risk assessments. These values have been almost universally adopted for other risk assessment uses, including RCRA evaluations and those risk assessments led by states. Any applicable state regulations or guidance should always be checked. The values are updated frequently, so current sources of toxicity information must always be consulted.The values are generally not derived from occupational experience, but from controlled animal studies, or accidental exposures to the general public (see Table 1). Values presented here are only examples.

For oral exposure, a specific dose (in milligrams of chemical per kilogram of person) or a cancer slope factor is given, instead of a concentration in water or soil. For inhalation exposure, a concentration or unit risk value is often given, instead of (or in addition to) the inhalation reference dose and cancer slope factor, partly because the toxicity information is collected in terms of exposure concentration rather than a measured dose relative to body weight. The concentration provided is expected to present a hazard index of 1.0, and the unit risk is the cancer risk per milligram of chemical per cubic meter of air. Both assume constant exposure. U.S. EPA defines unit risk, for example, as: "The upper bound excess lifetime cancer risk from continuous exposure to an agent at a concentration of 1 $\mu g/m^3$ in air." Thus, by definition, the unit risk concept is conservative and automatically overestimates risk both by using upper bound toxicity estimates and upper bound exposure estimates. It should be modified to reflect reduced exposure time, and the conservative nature of the toxicity component must be revealed to risk managers and to the public.

* The counter argument is that workers should be able to expect safe working conditions and not incur greater risk than their familes at home.
** The values are published in IRIS; the EPA's Integrated Risk Information System is available through several database services, and available on CD through Government Institutes. Other values are published in HEAST, the Health Effects Assessment Summary Tables (9200.6-303; EPA 540-R-94-020; call EPA for the most current information).

Table 1 Examples of U.S. EPA Inhalation Toxicity Values (note: E-2=10⁻²)

	Unit risk (cancer) concentration per µg/m³	Inhalation slope factor (kg-d/mg)	Reference dose (mg/kg/day)
Arsenic	4.3 E-3	1.5 E+1	—
Arsine	—	—	1.4 E-5
Benzene	8.3 E-6	2.9 E-2	1.7 E-3
Formaldehyde	1.3 E-5	4.6 E-2	—
Carbon Tetrachloride	1.5 E-5	5.3 E-2	5.71 E-4
TCDD (Dioxin)	—	1.2 E+5	—

Because the unit risk factor and the reference concentration combine exposure assumptions with toxicity information, the difference between a unit risk factor and a slope factor can be explained in terms of the risk equation in the following manner:

$$\text{Risk} = \text{Concentration} \times \text{Toxicity} \times \text{Exposure} \quad (1)$$

For noninhalation exposures:
$$\text{Risk} = \text{Concentration} \times \text{Slope Factor} \times \text{Exposure} \quad (2)$$

For inhalation exposures:
$$\text{Risk} = \text{Concentration} \times \text{Unit Risk} \quad (3)$$

Equations for reference doses and reference concentrations are written in the same basic form.

Doses may also be provided by EPA, although its current policy is to provide only unit risk values. Doses may be converted to exposure concentrations, and vice versa. Note in the examples given of toxicity values that the most carcinogenic chemicals have high slope factors (think of this as highest risk per milligram of exposure) and the most toxic gases or vapors have the lowest RfDs (think of the reference dose as the lowest acceptable exposure dose). So arsine gas is considered to be 1000 times more toxic than benzene (which we still breathe in gasoline fumes), and TCDD (the most potent of the dioxins, but not clearly a human carcinogen at low doses) is treated as though it is 10,000 times as carcinogenic as arsenic, which is known to be a human carcinogen. To use the unit risk factors or reference concentrations, multiply the appropriate value by the exposure concentration. For example, if there are 10 µg/m³ of benzene* in the air to which someone is exposed

---

* For many chemicals used in large quantities in the U.S., our experience with occupational or consumer exposure is at odds with the risk estimates projected under the conservative risk assessment approach used by U.S. EPA. Benzene is an example. According to the unit risk value, continuous exposure to just over 10 µg/m³ poses an upper bound risk of 10-4, the high end of the acceptable risk range under Superfund. However, the occupational limit is currently 10 ppm, about 30 mg/m³ (3000 times higher). Even adjusting for the length of the work week vs. a full 168-hour-7-day-week, and acknowledging that the relationship between exposure and cancer risk is not linear at high doses, a discrepancy remains. This has provoked many new studies and reevaluations in recent years and significant evolution in how cancer risk is evaluated. The results are partially reflected in the U.S. EPA's 1996 *Cancer Risk Guidelines*.

full time, the estimated cancer risk for lifetime exposure using this approach is 8.3×10^{-5} or about 8 in 100,000. To use the inhalation slope factors or reference doses, U.S. some exposure assumptions are necessary. An adult is assumed to inhale and absorb the chemicals from 20 cm³/day (the volume of air in a 9 × 10 foot room). Unless there is other information, it is assumed the person absorbs all of the chemical in air, even though humans absorb only a fraction of the oxygen we take in (otherwise artificial respiration would not work). Adults are assumed to weigh 70 kg and to live for 70 years. These assumptions are applied using the following basic equations:

$$\text{Intake} = \frac{C \times IR \times EF \times ED}{BW \times AT} \quad (4)$$

$$\text{Cancer risk} = \text{Intake} \times CSF \quad (5)$$

$$\text{Hazard quotient} = \text{Intake}/RfD \quad (6)$$

where C = concentration in air; IR = inhalation rate; EF = exposure frequency in days/year; ED = exposure duration in years; BW = body weight in kg; AT = averaging time in days (lifetime for cancer risk, equal to ED for noncancer hazard quotients); CSF = cancer slope factor; and RfD = reference dose.

That is all there is to doing a risk calculation for a single chemical along a single exposure pathway. The model is simple, linear, and easy to compute. It may or may not be accurate, but it is accepted, conservative, and widely used.

V. CONCLUSION: RISK CHARACTERIZATION AND INFORMING THE RISK MANAGER

As with all risk assessments, the final product is a statement of estimated risks under specific conditions along with the uncertainty surrounding the estimates. The risk characterization will provide an evaluation of the pathways identified in the conceptual site model, using appropriate exposure and toxicity information. Risk managers, including members of the public who influence the risk management decisions, will need help to understand the risk characterization. In the author's experience, their ability to understand is limited not by their intelligence or training, but by (1) the clarity of the communication, (2) their motivation to understand, and (3) the amount of time they have to spend understanding new technical information. Communication of risks should respect the limited time available to most people and focus on providing clear, jargon-free explanations of the most important points. An effectively executed and communicated risk assessment of air toxics or other chemical exposures will clarify why the selected action (or no action) is appropriate and protective of human health and the environment for all parties: agency staff, the public, and the industrial management (who may pay the bill).

REFERENCES

Burmaster, D.E. and Willson Appling, J., Introduction to human health risk assessment with an emphasis on contaminated properties, *Environ. Reporter*, 25, 48, 1995.

U.S. Environmental Protection Agency, *Addendum to the Methodology for Assessing Health Risks Associated with Indirect Exposure to Combustor Emissions*, Office of Research and Development, Washington, 1993.

U.S. Environmental Protection Agency, *Data Quality Objectives Process for Superfund: Interim Final Guidance*, Office of Emergency and Remedial Response, Washington, 1993b.

U.S. Environmental Protection Agency, *Exposure Assessment Guidance for RCRA Hazardous Waste Combustion Facilities: Draft*, Office of Solid Waste and Emergency Response, Washington, 1994.

U.S. Environmental Protection Agency, *Methodology for Assessing Health Risks Associated with Indirect Exposure to Combustor Emissions, Interim Final*, Environmental Criteria and Assessment Office, Cincinnati, 1990.

U.S. Environmental Protection Agency, *Proposed Guidelines for Carcinogen Risk Assessment*, National Center for Environmental Assessment, Washington, 1996.

U.S. Environmental Protection Agency, *Risk Assessment Guidance for Superfund, Volume I: Human Health Evaluation Manual (Part A), Interim Final*, Office of Emergency and Remedial Response, Washington, 1989.

U.S. Environmental Protection Agency, *Risk Assessment Guidance for Superfund, Volume I: Human Health Evaluation Manual (Part B, Development of Risk-Based Preliminary Remediation Goals), Interim Final*, Office of Emergency and Remedial Response, Washington, 1991.

CHAPTER 25

Radiation Risk Assessment

Nava C. Garisto and Donald R. Hart

CONTENTS

I. Introduction 479
II. Radiation Types and Sources 480
    A. Types of Radiation...480
    B. Radiation Units 481
    C. Radiation Sources 482
III. Risk Assessment for Radioactive Substances 482
    A. The Risk Assessment Process 482
    B. Problem Formulation 483
    C. Radiation Exposure Analysis..483
        1. Source Term Development 485
        2. Radionuclide Transport Analysis....................................486
        3. Food Chain Pathways Analysis 486
        4. Dose Rate Estimation 489
        5. Radiation Response Analysis 490
        6. Risk Characterization...492
IV. Conclusion 494
    References...494

I. INTRODUCTION*

Risk assessment for radioactive substances is a quantitative process that estimates the probability for an adverse response by humans and other biota to radiation

---

* The authors wish to thank Dr. D. Lush, Dr. F. Garisto, Ms. K. Fisher, and Mr. M. Walsh for critically reviewing early drafts of this manuscript. The graphics support of M. Green is greatly appreciated.

exposure. It has been used for a variety of regulatory purposes such as the derivation of site-specific radionuclide release limits, or the determination of the acceptability of proposed undertakings that may release radionuclides.

Radioactive substances, as compared to other chemical substances, have a long history of risk-based regulations. These regulations developed in reaction to early mismanagement of radiation risks. Today, the concept of site-specific risk assessment is fundamental to the regulation of radioactive substances and serves as a model for risk-based regulation of other chemicals.

The unique properties of radioactive substances, associated with their emissions of ionizing radiation, require specialized approaches to assessment of exposure, dose, and risk. For example, since a radiation dose can be received without physical contact with the radioactive substance, this external exposure, as well as internal exposure from radionuclides taken into the body, must be considered. Moreover, since radiation is the common agent of hazard for all radioactive substances, concentration and dose are usually expressed in radiation units (see below), and doses are additive across radionuclides, in contrast to the situation with chemical toxicants.

Whereas the fundamental concepts of risk assessment are the same for radioactive and other chemical substances, the unique properties of and approaches to radioactive substances must be understood in order to critically evaluate a consultant's work and integrate it into an overall risk assessment. The purpose of this chapter is to outline these unique properties and approaches to risk assessment of radioactive substances to better enable project managers to work with consultants in this technical area.

II. RADIATION TYPES AND SOURCES

A. Types of Radiation

Radiation consists of energetic particles or waves that travel through space. The less energetic wave types are said to be nonionizing because they do not cause atoms in biological tissue to become electrically charged. Familiar examples of nonionizing radiation are the visible light and heat that reach the earth from the sun. The more energetic wave types, such as ultraviolet rays, X-rays and gamma rays, are said to be ionizing, because they have enough energy to make electrons in biological tissues completely escape their atomic orbitals, forming electrically charged ions. In addition to wave energy, radioactive substances may emit sub-atomic particles such as beta or alpha particles. These particles also have sufficient energy to ionize biological tissues.

All types of ionizing radiation (both waves and particles) can produce damage to the biological tissues that they contact. Wave types can easily penetrate biological tissue. Some of the X or gamma rays that are directed towards the body will pass right through without being absorbed (i.e., without transferring energy to cause ionization). Others will be absorbed when they strike atoms in the tissue, forming charged ions. The charged ions are chemically reactive, and often react inappropriately. When this happens in the genetic material (DNA) that controls cell function, there is a chance that cell growth may eventually go out of control, causing cancer. If there is sufficient genetic damage in a reproductive tissue, there may also be some loss of reproductive function.

Particle radiations, because of their mass and electric charge, are less able to penetrate biological tissue. Their energy is absorbed and damage is concentrated closer to the point of biological contact. For example, if the radiation source is outside the body, most of the beta and alpha radiation will be absorbed in the skin. On the other hand, if the source is a radionuclide that has been incorporated into an internal tissue, most of the beta and alpha radiation will be absorbed inside that tissue. Alpha particles, because of their large mass, high charge, and high energy, produce more localized and intensive ionization effects than either waves or beta particles, and therefore tend to produce a greater amount of genetic damage. They also tend to produce a different spectrum of genetic damage (i.e., a higher proportion of chromosome breaks as opposed to point mutations) which makes accurate repair less likely.

Differences in the biological effectiveness of various radiation types are described by "quality factors" (QF). Gamma and beta radiations have quality factors of one (QF = 1), while alpha radiation has a much higher quality factor (QF = 20) based on its greater effectiveness in human cancer induction. Quality factors based on reproductive impairment have not been well defined, particularly for nonhuman species. This is a major source of uncertainty in assessment of ecological risks from alpha-emitting radionuclides.

B. Radiation Units

A radionuclide is designated by its atomic mass (isotope) number and its chemical element name. As it decays by atomic disintegration, its mass may change and it is transformed to a new element or a series of different "daughter" elements (a decay series). Alpha, beta, or gamma radiation is released with each disintegration over the course of this transition. Under secular equilibrium (i.e., undisturbed) conditions, each element in a decay series has the same activity.

Activity is a measure of radiation quantity in terms of atomic disintegration frequency. It is directly related to the amount of a radionuclide and its radiological half-life. Activity is expressed in becquerels (1 Bq = 2.7×10^{-11} Ci = one disintegration per second). Activity concentration in any medium is expressed in Bq per unit of mass, volume, or surface area.

The radiation energy absorbed by an organism is expressed as a dose in grays (1 Gy = 100 rad). The rate of energy absorption is expressed as a dose rate in Gy per unit of time. These units represent absorbed energy without regard to the radiation type or the effectiveness of the absorbed dose (1 Gy of alpha radiation is capable of causing more biological damage than 1 Gy of gamma radiation). Effective dose rates for humans are expressed as gamma dose equivalents in sieverts (Sv) per unit of time (1 Sv = 100 rem) after application of appropriate quality factors to account for radiation type.

C. Radiation Sources

All of us are exposed to ionizing radiation every day. The earth is continually bombarded by protons, X-rays, gamma rays, and ultraviolet radiation from cosmic sources. Approximately 67% of this radiation is absorbed by the earth's atmosphere and never reaches the earth's surface. Atmospheric gases such as ozone are particularly important in absorption of ultraviolet energy.

In addition to the cosmic sources of ionizing radiation, humans and other biota on earth are exposed to ionizing radiation from the decay of radioactive substances on earth. Ionizing radiation comes from such diverse sources as building materials in houses, glass and ceramics, water and food, tobacco, highway and road construction materials, combustible fuels, airport scanning systems, the uranium in dental porcelain used in dentures and crowns, diagnostic X-ray sources, and many others. Most of these substances contain radionuclides that are naturally present in the earth, although human activity has increased their production and/or the potential for human exposure. Other radionuclides, which are produced in nuclear reactors or accelerators, are geologically unknown or extremely rare.

The background radiation dose rate received by the average person from natural sources is approximately 2 mSv/a (UNSCEAR, 1988). Typical dose rates and doses from anthropogenic sources are as follows:

- Medical, average of all procedures = 1.0 mSv/a
- Fallout from nuclear weapons testing = 0.01 mSv/a
- Chernobyl accident, average first year commitment* in Bulgaria = 0.75 mSv
- Chest X-ray (one) = 0.1 mSv
- Dental X-ray (one) = 0.03 mSv
- Barium enema (one) = 8 mSv

Natural background varies geographically with altitude, latitude, and local geology. It is higher at high altitudes where the atmosphere is thinner and there is less atmospheric absorption of cosmic radiation. Fallout from long-range atmospheric transport varies mainly with latitude, due to global air circulation patterns, peaking at 40 – 70° north latitude.

III. RISK ASSESSMENT FOR RADIOACTIVE SUBSTANCES

A. The Risk Assessment Process

Risk assessment of radioactive substances should be conducted whenever radioactive substances are identified as contaminants of potential concern at a site. The process that is recommended by international agencies for risk assessment of radioactive substances (IAEA, 1989, 1992a) is consistent with the more recent U.S. EPA (1989, 1992) paradigms for human health risk assessment (HHRA) and ecological risk assessment (ERA) although there are minor variations in terminology. While the

---

* 50-year dose commitment from exposures over the first year.

process has historically been focused on the human receptor, there is increasing attention to nonhuman dose and risk estimation.

The radiological risk assessment process is outlined in Figures 1 and 2. The process is iterative as shown in Figure 1, with updating of methodology, models, and data between iterations. The risk assessment process includes the following basic components:

- Identification of events and processes which could lead to a release of radionuclides or affect the rates at which they are released and transported through the environment
- Estimation of the probabilities of occurrence of these release scenarios
- Calculation of the radiological consequences of each release (i.e., doses to individuals and populations and associated human cancer risks or ecological effects)
- Integration of probability and consequence over all scenarios to define the overall risk of human cancer or ecological effects
- Comparison of maximum doses and/or risks with current regulatory criteria

Deterministic estimates of maximum dose from each scenario are often made initially to evaluate whether further analyses are required. Probabilistic estimates are appropriate whenever maximum doses approach effect thresholds or acceptability criteria (IAEA, 1992a). The probabilistic methods explicitly consider the uncertainties in key parameters, but use best estimates as central values for each one. This produces a more realistic statement of risk.

B. Problem Formulation

Problem formulation is the scoping exercise which identifies the radionuclide sources, release scenarios, human and ecological receptors, exposure pathways, and response endpoints to be considered in the subsequent risk assessment. The spatial and temporal scales of analysis must also be defined. Collectively, these elements constitute a conceptual model of the system to be studied. They are included in the first two boxes on the main axis of Figure 1. It is important to ensure, at this stage, that all major stakeholder concerns are represented in the conceptual model.

There are few aspects of problem formulation that are unique to radioactive substances, although the gap between realistic concern and public perception is often particularly large for these substances. The scope of an assessment can easily escalate from local site-specific risk issues to encompass national energy policy issues. Without minimizing these public participation challenges, or the importance of problem formulation, this chapter focuses mainly on the subsequent stages of consequence analysis and risk characterization.

C. Radiation Exposure Analysis

Humans and other biota can be exposed to radiation by multiple routes. All environmental media must be considered as potential routes of exposure. For example, radionuclides may be carried into the atmosphere as aerosols or gases (e.g., radon),

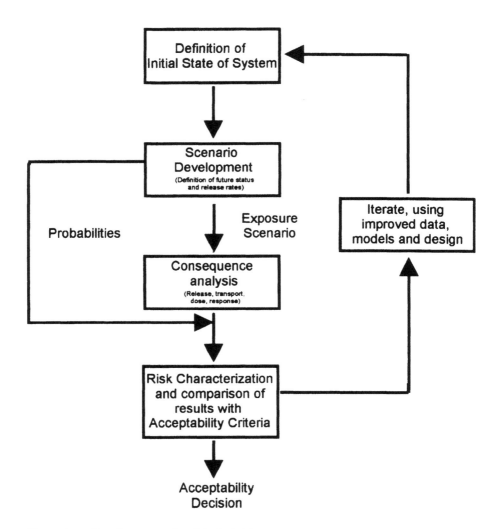

Figure 1 Overall process of radiological risk assessment.

and may fall onto the land and/or be leached into surrounding water bodies. As they disperse from the area of release, in either air or water, they are generally diluted and concentrations tend to decrease with distance from the source. Humans and biota near the source may take in larger quantities of radioactivity in the air they breathe, the water they drink, and the food they eat than organisms farther away. Figure 2 illustrates the major steps in exposure estimation within the overall risk assessment framework. These steps include source-term development, radionuclide transport analysis, food chain pathways analysis, and dose-rate estimation.

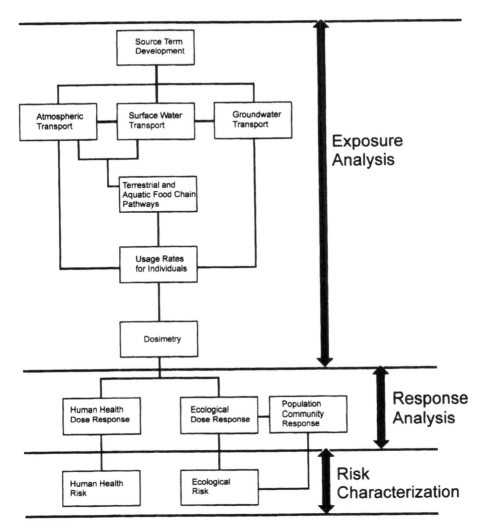

Figure 2 Major steps in radiological risk assessment as related to the framework for ecological risk assessment.

1. Source Term Development

The source term development will determine, through measurement or theoretical calculation, the type and quantity of radionuclides released in terms of activity per unit time. The chemical and physical form of the released radionuclides must also be considered. In the past, little emphasis was placed on accurately estimating source terms and considerable uncertainty still exists in this area for many assessments.

Source term models that have been developed specifically for radioactive waste management applications include, e.g., the AREST model (Liebetrau et al., 1987), the VAULT model (Johnson et al., 1994), and the RAMSIM model (BEAK, 1996a). These models take into account the evolution of geochemical and hydrological conditions in the source matrix, and the corresponding changes in radionuclide release rates over time.

2. Radionuclide Transport Analysis

The radionuclide transport calculations trace radionuclide movements through air, surface water, and groundwater. The objective here is to predict the activity concentrations of radionuclides to which humans and other biota are exposed. The contaminant transport models simulate physical transport due to processes such as advection and dispersion. The mechanisms of radionuclide movement through the natural environment are not dependent on the activity level of the radionuclide, except in a few cases (e.g., radiolysis of groundwater, the decomposition of groundwater caused by high levels of radiation, affects the oxidation states of radionuclides in groundwater and thereby affects radionuclide mobility). Since radioisotopes have chemical properties identical to those of their stable homologs, their movements will parallel those of stable elements. From the point of view of release and mobility, therefore, the important parameters are the physical state, the type of aggregation if any (e.g., colloidal), the chemical form, solubility, oxidation states, sorption properties, and volatility. The key product of a transport model is an estimate of radionuclide activity per unit volume of air, water, or soil as a function of time.

Processes that affect radionuclide transport through the atmosphere are shown schematically in Figure 3. In addition to the conventional dispersion processes, which are considered for all contaminants, radioactive decay and buildup have to be taken into account for radionuclides. For example, in modeling the transport of radon gas, it is important in some cases to consider its radioactive decay products and their deposition, especially within confined environments. The transformations that occur with degradation of some organic compounds add a similar level of complexity to their transport analyses. Atmospheric transport models include a whole range of models, from screening-level analytical (Gaussian plume) models to sophisticated numerical models that can take into account complex terrain, shoreline effects, building wake effects, and long-range transport. The more sophisticated models require more extensive input data. This often limits their usefulness.

Processes that affect contaminant transport through surface waters and groundwater are shown schematically in Figure 4. As with atmospheric transport, radioactive decay and buildup have to be taken explicitly into account. Numerous mathematical models, from simple to complex, have been developed to simulate the flow of water and the transport of radionuclides in surface waters and groundwater. It is important to understand the simplifying assumptions inherent in the simple models, in order to recognize the complex situations in which they are not applicable.

3. Food Chain Pathways Analysis

The food chain analysis traces radionuclide movements from surface water, soil, and atmosphere through a variety of internal exposure pathways to humans and other

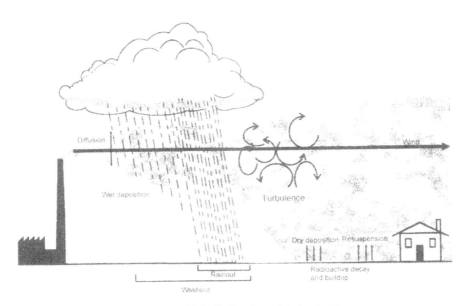

Figure 3 Atmospheric processes that affect radionuclide transport.

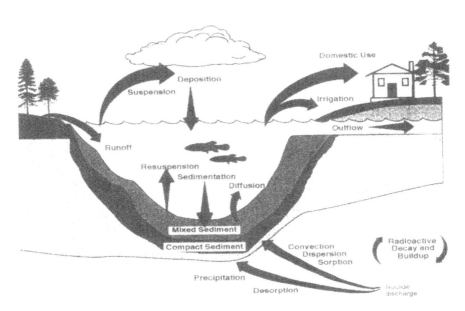

Figure 4 Radionuclide transport processes in surface waters and groundwater.

biota, in order to calculate radiation doses due to inhalation of air and ingestion of food, drinking water, and soil. Processes typically considered in food chain models include: atmospheric deposition to vegetation and soil, bioaccumulation from water

488 A PRACTICAL GUIDE TO ENVIRONMENTAL RISK ASSESSMENT REPORTS

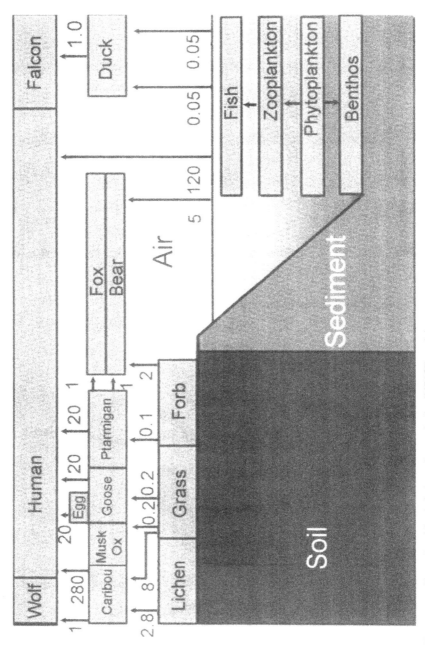

Figure 5 Example of food chain pathways in the IMPACT model

to fish and soil to vegetation, animal feed or forage ingestion, human and animal drinking water ingestion, and human ingestion of plant and animal food types (vegetables, fish, and meat). An example of a food chain is shown in Figure 5. Models such as RESRAD (Yu et al., 1993) and IMPACT* (BEAK, 1996b) have been designed for analysis of radionuclide transport and food chain exposure. The IAEA (1994) has tabulated food chain parameter values.

For human receptors, a "critical group" of individuals is identified as a defined group of people likely to receive the greatest radiation dose, based on location and lifestyle factors. Radionuclide incorporation into body tissues is usually represented either as a simple bioaccumulation factor (for fish and plants) or more explicitly in terms of food intakes and assimilation or transfer factors (for terrestrial vertebrates). Both approaches rely on steady-state assumptions. Detailed biokinetic models are available for use in short-term exposure situations where environmental concentrations change more rapidly than the time to achieve steady-state.

The long time frames that are often imposed on radionuclide risk assessments (e.g., 10,000 years) represent a particular challenge with respect to both exposure and response modeling. The environmental features that influence radionuclide transport, as well as the distributions, food chains, and radiosensitivities of receptor species, may well change with natural succession and radioadaptation. However, forecasting of these evolutionary processes involves large uncertainties.

Certain radionuclides, because of their ubiquitous nature, rapid biological exchange, or regulation in the body, may require alternate approaches to transport and food chain modeling. Radionuclides such as ^{3}H, ^{14}C, and ^{129}I require unique specific activity models. Till and Meyer (1983) discuss modeling approaches for these special cases.

4. Dose Rate Estimation

Calculation of radiation dose rates and cumulative doses to people and biota follow from measured or estimated activities of each radionuclide in each environmental medium, and from measured or estimated activities in the organisms themselves. The radiation dose is integrated over all contributing radionuclides and exposure pathways.

Once in the body, radionuclides continue to emit radiation, and even short-range emissions such as alpha and beta radiation can interact with body tissues. Radionuclides outside the body also emit radiation; however, for most large organisms, only the external gamma emissions have sufficient range to penetrate the body to a biologically significant depth. For humans, the external beta emissions of some radionuclides can be important, but their effects are confined to the skin, where effects other than cancer are limiting. In these cases, a separate skin dose is usually calculated. External doses arise mainly from air immersion, water immersion (swimming or bathing), and groundshine. Groundshine is the external gamma contribution

---

* IMPACT is a multiple source, multiple contaminant, multiple receptor risk assessment model which considers contaminant exposure through air, surface waters, and groundwater pathways. It estimates dose and risk for both radioactive and nonradioactive contaminants.

from radionuclides which have been deposited on the ground or otherwise incorporated into the soil.

The computation of radiation doses to various organisms from the radionuclide activities in their environment and their tissues, requires the use of a dosimetry model for each organism. Radiation dosimetry in human beings is well understood, resulting in a complex model of radionuclide distribution in the body, and integration of organ doses and radiosensitivities into a whole-body gamma-equivalent dose (i.e., sieverts). Dosimetry models for other organisms are less sophisticated and predict doses in terms of absorbed energy only (i.e., grays). Quality factors for integration of effective doses in nonhuman biota are lacking.

Standard human dose conversion factors (DCFs) are used to calculate the external radiation dose from radionuclide activities in the environment, and the internal radiation dose from radionuclide intake by inhalation and ingestion (ICRP, 1996). These DCFs incorporate all the complexities of human physiology and geometry, as represented by the ICRP (1975) reference man. They are generally greater for children than adults, although this can be offset to some extent by greater adult consumption rates. Dose conversion factors for nonhuman biota are less standardized.

5. Radiation Response Analysis

Radiation response analysis has a different focus in HHRA than in ERA. For humans, it is focused on protecting the individual. For other biota, it is focused on protection of populations and communities.

Certain value judgements are involved in determining the significance of a radiation dose. Generally, we consider stochastic effects, such as increased probability of cancer or hereditary disease, to be important to humans because of the value placed on quality of life for the individual. We assume that these effects may be produced at low-dose rates, based on linear extrapolation from high-dose rate data, but they tend to occur late in life or in the progeny of exposed individuals.

In other animal populations, stochastic effects are more accepted by society. The maintenance of animal population size or community diversity is usually our primary consideration. Stochastic effects may have little impact on such population and community endpoints. Higher dose rates are generally required to produce the nonstochastic (threshold) effects on survival and/or reproduction that are needed to impair a population or community.

Based on extrapolation from high-dose events, such as the Hiroshima and Nagasaki atomic explosions, we assume a risk factor of approximately 0.04 induced premature fatal cancers per Sv of radiation dose, and 0.01 induced hereditary effects. Thus, 30 years of exposure to 1 mSv/a (the ICRP [1991]) public dose limit) would produce a cumulative cancer risk of approximately 1×10^{-3}. The ALARA* policy

---

* ALARA Policy: compliance with dose limits ensures that working in a radiation laboratory is as safe as working in any other safe occupation. The goal of the radiation safety program is to ensure that radiation dose to workers, members of the public, and to the environment is as low as reasonably achievable (ALARA) below the limits established by regulatory agencies. The program also ensures that individual users conduct their work in accordance with university, state, and federal requirements.

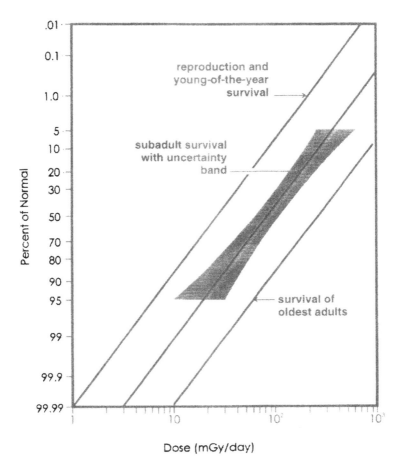

Figure 6 Example dose response curve for subsequent use in stressed population analysis.

in radiation protection states that public dose rates should be "as low as reasonably achievable" below this limit.

Threshold dose rates for survival and reproductive responses to radiation stress in nonhuman biota have been reviewed by many authors and several international agencies (e.g., IAEA, 1992b). Based on these documents, no-effect thresholds of approximately 1 mGy/day for mammals and 10 mGy/day for fish are defensible. Logistic (sigmoidal) response vs. dose relationships are usually assumed, although hormetic (stimulatory) responses to low doses are well known. In general, younger age classes and reproductive functions are most sensitive (see Figure 6). When no-effect threshold dose rates are exceeded, the possibility of population and/or community responses should be considered.

Population and community responses may be considered empirically by reference to relevant field studies of ecosystem exposure to radiation. However, since there are few such studies, empirical data relevant to the species and dose rates of interest are often lacking. The alternative is to model the population or community response. Population models can be used to translate survival and reproductive response functions (e.g., Figure 6) into a population response function (e.g., density reduction vs. dose rate) or a population response at a given dose rate.

6. Risk Characterization

Risk is the probability of a defined adverse effect arising from a defined set of chemical, physical, or biological stressors. Risk characterization is an integration of exposure and response analyses to provide a risk estimate. We are primarily concerned with cancer risks for humans and risks of radiotoxic (threshold) effects for other biota.

An estimate of cancer risk to humans can be derived directly from the estimated radiation dose rate. However, such a risk estimate is highly conditional on the accuracy of the estimated dose. A more meaningful risk estimate is one which incorporates all the uncertainties in both dose and response analyses.

Radiotoxicity risks to nonhuman organisms are sometimes expressed in terms of a hazard quotient (HQ = estimated dose rate/no-effect threshold dose rate). However, the HQ is not a probability and, therefore, not a true risk estimate. The risk of radiotoxicity (e.g., HQ >1), or of population reduction to x% of baseline, can only be determined by incorporation of uncertainties in exposure and response analyses.

Uncertainty analysis uses Monte Carlo or Latin hypercube methods to integrate the uncertainties in key exposure and response model parameters. This approach is illustrated in Figure 7. Distributions for each uncertain parameter are sampled repeatedly, and with each sampling the entire system of models is run to predict an effect. After many runs, a probability (risk) distribution for the effect is obtained. Sensitivity analysis is usually performed prior to uncertainty analysis to identify the key model parameters that most influence the effect prediction. These are the parameters for which uncertainty distributions must be defined.

Often the entire risk assessment is performed for a defined radionuclide release scenario, such as a waste container breach or a uranium tailings dam failure. It is important to realize that resulting risk estimates are conditional on scenario occurrence. When nonconditional (integrated) risk estimates are required (e.g., risk associated with a waste repository), it is critical to assign probabilities to all possible release scenarios, and to weight the risk for each scenario according to its probability of occurrence. Integrated risk estimates can then be generated by calculating a weighted sum across all scenarios.

Finally, it is important to realize that fundamental process uncertainty is not easily captured in any risk estimate. For example, if a population model incorrectly represents the mechanism of population regulation, the resulting risk estimate will be inaccurate, even when uncertainties in model parameters are fairly represented. Model validation and intercomparison exercises (e.g., BIOMOVS, 1995) can be used to test and build confidence in the tools of risk assessment.

RADIATION RISK ASSESSMENT

493

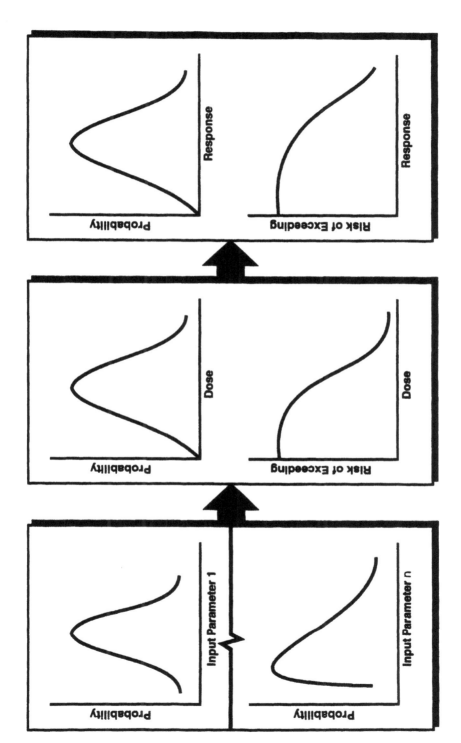

Figure 7 Examples of an uncertainty analysis.

IV. CONCLUSION

Compared to other chemical substances, radioactive substances have a long history of risk-based regulation. Risk assessment for radioactive substances is used to derive site-specific radionuclide release limits, for example, and to determine the acceptability of proposed undertakings that may release radionuclides. Although fundamental concepts common to all risk assessments apply also to radioactive substances, the unique physical properties of radioactive substances, and corresponding technical approaches, must be recognized. Such awareness will enable project managers to work with consultants and other professionals in this technical area.

REFERENCES

Amiro, B.D. and MacDonald, *Dose Conversion Factors for Non-Human Biota for Uranium Series Radionuclides*, Environmental Science Branch AECL Research, Whiteshell Laboratories, Pinawa, Manitoba, 1993.

Anspaugh, L.R., Catlin, R.J., and Goldman, M., The global impact of the Chernobyl reactor accident, *Science*, 242, 1513–1519, 1988.

Barnthouse, L.W., Suter, G.W., and Rosen, A.E., Risks of toxic contaminants to exploited fish populations: influence of life history, data uncertainty and exploitation intensity, *Environ. Toxicol. Chem.*, 9, 297, 1990.

BEAK, *Environmental Impact User Manual*, Beak Consultants Limited, Brampton, Ontario, 1996a.

BEAK, *A Reactive Acid Mine Drainage Simulation Model, User Manual*, Beak Consultants Limited, Brampton, Ontario, 1996b.

BIOMOVS, *Long-term Contaminant Migration and Impacts from Uranium Mill Tailings, Comparison of Computer Models Using a Hypothetical Dataset*, Tech. Report No. 4, Uranium Mill Tailings Working Group of BIOMOVS II, 1995.

Burmaster, D.E., Thompson, K.M., Crouch, E.A.C., Menzie, C.A., and McKone, T.E., Monte Carlo techniques for quantitative uncertainty analysis in public health risk assessments, *Proceedings of the 11th National Conference on Hazardous Wastes and Hazardous Materials*, Washington, D.C., 215, 1990.

Davis, P.A., Zach, R., Stephens, M.E., Amiro, B.D., Bird, G.A., Reid, J.A.K., Sheppard, M.I., Sheppard, S.C., and Stephenson, M., *The Disposal of Canada's Nuclear Fuel Waste: The Biosphere Model, BIOTRAC, for Post Closure Assessment*, Atomic Energy of Canada Limited Report AECL–10720, 1993.

Eckerman, K.F. and Ryman, J.C., External Exposure to Radionuclides in Air, Water and Soil, Federal Guidance Report No. 12, 1993.

Eckerman, K.F., Wohlbarst, A.B., and Richardson, A.C.B., *Limiting Values of Radionuclide Intake and Air Concentration and Dose Conversion Factors for Inhalation, Submersion and Ingestion*, Federal Guidance Report No. 11, 1988.

Garisto, N.C., LeNeveu, D.M., and Garisto, F., The mass transport of radionuclides in a multilayered medium, *Atomic Energy of Canada Limited Report*, AECL 10384, 1992.

Hart, D.R., Selection and adaptation in irradiated plant and animal population: a review, *Atomic Energy of Canada Limited*, AECL–6808, 1981.

International Atomic Energy Agency, *Generic Models and Parameters for Assessing the Environmental Transfer of Radionuclides form Route Releases, Exposures of Critical Groups*, Vienna, 1982.

International Atomic Energy Agency, *The Application of the Principles for Limiting Releases of Radioactive Effluents in the Case of the Mining and Milling of Radioactive Areas*, Vienna, 1989.
International Atomic Energy Agency, *Effects of Ionizing Radiation on Plants and Animals at Levels Implied by Current Radiation Protection Standards (draft)*, International Atomic Energy Agency, Vienna, 1991.
International Atomic Energy Agency, *Current Practices for the Management and Confinement of Uranium Mill Tailings, TRS 335*, Vienna, 1992a.
International Atomic Energy Agency, *Effects of Ionizing Radiation on Plants and Animals at Levels Implied by Current Radiation Protection Standards, TRS 332*, Vienna, 1992b.
International Atomic Energy Agency, *Handbook of Parameter Values for the Prediction of Radionuclide Transfer in Temperate Environments, TRS 364*, Vienna, 1994.
International Commission on Radiological Protection, *Age-Dependent Doses to Members of the Public from Intake of Radionuclides: Part I*, ICRP Publication 56, Ann. ICRP Vol. 20, No. 2, 1989.
International Commission on Radiological Protection, *Age-dependant Doses to Members of the Public from Intake of Radionuclides: Part 5, Compilation of Ingestion and Inhalation Dose Coefficients*, Pergamon Press, Elmsford, NY, 1996.
International Commission on Radiological Protection, *1990 Recommendations of the International Commission on Radiological Protection*, Pergamon Press, Elmsford, NY, 1991.
International Commission on Radiological Protection, *Report of the Task Group on Reference Man*, Pergamon Press, Elmsford, NY, 1975.
Johnson, H. and Tutiah, M., *Radiation is Part of Your Life*, Atomic Energy of Canada Limited (AECL) Report WNRE 1–501, 1993.
Johnson, L.H. et al., *The Vault Model for Post-closure Assessment*, Atomic Energy of Canada, Ontario, Canada, 1994.
Kocker, D.C. and Sjoreen, A.L., Dose rate conversion factors for external exposure to photon emitters in soil, *Health Physics*, 48(2), 193:205, 1985.
Lemire, R.J. and Garisto, F., *The Solubility of U, Np, Pu, Th and Tc in a Geological Disposal Vault for Used Nuclear Fuel*, Atomic Energy of Canada Limited Report AECL 10009, 1989.
Liebetrau, A.M. et al., *The Analytical Repository Source Term (AREST) Model: Description and Documentation*, Pacific Northwest Laboratory, Corvallis, OR, 1987.
Lush, D.L., Hart, D.R., and Acton, D.W., *Proceedings at the Fourth International Conference on High Level Radioactive Waste Management*, Las Vegas, 1993.
Luckey, T.D., *Radiation Hormesis*, CRC Press, Boca Raton, FL, 1991.
NCRP, *Effects of Ionizing Radiation on Aquatic Organisms*, Report No. 109, National Council on Radiation Protection, Bethesda, MD, 1991.
NRPB, *Committed Effective Organ Doses and Committee Effective Doses from Intakes of Radionuclides*, National Radiological Protection Board, Chilton, Canada, 1991.
Onishi, Y., Serne, R.J., Arnold, E.M., Cowan, C.E., and Thompson, F.L., Critical Review: Radionuclide Transport, Sediment Transport, and Water Quality Mathematical Modelling; and Radionuclide Adsorption/Desorption Mechanisms, WA, NUREG/CR-1322, PNL-2901, Pacific Northwest Laboratory, Richland, 1981.
Pinner, A.V. and Hill, M.D., *Radiological Protection Aspects of Shallow Land Burial of PWR Operating Wastes*, U.K. National Radiological Protection Board, 1982.
Science Applications, Inc. (SAI), *Tabulation of Waste Isolation Computer Models*, OH, ONWI-78, Office Nuclear Waste Isolation, Battelle Memorial Institute, Columbus, 1979.
Till, J. and Meyer, H.R., *Radiological Assessment*, U.S. Nuclear Regulatory Commission, Washington, 1983.

Turner, F.B., Effects of continuous irradiation on animal populations, *Advances in Radiation Biology*, 5, 83–144, 1975.

United National Scientific Committee on the Effects of Atomic Radiation, *Ionizing Radiation: Sources and Biological Effects*, United Nations, 1982.

United National Scientific Committee on the Effects of Atomic Radiation, *Sources, Effects and Risks of Ionizing Radiation*, United Nations Scientific Committee on the Effects of Atomic Radiation, United Nations, New York, 1988.

U.S. Environmental Protection Agency, *Risk Assessment Guidance for Superfund, Vol. 1. Human Health Evaluation Manual, Part A*, Washington, 1989.

U.S. Environmental Protection Agency, *Framework for Ecological Risk Assessment*, 1992.

U.S. Environmental Protection Agency, *Human Health Evaluation Manual, Supplemental Guidance: Standard Default Exposure Factors*, Washington, 1991.

U.S. Environmental Protection Agency, *Risk Assessment Forum, Framework for Ecological Risk Assessment*, Washington, 1992.

Whicker, F.W. and Fraley, L., Jr., Effects of ionizing radiation on terrestrial plant communities, *Advances in Radiation Biology*, 4, 317–366, 1974.

Woodhead, D.S., Methods of dosimetry for aquatic organisms, in *Methodology for Assessing Impacts of Radioactivity on Aquatic Ecosystems*, IAEA-TR-190, International Atomic Energy Agency, Vienna, 43–96, 1979.

Yu, C. et al., *Manual for Implementing Residual Radioactive Material Guidelines Using RESRAD, Version 5.0*, Argonne National Laboratory, Argonne, IL, 1993.

CHAPTER 26

Remediation Risk Assessment

William Phillips

CONTENTS

I. Introduction 497
II. Technical Review of Remediation Risk Assessment 498
    A. Evaluation of Remedial Alternatives..501
    B. Residual Risk 501
III. Conclusion 502
    References..503

I. INTRODUCTION

Remediation risk assessments calculate chemical specific numerical performance standards which must be achieved for a regulatory agency to find that medium or site acceptably clean. They are conducted to determine how much of one or more environmental contaminants must be removed from a site or medium to achieve acceptable risk levels sufficient to protect human and environmental health.

Cleanup decisions are not solely based on calculated risk levels and acceptable risk criteria. Risk managers are not captives of risk findings and must weigh several factors to arrive at a final equitable cleanup decision. For example, when formulating a cleanup decision for contaminated wetland sediments, risk managers must weigh reduced ecological risks against habitat destruction. Risk managers must balance the costs and benefits of chemical risk reduction and not rely solely on achieving a numerical performance standard.

Site or medium remediation efforts are often hampered by technological and financial constraints. Currently available technologies may not be able to reduce site contamination levels to regulatory agency numerical performance standards. As a remediation technology reaches its removal limits, its ability to efficiently and economically remove contaminants decreases. At the limits of remediation technology, each increment of pollutant removal increases cleanup costs steeply. For example, removal of a substance to 10 ppm could cost $1 million, 5 ppm, $10 million, and 1 ppm, $100 million. In an age of limited financial resources, risk managers must balance the need to achieve calculated acceptable risk levels with real economic and technological constraints.

Remediation risk assessments allow decision makers to balance public health concerns and cleanup goals with technological feasibility and cost-effective remedies. Such analyses support the selection of innovative technologies which can result in lower remedial costs, yet provide equivalent protection of human health and the environment. Remediation risk analyses can clearly define aspects of the costs and benefits to aid in decision making and can also potentially reduce financial liabilities associated with the cleanup. Considering responsible party cleanup costs at a Superfund site can range from $20–100 million, it is essential that rigorous remediation risk assessment be conducted and used effectively by risk managers in the remedy identification process.

II. TECHNICAL REVIEW OF REMEDIATION RISK ASSESSMENT

Remediation risk assessment practices in the U.S. have been driven by CERCLA and RCRA requirements. While a particular jurisdiction may have unique environmental laws and regulations, the basic processes and procedures are probably derived from these two national environmental statutes. The exact processes, procedures, and levels of regulatory flexibility can vary within and between government agencies. These differences are important because they provide an opportunity for the remediation risk assessor to negotiate risk assessment workplans and remedies that can be more economical while not reducing the level of human or environmental health protection offered by a site cleanup remedy.

Three important uses of risk assessment in environmental cleanup are: selection of numerical cleanup criteria, evaluating short and long term risks when using a cleanup technology, and residual risks following cleanup. While the discussion of these three items will be general in nature, examples from Superfund and RCRA will be used to illustrate key points.

Numerical cleanup criteria are typically derived for each environmental medium of interest. Called "preliminary remediation goals" under Superfund or "action levels" under RCRA, the cleanup criteria provide the remedial design staff with targets to use during the evaluation of alternative cleanup technologies. Chemical-specific cleanup criteria are concentrations based on applicable or relevant and appropriate requirements (ARARs) mandated by CERCLA, such as maximum contaminant levels promulgated under the Safe Drinking Water Act or Ambient Water Quality Criteria promulgated under the Clean Water Act, and risk-based calculations that set concentration limits using toxicity values under specific default exposure conditions. ARARs are rarely available for all chemicals and media of concern. When available, they provide a quick and convenient frame of reference for establishing the scope of site cleanup.

Guidance for calculating risk-based cleanup criteria is available from the CERCLA and RCRA programs. EPA publications provide general guidance, default exposure scenarios, and mathematical algorithms for developing risk-based preliminary remediation goals for the Superfund program. State regulatory agencies must be contacted for guidance regarding cleanup criteria for state-regulated sites or voluntary cleanup programs.

Cleanup criteria can be generic or site specific. Government developed generic risk-based cleanup criteria use a series of conservative assumptions that may not be appropriate for an individual site and are used as default cleanup values. U.S. EPA's development of national soil screening levels for compounds frequently detected at Superfund sites are typical of generic criteria. Generic criteria, ostensibly developed to shorten the time and resources needed to develop cleanup levels, use highly conservative methods to generate very conservative cleanup concentrations that may be technologically impossible to achieve and financially crushing. Inappropriate use of generic cleanup criteria as immutable standards is a common problem encountered by remediation risk assessors. In contrast, site specific solutions are tailored to the unique site conditions and can better reflect site risks and result in more equitable site contamination remedies.

For most sites, the environmental media requiring direct remediation (and development of cleanup criteria) are groundwater, soil, and sediment. Federal and state agencies have developed algorithms to back calculate environmental cleanup concentrations based on a desired risk level and generic or site specific scenarios. Agencies use target risk levels for cleanup purposes. Target carcinogenic and noncarcinogenic risks vary within and between government agencies and must be obtained from an appropriate responsible governmental unit. This hypothetical level of protectiveness is the typical point of departure for evaluating remedial technologies. Target risks for the final remedy may change, but are expected to achieve residual risks (cumulative across all exposure pathways) that lie within a government agency's acceptable risk range or equal to or less than a specific risk criteria.

The availability of generic risk assessment-based cleanup concentrations varies with media and locality. Since most waste sites and facilities involve contaminated groundwater, it is one medium where ARARs are most likely to be available. Cleanup criteria are often based on MCLs. When ARARs are not available or appropriate, site specific values can be generated. The key to developing a credible site specific cleanup value is using a high quality equation and input values. Since there are many possible input values that can be used in such equations (e.g., the amount of drinking water consumed per day for an adult or child), selection of equation input variables is extremely important. It is the job of the remediation risk assessment consultant to ensure that appropriate equations and input values are used to generate a credible cleanup concentration.

ARARs for surface water can be ambient water quality criteria developed under the Clean Water Act. For surface water bodies that are designated for drinking water supply, MCLs have been used as criteria. The potential for exposure through fishing should also be assessed. There is considerable inconsistency from site to site in the development of cleanup criteria for surface water. A practical consideration is to focus on source control in other media to prevent continuing discharges to surface water. Once the source is controlled, concentrations in surface water rapidly attenuate to nondetectable concentrations.

Cleanup criteria for soil have historically been risk-based criteria based on default future land use assumptions. Residential or commercial/industrial development scenarios are used to identify target receptors (residential child, residential adult, or adult worker) and the assumed frequency of exposure. For sites with volatile organic compounds (VOCs) in soil, the conservative default assumptions in the vapor release models make the inhalation exposure pathway the overriding determinant for soil cleanup criteria. Due to the excessive conservatism in the EPA's choice of vapor release models, site-specific modeling of volatile emissions may support less conservative cleanup criteria for soil.

Soil cleanup criteria often consider the potential for cross-media transfer of contaminants. Contaminated soils can be a source of groundwater contamination, and fate and transfer models have been developed to quantify the potential for leaching of soil contaminants to groundwater aquifers.

Soil-to-groundwater fate and transport models estimate cleanup criteria using the physical and chemical characteristics of the soil, the properties of the chemicals of concern, and the hydrogeological characteristics of the site. Generic cleanup criteria derived to protect groundwater resources are typically more conservative than criteria based on direct exposure to soil. It is usually advantageous to collect site-specific data, rather than to use default assumptions about the site characteristics.

Many of the models and default exposure assumptions favored by regulatory agencies in the development of cleanup values have not been validated. As a result, input numbers and final cleanup levels should be amenable to negotiations based on sound science. Remediation risk assessors bear the burden of proof for demonstrating that generic criteria or default assumptions are inappropriate for their particular site. Considerable effort and money often must be expended to meet that burden of proof, especially when negotiating with inexperienced regulators. However, the money spent on site characterization is usually a fraction of what must be spent on an overly conservative cleanup.

Software programs are available from vendors and can simplify calculations of cleanup criteria. Software programs are generally required for detailed statistical manipulation of monitoring data (e.g., kriging) or to perform Monte Carlo simulations of the distribution of input variables, exposures, and associated risks. Probabilistic methods like kriging and Monte Carlo simulations invariably provide a more accurate representation of potential risks than do deterministic methods.

In selected situations, it may also be necessary to evaluate indirect exposures that can occur through the aquatic and terrestrial food chains. Indirect exposures through the aquatic food chain can occur when contaminated sediments or surface water occur in sport fishing areas. Terrestrial food chain exposures are typically assessed when incinerators can enable downgradient dispersion of contaminants. However, for most sites, exposures through the food chain are unlikely.

Remember that the cleanup criteria are only preliminary and may change during the implementation of the cleanup. Site constraints such as fractured bedrock and the practical limits of remedial technologies make it difficult to achieve ideal criteria based on considerations of hypothetical exposure pathways.

REMEDIATION RISK ASSESSMENT

A. Evaluation of Remedial Alternatives

For most site cleanups, a number of potential technologies are evaluated for effectiveness, implementability, and cost. Those waste management options that are ineffective or too costly are eliminated from further consideration. Typical questions project managers should ask about the health risks posed by remedial technologies include:

- Which technologies are capable of achieving cleanup criteria in each environmental medium?
- Which alternative will *not* address the exposure pathways identified in the baseline risk assessment?
- Are the expected short-term risks or residual risks significantly different between alternatives?
- For each technology, what are the major uncertainties affecting implementation and performance?
- Are there other risk-based benefits, such as shorter time to completion, that are presented by particular technologies?
- Is there a need for engineering controls to mitigate risks during installation of a technology? If so, are the controls available and are they reliable?
- Will operating a specific technology create new chemicals of concern (e.g., products of incomplete combustion from incinerators) or new significant exposure pathways for the surrounding community? Are there appropriate engineering controls to mitigate the risks?
- Are containment technologies that leave contaminants at the site being used? If so, five-year reviews must be considered.

Programs such as Superfund and RCRA impose specific evaluation criteria and a labor-intensive selection process. In practical terms, technologies are selected because they accomplish the common sense considerations of cleaning up spills, controlling the source of the contaminants, and effectively managing the wastes that are generated by the cleanup activities. Considerations of protectiveness generally involve evaluating the short-term and long-term human health risks. Short term risks occur during implementation of the remedy or installation of a technology and include worker exposure to fugitive dusts or VOCs during soil excavation, or potential injury due to physical hazards, heat stress, and precarious work environments. Proper use of emergency response plans, engineering controls, work practices, and personal protective equipment can modify the magnitude of potential risks to workers. People who live and work in the vicinity of the site may also receive short-term exposures from, for example, fugitive emissions, emissions from an onsite air stripper or incinerator, runoff of water and sediments or leaching of water, and rupture of vessels containing treatment chemicals. Long-term risks are associated with a remedial alternative and involve evaluating permanence or protectiveness of the technology over time.

B. Residual Risk

Risks are associated with *in situ* treatment-based remedies such as bioremediation or soil vapor extraction, where total removal of a contaminant is not technically feasible. Because preliminary remediation goals and action levels are based on chronic human health risk and reasonable maximum exposure considerations, they usually provide the basis for evaluating residual risk. For many reports, it will be sufficient to indicate if an alternative has the potential to achieve the target numerical criteria rather than to quantify the risk that will remain after implementation of the alternative.

When selecting between otherwise similar alternatives, expert engineering judgement may be required to determine that a particular technology is more certain to achieve the targets or can achieve the target in a shorter schedule. A shorter remedial schedule is usually desirable; operating and maintenance costs are reduced and the potential for random, unforeseen developments is lessened.

The level of effort associated with a remediation risk assessment reflects a number of factors including proximity of populations to site, site chemical toxicity, releases by remedial technologies, unknown risks from innovative technologies, number of chemicals and exposure pathways, and frequency of releases can affect the costs of remediation risk assessments. These factors will determine the level of scientific rigor required by a risk analysis.

Environmental cleanup laws, such as CERCLA, may require that the effectiveness of site remedies be evaluated at regular intervals. Site-specific characteristics that should be considered during the evaluation of residual risks include contaminant source and release information, geological and hydrological information, contaminant fate and transport parameters, and exposure pathways. Periodic reviews also analyze changing legal requirements that have been promulgated by the federal or state government after the remedy was selected for a site. Based on this analysis, the risk assessor can determine whether the original remedial goals remain appropriate.

Depending on the complexity of the site, the level of effort for a residual risk assessment will vary widely. If the remedy performed as predicted, perhaps a qualitative assessment of the site would be sufficient. However, if the performance efficiency of the remedy was less than anticipated, a comprehensive risk assessment may be conducted to evaluate the residual risks to human health and the environment. In the event of remedy failure, the remediation risk assessor, in concert with the project team, will need to consider modifications to the original remedy.

Similar to the baseline risk assessment, a comprehensive residual risk assessment would quantitatively evaluate the levels of contaminants remaining at the site, assess exposure pathways associated with restricted and unrestricted land uses, and numerically characterize risks. The risk assessment may demonstrate that sites with remedies that achieve protectiveness for current use may require restrictions on future activities.

III. CONCLUSION

Changing political and societal goals may require reassessment of earlier remedial actions as technologies fail to meet lofty expectations. Given the disturbing current trend of increased cleanup costs, yet general ineffectiveness of the Superfund and

RCRA cleanup programs to clear the backlog of site cleanups, future statutory and policy changes in the programs are inevitable. These changes will eventually affect default exposure assumptions, target risks, cleanup criteria, and other determinants of protectiveness.

The uncertainties associated with quantifying the exposure and toxicity terms in the risk equation are one of the inherent scientific weaknesses in all risk assessments. The convention that has evolved is to accommodate the unknown by being conservative with respect to risks and by drastically cranking down on "safe" levels indicated by case histories.

Risk assessment based on statistical analysis, modeling, and common sense allows extrapolating beyond the available data and avoids the paralysis of protective action that would result from waiting for definitive data. However, risk assessment also creates a problem in that quantifying an uncertain process implies a level of precision that does not truly exist.

In terms of protecting human health and the environment, there is little meaningful difference between target cleanup levels for trichloroethylene of 20 mg/kg or 40 mg/kg in soil. In any risk assessment, the uncertainties of the analysis far exceed a multiple of two. Yet too much risk assessment effort and resources are currently committed to negotiations on such issues.

It is far more important to take early action to control the source of the contaminants and prevent additional releases. Perhaps 75% of the hypothetical risks to human health and the environment can be resolved by common sense cleanup activities that are immediately apparent to experienced project managers. Esoteric arguments involving "how clean is clean" can be resolved after the success of the early remedial efforts is evaluated.

REFERENCES

Habicht, H.F., *Guidance on Risk Characterization for Risk Managers and Risk Assessors: Memorandum*, Office of the Administrator, U.S. Environmental Protection Agency, Washington, 1992.

National Research Council, *Alternatives for Ground Water Cleanup*, Committee on Ground Water Cleanup Alternatives, National Academy Press, Washington, 1994.

U.S. Environmental Protection Agency, *Corrective Action for Solid Waste Management Units at Hazardous Waste Management Facilities, Proposed Rule*, 40 CFR Parts 264, 265, 270, and 271, *Federal Register* 55, 30798, 1990.

U.S. Environmental Protection Agency, *Guidance for Evaluating the Technical Impracticibility of Ground-water Restoration*, Office of Emergency and Remedial Response, Washington, 1993.

U.S. Environmental Protection Agency, *Risk Assessment Guidance for Superfund, Vol. I: Human Health Evaluation Manual (Part B, Development of Risk-based Preliminary Remediation Goals), Interim Final*, Office of Emergency and Remedial Response, Washington, 1991.

U.S. Environmental Protection Agency, *Risk Assessment Guidance for Superfund, Vol. I: Human Health Evaluation Manual (Part C, Risk Evaluation of Remedial Alternatives), Interim Final*, Office of Emergency and Remedial Response, Washington, 1992.

U.S. Environmental Protection Agency, *Role of the Baseline Risk Assessment in Superfund Remedy Selection Decisions*, Office of Solid Waste and Emergency Response, Washington, 1991.

U.S. Environmental Protection Agency, *Soil Screening Guidance*, Office of Emergency and Remedial Response, Washington, 1994.

CHAPTER 27

Facility Risk Assessment

George Anderson

CONTENTS

I. Introduction 506
II. RCRA Facility Assessments 506
    A. RCRA Compliance Assessment 506
        1. Determining the Type and Extent of the Assessment 506
        2. Writing a Request for Proposals..506
        3. Contractor Selection...507
        4. During the Assessment Audit 507
        5. Control and Review of Rough Drafts................................508
        6. Control of the Final Report 508
        7. Evaluating the Results..508
        8. Addressing Deficiencies...509
    B. RCRA Facility Assessment 509
        1. Determining the Type and Extent of the Assessment 509
        2. Writing the Request for the Proposal 509
        3. Contractor Selection...510
        4. Evaluation of Proposals 510
        5. Offer Acceptance..510
    C. Evaluation of Results 510
III. Preemptive Assessment..511
    A. Determining the Type and Extent of the Assessment...................511
        1. Writing the Request for Proposal 511
        2. Contractor Selection...512
        3. Evaluation of Proposals 512
        4. Accepting a Proposal 512
        5. The Assessment Audit..512
        6. Control of the Results 512
        7. Responding to the Results 513
IV. Conclusion 513
    References..513

I. INTRODUCTION

Other chapters of this book have discussed how to assess environmental media for environmental risk. This chapter discusses how to assess a facility for potential to generate environmental risk. Three forms of assessments will be described that are required under RCRA. RCRA compliance assessment evaluates a facility for compliance with RCRA requirements; RCRA facility assessment evaluates the liability assumed by sending waste to an RCRA treatment or disposal facility. Preemptive assessment of one's own property evaluates it under RCRA authority to avoid future complications under CERCLA. Compliance assessment, facility assessment, and preemptive assessment will be discussed separately because each is unique in the depth of the analysis required and the disciplines that should be involved.

II. RCRA FACILITY ASSESSMENTS

A. RCRA Compliance Assessment

An RCRA compliance assessment of a facility is directed at protecting the environmental manager and other top officials from criminal liability under RCRA. U.S. EPA and many states have increasingly personalized enforcement of environmental regulations by taking criminal enforcement actions against managers responsible for perceived violations. Conducting an RCRA risk assessment demonstrates intent to comply with the law. It reinforces the environmental manager's case for necessary changes in operations and it provides a different perspective on the facility's operations that may help identify problems that would otherwise be overlooked.

1. Determining the Type and Extent of the Assessment

The extent of the evaluation of a facility is determined by the complexity of operations at the site and the depth of assessment desired. However, compliance assessments are normally limited in their scope and cost.

The types of operations at the facility will also determine the disciplines required to develop the report. Basic types of operations to consider include: treatment, storage, disposal, recycling, fuel blending, transportation, and hazardous waste generator.

A facility can be evaluated from two perspectives: either according to the RCRA rules, or according to the conditions of the facility permit, if the facility is permitted. The permit conditions usually remain consistent throughout the permit period. The RCRA rules, however, change constantly as do the policies that agencies use to implement those rules.

2. Writing a Request for Proposals

Develop a scope of work by outlining the operations to be included in the compliance assessment and indicate the level of evaluation by listing items to be reviewed. For example an assessment might address:

FACILITY RISK ASSESSMENT

- Contingency plan and emergency procedures
- Waste analysis plan
- Closure plan/contingent long-term care plan
- Financial responsibility for liability coverage
- Personnel training/records
- Manifest records
- Inspection records
- Operating records
- Security, preparedness, and prevention
- Tank and container standards

When setting a deadline for proposals, allow enough lead time for consultants to schedule a site visit. This may be necessary to allow them to accurately estimate their costs (see Chapter 4).

3. Contractor Selection

Assemble a list of potential candidates who have demonstrated an understanding of RCRA rules and permitting. Consider how those rules apply to the facility and whether any of the contractors have specific knowledge in applicable areas of RCRA. Finally, contractor experience with the regulatory agency for the facility is a major factor.

Most compliance assessment contracts are minor, so project interviews are not normally warranted. Evaluation usually involves review of the proposal offer, and possibly a follow-up telephone call to deal with any questions. In evaluating proposals, consider both the firm's reputation and the project manager's qualifications. A reputable company name carries weight with a regulatory agency, but the project manager is the most important factor in determining the quality of the report. The project manager should be proficient in areas of RCRA pertaining to the facility operation. Experience with the agency or individuals that regulate the facility is also essential. If the operating systems are complex, such as a combustion unit, the project manager should also have an understanding of how the operating system's operation affects its performance. Although a number of firms have an understanding of RCRA, few have technical expertise on specific equipment. Additional factors to consider when evaluating the proposal include: similar project experiences, acceptable scope of work, confidentiality procedures, level of technical resources, and quality of writing. Consulting firms involved in this form of service should have a system in place to protect their clients' confidentiality.

Confirm acceptance of the consultant in a letter that restates the scope of work. Although many people do not notify the firms not selected, it is appropriate to send a short letter or make a telephone call to other candidates to notify them that they did not get the contract.

4. During the Assessment Audit

When the consultant makes arrangements to inspect the facility, it is important to determine which sets of records must be available for inspection and which personnel should be available for interviews. If possible, a facility management representative should accompany the auditors during the inspection. This will help insure that areas are not overlooked which pose the greatest concern to facility managers. Finally, the facility management should conduct an exit interview with the inspectors. An exit interview gives the managers an opportunity to learn of any significant areas of noncompliance and begin immediate correction of the problem. It also provides an opportunity to correct any misunderstanding of what was seen during the audit. Finally, it provides a last chance to review report confidentiality procedures with the auditors.

5. Control and Review of Rough Drafts

It is important to control access to rough drafts and to edit errors out of a draft report. An outside person auditing a facility for the first time may misinterpret what they see. Unless dealt with immediately and clearly, inaccuracies and misunderstandings develop a life of their own. Therefore, they should never be left in writing.

It is not the intent of the RCRA compliance assessment to allow the contracting company to edit out bad news, but false statements should be sought out, identified, and corrected as quickly as possible. Otherwise, they may someday appear in agency files as the truth. After inaccuracies have been identified and corrected, a final report should be generated and all drafts should be destroyed.

6. Control of the Final Report

An RCRA risk assessment can document potential violations, so it is important to protect its confidentiality. In some cases, it can warrant having all copies protected as confidential material under the attorney/client privilege. Then, if required, access to the information may be denied to regulators. In order for a document to be protected as attorney/client privileged material, it must be created by the attorney, or at his or her direction, in anticipation of litigation. It is not sufficient to work with the contractor to generate a final report and then present it to the attorney for safe keeping. In some states, such as Minnesota, a company's self-audits are protected by a limited privilege.

7. Evaluating the Results

Normally, it does not require a strong technical background to evaluate an RCRA compliance assessment. However, there are exceptions. For example, complex operations, such as a combustion unit, may require technical knowledge. Even in this case, however, the consultant's project manager should be capable of presenting the information at a level that managers can understand. Complex detail and specifics should be included in the report, but the explanation of this information should be conducted at a level that managers can comprehend.

If the facility operates under a permit, two separate sets of standards can apply: the original permit limits, and whatever current rules establish. If one standard is more stringent, the stricter standard applies. Therefore, both standards should be presented in series so that they are easy to interpret.

8. Addressing Deficiencies

The principal reason to conduct a RCRA risk assessment is to identify and correct problems that could lead to violations. Once problems have been identified, it is imperative to identify what is required to address each item, set a reasonable schedule for completion, and then clearly document progress toward correction.

Conducting the assessment, and then addressing problems in a timely fashion, is accepted as an expression of intent to comply with the law. This helps relieve the potential of criminal negligence. Failure to do so can negate the good accomplished by conducting the assessment. It can also create a trail of knowing noncompliance under RCRA and potentially lead to criminal penalties for the managers involved.

The consultant, as part of the final report, should suggest how to address the problems identified. These suggestions should be combined with alternatives generated by company managers. The person that regulates the facility can also be a great help if an open relationship exists between the management and the regulatory agency. However, the decision to involve a regulator is typically a very sensitive issue. It should occur only after consultation with all appropriate managers.

B. RCRA Facility Assessment

An RCRA facility assessment is conducted to evaluate the liability assumed by the company by using a treatment or disposal facility to handle waste. This type of analysis is an important part of protecting the company from potential CERCLA liability. Although a facility's RCRA compliance is important, this form of assessment can also include evaluations of a facility's environmental problems, general operations, financial strength, and customer base.

1. Determining the Type and Extent of the Assessment

The majority of off-site facility assessments are limited in scope. The basic intent is to evaluate the level of liability that will remain after a facility treats or disposes of your waste. This involves assessing the form of technology used, the expected longevity of the company, and if the company fails to survive, who will you share the CERCLA liability with?

The assessment scope of work can be limited to an information search of readily available regulatory and financial records. This is often done from the consultant's office. On the other hand, it may require an investigation that involves a team of auditors visiting the facility, interviewing appropriate regulatory personnel, and conducting an in-depth financial audit. The depth of this type of assessment should be related to the level liability that will be assumed.

2. Writing the Request for the Proposal

Outline the factors you want assessed. This should include at a minimum, a review of their technology, a compliance evaluation, and financial screening.

3. Contractor Selection

Following the guidelines outlined in the chapter on contractor selection, assemble a list of potential candidates. Conduct a phone survey to insure each candidate provides this form of service. For a project of this size, a limited number of candidates — three or four — would normally be adequate.

4. Evaluation of Proposals

Because of the limited scope of this form of assessment, oral presentations are normally not given. Evaluation of candidates is usually limited to written proposals, with a possible phone call as follow-up. Selection criteria should center on the consultants' understanding of the waste management industry, and their ability to communicate in written form.

5. Offer Acceptance

Confirm acceptance of your choice with a letter restating the scope of work. Although many people do not, a short letter or phone call notifying the other candidates that they did not get the bid is appropriate.

C. Evaluation of Results

The two factors to evaluate when assessing a treatment or disposal facility are:

- The soundness of their technology
- How long they will be around

Forms of treatment that reduce future liability, such as recycling, beneficial reuse, or destruction, hold the highest priority. The hazardous constituents are usually destroyed or transferred to the next consumer. Land disposal is usually considered as a last option. All operations should be well maintained and operated. Their equipment should also be current or state-of-the-art.

Determining how long the facility will be around includes a number of factors. Compliance is a critical issue. Facilities that have trouble complying with the regulations, also have trouble staying in business. If they have trouble in this area, it is a sign that they cut corners in other areas too. Knowingly using a company that has serious compliance problems transmits a certain level of liability to your company. You can also be drawn into some of their troubles.

A Dunn & Bradstreet, or other financial assessment, will give you a glimpse at their strength as a business. It will tell you how quickly they pay their bills, whether they have good credit ratings, if they are involved in any court actions, and if there are leans against their properties. Check with your accounting department — would they give them credit?

FACILITY RISK ASSESSMENT

Examine their list of customers. If they are used by a number of larger companies, they have probably already been closely assessed. If the company fails, you share the CERCLA liability with these companies.

Each of these is a potential red flag. If you see problems, you should either look further into their backgrounds, or drop them from consideration.

III. PREEMPTIVE ASSESSMENT

A preemptive assessment, to evaluate potential health or environmental problems of your site, can protect a company from future CERCLA liability. Although it often has the same cleanup standards as CERCLA, RCRA is much easier to work under and does not carry the stigma of the Superfund program. Evaluating and addressing potential problems while a facility is still active provides much greater control over the pace and extent of the evaluation and of any remediation that may be required. A preemptive assessment can also catch an environmental hazard early and reduce the extent of contamination to be addressed.

A preemptive RCRA assessment requires a consulting or engineering firm that offers general environmental investigation and remediation services. Two factors to consider are:

- How much experience does the firm have with the form of assessment requested?
- How well does that experience apply to the conditions at the site being assessed?

Additional assistance on selecting a consultant can be found in the chapter on contractor selection.

A. Determining the Type and Extent of the Assessment

To focus the scope of work determine by which potential pathways your facility could have affected the environment. Investigations usually center around groundwater, soil, and/or facilities. Investigations into residual contamination of soil and facilities are relatively straightforward and often have an acceptable cost. Groundwater investigations when contamination is detected, tend to be complicated, lengthy, and expensive. However, ignoring any of these can lead to even bigger problems.

Prior to initiating this form of assessment, management must be made an integral part of this process. They must be willing to accept the ramifications of the options you are considering. Without their long-term support, an investigation can leave you with documented problems, and little or no support for the solutions. There are serious personal and corporate liabilities that go along with failing to address known environmental problems.

1. Writing the Request for Proposal

Assemble a list of the areas, along with potential contaminants, that may require investigation. Use this as the base for writing a request for proposal.

2. Contractor Selection

There are numerous firms that provide this form of service. Because these evaluations can lead to additional phases, give special consideration to firms with experience remediating sites and contaminants similar to yours. Using the procedures outlined in the chapter on contractor selection, assemble a list of potential candidates. Because of the potential cost of evaluating and then remediating a site, consider as many firms as necessary to satisfy yourself that you have the best available candidates. Set a deadline for submittal that allows the consultants adequate time to develop a personalized presentation.

3. Evaluation of Proposals

Except for limited assessments, the breadth of this form of an evaluation warrants a combined written offer and oral presentation. Insist that the proposed project manager be part of the interview team. That person will be the focal point of the companies' resources. If they have good organizational and communication skills, they can elevate an average company to give you a good product. A poor project manager, even with a good company, will give you a weak product. Can they relay pertinent technical concepts of the report at a level that is useful to you and the other decision makers? How strong is their technical background and understanding of your site, or can they relay this from other team members?

As a company, is the presentation well organized and professional? What is their reputation? Do they have experience with a number of forms of remediation, or are they married to a select few? Do they have a good relationship with the regulatory community? Does their proposal list projects similar to yours? Does it accurately list all tasks that you expect in the scope of work? Is there a procedure to insure control of preliminary results?

4. Accepting a Proposal

Confirm acceptance of the primary proposal with a letter restating the scope of work. Although many people do not, a short letter or phone call notifying the other candidates that they did not get the bid is appropriate.

5. The Assessment Audit

Prior to the site investigation, insure that all pertinent information about past site activities and sampling have been made available to your consultant. Insure access to all appropriate personnel and locations.

6. Control of the Results

A review of preliminary results has two primary purposes: (1) to insure that inappropriate and inaccurate information is correctly dealt with; and (2) to give you as much lead time as possible to respond to critical decisions that arise from the investigation.

Once the report has been finalized, suppressing information that reveals environmental problems can expose both the company and responsible individuals to penalties under RCRA. Although it is important to control access to this information while appropriate direction is being determined, corrective actions must proceed in a timely manner. Your consultant should assist you in determining the distribution of those results.

7. Responding to the Results

The principal reason for conducting a preemptive assessment of your site is to identify and correct problems with as much company control as possible. Being proactive gives you the advantage of having additional control and lead time in evaluating the next appropriate step. You retain that advantage only as long as you remain proactive. Therefore, promptly addressing questions and problems highlighted by this assessment is critical. This requires documenting the steps taken to resolve any outstanding questions, how problems were addressed, and progress made. Failure to do so can erase the good accomplished by conducting the assessment. This could also leave a documented trail of knowing noncompliance that can lead to civil penalties for the company, and potential criminal penalties for the managers.

IV. CONCLUSION

The introductory chapters to this book discussed ways to determine whether a risk assessment should be performed. In addition to providing the legal benefits described above, and providing insight into the operation of the facility, a facility assessment provides another way to determine the need for a risk assessment.

REFERENCES

Anon., *Case Law Pertaining to the Personal Liability of Corporate Directors, Officers and Managers for Environmental Violations*, Bell, Boyd & Lloyd, Chicago, 1992.

Blaine, C., EPA issues interim policy on voluntary self audits, *1996 Executive File: Hot Environmental Issues*, M. Lee Smith Publishers & Printers, Nashville, 1995, 37-38.

Cox, D., *Handbook on Hazardous Materials Management*, 5th ed., Institute of Hazardous Materials Management, Rockville, MD, 1995.

Dreux, M., Legal affairs: building an effective program against corporate criminal liability, *Occup. Hazards*, 43, 1993.

Gold, K., Audit privilege prompts proactive approach to compliance issues, *1996 Executive File: Hot Environmental Issues*, M. Lee Smith Publishers & Printers, Nashville, 1995, 34.

Hartman, B., *EPA Enforcement Manual*, Thompson Publishing Group, New York, 1994.
Spacone, A., Individual criminal liability under environmental laws, *Hazardous Mater. Manage.*, 1993, 24.
U.S. Environmental Protection Agency, *Resource Conservation & Recovery Act, Inspection Manual*, Government Institutes Inc., Rockville, MD, 1993.

CHAPTER 28

CERCLA and RCRA Risk Assessments

Carol Baker

CONTENTS

I.	Introduction	515
II.	Overview of Risk Assessment	516
III.	Risk Assessment Process	517
	A. Purpose	517
	B. Approach	517
IV.	CERCLA	518
V.	Components of a Baseline Risk Assessment	520
VI.	RCRA	523
VII.	Comparison of RCRA and CERCLA Risk Assessment	525
VIII.	Conclusion	526
	References	526

I. INTRODUCTION

Risk assessment is used by federal and state regulatory agencies for implementation of Comprehensive Environmental Response, Compensation, and Liability Act of 1980 (CERCLA) and Resource Conservation and Recovery Act (RCRA) laws at contaminated sites. Effective use of a risk-based remediation strategy helps to focus environmental cleanup dollars on those areas of the site identified in the risk assessment as potentially presenting unacceptable risks. Risk assessments developed for CERCLA- and RCRA-regulated sites must be generally consistent with the various U.S. EPA guidance documents for conducting risk assessments to promote successful regulatory agency acceptance. However, strict adherence to U.S. EPA guidance is generally not conducive to developing site-specific information; deviation to address site-specific conditions and considerations is recommended and desirable.

While there are allegations that risk assessment as it exists today is over conservative, risk assessment is currently the best method available to predict the potential risks associated with contaminated sites. In January 1994 the Harvard Center for Risk Analysis released the results of a survey of 1000 Americans where 83% agreed "the government should use risk analysis to identify the most serious environmental problems." Many regulatory agencies consider risk assessment an acceptable method for identifying sites that present risks to human health and the environment and prioritizing them. Risk assessment offers a site-specific alternative to the application of generic cleanup standards. This chapter will address some of the major differences between preparation of a risk assessment for a site regulated by the U.S. EPA and state regulatory agencies under CERCLA and sites that fall under the jurisdiction of RCRA. Both of these laws require that remedial actions be taken to protect human health and the environment.

II. OVERVIEW OF RISK ASSESSMENT

There has been a growing discontent in the 1990s in the U.S. over the approximately $115 billion spent on environmental protection. Scientists, environmental policy makers, and the public question whether the money has been spent based on the best understanding of health and environmental risks (Stone, 1994).

Currently the general consensus is that remediation of many sites to background levels is inappropriate and technically infeasible. This is largely due to the limitations in remedial technologies. The traditional "cleanup to background or nondetectable levels" approach may not be necessary to protect human health but may still be required when a regulatory agency considers site remediation to be a function of regulatory standards. There is a growing movement, however, toward focusing remediation efforts (and funds) on a more practical goal — devising ways to minimize exposures based on the actual uses of a site, both currently and in the future. This process is called risk management. Exposures can be minimized through the use of remediation, control measures, or a combination of the two.

Risk assessment can provide information that is critical in making decisions concerning the following:

- The necessity of undertaking remedial action, if any, to protect human health
- The implementation of risk management alternatives at the site

Complete restoration and remediation to either background levels or generic standards are often impossible. Site-specific risk assessment can evaluate the health impacts at a site based on the intended use of the property, identify the potential for health risks associated with exposure, and determine risk-based, medium-specific cleanup targets based on realistic and intended use of the site. These targets are critical in selecting an appropriate remedial action to protect human health and the environment to the maximum extent practicable. Risk assessment can provide the logical framework for determining which actions might reduce or eliminate risk.

III. RISK ASSESSMENT PROCESS

A. Purpose

The risk assessment process is used to:

- Identify chemical- and medium-specific preliminary remediation goals (PRGs) early in the process that can be used to focus investigatory activities
- Determine the potential risks associated with site-specific exposure scenarios, assessing both the current and the hypothetical future exposures, without remediation
- Refine PRGs using site-specific data to generate health-protective cleanup targets for average and sensitive receptors, if potentially unacceptable risks are identified for a complete exposure pathway
- Estimate current and/or future risks (or reduction in risks) potentially associated with implementation of specific remedial alternatives or institutional/engineering controls
- Quantify risks potentially associated with exposure to residual chemicals following implementation of remediation and/or risk reduction measures

The information generated will depend on the direction of the project, the goals of responsible parties, and the regulatory program which governs the site.

The results of a risk assessment provide remediation engineers with site-specific cleanup target levels. These site- and medium-specific cleanup levels are useful in identifying appropriate cleanup alternatives.

B. Approach

Risk assessments developed for different sites are not always performed the same way, even under the same set of regulations and following the same guidelines. Either state or federal regulatory agencies may require slightly different evaluations and items in a risk assessment document. Guidelines vary from state to state, and some states have not yet developed state-specific guidance for risk assessment, but the impact of state guidance should be considered and addressed when it exists.

Risk assessors must participate in all phases of site work to assure the collection of appropriate data suitable for the development of a scientifically defensible risk assessment. If involvement of a risk assessor occurs later in the process, data needed to prepare a technically defensible risk assessment may not be available. This will either result in additional field work or greater uncertainty in the results because assumptions have to be made to represent the missing data. If a risk assessor is involved early on in a project, preferably as soon as data indicate that a release has occurred, then the needs of the risk assessor can guide investigatory activities and decision making strategies. Early risk assessor involvement applies to both regulated and nonregulated sites. Data required to complete a site-specific risk assessment or to develop medium- and chemical-specific remediation goals can then be identified, and efforts can be made to collect these data while field activities are being conducted.

When risk assessors do not obtain adequate site-specific data for use in a risk assessment, they must rely on conservative modeling input parameters as surrogate data. Those values may or may not be appropriate to evaluate conditions and subsequent potential risks associated with exposure, and do not allow for a site-specific assessment of conditions at the site. Risk assessments, default exposure assumptions, and toxicity values in particular, are designed to be conservative (i.e., overprotective). Conservatism is a value judgement introduced into the risk assessment to compensate for uncertainty. The default values constitute a deliberate slant on the side of safety and protection of human health and the environment (Covello and Merkhofer, 1993). Using conservative assumptions minimizes the probability that a risk estimate will underestimate the actual risks associated with exposure under a specific scenario by producing an upper-bound risk estimate. Thus, early involvement of a risk assessor should improve the collection and incorporation of site-specific data, result in less uncertainty in the results, and help offset the ultraconservative tendency of risk assessment.

IV. CERCLA

Risk assessments were first required for a site regulated under the CERCLA program in 1980; the original Superfund law required action at hazardous waste sites that posed "a substantial endangerment to public health or welfare or the environment" (Section 104[a]). U.S. EPA intended to use risk assessment to assist in making decisions regarding cleanup. U.S. EPA published guidance documents dealing with risk assessment for suspected carcinogens as early as 1976 (U.S. EPA, 1976), with the first relatively comprehensive risk assessment guidelines, *Superfund Public Health Evaluation Manual*, available in 1986 (U.S. EPA, 1986). In 1989 the U.S. EPA released *Risk Assessment Guidance for Superfund: Volume I — Human Health Evaluation Manual (Part A), Interim Final*. This guidance document describes what the U.S. EPA envisioned as the baseline risk assessment (BRA) procedure, the rationale for the process, and the components of a BRA. Although this guidance was never intended as a "cookbook" for preparing risk assessments, it has often been treated that way. Deviation from the U.S. EPA's BRA is often frowned upon by regulatory agencies. Additional and supplemental guidance has since been developed, and specific mathematical expressions have been refined; however, the 1989 Guidance (U.S. EPA, 1989) is still consulted for the framework of a BRA.

Once a site is listed on the National Priorities List (NPL), that site falls under CERCLA regulations and follows a prescribed process: remedial investigation (RI), BRA quantifying potential risks in the absence of remediation, feasibility study (FS), and remedial design/remedial implementation (RD/RI) (Figure 1). Immediate or interim measure actions can be implemented at any point in the process when data indicate it is necessary to take measures to protect human health and the environment. Risk assessment is a factor in each step of the CERCLA process. Following NPL identification, risk-related work occurs in at least four identified places in the CERCLA process.

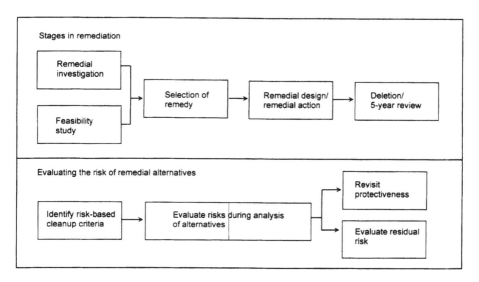

Figure 1 Risk evaluation of remedial alternatives in the Superfund process.

The first step in the RI process is to develop a sampling plan. At this time, risk assessors should be consulted to provide input on data needs and appropriate laboratory analytical techniques. For example, laboratory detection limits should not exceed health-based levels of concern. If they do, there is considerable uncertainty in the risk calculations, because even laboratory reported nondetects for certain constituents may result in unacceptable risks if exposure occurs. Due to the increasingly low detection limits that laboratories can routinely achieve, this has become less of a concern, but should still be considered.

Second, PRGs can be developed in the RI to focus the investigation. PRGs are typically developed early in the investigation using conservative default assumptions resulting in an upper-bound, worst-case remediation target protective of a sensitive receptor. In general, if sampling data demonstrate that constituent concentrations are below the PRGs, it is likely that the area will not require remediation to protect human health and the environment. In cases where there are numerous chemicals of potential concern, the PRGs may need to be refined to consider the potential for cumulative effects of the chemicals. However, even if all chemicals are reported at concentrations below their PRGs, an agency's interest in delineating impacts may require additional investigatory work.

Third, the BRA, prepared as part of the RI, is designed to evaluate potential risks to human health and the environment posed by the site without undertaking remediation. The results of the BRA should assist in decisions to:

- Identify areas of the site requiring remediation (i.e., those areas where exposure could result in potentially unacceptable health risks)
- Establish the level of remediation required in each of these areas

- Demonstrate the efficacy of implementing institutional and/or engineering controls
- Identify appropriate risk management decisions

Fourth, the FS follows the BRA. The FS is developed for the following reasons:

- To evaluate various remedial technologies that are effective and may be applicable at the site
- To organize applicable technologies into remedial alternatives
- To identify the usefulness of the various alternatives in conjunction with the media-, constituent-, and remediation-targets at a particular site.

The effectiveness of a specific alternative can be estimated through risk assessment, and the resulting information may eliminate an alternative from further consideration. For example, soil excavation may generate large amounts of fugitive dust that could pose an unacceptable risk both on and off site, limiting the viability of excavation as an alternative.

Finally, in the mandated 5-year review process of a CERCLA site, a risk assessment may be performed to assess and measure the potential risks associated with exposure to residual chemical levels remaining *in-situ* following implementation of the agreed upon remedial alternative.

V. COMPONENTS OF A BASELINE RISK ASSESSMENT

The human health BRA document prepared for a CERCLA site should consist of the following components: (1) site characterization; (2) constituent characterization; (3) exposure assessment; (4) toxicity assessment or hazard characterization; (5) risk characterization; and (6) uncertainties. Ecological evaluation should also be addressed. The entire site and all reportedly impacted media are typically included in one comprehensive risk evaluation unless circumstances allow for the logical division of the site. For instance, if significant information exists for groundwater, and there are soil data gaps that would weaken a risk assessment or contribute too much uncertainty, groundwater can be addressed. Soil can be dealt with after the necessary data are collected. This generally results in a smaller and more focused document.

The U.S. EPA issued guidance for the development of PRGs designed specifically for application to CERCLA sites (U.S. EPA, 1991b). U.S. EPA intended these PRGs to be refined through the CERCLA process as more site-specific information became available. PRGs are screening tools, not the final remediation target. They are used by engineers in the FS to evaluate alternatives. PRGs are designed to be conservative because they generally use upper-bound default exposure assumptions and assume prolonged exposure, and do not incorporate any chemical degradation over the exposure duration. The PRG guidance provides mathematical calculations to develop medium-specific goals for residential and commercial/industrial exposure scenarios. Although guidance addresses PRGs for residential or commercial exposure scenarios, relying on only these scenarios may not be in one's best interests. The mathematical equations could be adjusted to develop PRGs protective of less restrictive exposure scenarios. This option should be discussed with regulators if it appears feasible and appropriate.

While risk assessment guidance states that risk-based goals are to be developed for CERCLA sites, a review of 77 Records of Decision (ROD) for CERCLA sites in a study conducted by Walker et al. (1994) revealed that these risk-based goals were not always applied at a site. Fifty of 52 sites with goals for groundwater in the ROD relied on ARARs (existing federal standards or criteria) or "to be considered" guidance values alone or in combination with site-specific, risk-based goals. Only one ROD depended entirely on risk-based groundwater goals. Eighteen of 31 sites with goals for soil in the ROD relied on ARARs alone or in combination with risk-based goals; 10 sites used risk-based soil goals exclusively.

Whenever possible, data requirements of the BRA should be considered in developing the RI workplan to ensure that the appropriate data are obtained. Otherwise, the BRA will be less site-specific than desirable, will rely on default exposure assumptions to develop PRGs and estimate risks, and will tend to be more conservative than necessary.

Chemicals detected in a medium are considered chemicals of potential concern in the BRA unless at least one of the following conditions apply:

- Detection of the chemical is proved to result from laboratory contamination
- Reported concentrations are below site-specific or regional background levels (this is generally applicable for inorganic chemicals only)
- Chemicals are reported infrequently in one or a few specific media, or are not expected to be present considering operations previously conducted at the site
- The chemical is an essential human nutrient, reported at low concentrations (slightly elevated above naturally occurring concentrations), and is only considered to be toxic at very high doses (higher than those doses likely to occur at the site)

Chemicals identified as known or probable carcinogens (Class A or B) are generally retained for consideration in the BRA; chemicals identified by the U.S. EPA as possible carcinogens (Class C) are retained on a case-by-case basis (U.S. EPA, 1989). Class C carcinogens may be evaluated using their cancer slope factor or may be assessed using their RfDs or a modified RfD, which includes an additional uncertainty factor that reduces the RfD by an order of magnitude. Elimination of a chemical for other than the four reasons outlined above is often difficult to defend in CERCLA BRAs.

If the laboratory reports detectable concentrations of site-related chemicals, media typically investigated in a BRA include groundwater, surface water, sediment, leachate, air, surficial soil (typically 0 to 6 inches bellow ground surface, but this may be defined differently under a specific state's guidance), deeper soil (greater than 6 inches), and possibly biota.

A BRA developed under CERCLA often quantifies potential risks associated with a number of exposure scenarios, not just the one most likely to occur or the pathway with the most sensitive receptors. Exposure pathways that may be quantitatively evaluated in the BRA include:

- Exposure by trespassers and/or residents to surficial soil, air, surface water, sediment, and exposed leachate
- Worker exposure to surficial soil, deeper soil, air, surface water, sediment, and exposed leachate
- Fish ingestion exposure by potential receptors

Other exposure scenarios, such as exposure to chemicals of potential concern through ingestion of produce grown in impacted soil or irrigated with impacted groundwater, and exposure of potential receptors through ingestion of meat or milk from groundwater- or affected vegetation-fed livestock, are less commonly assessed. Exposure pathways or routes of exposure (i.e., oral, dermal, and inhalation) are not usually screened (except as detailed above in the determination of chemicals of potential concern) to determine which contribute significantly to the overall risks, but this step may be taken. Generally, all potentially applicable exposure scenarios, current and future hypothetical, involving different age groups, exposure routes, and media are included in a CERCLA BRA to characterize potential exposures.

It is unlikely that most CERCLA sites will be redeveloped as residential property in the next 10 years. A draft 1993 internal U.S. EPA memorandum stated that hypothetical future exposures for CERCLA sites should address potential site uses within the next 10 years and not necessarily beyond that time frame. Nevertheless, in the past it has been assumed with few exceptions that future hypothetical use of the site could be residential, and the risks associated with long-term residential exposure should be quantified. However, U.S. EPA is currently moving away from this position of assuming future residential use of the property if it can be demonstrated that a site is likely to remain industrial. Some factors influencing this determination include history of the site, zoning regulations, and anticipated future growth in the area. The willingness to use institutional controls to prevent residential redevelopment can assist in strengthening the "no residential exposure" stance.

There has been concern that unrealistic assumptions about future uses of a property may lead to an overestimate of potential risks. A study conducted by Walker et al. (1994) reported that the highest groundwater risks at CERCLA sites are generally associated with the assumption of residential use of groundwater. This is probably an unlikely pathway at most sites, especially given the availability of public water supplies. Resources can be needlessly expended in the baseline risk assessment evaluating all possible scenarios. Negotiating a tightly focused approach can save money and time, and still provide the information necessary to make informed remediation and risk management decisions.

The mathematical expressions used to calculate potential risks associated with human exposure to contaminated media are derived from U.S. EPA guidance documents developed for CERCLA sites (U.S. EPA, 1989 and 1991b). These two guidance documents are supplemented by other guidance documents and U.S. EPA directives. The equations set forth in these documents may be modified to accommodate site-specific conditions and updated methods. However, U.S. EPA (1991a) has provided conservative default exposure assumptions for use in calculations and U.S. EPA staff generally allow little site-specific deviation. This is probably because upper-bound assumptions assure a conservative estimate of risks, protecting even sensitive populations.

U.S. EPA (1989) states that an estimate of a scenario's reasonable maximum exposure (RME) should be quantified. The RME is the highest exposure that is reasonably expected to occur at a site. This is accomplished through the use of a mixture of reasonable and worst-case exposure assumptions, resulting in the estimate of a conservative exposure case that should still be within the range of possible exposures. In application, however, generally all conservative, upper-bound exposure assumptions are expected to be used.

For example, U.S. EPA (1989) states that the exposure point concentration (EPC) (that concentration of a chemical that a receptor may be exposed to over the exposure period) is the arithmetic average of the chemical concentration, but goes on to state that the uncertainty in EPC estimates dictates that the statistically derived 95% upper confidence limit (UCL) on the arithmetic average for a specific constituent be used as the EPC in the risk calculations. Additionally, the 95th percentile values for contact rate (amount of time contact with an impacted medium per unit time or event) and exposure frequency and duration (how often exposure occurs and the length of time for each exposure) should be used. While 9 years is considered to be a reasonable average length of residence in one home and 30 years is the upper-bound value, CERCLA risk assessments often use the 30-year value as the RME duration of a resident. Attention should be paid to the mixing of reasonable and worst-case assumptions in evaluating RME, and not defaulting to all worst-case assumptions, which would result in an overestimation of potential risk. Collection of site- and/or area-specific information can be useful in development of site-specific RME risks.

Ecological impacts potentially resulting from exposure must be addressed at CERCLA sites. Currently, however, the U.S. EPA has not developed guidance for quantifying potential impacts to ecological receptors, but has developed a qualitative approach generally used for ecological evaluation (U.S. EPA, 1989). The approach is a screening level assessment, which is useful in predicting whether biota have been affected by site-related chemicals or could be in the future. In this qualitative approach, U.S. EPA recommends comparing ambient environmental media concentrations with relevant criteria (including water quality) to determine whether the ecological receptors could potentially encounter EPCs exceeding these criteria. It is difficult to predict whether observed effects on individual populations will result in any real damage to the ecosystem because scientific understanding of ecosystem interactions is limited.

VI. RCRA

If a preliminary review of site information and noninvasive investigation of a site indicates that operational releases may have created environmental impacts, a site regulated under RCRA will be subject to corrective action. All subsequent actions must be performed in a manner consistent with the RCRA corrective action process until it can be demonstrated that corrective action is not required. This corrective action process consists of an RCRA Facility Assessment (RFA), RCRA Facility Investigation (RFI), Corrective Measures Study (CMS), and Corrective Measures Implementation (CMI). Performing the CMS and CMI is only necessary if impacts are identified in the RFI. Interim Corrective Measures (ICMs) can be implemented without agency authorization or penalty at any time in the process if human health or the environment is at immediate or imminent risk, or to minimize future chemical impacts or migration. Risk assessment can be used to support the utility of implementing the ICMs. ICMs may then be adopted as the final remedial measure.

Risk assessment provides decision making information at several points in the RCRA corrective action process.

1. Conservative and health-protective chemical-, medium-, and exposure-specific remediation goals, similar to PRGs for CERCLA sites, can be developed early in the RFI to focus additional investigatory activities.
2. Risk potentially associated with expected or possible exposure to contaminated media at a site can be quantitatively evaluated following delineation and identification of site-related chemical impacts in the RFI. If predicted risks fall below risk ranges identified by U.S. EPA as not requiring remediation, no action is required. Thus, depending on the results, the risk assessment can support a no action alternative. If the estimated risks fall within the range the U.S. EPA typically uses to evaluate the necessity of corrective action to protect human health and the environment, decisions on the necessity of remediation are made on a case-by-case basis. Most state regulatory agencies have adopted the U.S. EPA risk ranges to determine the need for remediation to protect human health and the environment, but this should be confirmed.
3. Risk assessment can be used to support or justify implementing an ICM. If risks quantified for a current scenario are unacceptable, an ICM can be used to reduce risk to an acceptable range without regulatory agency approval.
4. During the CMS, various remedial alternatives are evaluated for their usefulness and implementability at the site. Risks potentially associated with the implementation of each remedial alternative are assessed. Identifying risks associated with alternatives may be necessary to screen remedial options, or to eliminate alternatives from further evaluation.

The U.S. EPA has not developed risk assessment guidance specific to RCRA-regulated sites. However, U.S. EPA regional staff, as well as regulatory staff in many states which have RCRA program primacy, typically defer to U.S. EPA (1989), updated by recent risk guidance (U.S. EPA, 1991a and b). Due to the absence of a RCRA risk assessment guidance document, more variability exists in RCRA work products than in CERCLA risk assessments. However, without an RCRA-specific guidance document there may be more flexibility, and the option to avoid the use of default values and conservative assumptions, and rely instead on site-specific information. This adoption of a more tailored site-specific approach is most likely to occur when the future site use is known because the responsible parties are still operating the active site and are able to control current and future exposures through engineering or institutional controls.

VII. COMPARISON OF RCRA AND CERCLA RISK ASSESSMENT

RCRA and CERCLA risk assessments are similar in several ways. Chemicals of potential concern are selected for an RCRA-regulated site by a process similar to that used for a BRA at a CERCLA site. Risk assessments prepared for an RCRA site follow the same general format as those prepared for a CERCLA site. They use the same mathematical expressions. However, RCRA and CERCLA risk assessments also differ in certain important characteristics.

First, future hypothetical use of the RCRA-regulated site as a residence is not routinely required. In some cases, residential use may be evaluated because the future use of the entire site, or of a particular portion of an RCRA site, is not defined and it is decided that evaluation of potential risks associated with residential future use would provide information useful in making redevelopment decisions. Similarly, an RCRA risk assessment does not need to evaluate all potential exposure scenarios. It can quantify risks only for those pathways expected to occur or those where risks are expected to result in adverse health effects. For instance, risks potentially associated with a trespasser exposure are generally less than those associated with a worker- or residential-type exposure. Thus, if risks to a resident or worker are predicted to be acceptable, risks to a trespasser should also be acceptable.

Second, in contrast to CERCLA risk assessments, RCRA risk assessments do not routinely evaluate the site-related impacts throughout the entire RCRA facility in a single, comprehensive document, unless those operating the facility intend to discontinue its operations. Generally, RCRA risk documents focus instead on specific areas that facility owners or operators want to close or have removed from the RCRA corrective action process.

Third, RCRA risk assessments do not always adhere to the standard U.S. EPA risk assessment default-exposure assumptions. Site-specific data can, and should be, incorporated into the assessment. However, if site-specific data are difficult to collect, the U.S. EPA-identified default values may be used.

Fourth, RCRA risk assessments in the corrective action process differ somewhat from those developed to support a "clean closure," and both differ somewhat from CERCLA-regulated sites. Requirements for RCRA "clean closure," risk assessments are often more stringent than for a CERCLA-regulated site. The concentrations left in place should not present an unacceptable health risk, if exposure were to occur, and the use of institutional or engineering controls is not generally a part of the chosen remedy since these controls are not considered permanent.

Finally, RCRA calls for evaluation of impacts on biota. In practice, however, risk assessments of RCRA sites do not necessarily thoroughly assess the ecological impacts of current facility operations. Conditions surrounding most operating facilities do not generally sustain a complex natural ecosystem, even without site-related chemical impacts. At facilities with on-site, surface-water bodies or large facilities with areas on-site that are not currently in use, biota may exist that could be affected. If this is the case, the site can be assessed using a phased and focused approach to qualitatively evaluate ecological impacts cost-effectively.

VIII. CONCLUSION

Risk assessment serves as an integral part of the CERCLA and RCRA processes because it provides information critical to remediation evaluation decisions and other risk management decisions at sites. Even though the risk assessment guidances vary, RCRA generally relies on CERCLA guidance, and both processes are aimed at the same goal — developing quantitative and qualitative information on potential risks. This information serves as a basis for a risk-based approach to investigating and remediating contaminated sites and helps focus cleanup dollars on those areas of RCRA and CERCLA sites where exposure and associated risks are unacceptable.

REFERENCES

Covello, V.T. and Merkhofer, M., *Risk Assessment Methods: Approaches for Assessing Health & Environmental Risks,* Plenum Press, New York, 1993.

Stone, R., Can Carol Browner reform EPA?, *Science* , 263(21), 312, 1994.

U.S. Environmental Protection Agency, *Human Health Evaluation Manual, Supplemental Guidance: Standard Default Exposure Factors,* Washington, 1991a.

U.S. Environmental Protection Agency, *Interim Procedures and Guidelines for Health Risk and Economic Impact Assessments of Suspected Carcinogens,* Federal Register, 41, 24102, 1976.

U.S. Environmental Protection Agency, *Risk Assessment Guidance for Superfund,Volume I: Human Health Evaluation Manual (Part A), Interim Final,* Office of Emergency and Remedial Response, Washington, December 1989.

U.S. Environmental Protection Agency, *Risk Assessment Guidance for Superfund,Vol. I: Human Health Evaluation Manual (Part B, Development of Risk-based Preliminary Remediation Goals), Interim Final,* Office of Emergency and Remedial Response, Washington, 1991b.

U.S. Environmental Protection Agency, *Superfund Public Health Evaluation Manual,* Office of Emergency and Remedial Response, Washington, 1986.

Walker, K.D., Sadowitz, M. and Graham, J.D., *Confronting Superfund Mythology, the Case of Risk Assessment and Management,* Center for Risk Analysis, Harvard School of Public Health, Boston, 1994.

CHAPTER 29

International Health Risk Assessment Approaches for Pesticides

Colleen J. Dragula Johnson and Gary J. Burin

CONTENTS

I.	Introduction	527
II.	Sources of Information	528
III.	Differences in Policy and Objectives	528
IV.	Performing Assessments	531
V.	Issues Affecting Risk Assessment Decisions	534
VI.	Status of International Harmonization Efforts	533
VII.	Conclusion	534
	References	534

I. INTRODUCTION

The procedure for conducting risk assessments was originally developed by the National Academy of Sciences in the early 1980s and contained four steps: hazard identification, dose-response assessment, exposure assessment, and risk characterization. This same basic method continues to be used today as the basis for assessing chemicals. Despite the general scientific consensus on the individual components of a risk assessment, substantial differences exist in the numerical values calculated.

This chapter compares and contrasts the techniques used by various countries for assessing the risks associated with pesticide substances. Particular emphasis is placed on quantitating cancer risk, a contentious area within the scientific and regulatory communities for many years. In addition, the impact of mechanism of action (i.e., genotoxic and nongenotoxic) is emphasized, since this often drives the assessment procedures used by countries.

In order to provide more consistency between countries on the methods used to assess risk, a number of scientific meetings and conferences have discussed international harmonization efforts. To determine the status of this issue within the international community, the International Programme on Chemical Safety (IPCS) sponsored a project utilizing a survey questionnaire to obtain information on assessment techniques utilized by 21 Organisation for Economic Cooperation and Development (OECD) and select nonOECD countries. Several review articles and documents have also been prepared on this topic, especially in relation to carcinogenicity (GAO, 1993; OTA, 1993; WHO, 1993).

Finally, this chapter addresses a number of policy and technical issues relating to risk assessment. To the extent possible, updates on changes in policy have been included. The primary goal is to highlight differences that currently exist between the U.S. and other countries.

II. SOURCES OF INFORMATION

It is sometimes difficult to identify the appropriate governing body or group within a regulatory agency to contact with questions concerning risk assessment procedures. GAO (1993) provides a schematic breakdown of the regulatory agency structures in select OECD countries. In addition, the Pesticide Regulation Compendium (PRC, 1993) comprehensively describes the pesticide regulations in over 100 countries, including a description of each country's regulatory system, data requirements for residues, toxicology, ecotoxicology, and labeling. Table 1 lists the addresses for select regulatory agencies involved in conducting risk assessments.

III. DIFFERENCES IN POLICY AND OBJECTIVES

To investigate the intercountry differences in approaches to risk assessment, two components of the process have been examined: hazard identification and dose-response. Hazard identification examines all available data in humans and laboratory animals relating to a chemical's potential to induce toxicity. For carcinogenicity, the International Agency for Research on Cancer (IARC) uses an alphanumeric classification scheme to characterize whether the chemical is a known, probable, or possible human carcinogen. Although the U.S. EPA formerly used an alphanumeric system, in April 1996 it published new cancer risk assessment guidelines (U.S. EPA, 1996). These new guidelines proposed the use of three descriptive categories, similar to those used in the European Union (EU)* to characterize cancer risk (Directive 67/548/EEC).** Although assignment of these qualitative cancer ratings tends to be

---

* The European Community (now called the European Union) was established in 1958 by the Treaty of Rome and is responsible for developing governmental policy primarily through legislation known as regulations and directives. Much of the legislation affecting environmental issues involves directives. These directives are binding on the member nations with respect to the end result, but allow each state to individually decide the means by which they will implement the directive (OTA, 1993). This allows for considerable flexibility in fine-tuning the specifics of the directives.

** EU Directive 67/548/EEC addresses the classification, packaging, and labeling of dangerous substances and was last appended by the Seventh Amendment. Annex VI of this directive provides guidance for characterizing the carcinogenicity of chemicals in the EU.

Table 1 Regulatory Agencies Responsible for Conducting Pesticide Risk Assessments

Country	Agency
Australia	The Scientific Director Chemicals Safety Unit DHHLGCS, PO Box 9848 Canberra ACT 2601
Bulgaria	National Center of Hygiene Ecology and Nutrition D. Nestorov Str. 15 1431 Sofia, Bulgaria
Canada	Director General Food Directorate Health Protection Branch Health and Welfare Canada Health Protection Branch Bldg., Tunney's Pasture Ottawa, Ontario Canada K1A OL2
China	Institute for the Control of Agrochemicals, Ministry of Agriculture (ICAMA) Liang Maqiao, Chaoyang Qu, Beijing 100026, China
Czechoslovakia	National Institute of Public Health, National Reference Centre for Pesticides Srobarova 48, 100 42 Prana 10 Czech Republic
Egypt	Central Agricultural Pesticides Lab Ministry of Agriculture Dekki, Giza, Egypt
France	DGCCRF Commission of Toxicity 59 Boulevard Vincent Auriol, 75703 Paris cedex 13
Germany	Bundesgesundheitsamt 1. Eachgebeit C I 4 2. Postfach 33 00 13 3. B-1000 Berlin 33 4
India	Secretary, Central Insecticides Board & Registration Committee Directorate of P.P.Q & S NH-IV, Faridabad 121001
Korea	Agricultural Chemicals Research Institute 249 SeudoonDong SuweonSi KyunggiDo Republic of Korea
Thailand	Director of Agricultural Regulatory Division Department of Agriculture Bangkok 10900
United States	Office of Pesticide Programs, Health Effects Division Environmental Protection Agency Washington, DC

fairly consistent between countries, the situation for the dose-response evaluation is significantly different.

In the U.S., dose-response data are frequently assumed to be linear in the non-experimental low-dose region and are assessed through the derivation of cancer potency factors. Using mathematical models (i.e., the linearized multistage model, see below), upper limits of risk are calculated that yield a cancer potency factor. This factor is multiplied by the estimated exposure to yield a single risk value. In actuality, the "true" risk actually lies between this calculated upper limit (i.e., 95th percentile) and zero (EPA,1989).

The 15 EU member nations (Austria, Belgium, Denmark, Finland, France, Germany, Greece, Ireland, Italy, Luxembourg, The Netherlands, Portugal, Spain, Sweden, and the U.K.) use a significantly different approach than the U.S. for assessing the dose-response of carcinogenic pesticides. In these countries, cancer is thought to be a threshold response yielding a "safe" dose below which there is no risk. The EU also focuses on mechanism of action (e.g., genotoxic vs. nongenotoxic) in dealing with carcinogenic or potentially carcinogenic pesticides; registration of pesticides which are genotoxic is not permitted. Acceptable daily intakes (ADIs) are derived for nongenotoxic pesticides using NOELs or NOAELs* from animal carci-

---

* The NOAEL is the highest dose administered that produces "no statistically or biologically significant increases in frequency or severity of adverse effects." The NOEL is defined in the same manner, with the exception that there is no increase in the frequency or severity of effects (Hallenbeck and Cunningham, 1985).

nogenicity studies. This approach has occasionally been applied in the U.S. to assess pesticides for which the weight-of-evidence for carcinogenicity is not convincing. Thus, the use of a threshold-based approach in the U.S. is a function of the weight-of-evidence rather than whether or not genotoxicity is presumed to be responsible for tumor induction. Given the release of the new cancer risk assessment guidelines, however, the procedures used by the U.S. may change.

Health Canada reported employing different risk assessment techniques depending on the mechanism of tumor induction. For genotoxic pesticides, a quantitative risk assessment is performed whereas for nongenotoxic pesticides an ADI is derived in conjunction with a weight-of-evidence assessment (Dragula and Burin, 1994). Specific information on the type of quantitative risk assessment model used was not supplied. Typically, the federal government is responsible for these assessments, and not the individual provinces within Canada. Publicly available information on pesticide regulatory practices can be obtained through "Backgrounders" published by the Pesticides Directorate.

For occupational exposures, Denmark, The Netherlands, and the U.K. use quantitative risk assessment techniques (i.e., mathematical models) to generate a risk value which represents the probability of human cancer risk. This probability reflects the expected or the best estimate of the human cancer risk likely to occur in a population.

Finally, a significant portion of countries simply adopt the carcinogenicity assessment policies or evaluations developed by other countries or scientific groups. For example, Bulgaria and Thailand rely on the evaluations derived by MARC (which provides statements regarding the weight-of-evidence for carcinogenicity based on animal and epidemiological data); Korea relies on the risk assessment techniques and decisions developed by the EPA; Egypt utilizes the evaluations provided by the Joint FAO/WHO Meeting on Pesticide Residues; Czechoslovakia uses the risk assessment methods developed by WHO in their Environmental Health Criteria document (WHO, 1990); and Spain relies on criteria described in Annex VI of the EU Directive 67/548/EEC for dangerous substances (Dragula and Burin, 1994).

The most outstanding difference between the risk assessment goals or objectives in the U.S. vs. the EU is the default assumption in the U.S. that carcinogenicity is a nonthreshold, or linear process where every increase in dose is associated with an increased risk. The U.S. has utilized the linearized, multistage model for most potential human carcinogens, including pesticides, whereas the EU has generally used the ADI approach.

The U.S. also publishes extensive guidelines outlining the procedures for performing carcinogenic and other types of risk assessments. GAO (1993) referred to this policy as "transparent" and noted that such detailed procedures were not yet readily available in the EU and other OECD countries surveyed. However, the specific guidance provided by the EPA may have the unintended effect of forcing risk assessment decisions (i.e., through the application of default assumptions) in ways not consistent with expert scientific judgement. The interviewees considered the U.S. process to be less flexible than the procedures used in other OECD countries, a fact they viewed as a weakness when compared to their system's ability to flexibly address specific issues on a case-by-case basis (GAO, 1993).

IV. PERFORMING ASSESSMENTS

Two primary procedures exist for estimating the carcinogenic risks associated with exposure to pesticides (and other substances): mathematical models and ADIs. In general, quantitative risk assessment refers to the use of models since an actual risk value is calculated. This method is much more complex than the ADI approach in which the actual "risk" is presumed to be nonexistent. The ADI is derived from the highest dose in an animal study at which no adverse effects occur and is compared to the estimated human exposure.

The U.S. typically employs models using sophisticated computer software to estimate cancer risk. The Netherlands is the only other country that regularly performs quantitative risk assessment utilizing such models, and it only uses these models for occupational exposures, not for pesticide assessments. Quantitative assessments using models are performed on a limited basis in Canada, Denmark, Germany, and the U.K., but again not for pesticides.

In the U.S., the risk associated with carcinogenic substances is assessed using the linearized, multistage model. This model has two main constraints. First, it assumes no threshold for effects. In other words a single molecule of the pesticide/substance can induce the molecular events necessary to produce cancer (Barnes and Dourson, 1988). As a result, zero risk is only achieved when zero exposure occurs. Second, it assumes that the dose-response curve is linear in the low-dose region. This means that an increase in dose results in a proportional incremental increase in cancer risk. The decision to use this model was made in the 1970s and was based upon uncertainty regarding the shape of the dose-response curve in the nonexperimental, low-dose region. EPA believed that it was prudent to be conservative where the public was concerned and, thus, chose the linearized multistage model in order to provide the greatest protection. This model remains the default method for estimating the cancer risk of carcinogens. Alternative approaches may be implemented when the draft cancer risk assessment guidelines are formally adopted.

As noted above, most countries rely on ADIs for establishing exposure limits for carcinogenic pesticides. These limits are not the same as risk values because they represent levels at which no risk is predicted to occur. ADIs are calculated by dividing the NOAEL or NOEL by Safety Factors (SF), also called uncertainty factors. These factors reflect the reliability and consistency of the experimental animal data. Generally, the more SFs applied the less confidence is placed in the data. The majority of the nations polled in the IPCS survey (Australia, Belgium, Bulgaria, Canada, China, Czechoslovakia, Denmark, France, Germany, India, Japan, Korea, The Netherlands, Spain, and the U.S.) routinely utilized the NOAEL/SF approach for estimating the risks associated with pesticide exposure (Dragula and Burin, 1994). As previously noted, the U.S. often uses quantitative models for estimating cancer risks.

The reliance on the NOAEL/SF approach is easy to understand: it is simple to use and provides a clear, limit value below which exposures are considered acceptable. Although many countries currently rely on this approach for assessing cancer risks there are practical advantages to the risks calculated from quantitative models.

These include the ability to calculate "cost" per tumor in risk benefit comparisons and to develop "bright lines" or clear risk values upon which regulatory decisions can be based (e.g., a cancer risk of 1 in 1 million).

V. ISSUES AFFECTING RISK ASSESSMENT DECISIONS

The mechanism by which substances induce cancer (e.g., genotoxic vs. nongenotoxic) is the single most important factor affecting the approach many countries take to assessing human cancer risk.* Terms used to describe the mechanism of nongenotoxic and genotoxic substances are threshold and nonthreshold, respectively. There appears to be general international consensus that most biologic effects occur through a threshold mechanism (i.e., there is a dose, which is unique for each substance and endpoint, below which no adverse effect or response is observed); however, there is some debate about whether this is also true for carcinogenicity.

According to U.S. policy, cancer can develop from a single event with no threshold. Yet, several nongenotoxic mechanisms of action have been identified and studied extensively. These include thyroid tumors induced by hormone imbalance and kidney tumors associated with $\alpha-2\mu$ globulin.

Two other issues affect the conduct of risk assessments and represent substantial differences in international scientific opinion: the significance of tumors arising at the maximum tolerated dose (MTD); and the significance of mouse liver tumors (MLTs) in extrapolating carcinogenic risk to humans. Not only do these issues symbolize a significant divergence of opinion and policy between the U.S. and other countries, they also represent barriers between countries with respect to data interpretation in carcinogenicity studies.

Most countries believe that tumors only occurring above the MTD result from physiologic changes that cannot be directly associated with the substance administered. Such changes include, but are not limited to, cell death and concomitant cellular repair processes, metabolic overload, pharmacokinetic alterations, and hormonal changes. Australia, France, Bulgaria, and Canada specifically concluded that tumor data collected at the MTD were either of limited or no usefulness for assessing carcinogenic risks in humans (Dragula and Burin, 1994).

In comparison, the 1986 EPA cancer risk assessment guidelines state that data must be collected at the MTD in order to provide adequate statistical power for assessing the carcinogenicity of a substance (U.S. EPA, 1986). An agency position document stated that the purpose of the MTD was to "vigorously" test a substance "for oncogenic potential at levels somewhat below test levels which might compromise survivability" (Farber, 1987). Interestingly, EPA did not specifically use the term "MTD" in the proposed new cancer risk assessment guidelines, but it did state that the high dose in a carcinogenicity bioassay should produce some toxicity (not

---

* Genotoxic compounds directly interact with DNA and alter it permanently such that the change is passed on to future generations of cells. This is referred to as a heritable change and the substances that produce this effect are often termed "initiators" to describe the initiation of cancer. In comparison, nongenotoxic substances simply enhance or promote the growth of cells already genetically altered or initiated; they do not interact directly with the DNA to produce any alterations. These substances are referred to as "promoters."

to exceed a 10% reduction in body weight gain during the lifespan of the animal) without unduly affecting the survival or the nutrition and health of the test species (U.S. EPA, 1996).

Of the 16 countries responding to related questions in the IPCS survey, eight (Bulgaria, Denmark, Germany, India, Japan, The Netherlands, Spain, and the U.S.) stated that dose selection "frequently" affected carcinogenicity study results and another two (Belgium and China) stated that it "sometimes" affected study results (Dragula and Burin, 1994). Furthermore, seven of the countries concluded that the highest dose used was too high (Belgium, Bulgaria, Czechoslovakia, Denmark, Germany, Japan, and Spain) and five felt that it was inadequately justified (Canada, France, Germany, India, and the U.S.). These survey results point to the need for international consensus on the highest dose to be used in animal carcinogenicity studies and clarification as to whether this dose should be the MTD. Clearly, many countries disagree with the U.S. criteria for determining whether dosing is adequate.

The U.S. has also developed a policy/regulatory position on the significance of MLTs for estimating carcinogenic risk to humans. EPA (1986) indicated that MLTs, under particular conditions, provide sufficient evidence of carcinogenicity in animals, even if they are the only type of tumor response observed. This "sufficient" classification may be withdrawn under particular conditions but the policy is in direct conflict with the opinions and policies of other countries (Dragula and Burin, 1994). In India, for example, MLTs typically receive little "weight" as indicators of carcinogenic potential in humans. Bulgaria also assumes that these tumors are of little relevance in estimating human cancer risks. In Germany, the significance of MLT data is assessed on a case-by-case basis. Because MLTs are a frequent finding in carcinogenicity studies, the different views on their significance represent a barrier to risk assessment harmonization.

VI. STATUS OF INTERNATIONAL HARMONIZATION EFFORTS

Significant interest in harmonizing risk assessment techniques has culminated in a number of international scientific conferences as well as national and international sponsorship of data gathering activities on the subject. At present, scientists are attempting to determine the status of risk assessment procedures in different nations and have discovered that some countries have had the opportunity to focus efforts on it far more than others. This information represents a first step towards identifying the areas most likely to benefit from harmonization efforts.

In the IPCS survey, 17 of the 18 nations clearly indicated that they were interested in harmonizing risk assessment procedures (Dragula and Burin, 1994). However, the area(s) that each country identified as being most in need of harmonization activities varied considerably and included virtually every topic related to toxicology. This diversity of responses truly reflects the varying interests of the countries polled.

Involvement in international harmonization endeavors is not limited to pesticide chemicals. In fact, the pharmaceutical industry has recently worked to harmonize various topics of concern. Several consensus documents have been published under the sponsorship of the International Congress on Harmonization. It would appear that some of the lessons learned by pharmaceutical manufacturers may also be applied to pesticide manufacturers.

VII. CONCLUSION

The four-step paradigm originally proposed in 1983 by the National Academy of Sciences continues to be the most widely used methodology for presenting the results of risk assessments. However, as this chapter shows, significant differences exist with respect to individual country's policies for interpreting toxicity data and assessing carcinogenic risk. These differences are particularly interesting given the fact that toxicity data requirements for pesticide registration are fairly standardized across most countries (GAO, 1993 and PRC, 1993). They appear to arise largely in the area of data interpretation reflecting each country's autonomy in regulatory decision-making. This desire for individuality complicates efforts to enter international markets, since a pesticide developed in one country and regarded as noncarcinogenic may be considered carcinogenic in another country. Johnson (1989) captured this reality succinctly when stating that although significant agreement exists with respect to the scientific principles for regulating pesticides, the application of these principles is based on various factors, not the least of which are "national, social, and institutional structures and values."

To date, the U.S. has taken the international lead in developing risk assessment methodologies. More countries are becoming involved in developing approaches to risk assessment, such as The Netherlands. International scientific bodies, such as WHO, IARC, IPCS, OECD, and the Pan American Health Organization, are also playing an increasingly important role in advancing this process, especially for developing countries (OTA, 1993).

In conclusion, the following two general statements can be made regarding the current state of risk assessment and its future direction. First, the status quo results in an unnecessary duplication of resources and the generation of potential trade barriers. Second, harmonization efforts are underway to establish a uniform approach to data interpretation, especially for carcinogenicity data, in risk assessments. It is clear that international harmonization of risk assessment procedures is an objective which will result in more consistent international risk decisions.

REFERENCES

Anon., *Pesticide Regulation Compendium*, 4th ed., Editions Agrochimie, Switzerland, 1993.

Barnes, D.G. and Dourson, M., Reference dose (RfD): description and use in health risk assessments, *Regul. Toxicol. Pharmacol.*, 8, 471, 1988.

Dragula, C. and Burin, G., International harmonization for the risk assessment of pesticides: results of an IPCS survey, *Regul. Toxicol. Pharmacol.*, 20, 337, 1994.

Farber, T.M., *A Position Document of the U.S. Environmental Protection Agency Office of Pesticide Programs: Selection of a Maximum Tolerated Dose (MTD) in Oncogenicity Studies*, Office of Pesticide Programs, Washington, 1987.

Government Accounting Office, *Pesticides: A Comparative Study of Industrialized Nations' Regulatory Systems, Report to the Chairman of the Committee on Agriculture, Nutrition, and Forestry*, U.S. Senate, Washington, 1993.

Hallenbeck, W.H. and Cunningham, K.M., Qualitative evaluation of human and animal studies, in *Quantitative Risk Assessment for Environmental and Occupational Health*, Lewis Publishers, Ann Arbor, MI, 4041, 1988.

Johnson, E.L., Pesticide residues, In *International Food Regulation Handbook: Policy, Science, Law*, Middlekauff, R.D., and Shubik, P., Eds., 253, 1989.

Office of Technology Assessment, Appendix A: International risk assessment, in *Researching Health Risks*, U.S. Government Printing Office, Washington, 187, 1993.

U.S. Environmental Protection Agency, *Guidelines for Carcinogen Risk Assessment*, Federal Register, 51, 33992, 1986.

U.S. Environmental Protection Agency, *Proposed Guidelines for Carcinogen Risk Assessment*, Federal Register, 61(79), 17959, 1996.

U.S. Environmental Protection Agency, *Risk Assessment Guidance for Superfund, Vol. I: Human Health Evaluation Manual (Part A), Interim Final*, Office of Emergency and Remedial Response, Washington, 1989.

World Health Organization, *Classification Systems on Carcinogens in OECD Countries — Similarities and Differences*, prepared by Norway and The Netherlands, 1993.

World Health Organization, *Environmental Health Criteria, Principles for the Toxicological Assessment of Pesticide Residues in Food, Vol. 104*, International Programme on Chemical Safety, Geneva, Switzerland, 1990.

CHAPTER 30

Historical Toxicology and Risk Assessment

David A. Belluck, Mark W. Rattan, and Sally L. Benjamin

CONTENTS

I.	Introduction	537
II.	Historical Toxicology and Risk Assessment	538
	A. Selecting the Proper Analysis	538
	B. Historical Toxicology	539
	C. Alcohol Consumption: How Simple Observations Lead to Complex Toxicological Understandings	539
III.	Application of Historical Toxicology in Insurance Law	541
IV.	Historical Toxicology in Litigation	542
	A. Analysis: Manufactured Gas Plant Industry	543
	1. The Legal Setting	543
	2. A Historical Toxicology Review of the MGP Industry	543
	B. Using a Historical Toxicology Review in Litigation	546
V.	Conclusion	547
	References	548

I. INTRODUCTION*

Since the invention of the risk assessment paradigm in the early 1980s, the risk assessment process and its products have become key elements in public and environmental health decision making. Although risk assessment methods and policies have changed since the early 1980s, the basic risk assessment paradigm has remained relatively stable. This makes it possible to generate a risk assessment appropriate

---

* The authors are indebted to Desiree Savage, Esq., who researched the technical information that forms the basis for this chapter.

for any given time period since the 1980s by using risk assessment methods and data inputs from that time and, thus, to reconstruct what risk assessors and other professionals of that period knew or should have known about the human health or environmental risks posed by a facility, activity, or site.

Court cases can be won or lost on the basis of characterizations of what a litigant knew or should have known about the environmental or human health effects of their activities. Attorneys and their technical experts expend considerable time and resources to make this determination.

Risk assessment is a useful tool in this endeavor. Since modern risk assessment did not exist as a regulatory construct until the early 1980s, however, it can be argued that modern risk assessment principles and methods are not useful to infer knowledge of risks before that time.

II. HISTORICAL TOXICOLOGY AND RISK ASSESSMENT

If it is anachronistic to use modern risk assessment-based methods to gauge the understanding of human health and environmental impacts of environmental releases before the early 1980s, is there another way? Instead of using modern risk assessment, we suggest an in-depth analysis using disciplines that existed at the time of an environmental release to ascertain what individuals or organizations did know, or should have known, about the effects of such releases. We term this analysis "historical risk assessment and toxicology."

A. Selecting the Proper Analysis

Before starting a project it is essential to determine which research standard applies to the analysis; "historical risk assessment and toxicology" or "modern" risk assessment. While the historical method described in detail below is appropriate for reconstructing what a technical person in a given discipline should have known at a critical point in time, certain laws use a different standard. Although CERCLA, for example, imposes liability for pollution resulting from the polluter's activities in the past, the cleanup standards it requires the polluter to meet are based upon current scientific knowledge of the hazardous properties of the chemicals involved and their behavior in the environment.

Persons working on modern risk assessments will recognize that historical risk assessment and toxicology involves several fundamental tools of their profession. Historical risk assessment and toxicology reviews are essential for cutting edge risk assessments, to ensure that outdated toxicological values and methods are not used in their reports. Data obtained by historical risk assessment and toxicology reviews are also used in modern risk assessment reports (i.e., 1980s onward). Searching records and reviewing the literature and databases are part of modern risk assessment, as well as historical risk assessment and toxicology.

The rest of this book discusses how to obtain and use current risk assessment resources and tools. This primer describes the use of historical toxicology. Although historical toxicology reviews are valid for all periods of time in which toxicology has been an established profession, as discussed above, this primer deals with uses of historical toxicology in the period prior to the current risk assessment paradigm.

B. Historical Toxicology

How can researchers reasonably determine risk understandings of the past for use in current litigation? Historical toxicologists do so by asking the following three simple questions. What did they know? What should they have known? When should they have known it?

Although the questions are simple, answering them can require in-depth historical research. Rigorous review of media reports, letters, memoranda, books, journals, and technical reports of the time period in question can (through what we term a "weight-of-evidence" approach) marshall logically consistent, detailed, and compelling proof concerning what a given industry knew at critical points in time about the hazardous propensities of byproducts from its manufacturing operations, or what the industry should have known if it had made reasonable, prudent, and responsible inquiry.

The application of historical toxicology to a given industry, therefore, involves painstaking investigation of the medical, scientific, and technical knowledge concerning the byproducts from the manufacturing operations and their chemical constituents. When evaluating the three pivotal questions, researchers must resist the temptation to evaluate past actions and understandings in light of modern methods and standards of professional behavior.

After contaminants have been identified and their impact on the environment evaluated, the parties must resolve the legal question of who is responsible for paying the cleanup costs. Polluters who have been held liable for cleanup costs by regulatory agencies often assert that the loss should be covered by their liability insurance policies. The insurance companies in turn assert that the cleanup costs are either not covered by the insuring language of the policy, or are excluded from coverage by various policy exclusions. The resolution of the dispute often turns on the factual information obtained in the historical toxicology review concerning what the polluter knew or should have known during the time period when the contaminants were released into the environment.

There are a wide range of observation types that can be used to determine the effects of chemical exposures to humans or the environment from a process, activity, or a facility. They can range from a simple observation recorded in a diary, or noted in a file memo, to the traditional scientific method of data collection, interpretation, and hypothesis testing. How simple observations can lead to complex toxicological understandings is illustrated by the decision to consume alcohol.

C. Alcohol Consumption: How Simple Observations Lead to Complex Toxicological Understandings

People conduct qualitative risk assessments concerning the risks of alcohol consumption without formal risk assessment and toxicology training. Persons consuming alcoholic beverages know that there are potential health problems associated with consuming alcoholic beverages (i.e., hazard evaluation); that the hazards are related to the amount consumed in a given period of time (i.e., exposure assessment); that the various chemicals and process impurities can be toxic by themselves or in combination with pharmaceuticals or other drugs (i.e., toxicity assessment); and that the combination of these factors results in a certain level of risk (i.e., risk characterization).

This type of analysis, conducted by many people on Friday evenings on the way to Happy Hour, illustrates that almost everyone has a basic grasp of the main components of risk assessment and that the results of individual assessments can actually influence individual behavior. Characterizing risks allows a person to perform a cost-benefit analysis and decide what level of risk to accept and what risk-reduction measures (e.g., take a cab home) to employ to reduce short- or long-term risks.

Knowledge of the toxic effect of a component of a complex mixture (e.g., alcohol in alcoholic beverages) can be successfully applied to all other alcoholic beverages. A person might not be able to express the exact toxicity of a given alcoholic beverage, but common sense indicates that any alcoholic beverage will, at some point, induce the toxic effects, even though the point differs according to the type of alcoholic drink. Personal experience with every alcoholic beverage is not required to know this.

The above risk analysis reveals simple truths about alcohol consumption. These simple truths enable the individual to make choices to control the risks. In a similar fashion, historical toxicology reviews reveal awareness of simple scientific correlations (e.g., exposure to chemical A causes effect B) long before a definitive treatise is published quantifying the fact and explaining its precise mechanism of action. Historical toxicology reviews usually reveal the same general pattern of awareness and scientific analysis:

- An ongoing exposure is identified as the possible cause of a toxic effect that is observed repeatedly in humans.
- This awareness leads to data collection on the phenomenon (case studies) and initiation of testing (epidemiological or animal studies).
- Results of systematic information gathering, in turn, spur more detailed research into the phenomenon.
- A cause and effect relationship is established.
- Research turns to quantifying the dose-response relationship and to determining the mechanism of action.

The point is that human awareness grows and becomes more refined over time, but even simple observations are valuable clues for the historical toxicologist's quest to understand what could or should have been known at a given point in time about human or environmental health effects of some human action.

III. APPLICATION OF HISTORICAL TOXICOLOGY IN INSURANCE LAW *

In 1980 the passage of Superfund** fostered an explosion of litigation between insurers and their policyholders concerning the issue whether the expense associated with environmental clean-ups, often referred to as "response costs" or "remediation," is covered under one or more of the insurance industry standard form Comprehensive General Liability (CGL) insurance policies issued to businesses since the 1950s. As the Supreme Court of Minnesota noted in the recent case, *Northern States Power Co. v. Fidelity and Gas. Co. of New York*, 521 N.W.2d 21, 31 (1994), "The stakes in these cases are extremely high." Quoting law review articles by Kenneth S. Abraham, the Supreme Court of Minnesota further noted:

> The average cost of remedying hazardous conditions at a site on the Superfund "National Priority List" now exceeds $30 million (Abraham, 1993). The cost of hazardous waste cleanup under the federal superfund program and analogous state regimes ... is likely to be several hundred billion dollars before these programs are completed (Abraham, 1991) (521 N.W.2d at 31).

Given the ruinous liability which can be imposed on a polluter under CERCLA and comparable state statutes, the polluter looks to its insurers to pay or at least share in the remediation cost. The language of a standard CGL policy typically provides that the insurer agrees to pay on the policyholder's behalf the amounts that a policyholder is legally obligated to pay in "damages" as the result of unintentional conduct by the policyholder which causes property damage to a third person. Insurers contend that the amounts government regulators order a policyholder to pay under CERCLA to cover the "response costs" necessary to cleanup pollution to the environment are not "damages" within the meaning of the policy language, and were never intended to be covered by a CGL policy

Insurers further assert that, to the extent the courts construe CERCLA "response costs" to constitute "damages" within the meaning of the insuring agreement, an exclusion introduced into CGL insurance policies in about 1973 precludes coverage for all liability associated with discharges of contaminants into the environment unless the discharges were sudden and accidental. That exclusion is known as the "sudden and accidental" pollution exclusion. Since contamination on most of the CERCLA cleanup sites results from repeated and gradual discharges of pollutants over the course of several years, as opposed to a "sudden and accidental" event, insurers contend that the sudden and accidental pollution exclusion precludes coverage for the cost necessary to remedy the contamination.***

---

* This section is not intended and does not pretend to be a complete or exhaustive discussion of the law of insurance coverage in the environmental context. Rather, it is designed to give the reader an overview of some environmental insurance law issues and how they relate to historical risk assessment and toxicology reviews.
** The Comprehensive Environmental Response, Compensation, and Liability Act, §9601 (CERCLA).
*** As the result of court decisions in some states which weakened the sudden and accidental pollution exclusion (see, for example, *Just v. land Reclamation*, 155 Wis.2d 737, 456 N.W.2d 570 [1990], *amended, recons. den'd.*, 157 Wis.2d 507 [1990], the insurance industry introduced the "absolute pollution exclusion" into CGL policies in 1985 which, among other things, excluded coverage for even sudden and accidental polluting events.

Insurers also believe they drafted and rated their policies to cover only liability resulting from accidents, not from intentional conduct on the part of the policyholder. In the pollution context, the insurers endeavor to deny coverage on the basis that repeated intentional and gradual discharges of pollutants by the policyholder are not the result of unintentional conduct. The language on which insurers rely to deny coverage for discharges of pollutants provides that the policy covers "an accident . . . which is neither expected nor intended from the standpoint of the insured." Under that language, insurers contend, there is no coverage for pollution liability if the insured knew or should have known that the substances discharged into the environment would likely be hazardous to human beings or the environment (i.e., if the insured "expected and intended" the results). In the mind of the insurer, there was no accident under these circumstances because the insured intentionally and continually released pollutants into the environment knowing them to be harmful.

As mentioned earlier, analysis of whether the insured "expected and intended" any damage to property often devolves into three areas of factual inquiry: what did the policyholder/polluter know, what should he have known, and when did he know it? In some cases, that factual inquiry is resolved by the information revealed in a historical toxicology review of the nature of the pollutants and the industry's knowledge of its hazardous propensities. A toxicologist and his risk assessment team can research the knowledge during the time period the policyholder was releasing pollutants into the environment to determine whether the "weight-of-evidence" shows that the policyholder knew or should have known of its hazardous propensities. An attorney will use the information garnered by the risk assessment team to present a clear and compelling case to the jury that the response costs are not covered because they did not result from "an accident . . . neither expected nor intended from the standpoint of the insured."

The following case history illustrates the use of historical toxicology reviews to determine what the Manufactured Gas Plant (MGP) industry should or could have known about the human and environmental health effects and impacts of its operations during the time it was operating.

IV. HISTORICAL TOXICOLOGY IN LITIGATION

In litigation between an insurance company and its policyholder over whether remediation costs are covered by a CGL policy, a litigator representing the insurance company requests historical toxicology reviews to answer three pivotal questions:

- What did the policyholder/polluter know about the hazardous propensities of the contaminants discharged into the environment?
- What should he have known?
- When did he know it?

A. Analysis: Manufactured Gas Plant Industry

1. The Legal Setting

Throughout their history, from the early 1800s to the late 1950s, manufactured gas plants generated tremendous amounts of coal tar. This byproduct was often pumped into coal tar "lagoons" on a plant's premises. The coal tar and its constituents, primarily benzene and naphthalene, often saturated the soil under the lagoon and contaminated the groundwater.

As authorized under CERCLA and comparable state statutes, government entities are now requiring gas utilities to remediate coal tar residuals in the soil and groundwater. The CERCLA response costs or cleanup costs for a single site can amount to tens of millions of dollars. Faced with this extensive liability, the gas utilities seek reimbursement for these response costs from their insurers. Information obtained in a historical toxicology review concerning what the gas industry knew at various points in time can be pivotal in determining whether an insurer must pay the response costs on behalf of a gas utility.

2. A Historical Toxicology Review of the MGP Industry

Various technical and scientific developments in the early to mid-1800s led to the operation of MGPs in many towns and cities across the U.S. by 1860. Gas from these facilities was obtained through various physical and chemical processes and was used for light and heat in homes and factories. Processing coal and other feedstocks produced many byproducts and wastes. Byproducts could be further refined into important chemical feedstocks on- or off-site. Wastes were routinely released into the air, adjacent waterways, sewers, and land. Facility practices and government requirements for manufacturing operations and disposal of wastes varied with locality and year of operation.

The first question to be answered in a historical toxicology review of a given MGP facility is "what scientific disciplines were recognized during its operation?" For example, was toxicology a recognized discipline at the time? A review of available literature indicates that numerous fundamental works on toxicology were available to the technical community starting in the years 1814-1815 (from France), 1884 (from England), and 1902 (from the U.S.) (Teleky, 1948).

If a discipline exists, it is important to next understand when fundamental principles of that discipline were known. Since toxicology was a recognized discipline during the operation of a given facility, then awareness of potential human health and environmental effects of a given release could have been informed by concepts used by toxicologists and those who use toxicological information, such as dose-response. As the dose of a chemical increases, an exposed organism exhibits some toxic response. A review of available literature found one author stating that the 16th century scientist Paracelsus understood that as the dose of a chemical increases the response or the effect from that exposure will also increase (Stacey, 1993). A report in a 1901 edition of the *Lancet* noted that the poisonous effects of certain chemicals at different dosages were known (White and Hay, 1901).

If dose-response relationships are known, then how was toxicological testing used during the time MGPs were operating? A review of available literature found that toxicological testing for industrial hygiene purposes can be traced back to 1886 (Teleky, 1948). Based on this understanding, it could be concluded that the methods were available to persons during the MGP era to test for, observe, and evaluate the potential toxic effects of MGP operations.

Coal tar was a major byproduct of coal gassification processes and was produced in great quantities during the MGP era. The amount and chemical composition of coal tar varied from facility to facility and was, in large part, dependent on the type of physical/chemical processes used to remove coal gas from coal, and the type of coal used. Coal tar was used as a source of numerous organic chemicals that could be used alone or as feedstocks for other synthetic chemical manufacturing. A historical toxicology review provides an understanding of what persons during the MGP era should or could have known about the chemical composition and toxicological effects of coal gassification byproducts.

A historical toxicology review of available literature found that the toxic properties of coal tar have been known since 1775. In 1775 Potts observed scrotal cancer among chimney sweeps who were exposed to coal combustion residue products (ATSDR, 1994). According to the International Agency for Research on Cancer, for over 200 years skin cancers have been recognized to be associated with work exposures to coal pitches and tars (IARC, 1984). An 1859 paper reported dermatitis induced by coal tar (Mathias, 1988). The precancerous and cancerous conditions resulting from exposure to industrial tars were reported in 1875 (O'Donovan, 1920). In 1875, the first recorded cases of tar cancer were reported. Subsequently, ten cases involving cancer of the arm or scrotum were recorded between 1873 and 1890 (Kennaway, 1923). A comprehensive review of surgical and dermatological cases of tar cancer of the skin seen at London Hospital from 1903–1920 resulted in the first paper where an expert clinical description of the development and varieties of tar cancer was published (O'Donovan, 1928).

By 1918, the basis of modern hydrocarbon carcinogenesis testing for coal tar substances was in place. In 1915 researchers succeeded in inducing skin tumors in the ears of rabbits by repeated applications of coal tar, firmly establishing the modern era of hydrocarbon carcinogenesis research. In 1918, tumors were produced in mice by repeated application of tar to the skin. This methodology was quickly adopted and is still in use in carcinogenicity research (Harvey, 1919).

By the mid 1920s to early 1930s, there was little doubt that coal tars caused cancer. Research during this period was directed at determining which tars caused greater or lesser morbidity (e.g., cancer) in exposed organisms. Important findings include:

- The ability of coal tar to produce cancer was known for about 50 years (Kennaway, 1924).
- Researchers comparing the carcinogenic effects of horizontal retort tar and vertical retort tar of the Amsterdam gasworks found that the latter produced cancer much more slowly than the former (Kennaway, 1924).

- Of the coal tars produced at controlled temperatures, those created at temperatures most closely corresponding to gasworks production temperatures generated the most carcinomas in mice (Kennaway, 1925).
- Tar and pitch from gasworks was more carcinogenic than tar and pitch from coke ovens (Patty, 1948).
- The higher boiling fractions of gasworks tar contains a carcinogenic factor (Kennaway, 1930).
- Death-rate from industrial skin cancer from 1920-1931. Gasworks: 125 cases, 58 deaths, 46.4% death rate. Tar distilling: 218 cases, 22 deaths, 10% death rate (Legge, 1934).
- A reported 324 cases of occupational cutaneous cancers in 309 persons (i.e., managers, retort-stokers, retort-setters and repairers, workers on mains and pipes, fitters, pipe-laggers, carpenters and other maintenance men, and yard-laborers) employed in the coal gas industry (Henry, 1947).
- Coal gas and tar workers show an increased prevalence of cancer of the lung (Kennaway and Kennaway, 1947).
- Coal tar is the main carcinogenic agent with which the gasworker comes into contact (Henry, 1947).
- In the U.S. heavy coal-tar distillates and pitch are the most frequent causes of occupational cancer causing more than 90% of all reported occupational cancers (Patty, 1948).

It is clear from this discussion that an understanding of the human health effects of coal gassification products and byproducts has been known since the late 1700s. The next step in this analysis is to identify each coal tar constituent, determine when it was identified as being present in coal tar, and establish the earliest reported toxic effects for each constituent. For example, coal tar contains anthracene, arsenic, benzene, cresol, naphthalene, PAHs, toluene, and xylene, as well as numerous other constituents. Of these constituents:

- Anthracene reported in coal tar as early as 1832 (Partington, 1964). Health effects related to exposure to anthracene, such as epithelioma of the hand, cheek, and wrist, were reported as early as 1921 (Kennaway, 1924).
- Arsenic was reported in coal tar as early as 1901 (Weyman, 1922). Health effects such as palmar and plantar hyperkeratoses and a variety of cancerous and precancerous lesions on the hands, feet, and trunk were reported as early as 1921 (NAS, 1977).
- Benzene was reported in coal tar as early as 1783 (Lunge, 1916). Health effects such as unpleasant side- and after-effects from benzene use in anesthesia were reported as early as 1810 (Clayton and Clayton, 1981).
- Cresol was reported as a constituent of coal tar as early as 1868 (Gardner, 1915). Health effects such as renal toxicity were reported as early as 1922 (ATSDR, 1992).
- Naphthalene was reported as a constituent of coal tar as early as 1819 (Rhodes, 1945). Health effects such as cataract induction were reported as early as 1886 (Grant, 1962).
- While it is known that PAHs were probably responsible for chimney sweep cancers reported in 1775, clear identification of PAH toxicity was reported as early as 1918 (ATSDR, 1994).

- Toluene was reported as a constituent of coal tar as early as 1888 (Haynes, 1954). Health effects were reported as early as 1921 (Hamilton, 1921).
- Xylene was reported as a constituent of coal tar as early as 1888 (Haynes, 1954). Health effects were reported as early as 1921 (Hamilton, 1921).

The above is a summary of painstaking literature review of documents that existed at the time the MGP industry was active. This weight of evidence approach can be used to provide a broad view of what was or should have been known to persons in charge of operating, maintaining, or managing facilities. This review revealed that the identity and toxic properties of coal tar and its constituents were known to the scientific community during the MGP era. It is possible to reach this conclusion based on the fact that chemical identity and toxicology information was available from sources accessible to individuals at the time of facility operation had they used obtainable library resources.

As this review illustrates, historical toxicology is an effective method for determining the understanding of risks and impacts from environmental releases. It is especially useful for industrial practices that predate modern risk assessment.

B. Using a Historical Toxicology Review in Litigation

An attorney who represents an insurance company in litigation would retain a historical toxicologist to conduct extensive research into the state of knowledge concerning coal tar and its constituents at various critical points in time. As demonstrated above, that research would include a historical review of medical, technical, and other scientific concerning coal tar. It would also include a meticulous review of gas industry publications and journals from the period to ascertain whether gas industry organizations warned or alerted individual gas utilities of harmful effects coal tar could have on human beings and the environment. Depending on information obtained by the risk assessment team, the attorney may decide to have the toxicologist testify as an expert witness.

Taking the above gas plant example a step farther, suppose a gas plant started operations in 1925, and continually pumped coal tar into a "lagoon" on its manufacturing premises, until it ceased gas manufacturing operations in the 1950s and started using natural gas. Then, in 1983, 3 years after CERCLA became law, U.S. EPA ordered the gas utility to remediate the soil and groundwater beneath the "lagoon" which had been saturated with coal tar from 30 years of repeated and systematic dumping. U.S. EPA ordered the cleanup because carcinogenic constituents of coal tar, such as benzene and naphthalene, have appeared in drinking water wells located down gradient from the former coal tar lagoon.

Suppose further that the toxicologist concludes that, based upon extensive research by the risk assessment team, the weight-of-evidence indicates that, as early as 1910, coal tar and its constituents were known human carcinogens, and that this information was circulated to gas utilities through industry journals prior to 1925, the year the plant started operations.

The insurer's attorney under this hypothetical would probably ask the toxicologist to testify in court to explain the risk assessment teams' findings to the jury in plain, simple English. The attorney would then argue to the jury in his summation that they should find that the government-ordered CERCLA response costs are not covered by the gas utility's CGL insurance policy because, given that the utility knew or should have known of the carcinogenic propensities of coal tar prior to initiating operations in 1925, any "damage" to property (i.e., groundwater and wells used for drinking) caused by the coal tar carcinogens was not an accident. Rather, it was "expected and intended" within the meaning of the applicable policy language. The research of the risk assessment team as presented by the testimony of the toxicologist would be compelling proof that CERCLA response costs are not covered by the CGL insurance policy.

The manufactured gas plant discussion is just one example of what litigators seek in historical toxicology reviews and how the information obtained in a historical toxicology review is useful in the litigation process to aid the court and jury in the difficult determination of whether the policyholder's liability is covered by the CGL policy.

V. CONCLUSION

The role of the risk assessment team is to determine, through intensive historical research, an assessment of what the weight-of-evidence shows about what the polluter knew and when he knew it, and to develop a clear presentation of these findings for the court and jury. Given this role, the ideal consultant will be a highly trained risk assessor with an advanced degree in one of the technical fields associated with risk assessment preparation (e.g., toxicology, hydrology, medical sciences) and training in the history of science. Finding one individual with cross-training in these fields may be very difficult. When this is the case, hiring a team possessing these skills is the next best alternative.

While it is more efficient to hire a person or team with these skills and experience in a particular type of case, the probability of this occurring can be quite small. Look for evidence that this individual or team is able to conduct highly complex cross-discipline research, can integrate this information into a useful report or strategy, and can then explain the findings to the client and courts. Team members should be skilled in creative use of library resources and electronic databases. Since the information they will need may be in many different locations (e.g., scattered in libraries and government offices around the U.S.), the individual or team must demonstrate flexibility in the environments in which they work effectively.

The consultant or consultant team must have excellent verbal communication skills. First, the team will have to be able to review, digest, and synthesize massive amounts of information, often within an extremely short time frame. Second, the team will have to organize the material into a short, concise presentation which will be readily understandable by lawyers, judges, and juries who do not possess the same scientific education and technical training of the members of the risk assessment team. The ability to converse in technical terms may impress a professional colleague, but it will not likely sway a jury because they may not understand it. One of the greatest downfalls of expert witnesses is their inability to abandon their technical jargon, and communicate highly scientific concepts into plain, simple English lay people can understand. In evaluating the testimony of an expert, for example, a juror will consciously or unconsciously ask two fundamental questions: did I understand what the witness said? Was it believable?

If the juror does not understand the risk assessor's testimony, the team's project quickly degenerates into an extremely expensive and wholly academic exercise.

REFERENCES

Abraham, K.S., *Cleaning Up the Environmental Liability Insurance Mess*, 27 Val. U.L. Rev. 601, 603, 1993.

Abraham, K.S. *Environmental Liability Insurance Law: An Analysis of Toxic Tort and Hazardous Waste Insurance Coverage Issues*, 1, 1991.

Agency for Toxic Substances and Disease Registry (ATSDR), *Toxicological Profile for Cresols: o Cresol, p-Cresol, m-Cresol*, U.S. Dept. of Health and Human Services, Public Health Service, Atlanta, 1992.

Agency for Toxic Substances and Disease Registry (ATSDR), *Toxicological Profile for Polycyclic Aromatic Hydrocarbons (PAHs), Update*, U.S. Dept. of Health and Human Services, Public Health Service, Atlanta, 1994.

Assembly of Life Sciences (U.S.), Committee on Medical and Biologic Effects of Environmental Pollutants, Arsenic, National Academy of Sciences, Washington, 1977.

Clayton, G.D. and F.E. Clayton, Eds., *Patty's Industrial Hygiene and Toxicology, Third Revised Edition*, Vol. 2B, Toxicology, John Wiley & Sons, New York, 1981.

Gardner, W.M., *The British Coal-Tar Industry, its Origin, Development and Decline*, Williams and Norgate, London, 1915.

Grant, W.M., *Toxicology of the Eye*, Charles C. Thomas Publisher, Springfield, 1962.

Hamilton, A., *Industrial Poisoning in Making Coal-Tar Dyes and Dye Intermediates*, Bulletin of the U.S. Bureau of Labor Statistics, No. 280, Washington, 1921.

Harvey, G., *Polycyclic Aromatic Hydrocarbons: Chemistry and Carcinogenicity*, Cambridge University Press, Great Britain, 1919.

Haynes, W., *American Chemical Industry, Background and Beginnings*, D. Van Nostrand, New York, 1954.

Henry, S. A., Occupational cutaneous cancer attributable to certain chemicals in industry, *Br. Med. Bull.*, 4, 389, 1947.

IARC, *IARC Monographs on the Evaluation of the Carcinogenic Risk of Chemicals to Humans, Polynuclear Aromatic Compounds, Part 3*, Industrial Exposures, in Aluminum Production, Coal Gassification, Coke Production, and Iron and Steel Founding, Vol. 34, 1984.

Kennaway, E.L., On cancer-producing tars and tar-fractions, *J. Industrial Hygiene*, 5, 462, 1923.

Kennaway, E.L., On the cancer-producing factor in tar, *Br. Med. J.*, March 29, 564, 1924.

Kennaway, E.L., Experiments on cancer-producing substances, *Br. Med. J.*, 2, 3366, 1925.

Kennaway, E.L., LVII, Further experiments on cancer-producing substances, *The Biochemical J.*, 24, 497, 1930.

Kennaway E.L. and Kennaway, N. M., A further study of the incidence of cancer of the lung and larynx, *Br. J. Cancer*, 1, 260, 1947.

Legge, T., *Industrial Maladies*, Oxford University Press, London, 1934.

Lunge, G., *Coal-Tar and Ammonia*, D. Van Nostrand, New York, 1916.

Mathias, C.G.T., Occupational dermatoses, in *Occupational Medicine, Principles and Practical Applications*, 2nd ed., Zenz, C., Ed.,Year Book Medical Publishers, Chicago, 1988.

O'Donovan, W.J., Epitheliomatous ulceration among tar workers, *Br. J. Dermatology and Syphilis*, 1920, 219.

O'Donovan, W.J., Cancer of the skin due to occupation, report of the Internal Conference on Cancer, London — July 17–20, 1928, William Wood and Co., New York, 1928.

Partington, J.R., *A History of Chemistry*, Vol. 4, Macmillan & Co., New York, 1964.

Patty, F.A., Ed., *Industrial Hygiene and Toxicology*, Vol. I, InterScience, New York, 1948.

Rhodes, E.O., The chemical nature of coal tar, in *Chemistry of Coal Utilization*, vol. II, Lowrey, H.H., Ed., 1945.

Stacey, N.H., *Occupational Toxicology*, Taylor and Francis, Ltd., London, 1993.

Teleky, L., *History of Factory and Mine Hygiene*, Columbia University Press, New York, 1948.

Weyman, G., *Modern Gasworks Chemistry*, Benn Brothers, Ltd., London, 1922.

White, R.P. and Hay, J., Some recent inquiries and researches into the poisonous properties of naphthalene and the aromatic compounds, *The Lancet*, August 31, 582, 1901.

CHAPTER 31

Special Topics in Risk Assessment: Models and Uncertainties

Stephen G. Zemba and Laura C. Green

CONTENTS

I.   Introduction 551
II. Use of Models in Risk Assessment 552
     A.   Consider the Relevance of the Model 553
     B.   Review Input and Output Parameters 554
     C.   Check Equations and Calculations 555
     D.   Perform Reality Checks 556
III. Uncertainty in Risk Assessment 557
IV. Conclusion 560
     References 561

I. INTRODUCTION

Risk assessment* is an analytic tool intended to quantify possible threats to the environment and/or the public health. Once an academic instrument played by relatively few analysts, risk assessments are now routinely performed by hundreds of professionals and for many regulatory purposes, at least within the U.S.** Risk-based decisions and regulations abound, and given the current political climate, the use of risk assessment is likely to continue to expand.

---

* As used here, risk assessment means the quantitative assessment of risks to both human health and the environment due to exposure to chemical contaminants present in air, soil, water, and/or food.
** Formal, quantitative risk assessment is somewhat unique to the U.S. In Europe and elsewhere, analyses typically rely less on detailed modeling and extrapolations and more on the semiquantitative judgements of toxicologists and other scientists and engineers.

Risk assessment is distinguished from other environmental disciplines by its integration of the physical and biological sciences. A thorough understanding of a risk assessment typically requires detailed expertise in a variety of fields. In addition, risk assessment procedures have burgeoned in complexity. Ten or 15 years ago, a typical assessment included at most a few basic pathways — routes from a source to a person — and risk estimates were typically constructed as simple, order-of-magnitude estimates. Nowadays, risk assessments endeavor to account for all relevant avenues of exposure, to model in detail the environmental transport and fate of contaminants, and to describe and quantify the inherent variabilities and uncertainties in crucial variables.

In this chapter, we focus on two current issues in risk assessment. The first topic — the use of models in risk assessment — is motivated by the recent regulatory emphasis on multipathway risk assessment. The desire to quantify the movement of contaminants within the environment has been accompanied by proliferation of fate-and-transport models. For example, by utilizing only a few chemical-specific partitioning coefficients, a chain of models can be constructed to trace pollutants from air into water, soil, vegetation, and foodstuffs. Unfortunately, this propagation of models has not typically been accompanied by a commensurate level of testing and validation of their predictions. In addition, a wider audience of (and for) model users has, in some cases, led to improper or at least questionable applications of models. We thus discuss here some critical characteristics of models and suggest procedures that can be used in their selection and review.

The second topic we address is uncertainty. Probabilistic methods are an important advancement in risk assessment methodology; an illustration of a Monte Carlo assessment is included to demonstrate advantages and caveats of the method. Proper characterization of uncertainty is a focus of recent risk characterization policies released by the U.S. EPA (1995).

II. USE OF MODELS IN RISK ASSESSMENT

Broadly defined, a model is an abstraction used to mimic, describe, and/or predict some aspect of reality. Models may be used to extrapolate from data sets, to interpolate between data points, or to provide estimates where few or no measurements exist. Models permeate all facets of risk assessment. For example, the characterization of a series of measurements of contaminant concentrations in groundwater may assume an underlying statistical model. The definition of an exposure pathway requires the conceptualization of the process whereby a contaminant reaches a human or environmental receptor. The linearized multistage method, as another example, is the extrapolation model used by EPA to estimate the carcinogenic potency of a chemical in humans, given, typically, dose-response data from laboratory animal bioassays at doses vastly greater than those of interest for the risk assessment.

Most risk assessors think of models in terms of contaminant fate-and-transport models. Within this category, models range in complexity from simple analytical expressions that consider a few parameters to sophisticated "black box" algorithms

Table 1 Factors to Consider When Renewing Model Usage and Results

RELEVANCE	Is the model appropriate for the physical situation?
	Does the model meet regulatory requirements?
INPUT/OUTPUT	Do parameter values seem unusually large or small?
	Are parameters easily checked against standard, common values?
	Are units specified, and are they consistent among parameters?
	Are site-specific values used where possible?
CALCULATIONS	Can the results be reproduced from given equations and parameters?
REALITY CHECKS	Do model results exceed real-world constraints?
	Are there idiosyncrasies between model predictions and physical expectations?
	Do model predictions violate conservation of mass?

that simulate complex mathematical relationships. Models may be theoretically derived from underlying physical principles, empirically based on statistical inferences, or both.

The vast number, variety, and complexity of models available can make it difficult to:

- Select appropriate models for use in risk assessments
- Choose appropriate input values
- Review modeling results

There are many modeling pitfalls, and even the most experienced users and reviewers must exercise considerable caution. Analysts must be ever mindful both that no model is a perfect representation of the real world, and that considerable expertise (sometimes different from one's own) may be needed to differentiate between meaningful and meaningless results. Having reviewed many erroneous applications of models in risk assessments, we have developed or used a number of techniques to identify errors and/or inappropriate applications. We offer the following advice to modelers and reviewers (Table 1).

A. Consider the Relevance of the Model

Models should be applied only for situations for which they have been designed. Though it is obvious advice, we find it often ignored; perhaps because model users do not always review the derivation or limitations of the model before employing it. For example, a model designed to simulate groundwater flow should not be applied to the unsaturated zone. More often, however, poor judgement in model application involves more subtle errors. For example, Gaussian plume (GP) models are commonly applied to estimate the dispersion of contaminants in air. Most GP models simulate a plume that proceeds in a straight line from the point of pollutant release.

In many cases, such an assumption is appropriate. Application of a GP model in settings where winds change direction (such as a valley), however, can lead to the prediction of impacts at erroneous locations.

Models should include the essential physics needed to simulate a particular environmental situation. We prefer to use the simplest model possible that contains the basic factors that influence contaminant transport. Care must be taken, however, to insure that all relevant mechanisms have been included (of course, there are many settings in which all relevant mechanisms are unknown or unquantified; only additional research can help remedy such defects). For example, a soil model that neglects water-phase transport may grossly overpredict vapor diffusion rates.

Model selection also requires consideration of regulatory requirements. In many situations, agencies recommend the use of specific models. It can be easier (from a political perspective) to apply a less-than-ideal model to avoid costly regulatory review. For example, in permitting of air pollution sources, EPA provides a list of "approved" (though not necessarily fully validated) models that can be used for specific purposes. Use of alternate models instead may involve extensive justification; depending upon the discretion and/or tastes of the regulators, such alternates may or may not win approval.

Models vary greatly in complexity. As a general rule, simple screening models produce less accurate results than more elaborate (refined) models. This does not mean, however, that the most refined model should always be selected. Instead, model sophistication should be matched to the level (and certainty) of available information. Use of overly sophisticated models can produce misleading or inaccurate results if they are based on generic default parameters that may bear little or no similarity to site-specific conditions.

In selecting and evaluating a model for appropriateness, the best advice we can offer is to gather, read, and assimilate documentation regarding the model's basis and development. Such documentation may be found in users' manuals, technical reports, and the scientific literature. From these, one may glean a sense of whether the model's purposes and strengths are matched by the application at hand. One may also find in the literature additional or alternate models; and it is sometimes instructive to run these and compare results with those of the proffered model.

B. Review Input and Output Parameters

Models produce one or more outputs given one or more inputs. The GIGO (garbage in/garbage out) principle* is one of the cardinal rules of modeling, and has never been more relevant given the proliferation of user-friendly, menu-driven models that provide default parameters and run with little or no user interaction.

While we know of no systematic ways to avoid input/output errors, there are several measures that can be taken to reduce the possibility of errors. First and foremost, check the units. Since models represent mathematical equations, they require consistency among parameters, and generally demand precise specification of parameters. Factors of 1000 errors in using metric units are remarkably common.

* Also known as, "you can't make good applesauce with bad apples;" and note, that while some may find the applesauce made with bad apples to be tasty enough, gourmets won't be fooled.

Also, be careful in applying conversion factors, since there are unusual (but conventional within specified settings) parameters in use in various disciplines. For example, pollutant concentrations in stack gases are often expressed as mass per dry standard volume, which permits comparisons among facilities operating under wide ranges of conditions, but can lead to erroneous calculations of emission rates if used without the necessary conversions.

Methods to check units vary according to the type of model. In applying simple algebraic equations in spreadsheets, explicitly write out units by hand using the factor-label method to verify consistency. For computer algorithms, carefully review the users' guide to make sure that all inputs are specified in the units demanded by the program; be aware of the units specified for output parameters and use them accordingly. In reviewing reports, look for values expressed in suspicious units. For example, rates (flow, emission, etc.) should always be expressed per unit time (although they are frequently not).*

As a second step, review input parameters for consistency with your intuition. In some cases, this may require a greater (but useful) assimilation of the metric system. As examples, the sizes of physical objects such as farm fields and surface water bodies should be reasonable. Groundwater should move more slowly than surface water. Densities of liquids and gases should be of the same orders of magnitude as those of water (1000 kg/m^3) and air (1.2 kg/m^3), respectively. Values should not be outside allowable limits (e.g., 10,000 g/kg signals an error). Although these comparisons cannot be done for all parameters, one can learn with practice to recognize a wider range of outliers. For example, by reviewing only a few studies of subsurface transport, it becomes readily apparent that molecular diffusivities of contaminants in air and water are always of the order of 10^{-5} and 10^{-9} m^2/s, respectively.

Third, take steps to ensure that the most appropriate parameters have been selected. Where possible, choose site-specific values that reflect on-site measurements or regional characteristics. Be wary of default values and parameters apparently chosen arbitrarily from the literature. When assigning parameters, have another person peer review your choices — a second perspective is always useful. As an example, one of our colleagues was charged with selecting half-lives for various organic pollutants in a groundwater modeling study. For one of the pollutants, a half-life of 5 days, as reported in a handbook, was selected. It was readily apparent to another of us, however, that this value was unrealistic in our application, since the pollutant had persisted at the site for many years. Consequently, a longer half-life was chosen and justified.

C. Check Equations and Calculations

Peer-review is the best method of checking model calculations. Given the volume of calculations they encompass, numerical mistakes in risk assessments are common. We often identify mistakes through reviewing our own and others' work, and cannot overemphasize the need for checking. We find that electronic spreadsheets, which

---

* Atypical units sometimes reflect conventions and not errors. For example, hydrogeologists express pressures in terms of the equivalent feet of water, while meteorologists express pressures as inches of mercury.

might have been expected to reduce the chances of error, instead introduce opportunities for mistakes through the incorrect transcription of values, implementation of formulae, and copying and rearranging of cells.

We recommend two methods to check results. First, reproduce the calculations of a representative example (i.e., a single chemical) by hand. Often it is prudent to focus on the constituent most critical to the risk assessment.* Second, check the spreadsheet implementation of each equation to confirm correspondence. In addition, in cases where a number of similar calculations is performed, check that the values that differentiate the calculations have been used in the appropriate places. In large, complex spreadsheets, check one of a similar lot of calculations, and look through columns and rows to ensure that formulae and values have been correctly propagated. The use of named parameters for constants can facilitate the verification of equations and calculations.

D. Perform Reality Checks

The most difficult errors to identify are those that are numerically correct but violate physical limits or other characteristics of the real world. For example, we recently reviewed a model in which vapors of semivolatile organic compounds were assumed to diffuse from soil into the basement of an enclosure and be removed by dilution from outdoor air. In and of itself, vapor diffusion was modeled correctly, as was the simple box model used to predict indoor air concentrations. Coupled together, however, the two models produced indoor air concentrations that exceeded the soil-gas concentrations diffusing into the room.**

We promote two categories of reality checks. The first is common sense and the ability to apply insight to model results. Put another way, one should attempt to analyze one's (or others') results in terms of constraints. As examples:

- Concentrations of contaminants predicted in air should not exceed vapor pressures
- Dissolved concentrations of pollutants predicted in water should not exceed solubility limits
- Pollutants should dilute in concentration (unless there is a valid means of bioconcentration)

The second reality-checking technique involves developing an awareness of the implications of averaging-time. Most environmental fate-and-transport problems are time-dependent — meaning that concentrations change (either slowly or rapidly) with time. Paradoxically, most mathematical models are easier to solve if they are assumed not to depend upon time.

Problems in which the transient (time-dependent) portion of the solution is ignored are described as steady-state solutions.*** As a general rule, the plausibility

---

* Conversely, if a constituent was expected to be important in the risk assessment, but the results suggested otherwise, the modeling for that constituent should also be checked.
** Under the assumptions of the model, this result violates the second law of thermodynamics.
*** Most models based on analytical solutions are of the steady-state variety.

of steady-state solutions are best checked with a simple mass balance.* Two basic mass balance methods are available and easy to apply.

First, fluxes predicted by any model can be used to estimate depletion rates of contaminants as they leave a source of emission. For example, a modeled soil-gas emission rate can be used in conjunction with the contaminant concentration in soil to estimate the depth of contamination that would be depleted in a given amount of time. With this method, unrealistically high depletion values may suggest the need to use a different model, or could be used to interpret the results properly (e.g., the high rate of emission may be assumed to occur only over a short period of time in calculating potential exposure).

Second, mass balance checks can be useful in identifying fallacious results that otherwise appear correct. As an example, consider equilibrium partitioning models that are often used to estimate the distribution of contaminants (sorbed vs. water phases) in sediments.** Typically, these models assume that, within the water column of a lake or river, a contaminant distributes between a dissolved phase and a fraction that is attached to suspended particles. Some models extrapolate particle-bound concentrations in the water column to those in the sediment layer, in which the particle density is much greater. For compounds that partition heavily to particles, this extrapolation implies a substantial pollutant concentration in sediments. In such cases, it can be useful to use a mass balance model to calculate the implied loading to the sediment layer. To do so, one can estimate mass of the contaminant contained in the sediment layer as the concentration in sediment times the volume of the sediments (area times a given depth). By then dividing this mass by the contaminant loading to the water body, one obtains an estimate of the minimum time over which loading would have to occur to deposit that level of the contaminant to the sediment.*** Often these calculations demonstrate hundreds of years of contaminant loading to be present in sediments, which equilibrium partitioning models assume to be established immediately as the contaminant is introduced to the water body.

III. UNCERTAINTY IN RISK ASSESSMENT

Until recently, uncertainty of risk estimates received cursory attention. Ten years ago, uncertainty analyses typically consisted of a few paragraphs' discussion of the level of conservatism embodied in the risk estimates (and possible nonconservative or anticonservative assumptions were barely mentioned at all). The increased complexity of risk assessments, combined with a shift away from worst-case scenarios

---

* In most cases, mass balances are not automatically performed within a risk assessment since they are not the objective of modeling.
** The results of equilibrium partitioning models should always be checked carefully, since they frequently violate constraints imposed by mass transfer (e.g., the assumed movement of pollutants needed to reach equilibrium cannot be achieved on practical time scales).
*** The estimated time is a minimum because the calculation assumes that all of the pollutant loading in the water column is deposited to the sediment, and that none is discharged at the outlet to the water body.

towards "reasonable maximum" and "central tendency"* estimates, has magnified the potential importance of assessing uncertainties.

When considering uncertainty, it is important to distinguish it from variability, which is the measured (or at least expected and reasonably well-known) variation among members of a defined population that leads to potential differences in risks. Uncertainty is the combination of all other effects — those about which we are genuinely or at least largely ignorant — that lead to variations in risk estimates for the defined population. These definitions are necessarily somewhat arbitrary, since the distinction between the two can become blurred. As an example, the variation in individual bodyweight contributes to variability of a risk estimate in a population if the distribution of bodyweights in that population is known. For a given individual of a known bodyweight in the population, this variabililty is removed. However, for individuals of unknown bodyweights, the distribution of bodyweights contributes to the uncertainty in risk estimates.

The conventional way of accounting for both uncertainty and variability has been to choose point estimates for some parameters that come from relatively extreme ("conservative") values of the variability or uncertainty distributions characterizing some of those parameters, and central values for others (U.S. EPA, 1989). This method leads to the concept of selection of individuals or populations who are reasonably maximally exposed receptors. Problematically, the definition of reasonable maximum exposure is to a large degree arbitrary.

The limitations of conventional point risk estimates can be transcended (it is hoped) by probabilistic risk assessment methods, which are becoming increasingly popular. These techniques differ from conventional algorithms by explicitly considering and quantifying variability and uncertainty in parameter values and models. In doing, the results of the risk assessment are no longer limited to point estimates, but rather are a distribution of possible values. Compared with a point estimate, a distribution of values provides a greater amount of information. Percentile values, ranges, and other statistical measures can be used to characterize likelihood and uncertainty. For example, 95th percentile values may be used to characterize the risks due to reasonable maximum exposure; the risk due to average-case exposure can be gauged by the median, mean, or other "central tendency" portion of the distribution. Properly combined with demographic information, probabilistic methods can be used to derive estimates of population-weighted risk and distributions of risk to specific segments of the population.

The Monte Carlo method is perhaps the best-known probabilistic technique. In a Monte Carlo simulation, distributions are specified for each parameter that account for both variability and uncertainty.** A single estimate of risk is computed by: (1) selecting random values for each parameter from the distributions and accounting for any correlations between variables (e.g., food consumption rates may be partially

---

* Neither "reasonable maxima" nor "central tendencies" have unique or unambiguous definitions; one analyst's reasonable maximum" is another's extreme values and still another's moderate assumption. Even central tendency estimates vary widely among analysts given the same nominal constraints.
** Of course, it is difficult to fully characterize uncertainty since, by definition, one is uncertain about one's uncertainties. In practice, one guesses at a practical range of uncertainty about various parameters. Risk assessments performed iteratively, perhaps after measurements have been taken or research completed, may be characterized by smaller ranges of uncertainty than those at present.

related to body weight); and (2) calculating the value of risk (using deterministic models and relationships) with the set of random values for that instance. By conducting a large number of determinations — on the order of 20,000 — a stable distribution of risk values is generated from which it is possible to determine relevant statistical measures. Of course, these determinations are performed via computer.

An example of a practical Monte Carlo application is detailed in Zemba et al. (1994), which estimates risk of contracting cancer incurred by any of the 400,000 California residents who ingest water contaminated with the pesticide dibromochloropropane (DBCP). Individuals using the contaminated water may be exposed to DBCP principally through drinking, showering (DBCP escapes into the air, and may be inhaled), and through dermal contact (principally in showers and baths). Different individuals drink different quantities of water, and have differing showering habits, so that their exposures differ even for similar concentrations in the water supply.

Distributions were constructed to consider variabilities and uncertainties in each important parameter, including:

- DBCP concentrations in the water supply
- Rates and frequencies of contact (water ingestion, showering/bathing habits, etc.)
- Physiological parameters (body weight, skin area, etc.)
- Dose-response characteristics of the carcinogenic potency of DBCP

In this example, the main source of variability between individuals arises from the distribution of concentration in the water supply, and the main source of uncertainty is in the estimate of human carcinogenic potency.* Figure 1 shows the results of our example simulation. The cumulative distribution for individual risk is shown as a solid line (left scale), and the differential distribution as a dotted line (arbitrary scale). The location of the point estimate (as calculated by conventional deterministic methods) is also shown (it is at approximately the 99th percentile of the distribution). The 50th percentile value, which represents the median estimate of cancer risk to any individual in the specified population, is more than two orders of magnitude lower than the point risk estimate.

The distribution shown in Figure 1 must be interpreted carefully within the assumptions of the simulation — a caveat that applies to all Monte Carlo analyses. For each value of risk (x-axis value) the cumulative curve shows the probability (y-axis value) that the lifetime risk to a randomly chosen exposed individual is smaller than the given value. The 95th percentile, for example, is a 95% upper confidence limit in this sense: we are 95% certain that any of the 400,000 Californians modeled here are at no more than an 8 in 10,000 excess lifetime risks of developing cancer by dint of their waterborne exposures to DBCP. The distribution of risk estimates shown applies only to

---

* Uncertainties in risk assessments for chemical carcinogens are essentially always dominated by uncertainties in the potency of those chemicals as human carcinogens at the typical low levels of interest for environmental risk assessment. In this example, the specified uncertainty in our estimation of the carcinogenic potency of DBCP is intentionally understated since we follow certain U.S. EPA conventions for the interpretation and extrapolation of dose-response information. The actual dose-response relationship between low-level exposures to DBCP and effect on human carcinogenesis is essentially completely unknown. Risk assessments for occupational exposures to established carcinogens, on the other hand, will typically involve a tighter distribution of uncertainty about the carcinogenic potency term.

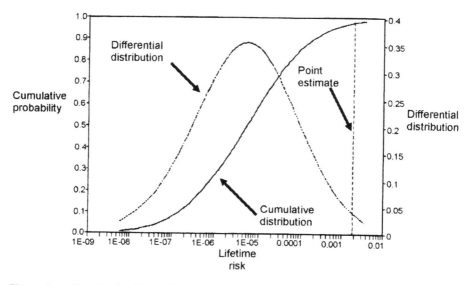

Figure 1 Results of a Monte Carlo simulation for individual risk from DBCP.

randomly chosen individuals within the population — it does not correspond to the uncertainty distribution for an individual exposed to a known concentration, for example, or an individual with other specified relevant characteristics.

Consideration of the uncertainties inherent in the cancer potency is a critical feature of this example that has rarely been considered in Monte Carlo assessments. Since most cancer potency estimates are derived in a conservative manner, the failure to consider the potency explicitly in an uncertainty analysis introduces a conservative bias in the simulation. Even in the context of conventional point-risk estimation, central tendency exposure estimates coupled with conservatively derived dose-response parameters have been erroneously interpreted as central tendency risk estimates. Given how little we still know about the causes of human cancer (with a few notable exceptions), using point-estimate values for carcinogenic potency strikes many observers as audacious.

IV. CONCLUSION

The uses and misuses of models and the proper treatment of uncertainty are two topics of great importance to risk assessment. Moreover, the two topics are related. In many risk assessments, models are a significant (perhaps the greatest) source of uncertainty. The importance of uncertainty will be magnified (and must be recognized) as model-intensive, multipathway risk assessments become more commonplace. Coincidentally, the best way to assess these uncertainties is through the use of probabilistic techniques such as the Monte Carlo analysis described here.

We expect that future risk assessments will become even more complex, but will (perhaps after a few years of working things through) provide a significantly greater amount of usable information to regulators, decision makers, and other managers of risk. Uncertainties and variabilities are daunting, but need not be paralyzing. It is hoped that careful, holistic, and probabilistic assessments of risk will afford more equitable and efficient means of minimizing threats to the environment and the public health.

REFERENCES

U.S. Environmental Protection Agency, *Policy for Risk Characterization*, included as part of a memorandum from Carol Browner to various EPA administrators, dated March 21, 1995.

U.S. Environmental Protection Agency, *Risk Assessment Guidance for Superfund, Vol. I: Human Health Evaluation Manual (Part A), Interim Final*, Office of Emergency and Remedial Response, Washington, 1989.

Zemba, S.G., Green, L.C., and Crouch, E.A.C., Risk assessment of combuster stack emissions, In *National Waste Processing Conference Proceedings, American Society of Mechanical Engineers*, Boston, 1994.

APPENDIX A

Risk Assessment Resources Guide*

AIR

Anderson, E.L. and Albert, R.E., *Risk Assessment and Indoor Air Quality*, Lewis Publishers, Boca Raton, FL, 1999.

Belluck, D.A. and Criswell, R., Comparative Risk Analysis Framework for Ash Utilization, in *Proceedings of the Fourth U.S.-Dutch International Symposium "Comparative Risk Analysis and Priority Setting for Air Pollution Issues,"* Keystone, CO, 1993.

Booz-Allen and Hamilton, Inc., *RCRA/UST, Superfund, and Epcra Hotline Training Module, Introduction to Accidental Release Prevention Program (CAA Section 112(R); 40 CFR Part 68)*, Bethesda, MD; Office of Solid Waste and Emergency Response, U.S. Environmental Protection Agency, Washington, 1996, EPA-68-W0-0039; EPA/550/B-96/005.

California Air Pollution Control Officers Association, *Capcoa Air Toxics "Hot Spot" Program*, Cameron Park, CA, 1990.

California Air Resources Board, *Adoption of a System for the Classification of Organic Compounds According to Photochemical Reactivity*, Staff Report 76-3-4, Sacramento, CA, 1986.

Code of Federal Regulations of the United States, Air Quality Checklist, Appendix W, 40 CFR, Part 51.

Committee on Health Effects of Waste Incineration, *Waste Incineration and Public Health*, National Research Council, 2000.

Darnall, K.R., Reactivity scale for atmospheric hydrocarbons based on reaction with hydroxyl radicals, *Environ. Sci. Technol.*, 10, 692, 1986.

Gad, S.C., *Combustion Toxicology*, CRC Press, Boca Raton, FL, 1990.

Hendry, D.G. and Kenley, R.A., *Atmospheric Reaction Products of Organic Compounds*, Office of Toxic Substances, U.S. Environmental Protection Agency, Washington, 1979, EPA-560/12-79-001.

Howitt, A.M., *Linking Transportation and Air Quality Planning: Implementation of the Transportation Conformity Regulations in 15 Nonattainment Areas*, Taubman Center for State and Local Government, John F. Kennedy School of Government, Harvard University, Office of Air and Radiation, U.S. Environmental Protection Agency, Washington, 1999, EPA/20-R-99-011.

---

* Government, private industry, academic institutions, nonprofit organizations, international organizations, and others have Internet sites that may contain extensive listings of published documents, documents found only on Internet sites, links to other sites, and searchable databases.

Jarabek, A.M. et al., *Interim Methods for Development of Inhalation Reference Doses*, U.S. Environmental Protection Agency, Research Triangle Park, 1989, EPA/600/8-88/006f.

Lee, C.W. et al., *Research on Emissions and Mitigation of Pop's from Combustion Sources*, Air Pollution Prevention and Control Div., U.S. Environmental Protection Agency, Research Triangle Park, NC, 1999, EPA/600/A-99/081.

Meardon, K., *Cancer Risk from Outdoor Exposure to Air Toxics, Vol. 1*, Pacific Environmental Services, Inc., Durham, NC, Office of Air Quality Planning and Standards, U.S. Environmental Protection Agency, Research Triangle Park, NC, 1990, EPA-68-02-4393; EPA/450/1-90/004a.

National Research Council, *Human Exposure Assessment for Airborne Pollutants: Advances and Opportunities*, National Academy of Sciences, Board on Environmental Studies and Toxicology, Committee on Advances in Assessing Human Exposure to Airborne Pollutants,Washington, 1991.

Neuman H.G. et al., Markers of Exposure to Aromatic Amines and Nitro-Pah, *Arch. Toxicol. Suppl.*, 20, 179, 1998.

Pacific Environmental Services, Inc., *Cancer Risk from Outdoor Exposure to Air Toxics, Vol. 2, Appendices*, Pacific Environmental Services, Inc., Durham, NC, Office of Air Quality Planning and Standards, U.S. Environmental Protection Agency, Research Triangle Park, NC, 1990, EPA-68-02-4393; EPA/450/1-90/004b.

Peach, H.G., *Air Quality and Human Health*, Environment Australia, Canberra, 1997.

Pierce, T.E. et al., *PTPLU — A Single Source Gaussian Dispersion Algorithm User's Guide*, Aerocomp, Inc., Costa Mesa, CA; Environmental Sciences Research Laboratory, U.S. Environmental Protection Agency, Research Triangle Park, NC, 1982, EPA-68-02-3442; EPA-600/8-82-014.

Research Triangle Institute, *Risk Assessment to Support the Development of Technical Standards for Emissions from Combustion Units Burning Hazardous Wastes: Background Information Document, Final Report*, Industrial Economics Incorporated, Cambridge, MA, 1996.

Rimer, K., *Guidance for Review of High Risk Point Sources under Section 112 of the 1990 Clean Air Act Amendments*, Office of Air Quality Planning and Standards, U.S. Environmental Protection Agency, Research Triangle Park, NC, EPA-453/R-93-039.

Roberts, S.M. et al., *Hazardous Waste Incineration: Evaluating the Human Health and Environmental Risks*, Lewis Publishers, Boca Raton, FL, 1999.

Sample, B.E., *Methods for Field Studies of Effects of Military Smokes, Obscurants, and Riot-Control Agents on Threatened and Endangered Species; Vol. 2: Methods for Assessing Ecological Risks*, U.S. Army Corps of Engineers, Construction Engineering Research Laboratories, Champaign, IL, 1997.

U.S. Environmental Protection Agency, *A Descriptive Guide to Risk Assessment Methodologies for Toxic Air Pollutants*, Office of Air Quality Planning and Standards, Research Triangle Park, NC, 1993, EPA-453/R-93-038.

U.S. Environmental Protection Agency, *Air/Superfund National Technical Guidance Study Series*, Office of Air Quality Planning and Standards, Research Triangle Park, NC, Multi-Vol. Series, 1989.

U.S. Environmental Protection Agency, *Air Pathway Analysis*, Office of Emergency and Remedial Response, Washington, 1992, EPA/540/S-92/013.

U.S. Environmental Protection Agency, *Air RISC, a Short Course in Exposure for Air Toxics*, Air Risk Information Support Center, Research Triangle Park, NC, 1995.

U.S. Environmental Protection Agency, *Cancer Risk from Outdoor Exposure to Air Toxics: Final Report*, Office of Air Quality Planning and Standards, Washington, 1990, EPA-450/1-90-004a;EPA-450/1-90-004b.

RISK ASSESSMENT RESOURCES GUIDE

U.S. Environmental Protection Agency, *Contingency Analysis Modeling for Superfund Sites and Other Sources*, Office of Air Quality Planning and Standards, Research Triangle Park, NC, 1993, EPA-454/R-93-001.

U.S. Environmental Protection Agency, *EPA Risk Management Programs, 40 CFR Part 68*, Office of Emergency and Remedial Response, Washington, 1998, EPA/540/R-97/039.

U.S. Environmental Protection Agency, *Hazardous Air Pollutants, Profiles of Noncancer Toxicity from Inhalation Exposures*, Office of Research and Development, Washington, 1993, EPA/600/R-93/142.

U.S. Environmental Protection Agency, *Indoor Air Assessment: a Review of Indoor Air Quality Risk Characterization Studies*, Environmental Criteria and Assessment Office, Research Triangle Park, NC, 1991, EPA/600/8-90/044; ECAO-R-0314.

U. S. Environmental Protection Agency, *Integrated Human Exposure Committee Commentary on Indoor Air Strategy*, Office of the Administrator, Science Advisory Board, Washington, 1998, EPA SAB-IHEC-COM-98-001.

U.S. Environmental Protection Agency, *Inventory of U.S. Greenhouse Gas Emissions and Sinks: 1990-1996*, Office of Policy, Planning and Evaluation, Washington, 1998, EPA 236-R-98-006.

U.S. Environmental Protection Agency, *Methodology for Assessing Health Risks Associated with Indirect Exposure to Combustor Emissions: Project Summary*, Environmental Criteria and Assessment Office, Cincinnati, 1991, EPA/600/S6-90/003.

U.S. Environmental Protection Agency, *Methodology for Assessing Health Risks Associated with Indirect Exposure to Combustor Emissions, Interim Final*, Office of Health and Environmental Assessment, Washington, 1990, EPA/600/6-90/003.

U.S. Environmental Protection Agency, *Methods for Derivation of Inhalation Reference Concentrations and Application of Inhalation Dosimetry*, Environmental Criteria and Assessment Office, Office of Health and Environmental Assessment, Office of Research and Development, Washington, 1994, EPA/600/8-90-066f.

U.S. Environmental Protection Agency, *Risk Assessment for Toxic Air Pollutants: a Citizen's Guide*, Air Risk Information Support Center, Research Triangle Park, NC, 1991, EPA 450/3-90-024.

U.S. Environmental Protection Agency, *Risk Management Program Guidance for Ammonia Refrigeration (40 CFR Part 68)*, Office of Solid Waste and Emergency Response, Chemical Emergency Preparedness and Prevention Office, Washington, 1998, EPA 550-B-98-014.

U.S. Environmental Protection Agency, *Risk Management Program Guidance for Propane Users and Small Retailers (40 CFR Part 68)*, Office of Solid Waste and Emergency Response, Chemical Emergency Preparedness and Prevention Office, Washington, 1998, EPA 550-B-98-022.

U.S. Environmental Protection Agency, *Risk Management Program Guidance for Warehouses (40 CFR Part 68)*, Office of Solid Waste and Emergency Response, Chemical Emergency Preparedness and Prevention Office, Washington, 1999, EPA 550-B-99-004.

U.S. Environmental Protection Agency, *Screening Level Ecological Risk Assessment Protocol for Hazardous Waste Combustion Facilities; Peer Review Draft, Vol. 1*, Office of Solid Waste and Emergency Response, Washington, 1999, EPA/530/D-99/001a.

U.S. Environmental Protection Agency, *SAB Report: Review of the RFC Methods Case Studies; Review of Case Studies Associated with the Document Methods for Derivation of Inhalation Reference Concentrations and Application of Inhalation Dosimetry*, Science Advisory Board, Washington, 1998, EPA-SAB-EHC-99-002; EPA/600/8-90/066f.

U.S. Environmental Protection Agency, *SAB Report: Review of the USEPA'S Report to Congress on Residual Risk*, Science Advisory Board, Washington, 1998, EPA-SAB-EC-98-013.

U.S. Environmental Protection Agency, *The Total Integrated Methodology, Implementation of the Trim Conceptual Design Through TRIM.FaTE Module, a Status Report*, Office of Air Quality Planning and Standards, Research Triangle Park, NC, 1998, EPA-452/R-98-001.

AQUATIC BIOTA

Environmental Management & Enrichment Facilities, *Radiological Benchmarks for Screening Contaminants of Potential Concern for Effects on Aquatic Biota at Oak Ridge National Laboratory, Oak Ridge, Tennessee*, Bechtel Jacobs Company LLC, Oak Ridge, TN, 1998, BJC/OR-80.

Hull, R.N. and Suter, G.W., *Toxicological Benchmarks for Screening Contaminants of Potential Concern for Effects on Sediment-Associated Biota: 1994 Revision*, Oak Ridge National Laboratory, Oak Ridge, TN, 1994, ES/ER/TM-95/R1.

Lackey, R.T., *Fisheries Management: Integrating Societal Preference, Decision Analysis, and Ecological Risk Assessment*, Western Ecology Division, National Health and Environmental Effects Research Laboratory, Corvallis, OR, 1997, EPA/600/A-97/004.

Laws, E.A., *Aquatic Pollution, an Introductory Text*, John Wiley & Sons, New York, 1981.

Olsen, A.R. et al., *Frequency Analysis of Pesticide Concentrations for Risk Assessment (FRANCO MODEL)*, Battelle Pacific Northwest Laboratories, Richland, WA; Environmental Research Laboratory, Athens GA, 1982, EPA-68-03-2613; EPA-600/3-82-044.

Onishi, Y. et al., *Methodology for Overland and Instream Migration and Risk Assessment of Pesticides*, Battelle Pacific Northwest Laboratories, Richland, WA; Environmental Research Laboratory, Athens, GA, 1982, EPA-68-03-2613; EPA-600/3-82-024.

Rand, G.M. and Petrocelli, S.R., Eds., *Fundamentals of Aquatic Toxicology, Methods and Applications*, Hemisphere Publishing, Washington, 1985.

Suter, G.W. II and Tsao, C.L., *Toxicological Benchmarks for Screening of Potential Contaminants of Concern for Effects on Aquatic Biota on Oak Ridge Reservation: 1996 Revision*, Oak Ridge National Laboratory, Oak Ridge, TN, 1996, ES/ER/TM-96/R2.

BIOACCUMULATION

Arts, M.T. and Wainman, B.C., Eds., *Lipids in Freshwater Ecosystems*, Springer, New York, 1999.

Battelle, *Risk Analysis to Support Standards for Lead in Paint, Dust, and Soil, Vol. 1 & 2*, Columbus, OH, Office of Pollution, Prevention, and Toxics, U.S. Environmental Protection Agency, Washington, 1998, EPA/747/R-97/006.

Crompton, T. R., *Occurrence and Analysis of Organometallic Compounds in the Environment*, John Wiley & Sons, Chichester, England; New York, 1998.

De Rosa, C.T. et al., Dioxin and dioxin-like compounds in soil, part 1: ATSDR interim policy guideline, *Toxicol. and Industrial Health*, 13(6), 759, 1997.

European Centre for Ecotoxicology and Toxicology of Chemicals, *Role of Bioaccumulation in Environmental Risk Assessment of the Aquatic Environment and Related Food Webs*, Brussels, Belgium, 1995, OCLC 35703592.

Hall, L.W., *Acute and Chronic Toxicity of Copper to the Estuarine Copepod Eurytemora Affinis: Final Report*, U.S. Environmental Protection Agency, Chesapeake Bay Program, Annapolis, 1998, EPA 903/R/98/005.

Hwang, S.T., *Multimedia Approach to Risk Assessment for Contaminated Sediments in a Marine Environment*, Office of Health and Environmental Assessment, U.S. Environmental Protection Agency, Washington, 1987, EPA/600/D-87/350.

Ingersoll, C.G. et al., Chapter 18: Workgroup summary report on an uncertainty evaluation of measurement endpoints used in sediment ecological risk assessments, in *Proc. Pellston Workshop (22nd) Soc. Environ. Toxicol. and Chem. (SETAC)*, Pacific Grove, CA, 1995.

Ingersoll, C.G., Ankley, G.T., Buado, R., Burton, G.A., and Lick, W., Workgroup Summary Report on an Uncertainty Evaluation of Measurement Endpoints Used in Sediment Ecological Risk Assessments, Corvallis Environmental Research Laboratory, Corvallis, OR, National Biological Service, Onalaska, WI, 1996, EPA/600/A-96/097.

Lee, H. et al., *Computerized Risk and Bioaccumulation System (Version 1.0)*, Environmental Research Laboratory, Narragansett, RI, 1991, EPA/600/3-91/069.

Marquis, P.J. et al., *Analytical Methods for a National Study of Chemical Residues in Fish, I, Polychlorinated Dibenzo-P-Dioxins/Dibenzofurans*, Environmental Research Laboratory, Duluth, MN, Center for Lake Superior Environmental Studies, Wisconsin University, Superior, WI, 1994, EPA/600/J-94/431.

Marquis, P.J. et al., *Analytical Methods for a National Study of Chemical Residues in Fish, II, Pesticides and Polychlorinated Biphenyls*, Environmental Research Laboratory, Duluth, MN, Center for Lake Superior Environmental Studies, Wisconsin University, Superior, WI, 1994, EPA/600/J-94/432.

Maukerjee, D. and Cleverly, D.H., *Risk from Exposure to Polychlorinated Dibenzo-P-Dioxins and Dibenzofurans Emitted from Municipal Incinerators*, Environmental Criteria and Assessment Office, U.S. Environmental Protection Agency, Cincinnati, 1987, EPA/600/J-87/516.

Mcfarland, V.A., *Evaluation of Field-Generated Accumulation Factors for Predicting the Bioaccumulation Potential of Sediment-Associated Pah Compounds*, U.S. Army Corps of Engineers, Waterways Experiment Station, Vicksburg, MS, 1996, Technical Report D-95-2.

Mukerjee, D., Ris, C.H., and Schaum, J., *Health Risk Assessment Approach for 2,3,7,8-Tetrachlorodibenzo-P-Dioxin*, Environmental Criteria and Assessment Office, U.S. Environmental Protection Agency, Cincinnati, OH, 1985, EPA/600/8-85/013.

Palmer, K.R.S., *Bioaccumulation of Trans- and Cis-Chlordane and Trans- and Cis-Nonachlor in Lake Michigan Plankton*, thesis (M.S.), University of Minnesota, Minneapolis, 1998.

Sample, B.E. et al., *Development and Validation of Bioaccumulation Models for Small Mammals*, Oak Ridge National Laboratory, Oak Ridge, TN, 1998, ES/ER/TM-219.

Sample, B.E. et al., *Development and Validation of Bioaccumulation Models for Earthworms*, Oak Ridge National Laboratory, Oak Ridge, TN, 1998, ES/ER/TM-220.

Skoglund, R.S., *Modeling the Bioaccumulation of PCBS in Phytoplankton*, Ph.D. thesis, University of Minnesota, Minneapolis, 1997.

Smith, D.J., Gingerich, W.H., and Beconi-Barker, M.G., Eds., *Xenobiotics in Fish*, Kluwer Academic/Plenum Publishers, New York, 1999.

University of Maryland Agricultural Experiment Station, *The Influence of Salinity on the Chronic Toxicity of Atrazine to Sago Pondweed: Filling a Data Need for Development of an Estuarine Chronic Condition*, U.S. Environmental Protection Agency, Chesapeake Bay Program, Annapolis, 1998, EPA 903-R-98-020.

U.S. Army Corps of Engineers, *Summary of a Workshop on Interpreting Bioaccumulation Data Collected During Regulatory Evaluations of Dredged Material*, Waterways Experiment Station, Vicksburg, MS, 1998, D-96-1.

U.S. Department of Commerce, *Toxicokinetics of Polychlorinated Biphenyl Congeners by Diporeia Spp.: Effects of Temperature and Organism Size*, National Oceanic and Atmospheric Administration (NOAA), Environmental Research Laboratories, Great Lakes Environmental Research Laboratory, Ann Arbor, MI, 1998, ERL GLERL, 106.

U.S. Environmental Protection Agency, *Guidance for Assessing Chemical Contaminant Data for Use in Fish Advisories, Vol. 2, Risk Assessment and Fish Consumption Limits*, 2nd Edition, Office of Water, Washington, 1997, EPA/823/B-97/009.

U.S. Environmental Protection Agency, *Guidance for Assessing Chemical Contaminant Data for Use in Fish Advisories, Vol. 3, Overview of Risk Management*, Office of Water, Washington, 1996, EPA/823/B-96/006.

U.S. Environmental Protection Agency, *Guidelines for Reproductive Toxicity Risk Assessment*, Office of Research and Development, Washington, 1996, EPA/630/R-96/009a.

U.S. Environmental Protection Agency, *Report on the Peer Consultation Workshop on Selenium Aquatic Toxicity and Bioaccumulation*, Office of Water, Washington, 1998, EPA-822-R-98-007.

Willard, R.E., *Assessment of Cadmium Exposure and Toxicity Risk in an American Vegetarian Population*, Loma Linda University Medical Center, CA; Health Effects Research Laboratory, U.S. Environmental Protection Agency, Research Triangle Park, NC, 1985, EPA-R-806006; EPA/600/1-85/009.

BIOLOGICAL CRITERIA

Battelle, *Biological Criteria: Guide to Technical Literature*, Office of the Assistant Administrator for Water, U.S. Environmental Protection Agency, Washington, 1991, EPA-68-03-3534; EPA 440/5-91-004.

Brooks, L.T. and Carr, R.S., *Ambient Aquatic Life Water Quality Criteria for Di-2-Ethylhexyl Phthalate*, Wisconsin University, Superior, WI, Battelle Ocean Sciences, Duxbury, MA, Environmental Research Laboratory, Duluth, MN, 1987, EPA/440/5-87/013.

Environmental Research Laboratory, *Ambient Aquatic Life Water Quality Criteria for Aniline (CAS Registry Number 62-53-3)*, Narragansett, RI; Wisconsin University, Superior, WI, Duluth, MN, 1993.

Environmental Research Laboratory, *Ambient Aquatic Life Water Quality Criteria for Hexachlorobenzene*, Duluth, MN, 1988.

Environmental Research Laboratory, *Ambient Aquatic Life Water Quality Criteria for Selenium*, Office of Water Regulations and Standards, Criteria and Standards Division, U.S. Environmental Protection Agency, Duluth, MN, Narragansett, RI, 1987, EPA-440/5-87-006.

Environmental Research Laboratory, *Ambient Aquatic Life Water Quality Criteria for Silver*, Environmental Research Laboratory, Narragansett, RI, U.S. Environmental Protections Agency, Washington, 1987, EPA/440/5-87/011.

Gibson, G.R. et al., *Biological Criteria: Technical Guidance for Streams and Small Rivers*, Office of Science and Technology, Health and Ecological Criteria Division, U.S. Environmental Protection Agency, Washington, 1996, EPA/822/B-96/001.

Griffith, G.E., Omernik, J.M., Azevedo, S.H., *Ecoregions of Tennessee*, National Health and Environmental Effects Research Laboratory, Corvallis, OR, 1997, EPA/600/R-97/022.

Griffith, G.E. et al., *Ecoregions and Subregions of Iowa: a Framework for Water Quality Assessment and Management*, Corvallis Environmental Research Laboratory, Corvallis, 1994, EPA/600/J-94/466.

Griffith, G.E. et al., *Massachusetts Ecological Regions Project*, U.S. Environmental Protection Agency; Corvallis Environmental Research Laboratory, 1994, EPA/600/A-94/111.

Kepner, W.G., *Arid Ecosystems 1992 Pilot Report, Environmental Monitoring and Assessment Program*, Environmental Monitoring Systems Laboratory, Office of Research and Development, U.S. Environmental Protection Agency, Research Triangle Park, NC, 1994, EPA/620/R-94/015.

Larson, L. and Hyland, J., *Ambient Aquatic Life Water Quality Criteria for Zinc*, Office of Research and Development, Environmental Research Laboratories, U.S. Environmental Protection Agency, Washington, 1987, EPA-440/5-87-003.

Oklahoma Water Resources Board, *Standard Operating Procedures for Stream Assessments and Biological Collections Related to Biological Criteria in Oklahoma*, Oklahoma Water Resources Board, Oklahoma City, OK, 1999, OWRB Technical Report 99-3.

Simon, T.P., *Development of Index of Biotic Integrity Expectations for the Ecoregions of Indiana: II, Huron-Erie Lake Plain*, Water Quality Surveillance and Standards Branch, U.S. Environmental Protection Agency, Chicago, 1994, EPA/905/R-92/007.

Simon, T.P., *Development of Index of Biotic Integrity Expectations for the Ecoregions of Indiana: III, Northern Indiana Till Plain*, U.S. Environmental Protection Agency, Chicago, 1997, EPA-905-R-96-002.

Stephan, C.E. et al., *Interim Guidance on Determination and Use of Water-Effect Ratios for Metals*, Office of Water, U.S. Environmental Protection Agency, Washington, 1994, EPA/823/B-94/001.

Telliard, W.A., *Monitoring Trace Metals at Ambient Water Quality Criteria Levels*, Engineering and Analysis Division, U.S. Environmental Protection Agency, Washington, 1995, EPA/821/R-94/027.

University of Wisconsin-Superior, *Ambient Aquatic Life Water Quality Criteria 'Tributyltin,'* Great Lakes Environmental Center, Traverse City, MI, Office of Water, U.S. Environmental Protection Agency, Washington, 1997, EPA-68-C6-0036; EPA-822-D-97-001.

U.S. Environmental Protection Agency, *Ambient Water Quality Criteria Derivation Methodology Human Health: Technical Support Document, Final Draft*, Washington, 1998, EPA/822-B-98-005.

U.S. Environmental Protection Agency, *Ambient Aquatic Life Water Quality Criteria for 2,4-Dimethylphenol: Fact Sheet*, Office of Water, Washington, 1993, EPA-822-F-93-002.

U.S. Environmental Protection Agency, *Ambient Aquatic Life Water Quality Criteria for Aniline: Fact Sheet*, U.S. Environmental Protection Agency, Office of Water, Washington, 1993, EPA-822-F-93-001.

U.S. Environmental Protection Agency, *Ambient Aquatic Life Water Quality Criteria for 2,4-Dimethylphenol (Cas Registry Number 105-67-9)*, Office of Water, Washington, 1993.

U.S. Environmental Protection Agency, *Water Quality Criteria Documents for the Protection of Aquatic Life in Ambient Water*, Office of Water, Washington, 1996, EPA 820-B-96-001.

U.S. Environmental Protection Agency, *Ambient Aquatic Life Water Quality Criteria for Ammonia (Saltwater)*, Office of Water Regulations and Standards, Criteria and Standards Division, Washington, 1989, EPA 440/5-88-004.

U.S. Environmental Protection Agency, *Ambient Aquatic Life Water Quality Criteria for Aluminum*, Office of Water Regulations and Standards, Criteria and Standards Division, Washington, 1988, EPA 440/5-86-008.

U.S. Environmental Protection Agency, *Ambient Aquatic Life Water Quality Criteria for Antimony (III) Draft*, Office of Research and Development, Environmental Research Laboratories, Washington, 1988.

U.S. Environmental Protection Agency, *Ambient Aquatic Life Water Quality Criteria for Chloride*, Office of Research and Development, Environmental Research Laboratory, Washington, 1988, EPA 440/5-88-001.

U.S. Environmental Protection Agency, *Ambient Aquatic Life Water Quality Criteria for Hexachlorobenzene Draft*, Office of Research and Development, Environmental Research Laboratories, Washington, 1988.

U.S. Environmental Protection Agency, *Ambient Aquatic Life Water Quality Criteria for Phenanthrene Draft*, Office of Research and Development, Environmental Research Laboratories, Washington, 1988.

U.S. Environmental Protection Agency, *Ambient Aquatic Life Water Quality Criteria for 2,4,5-Trichlorophenol Draft*, Office of Research and Development, Environmental Research Laboratories, Washington, 1987.

U.S. Environmental Protection Agency, *Ambient Aquatic Life Water Quality Criteria for Di-2-Ethylhexyl Phthalate Draft*, Office of Research and Development, Environmental Research Laboratories, Washington, 1987.

U.S. Environmental Protection Agency, *Ambient Aquatic Life Water Quality Criteria for Nickel*, Office of Water Regulations and Standards, Criteria and Standards Division, Washington, 1986, EPA 440/5-86-004.

U.S. Environmental Protection Agency, *Ambient Aquatic Life Water Quality Criteria for Parathion*, Office of Water Regulations and Standards, Criteria and Standards Division, Washington, 1986, EPA 440/5-86-007.

U.S. Environmental Protection Agency, *Ambient Aquatic Life Water Quality Criteria for Pentachlorophenol*, Office of Water Regulations and Standards, Criteria and Standards Division, Washington, 1986, EPA 440/5-86-009.

U.S. Environmental Protection Agency, *Ambient Aquatic Life Water Quality Criteria for Toxaphene*, Office of Water Regulations and Standards, Criteria and Standards Division, Washington, 1986, EPA/440/5-86-006.

U.S. Environmental Protection Agency, *Ambient Aquatic Life Water Quality Criteria for Chlorpyrifos*, Office of Water Regulations and Standards, Criteria and Standards Division, Washington, 1986, EPA/440/5-86/005.

U.S. Environmental Protection Agency, *Ambient Aquatic Life Water Quality Criteria for Dissolved Oxygen: (Freshwater)*, Office of Water Regulations and Standards, Criteria and Standards Division, Washington, 1986, EPA 440/5-86-003.

U.S. Environmental Protection Agency, *Ambient Aquatic Life Water Quality Criteria for Cyanide*, Office of Water Regulations and Standards, Criteria and Standards Division, Washington, 1985, EPA 440/5-84-028.

U.S. Environmental Protection Agency, *Ambient Aquatic Life Water Quality Criteria for Chromium*, Office of Water Regulations and Standards, Criteria and Standards Division, Washington, 1985, EPA 440/5-84-029.

U.S. Environmental Protection Agency, *Ambient Aquatic Life Water Quality Criteria for Mercury*, Office of Water Regulations and Standards, Criteria and Standards Division, Washington, 1985, EPA 440/5-84-026.

U.S. Environmental Protection Agency, *Ambient Aquatic Life Water Quality Criteria for Copper*, Office of Research and Development, Environmental Research Laboratories, Washington, 1985, EPA 440/5-84-031.

U.S. Environmental Protection Agency, *Ambient Aquatic Life Water Quality Criteria for Arsenic*, Office of Water Regulations and Standards, Criteria and Standards Division, Washington, 1985, EPA 440/5-84-033.

U.S. Environmental Protection Agency, *Ambient Aquatic Life Water Quality Criteria for Cadmium*, Office of Water Regulations and Standards, Criteria and Standards Division, Washington, 1985, EPA 440/5-84-032.

U.S. Environmental Protection Agency, *Ambient Aquatic Life Water Quality Criteria for Chlorine*, Office of Water Regulations and Standards, Criteria and Standards Division, Washington, 1985, EPA 440/5-84-030.

U.S. Environmental Protection Agency, *Ambient Aquatic Life Water Quality Criteria for Ammonia*, Office of Water Regulations and Standards, Criteria and Standards Division, Washington, 1985, EPA 440/5-85-001.

U.S. Environmental Protection Agency, *Ambient Aquatic Life Water Quality Criteria for Lead*, Office of Water Regulations and Standards, Criteria and Standards Division, Washington, 1985, EPA 440/5-84-027.

U.S. Environmental Protection Agency, *Biological Criteria: National Program Guidance for Surface Waters*, Office of Water/Office of Science and Technology, Washington, 1990, EPA/440/5-90/004.

U.S. Environmental Protection Agency, *Development of Index of Biotic Integrity Expectations for the Ecoregions of Indiana I, Central Corn Belt Plain*, Water Quality Surveillance and Standards Branch, Chicago, 1991, EPA/905/9-91/025.

U.S. Environmental Protection Agency, *Great Lakes Water Quality Initiative Criteria Documents for the Protection of Aquatic Life in Ambient Water*, Office of Water, Washington, 1995, EPA-820-B-95-004.

U.S. Environmental Protection Agency, *Guidelines for Deriving Site Specific Sediment Quality Criteria*, Office of Water, Washington, 1993, EPA 822-R-93-017.

U.S. Environmental Protection Agency, *Procedures for Initiating Narrative Biological Criteria, Office of Science and Technology*, Office of Water, 1992, EPA-822-B-92-002.

U.S. Environmental Protection Agency, *SAB Report Review of Rationale for Development of Ambient Aquatic Life Water Quality Criteria for TCDD (Dioxin)*, Science Advisory Board, Washington, 1992, EPA-SAB-EPEC-92-024.

CHEMICAL PROPERTIES

Anderson, E., Veith, G.D., and Weininger, D., *SMILES (Simplified Molecular Identification and Line Entry System): a Line Notation and Computerized Interpreter for Chemical Structures*, Environmental Research Laboratory, Duluth, MN, Computer Sciences Corporation, Falls Church, VA, 1987, EPA/600/M-87/021.

Beilstein, F.K., *How to Use Beilstein Handbook of Organic Chemistry*, Beilstein Institute, Frankfurt/Main, Germany, 1990, OCLC 23590912.

Boethling, R. and Tirado, N., *Environmental Fate Data Base (ENVIROFATE): Journal Citations (XREF), April 1990*, Office of Pesticides and Toxic Substances, U.S. Environmental Protection Agency, Washington, 1990, EPA/DF/MT-90/032.

Boethling, R.S., Structure activity relationships for evaluation of biodegradability in the Environmental Protection Agency's Office of Pollution Prevention and Toxics, in *Environmental Toxicology and Risk Assessment, Vol. 2*, Gorsuch, J.W. et al., Eds., American Society for Testing and Materials, Philadelphia, 1993, ASTM STP 1216.

Boethling, R.S. and Mackay, D., *Handbook of Property Estmation Methods for Environmental Chemicals: Environmental and Health Sciences*, Lewis Publishers, Boca Raton, FL, 2000.

Boethling, R.S. and Mackay, D., *Handbook of Property Estimation Methods for Chemicals: Environmental and Health Sciences*, Ann Arbor Press, Chelsea, MI, 1998.

Bolz, R.E. and Tuve G.L., *CRC Handbook of Tables for Applied Engineering Science*, Chemical Rubber Company, Cleveland, 1970.

Briggs, G.C., Rheoretical and experimental relationships between soil adsorption, octanol-water partition coefficients, water solubilities, bioconcentration factors, and the parachor, *J. Agric. Food Chem.*, 29, 1050, 1980.

Buckingham, J., *Dictionary of Organic Compounds*, 5th ed., Chapman & Hall, New York, 1982.

Budavari, S., *Merck Index: an Encyclopedia of Chemicals, Drugs, and Biologicals*, Merck, Whitehouse Station, NJ, 1996.

Clansky, K.B., *Suspect Chemicals Sourcebook a Guide to Industrial Chemicals Covered under Major Federal Regulatory and Advisory Programs*, Roytech Publications, Burlingame, CA, 1988.

Clements, R.G., Ed., *Estimating Toxicity of Industrial Chemicals to Aquatic Organisms Using Structure-Activity Relationships*, Office of Pollution Prevention and Toxics, Health and Environmental Review Division, Environmental Effects Branch, U.S. Environmental Protection Agency, Washington, 1994, EPA-748-R-93-001.

Clements, R.G. et al., The use and application of QSARS in the office of toxic substances for ecological hazard assessment of new chemicals, in *Environmental Toxicology and Risk Assessment*, Landis,W.G. et al., Eds., American Society for Testing and Materials, Philadelphia, 1993, ASTM STP 1179.

Clements, R.G. et al., The use of quantitative structure-activity relationships (QSARS) as screening tools in environmental assessment, in *Environmental Toxicology and Risk Assessment*, Gorsuch, J.W. et al., Eds., American Society for Testing and Materials, Philadelphia, 1993, ASTM STP 1216.

Clements, R.G. and Nabholz, J.V., *ECOSAR: a Computer Program for Estimating the Ecotoxicity of Industrial Chemicals Based on Structure Activity Relationships: User's Guide*, U.S. Environmental Protection Agency, Office of Pollution Prevention and Toxics, Washington, 1994, EPA-784-R-93-002.

Devillers, J., *Comparative QSAR*, Taylor & Francis, Washington, 1998.

Dow Chemical Company, *Dow's Fire and Explosion Index Hazard Classification Guide*, 6th ed., American Institute of Chemical Engineers, Midland, MI, 1987.

Eastern Research Group, *Inventory of Exposure-Related Data Systems Sponsored by Federal Agencies*, Eastern Research Group, Lexington, MA, 1992, EPA/600/R-92/078.

Elvers, B., Hawkins, S., and Ullmann, F., *Ullmann's Encyclopedia of Industrial Chemistry*, VCH Publishers, Deerfield Beach, FL, 1996.

Farley, F., *Photochemical Reactivity Classification of Hydrocarbons and Other Organic Compounds*, U.S. Environmental Protection Agency, Research Triangle Park, NC, 1977, EPA-600-R-92-078.

Foster Wheeler Enviresponse, Inc., *Eco Logic International Gas-Phase Chemical Reduction Process — the Thermal Desorption Unit: Applications Analysis Report*, Office of Research and Development, Risk Reduction Engineering Laboratory, Center for Environmental Research Information, U.S. Environmental Protection Agency, Cincinnati, 1994, EPA/540/AR-94/504.

Gill, B.S. et al., *Application of a Plant Test System in the Identification of Potential Genetic Hazards at Chemical Waste Sites*, Health Effects Research Laboratory, Research Triangle Park, NC, Genetic Toxicology Div. Environmental Health Research and Testing, Inc., Research Triangle Park, NC, 1991, EPA/600/D-91/275.

Hamrick, K.J., Kollig, H.P., and Bartell, A., Computerized extrapolation of hydrolysis rate data, *J. Chem. Inf. Comp. Sci.*, 32, 511, 1992.

Hansch, C. and Fujita, T., *Classical and Three-Dimensional QSAR in Agrochemistry*, American Chemical Society, Washington, 1995.

Hansch, C., Leo, A., and Hoekman, D.H., *Exploring QSAR*, American Chemical Society, Washington, 1995.

Hansch, C. and Leo, A., *The Log P Data Base*, Pomona College, Claremont, CA, 1987.

Hassett, J.J., Correlation of compound properties with sorption characteristics of nonpolar compounds by soils and sediments: concepts and limitations, in *Environment and Solid Wastes: Characterization, Treatment, and Disposal*, Francis, C.W. and Auerbach, S.I., Eds., Butterworths, Boston, MA, 1981.

Hawley, G.G., Sax, N.I., and Lewis, R.J., Eds., *Condensed Chemical Dictionary (Computer File)*, Van Nostrand Reinhold Co., New York, 1998, OCLC 40569260.

Hawley, G.G., Sax, N. I., and Lewis, R.J., Eds., *Condensed Chemical Dictionary*, Van Nostrand Reinhold Co., New York, 1997, OCLC 37370603.

Helfgott, T.B., Hart, F.L., and Bedar, R.G., *An Index of Refractory Organics*, Office of Research and Development, U.S. Environmental Protection Agency, Ada, OK, 1977, EPA-600/2-77-174.

Heller, S.R., *Online Searching on STN Beilstein Reference Manual*, Springer, New York, 1989, OCLC 20498550.

Himmelblau, D.M., *Basic Principles and Calculations in Chemical Engineering*, 4th ed., Prentice-Hall, Englewood Cliffs, NJ, 1990.

Howard, P.H. et al., *Handbook of Environmental Degradation Rates*, Lewis Publishers, Chelsea, MI, 1991.

Hunter, R.S. and Culver, F.D., *Microqsar Version 2.0: a Structure-Activity Based Chemical Modeling and Information System*, U.S. Environmental Protection Agency and Montana State University, Institute for Biological and Chemical Process Analysis, Duluth, MN, 1992.

Hunter, R.S. and Culver, F.D., *QSAR System User Manual: a Structure-Activity Based Chemical Modeling and Information System*, Institute for Biological and Chemical Process Analysis, Montana State University, Bozeman, MT, 1988.

Jaber, H.M. et al., *Data Acquisition for Environmental Transport and Fate Screening*, Sri International, Menlo Park, CA; Office of Health and Environmental Assessment, U.S. Environmental Protection Agency, Washington, 1984, EPA-68-03-2981; EPA/600/6-84/009.

Jochum, C., Hicks, M.G., and Sunkel, J., *Physical Property Prediction in Organic Chemistry Proceedings of the Beilstein Workshop, 16-20th May, 1988, Schlob Korb, Italy*, Springer-Verlag, New York, 1988, OCLC 18558261.

Klecka, G.M. and Mackay, D., *Evaluation of Persistence and Long-Range Transport of Organic Chemicals in the Environment: Summary of a SETAC Pellston Workshop*, SETAC Press, Pensacola, FL, 1999.

Kroschwitz, J.I., *Kirk-Othmer Concise Encyclopedia of Chemical Technology*, John Wiley & Sons, New York, 1999.

Kubinyi, H., *3D QSAR in Drug Design: Theory, Methods and Applications*, Escom, Leiden, The Netherlands, 1993.

Landrum, P.F. et al., Synopsis of discussion session on the kinetics behind environmental bioavailability, in *Bioavailability: Physical, Chemical and Biological Interactions*, Environmental Research Laboratory, U.S. Environmental Protection Agency, Narragansett, RI, 1994, EPA/600/A-94/235.

Lee, S.C., Shiu, W.Y., and Mackay, D., *Study of the Long Term Fate and Behaviour of Heavy Metals*, Department of Chemical Engineering and Applied Chemistry, University of Toronto, 1992.

Lide, D.R., *CRC Handbook of Chemistry and Physics: a Ready-Reference Book of Chemical and Physical Data*, CRC Press, Boca Raton, FL, 1999.

Lipnick, R.L., Baseline toxicity QSAR models: a means to assess mechanism of toxicity for aquatic organisms and mammals, in *Environmental Toxicology and Risk Assessment, Vol. 2*, Gorsuch, J.W. et al., Eds., American Society for Testing and Materials, Philadelphia, 610, 1993, ASTM STP 1216.

Luyben, W. and Wenzel, L., *Chemical Process Analysis: Mass and Energy Balances*, Prentice-Hall, Englewood Cliffs, NJ, 1988.

Lyman, W.J., Reehl, W.F., and Rosenblatt, D.H., *Handbook of Chemical Property Estimation Methods*, American Chemical Society, Washington, 1990.

Mabey, W.R. et al., *Aquatic Fate Process Data for Organic Priority Pollutants*, SRI International, Menlo Park, CA, Monitoring and Data Support Division, U.S. Environmental Protection Agency, Washington, 1982, EPA-68-01-3867; EPA-68-03-2981; EPA/440/4-81/014.

Mabey, W.R. and Mill, T., *Chemical Transformations in Groundwater*, SRI International, Menlo Park, CA, 1984, OCLC 33115461.

Mabey, W.R. and Mill, T., Critical review of hydrolysis of organic compounds in water under environmental conditions, *J. Phy. Chem. Reference Data*, 7, 383, 1978.

Mabey, W.R., Mill, T., and Podoll, R.T, *Estimation Methods for Process Constants and Properties Used in Fate Assessments*, SRI International, Menlo Park, CA, Environmental Research Laboratory, U.S. Environmental Protection Agency, Athens, GA, 1984, EPA-68-03-2981; EPA-600/3-84-035.

Mackay, D., *Multimedia Environmental Models the Fugacity Approach*, Lewis Publishers, Chelsea, MI, 1991.

Mackay, D. et al., *Volatilization of Organic Pollutants from Water*, Toronto University, Ontario; Environmental Research Laboratory, U.S. Environmental Protection Agency, Athens, GA, 1982, EPA-R-80515; EPA-600/3-82-019.

Mackay, D., Shiu, W.Y., and Ma, K.C., *Illustrated Handbook of Physical-Chemical Properties and Environmental Fate for Organic Chemicals, Vols. 1-4*, Lewis Publishers, Chelsea, MI, 1992.

Mackay, D., Shiu, W.Y., and Ma, K.C., *Physical-Chemical Properties and Environmental Fate Handbook (Computer File)*, Chapman & Hall/CRCnet Base, Boca Raton, FL, 1999.

Mekenyan, O.G. and Veith, G.D., *Electronic Factor in QSAR: MO-Parameters, Competing Interactions, Reactivity and Toxicity*, Environmental Research Laboratory, U.S. Environmental Protection Agency, Duluth, MN, Department of Physical Chemistry, Burgas University of Technology (Bulgaria), 1994, EPA/600/J-94/362.

Mill, T. et al., *Laboratory Protocols for Evaluating the Fate of Organic Chemicals in Air and Water*, SRI International, Menlo Park, CA, Environmental Research Laboratory, U.S. Environmental Protection Agency, Athens, GA, 1982, EPA-68-03-2227; EPA-600/3-82-022.

Mill, T. et al., *Validation of Estimation Techniques for Predicting Environmental Transformation of Chemicals*, Office of Pesticides and Toxic Substances, U.S. Environmental Protection Agency, Washington, SRI International, Menlo Park, CA, 1982, EPA 560/1982 SRI/003; 68-01-6269.

O'Bryan, T. and Ross, R.H., *Chemical Scoring System for Hazard and Exposure Identification*, Hemisphere Publishing, Washington, 1988.

Peijnenburg, J.G.M. and Damborsky, J., *Biodegradability Prediction*, Kluwer Academic Publishers, Boston, 1996.

Perry, R.H., Green, D.W., and Maloney, J.O., *Perrys' Chemical Engineers' Handbook*, McGraw-Hill, New York, 1997.

Pitter, P., Determination of biological degradability of organic substances, in *Water Resources*, 10, 231, 1976.

STN International, *Online Searching on STN Beilstein Workshop Manual*, Springer, New York, 1989, OCLC 20709984.

Stull, D.R., Fundamentals of fire and explosion, *Amer. Inst. Chem. Eng. Monog. Ser.*, 73, 10, 1977.

Weiss, G., *Hazardous Chemicals Data Book*, Noyes Data Corp., Park Ridge, NJ, 1986.

CONTRACTOR MANAGEMENT

Batelle Memorial Institute, *Project Management*,
http://sbms.pnl.gov:2080/standards/61/6100p010.htm, 1995-1998.

Gil, E. et al., *Working with Consultants*, American Planning Association, Chicago, 1983, Planning Advisory Service Report Number 378.

U.S. Environmental Protection Agency, *Survey Management Handbook, Vol. II: Overseeing the Technical Progress of a Survey Contract*, Office of Policy, Planning, and Evaluation, Washington, 1984, EPA 230/12-84-002.

DATA ANALYSIS

Abramson, J.H., *Making Sense of Data: a Self-Instruction Manual*, Oxford University Press, New York, 1988.

Adkins, N.C., *Framework for Development of Data Analysis Protocols for Ground Water Quality Monitoring*, Colorado Water Resources Research Institute, Fort Collins, CO, 1993.

Graham, J. A. et al., *Proceedings of the NHEXAS Data Analysis Workshop*, National Exposure Research Laboratory, U.S. Environmental Protection Agency, Cincinnati, 1999, EPA/600/R-99/077.

Hattis, D., Erdreich, L., and Ballew, M., *Human Variability in Susceptibility to Toxic Chemicals: a Preliminary Analysis of Pharmacokinetic Data from Normal Volunteers*, Environmental Criteria and Assessment Office, U.S. Environmental Protection Agency, Cincinnati, Massachusetts Institute of Technology, Center for Technology, Policy and Industrial Development, Cambridge, MA, 1987, EPA/600/J-87/320.

ICF Kaiser, Inc., *Summary and Analysis of Available Air Toxics Health Effects Data*, ICF Kaiser International, Inc., Fairfax, VA, Office of Air Quality Planning and Standards, U.S. Environmental Protection Agency, Research Triangle Park, NC, 1995, EPA/456/R-97/001.

Kilaru, V., *Screening Analysis of Ambient Monitoring Data for the Urban Area Source Program, Final,* Office of Air Quality Planning and Standards, U.S. Environmental Protection Agency, Research Triangle Park, NC, 1994, EPA/453/R-94/075.

Mcbean, E.A., *Statistical Procedures for Analysis of Environmental Monitoring Data and Risk Assessment,* Prentice-Hall PTR, Upper Saddle River, NJ, 1998.

Nesnow, S. and Bergman, H., *Analysis of the Gene-Tox Carcinogen Data Base (Journal Version),* Health Effects Research Laboratory, Research Triangle Park, NC, 1987, EPA/600/J-88/096.

Pfetzing, E. and Allen, B., *Guidelines for Statistical Analysis of Occupational Exposure Data,* IT Environmental Programs, Inc., Cincinnati, ICF Kaiser, Inc., Fairfax, VA, Office of Pollution, Prevention, and Toxics, U.S. Environmental Protection Agency, Washington, 1994, EPA-68-D2-0064; EPA/744/B-94/003.

Rubenstein, S. and Horn, R.L., *Risk Analysis in Environmental Studies: I. Risk Analysis Methodology: a Statistical Approach; II. Data Management for Environmental Studies,* Rockwell International, Atomics International Division, Richland, Washington, 1978.

Sparks, L.E., Molhave, L., and Dueholm, S., *Source Testing and Data Analysis for Exposure and Risk Assessment of Indoor Pollutant Sources,* Air and Energy Engineering Research Laboratory, U.S. Environmental Protection Agency, Research Triangle Park, NC, Aarhus University, Wesser and Dueholm, Copenhagen, 1995, EPA/600/A-95/104.

Stockwell, J.R., *TRI Data Analysis, Ashland, KY, Area,* U.S. Environmental Protection Agency, Atlanta, 1991, EPA 904/R-91/107.

Stockwell, J.R., *TRI Data Analysis, Calvert City, KY, Area,* U.S. Environmental Protection Agency, Atlanta, 1991, EPA 904/R-91/106.

TRC Environmental Corporation, *Analysis of Ambient Monitoring Data in the Vicinity of Open Tire Fires,* TRC Environmental Corporation, Chapel Hill, NC, Air Risk Information Support Center, U.S. Environmental Protection Agency, Research Triangle Park, NC, 1993, EPA-68-D0-0121; EPA/453/R-93/029.

Tsang, A.M. and Klepeis, N.E., *Descriptive Statistics Tables from a Detailed Analysis of the National Human Activity Pattern Survey (NHAPS) Data,* Lockheed Martin Environmental Systems and Technologies, National Exposure Research Laboratory, U.S. Environmental Protection Agency, Las Vegas, 1996, EPA-68-01-7325; EPA/600/R-96/148.

U.S. Environmental Protection Agency, *Influence of Plumbing, Lead Service Lines, and Water Treatment on Lead Levels at the Tap: Analysis of Available Data,* Office of Drinking Water, Washington, 1990.

U.S. Environmental Protection Agency, *National Sediment Contaminant Point Source Inventory: Analysis of Facility Release Data,* Office of Water, Washington, 1996, EPA/823/D-96/001.

DATA COLLECTION

American Society for Testing and Materials, *ASTM 1998 Annual Book of ASTM Standards, Water and Environmental Technology,* Philadelphia, 1998.

Barcelona, M.J. et. al., *Practical Guide for Ground-Water Sampling,* Illinois State Water Survey, Department of Energy and Natural Resources, Champaign, IL, Robert S. Kerr Environmental Research Laboratory, U.S. Environmental Protection Agency, Ada, OK, 1985, EPA/600/2-85/104.

Bayard, S. et al., *Technical Analysis of New Methods and Data Regarding Dichloromethane Hazard Assessments, External Review Draft,* Office of Health and Environmental Assessment, U.S. Environmental Protection Agency, Washington, 1987, EPA/600/8-87/029a.

Bigler, J., *Guidance for Assessing Chemical Contaminant Data for Use in Fish Advisories: Vol. I, Sampling and Analysis*, Office of Water, U.S. Environmental Protection Agency, Washington, 1994, EPA-823-B-94-004.

Budde, W.L. and Munch, J.W., *Methods for the Determination of Organic Compounds in Drinking Water, Suppl. 3*, National Exposure Research Lab, U.S. Environmental Protection Agency, Cincinnati, 1995, EPA/600/R-95/131.

Columbari, V., Reliability data collection and use in risk and availability assessment, *Proc. 6th EuReData Conf.*, Springer-Verlag, New York, 1989.

de Vera, E.R., *Samplers and Sampling Procedures for Hazardous Waste Streams*, Municipal Environmental Research Laboratory, Office of Research and Development, U.S. Environmental Protection Agency, Cincinnati, 1980, EPA-600/2-80-018.

Fisher, N.A., *Volunteer Estuary Monitoring: a Methods Manual*, Office of Water, Office of Wetlands, Oceans and Watersheds, Oceans and Coastal Protection Division, U.S. Environmental Protection Agency, Washington, 1993, EPA/842-B-93/004.

Gibson, G. R. et al., *Biological Criteria: Technical Guidance for Streams and Small Rivers, Revised Edition*, Office of Science and Technology, Health and Ecological Criteria Division, U.S. Environmental Protection Agency, Washington, 1996, EPA/822-B-96/001.

Horning, W.B. and Weber, C.I., *Short-Term Methods for Estimating the Chronic Toxicity of Effluents and Receiving Waters to Freshwater Organisms*, Biological Methods Branch, Environmental Monitoring and Support Laboratory, U.S. Environmental Protection Agency, Cincinnati, 1985, EPA-600/4-85-014; PB 86-158474.

Klemm, D.J., *Fish Field and Laboratory Methods for Evaluating the Biological Integrity of Surface Waters*, Environmental Monitoring Systems Laboratory, Office of Modeling, Monitoring Systems, and Quality Assurance, Office of Research and Development, U.S. Environmental Protection Agency, Cincinnati, 1993, EPA/600/R-92/111.

Klemm, D.J. et al., *Macroinvertebrate Field and Laboratory Methods for Evaluating the Biological Integrity of Surface Waters*, Environmental Monitoring Systems Laboratory, Office of Modeling, Monitoring Systems, and Quality Assurance, Office of Research and Development, U.S. Environmental Protection Agency, Washington, 1990, EPA/600/4-90/030.

Kopp, J.F. and McKee, G.D., *Methods for Chemical Analysis of Water and Wastes*, Office of Research and Development, U.S. Environmental Protection Agency, Washington, 1983, EPA/600/4-79/020.

Mahy, B.W.J. and Kangro, H., *Virology Methods Manual*, Academic Press, San Diego, 1996.

McGuire, J.M., *Multispectral Identification and Confirmation of Organic Compounds in Wastewater Extracts*, Environmental Research Laboratory, Athens, GA, U.S. Environmental Protection Agency, Washington, 1990, EPA/600/S4-90/002.

Mills, W.B., *Handbook: Stream Sampling for Waste Load Allocation Applications*, Office of Research and Development, U.S. Environmental Protection Agency, Washington, 1986, EPA/625/6-86/013.

Mills, W.B. et al., *Water Quality Assessment: Screening Procedure for Toxic and Conventional Pollutants in Surface and Ground Water, Part 1, Revised 1985*, Environmental Research Laboratory, Athens, GA, U.S. Environmental Protection Agency, Washington, 1985, EPA/600/6-85/002a.

Mills, W.B. et al., *Water Quality Assessment: Screening Procedure for Toxic and Conventional Pollutants in Surface and Ground Water, Part 2, Revised 1985*, Environmental Research Laboratory, Athens, GA, U.S. Environmental Protection Agency, Washington, 1985, EPA/600/6-85/002b.

Montgomery, D.C., *Design and Analysis of Experiments*, John Wiley & Sons, New York, 1991.

Mount, D.I., *Methods for Aquatic Toxicity Identification Evaluations, Phase III Toxicity Confirmation for Samples Exhibiting Acute and Chronic Toxicity*, Office of Research and Development, U.S. Environmental Protection Agency, Washington, 1993, EPA 6/600/R-92/081.

Mueller, C., *Standard Operating Procedures and Field Methods Used for Conducting Ecological Risk Assessment Case Studies*, Naval Command, Control and Ocean Surveillance Center, RDT&E Division, San Diego, 1992, EPA Tech. Doc. 2296.

Plafkin, J.L., *Rapid Bioassessment Protocols for Use in Streams and Rivers: Benthic Macroinvertebrates and Fish*, Office of Water, U.S. Environmental Protection Agency, Washington, 1989, EPA/440/4-89/001.

Ray, M.S., *Engineering Experimentation: Ideas, Techniques, and Presentations*, McGraw-Hill, New York, 1988.

Reckhow, K.H., *Quantitative Techniques for the Assessment of Lake Quality*, Office of Water Planning and Standards, U.S. Environmental Protection Agency, Washington, 1979, EPA 440-5-79-015.

Schlekat, C.E. and Scott, K.J., *Methods for Measuring the Toxicity of Sediment-Associated Contaminants with Estuarine and Marine Amphipods*, Office of Research and Development, U.S. Environmental Protection Agency, Narragansett, RI, 1994, EPA 600/R-94/025.

Sherma, J., *Manual of Analytical Quality Control for Pesticides and Related Compounds in Human and Environmental Samples: a Compendium of Systematic Procedures Designed to Assist in the Prevention and Control of Analytical Problems*, Office of Research and Development, Health Effects Research Laboratory, U.S. Environmental Protection Agency, Research Triangle Park, NC, 1981, EPA-600/2-81-059.

Simpson, J.T., *Volunteer Lake Monitoring: a Methods Manual*, Office of Wetlands, Oceans, and Watersheds, Assessment & Watershed Protection Division, U.S. Environmental Protection Agency, Washington, 1991, EPA 440/4-91-002.

Smoley, C.K., *Methods for the Determination of Metals in Environmental Samples, Environmental Monitoring Systems Laboratory*, U.S. Environmental Protection Agency, Cincinnati, 1992.

U.S. Environmental Protection Agency, *Analytical Procedures and Quality Assurance Plan for the Determination of Xenobiotic Chemical Contaminants in Fish*, Environmental Research Laboratory, Duluth, MN, 1990, EPA/600/3-90/023.

U.S. Environmental Protection Agency, *Analytical Procedures and Quality Assurance Plan for the Determination of Xenobiotic Chemical Contaminants in Fish, National Dioxin Study, Phase 2*, Environmental Research Laboratory, Duluth, MN, 1989, EPA/600/3-90/023.

U.S. Environmental Protection Agency, *Bibliography of Methods for Marine and Estuarine Monitoring*, Office of Water Office of Wetlands, Oceans, and Watersheds, Ocean and Coastal Protection Division, Office of Water, Washington, 1995, EPA 842-B-95-002.

U.S. Environmental Protection Agency, *CWA Section 403: Procedural and Monitoring Guidance*, Washington, 1994, EPA 842-B-94-003.

U.S. Environmental Protection Agency, *Evaluation of Dredged Material Proposed for Ocean Disposal Testing Manual*, Office of Marine and Estuarine Protection, U.S. Army Corps of Engineers, Washington, 1991, EPA 503/8-91/001.

U.S. Environmental Protection Agency, *Guidance for Assessing Chemical Contaminant Data for Use in Fish Advisories, Vol. 2, Risk Assessment and Fish Consumption Limits*, Office of Water, Washington, 1994, EPA/823/B-94/004.

U.S. Environmental Protection Agency, *Guidance for Data Useability in Risk Assessment (Part A)*, Office of Emergency and Remedial Response, Washington, 1992, OSWER-9285.7-09a.

U.S. Environmental Protection Agency, *Guidance for State Water Monitoring and Wasteload Allocation Programs*, Monitoring and Data Support Division, Washington, 1985, EPA 440/4-85-031.

U.S. Environmental Protection Agency, *Handbook for Sampling and Sample Preservation of Water and Wastewater*, Environmental Monitoring and Support, Laboratory Office of Research and Development, Cincinnati, 1982, EPA-600/4-82-029.

U.S. Environmental Protection Agency, *Innovations in Site Characterization: Interim Guide to Preparing Case Studies*, Office of Solid Waste and Emergency Response, Washington, 1998, EPA/542/B-98/009.

U.S. Environmental Protection Agency, *Manual for the Certification of Laboratories Analyzing Drinking Water: Criteria and Procedures Quality Assurance*, 3rd ed., Office of Ground Water and Drinking Water, Technical Support Division, Cincinnati, 1992, EPA-814B-92-002.

U.S. Environmental Protection Agency, *Methods for Measuring the Toxicity of Sediment-Associated Contaminants with Freshwater Invertebrates*, Office of Research and Development, Duluth, MN, 1994, EPA 600/R-94/024.

U.S. Environmental Protection Agency, *Methods for the Determination of Inorganic Substances in Environmental Samples*, Environmental Monitoring Systems Laboratory, Cincinnati, 1993, EPA/600/R-93/100.

U.S. Environmental Protection Agency, *Methods for the Determination of Metals in Environmental Samples, Supplement I*, Environmental Monitoring Systems Laboratory, Cincinnati, 1993, EPA-600/R-94/111.

U.S. Environmental Protection Agency, *Methods for the Determination of Organic Compounds in Drinking Water, Suppl. 1*, Office of Research and Development, Washington, 1990, EPA/600/4-90/020.

U.S. Environmental Protection Agency, *Methods for the Determination of Organic Compounds in Drinking Water, Suppl. 2*, Environmental Monitoring Systems Laboratory, Office of Research and Development, Cincinnati, 1995, EPA-600/R-92/129.

U.S. Environmental Protection Agency, *Monitoring Guidance for the National Estuary Program*, Office of Water, Washington, 1992, EPA 842 B-92-004.

U.S. Environmental Protection Agency, a *Review of Methods for Assessing Nonpoint Source Contaminated Ground Water Discharge to Surface Water*, Office of Water Protection, Washington, 1991, EPA 570/9-91-010.

U.S. Environmental Protection Agency, *Risk Assessment Guidance for Superfund (Training Course)*, Office of Emergency and Remedial Response, Washington, 1995, EPA/540/R-95/132; OSWER-9285.9-22a.

U.S. Environmental Protection Agency, *SAB Report: Review of a Testing Manual for Evaluation of Dredged Material Proposed for Ocean Disposal*, Science Advisory Board, Washington, 1992, EPA-SAB-EPEC-92-014.

U.S. Environmental Protection Agency, *Statistical Methods for the Analysis of Lake Water Quality Trends: Technical Supplement to the Lake and Reservoir Restoration Guidance Manual*, Office of Water, Washington, 1993, EPA 841-R-93-003.

U.S. Environmental Protection Agency, *USEPA Manual of Methods for Virology*, Computer Sciences Corp.; Cincinnati, Office of Research and Development, Washington, 1993, EPA/600/4-84/013.

Weber, C.I., *Methods for Measuring Acute Toxicity of Effluents and Receiving Waters to Freshwater and Marine Organisms*, 4th ed., Environmental Monitoring Systems Laboratory, Cincinnati, 1993, EPA-600/4-90-027f.

Weber, C.I. et al., *Short-Term Methods for Estimating the Chronic Toxicity of Effluents and Receiving Waters to Freshwater Organisms*, Biological Methods Branch, Environmental Monitoring and Support Laboratory, Cincinnati, 1988, EPA-600/4-87-028.

DATA QUALITY

Arnold, J.R., Dennis, R.L., and Tonnesen, G.S., *Advanced Techniques for Evaluating Eulerian Air Quality Models: Background and Methodology*, U.S. Environmental Protection Agency, Research Triangle Park, NC; National Exposure Research Laboratory, National Oceanic and Atmospheric Administration, Research Triangle Park, NC, 1998, EPA/600/A-97/098.

Barnthouse, L.W., *Assessing the Transport and Fate of Bioengineered Microorganisms in the Environment*, Oak Ridge National Laboratory, Oak Ridge, TN, 1990.

Barnthouse, L.W., *Methodology for Environmental Risk Analysis, United States*, Office of Research and Development, U.S. Environmental Protection Agency, Oak Ridge National Laboratory, Oak Ridge, TN, 1982.

Barnthouse, L.W., *Unit Release Risk Analysis for Environmental Contaminants of Potential Concern in Synthetic Fuels Technologies*, Oak Ridge National Laboratory, Oak Ridge, TN, 1985.

Barnthouse, L.W. et al., *Risks of Toxic Contaminants to Exploited Fish Populations: Influence of Life History, Data Uncertainty and Exploitation Intensity*, Oak Ridge National Laboratory, TN, Martin Marietta Energy Systems, Inc., Oak Ridge, TN, Department of Energy, Washington, Environmental Research Laboratory, U.S. Environmental Protection Agency, Gulf Breeze, FL, 1990.

Barnthouse, L.W. and Suter, G.W., *United States, User's Manual for Ecological Risk Assessment*, Office of Research and Development, U.S. Environmental Protection Agency, Oak Ridge National Laboratory, Oak Ridge, TN, 1986.

Barth, D.S., *Sediment Sampling Quality Assurance User's Guide*, Environmental Monitoring Systems Laboratory, Las Vegas; U.S. Environmental Protection Agency, Washington, 1985, EPA/600/4-85/048.

Barth, D.S., *Soil Sampling Quality Assurance User's Guide*, Office of Emergency and Remedial Response, U.S. Environmental Protection Agency, Washington, 1984, EPA 600/4-84-0043.

Burgess, R.M. et al., *Water Column Toxicity from Contaminated Marine Sediments: Effects on Multiple Endpoints of Three Marine Species*, U.S. Environmental Protection Agency, Narragansett, RI, 1993, EPA-68-C1-0005; EPA-68-03-3529; EPA/600/J-93/311; ERLN-1191.

Cadmus Group, Inc., *Testing of Meteorological and Dispersion Models for Use in Regional Air Quality Modeling*, Waltham, MA, National Park Service, Denver, 1995, EPA-68-D0-0095; EPA/454/R-95/005.

Cline, S.P. et al., *Environmental Monitoring and Assessment Program Forest Health Monitoring Quality Assurance Project Plan for Detection Monitoring Project*, U.S. Environmental Protection Agency, Research Triangle Park, NC, 1995, EPA/620/R-95/002.

Crockett, A.B. et al., *Federal Facilities Forum Issue: Field Sampling and Selecting On-Site Analytical Methods for Explosives in Soil*, U.S. Environmental Protection Agency, Las Vegas, National Exposure Research Laboratory; Idaho National Engineering and Environmental Laboratory, Idaho Falls Army Environmental Center, Aberdeen Proving Ground, MD; Department of Energy, Washington, 1996, EPA/540/R-97/501; NERL-LV-97-037; DE-AC07-94ID13223.

Cross-Smiecinski, A., *Quality Planning for the Life Science Researcher Meeting Quality Assurance Requirements*, CRC Press, Boca Raton, FL, 1994.

Eskin, R.A., Rowland, K.H., and Alegre, D.Y., *Contaminants in Chesapeake Bay Sediments, 1984-1991*, Chesapeake Bay Program, U.S. Environmental Protection Agency, Annapolis, 1996, CBP/TRS-145/96.

Freas, W.P., *Guideline on Data Handling Conventions for the 8-Hour Ozone NAAQS*, Office of Air Quality Planning and Standards, U.S. Environmental Protection Agency, Washington, 1998, EPA-454/R-98-017.

Gilbert, R.O. and Simpson, J.C., *Statistical Methods for Evaluating the Attainment of Cleanup Standards, Vol. 3, Reference-Based Standards for Soils and Solid Media*, Office of Policy, Planning and Evaluation, U.S. Environmental Protection Agency, Washington, 1994, EPA/230/R-94/004.

Heck, W.H. et al., *Environmental Monitoring and Assessment Program: Agroecosystem 1992 Pilot Project Plan (April 3, 1992)*, Environmental Monitoring and Assessment Program, U.S. Environmental Protection Agency, Washington, 1993, EPA-68-C0-0049; EPA-68-C8-0006.

Hoos, A.B., *Procedures for Adjusting Regional Regression Models of Urban-Runoff Quality Using Local Data*, U.S. Government Printing Office, Washington, 1996.

Kepner, W.G. et al., *Mid-Atlantic Landscape Indicators Project Plan*, Environmental Monitoring and Assessment Program, U.S. Environmental Protection Agency, Research Triangle Park, NC, 1995, EPA/620/R-95/003.

Lapakko, K. et al., *Long Term Dissolution Testing of Mine Waste*, Minnesota Department of Natural Resources, Minerals Division, St. Paul Office of Solid Waste, U.S. Environmental Proction Agency, Washington, 1995, EPA-X-8200322-01-0; EPA/530/R-95/040.

Longest, H.L., *Superfund Data Quality Objectives: Fact Sheet, Interim Final Guidance, and Workbook*, Office of Solid Waste and Emergency Response, U.S. Environmental Protection Agency, Washington, 1993, OSWER 9355.9-02.

Mason, B.J., *Preparation of Soil Sampling Protocols: Sampling Techniques and Strategies*, Environmental Monitoring Systems Laboratory, Office of Research and Development, U.S. Environmental Protection Agency, Las Vegas, 1993, EPA/600/R-92/128.

Mcnider, R.T. et al., *Assimilation of Satellite Data in Regional Air Quality Models*, National Exposure Research Laboratory, U.S. Environmental Protection Agency, Research Triangle Park, NC, 1997, EPA/600/A-97/058.

Myers, D.N., *Handbooks for Water-Resources Investigations National Field Manual for the Collection of Water-Quality Data*, U.S. Government Printing Office, Washington, 1997.

National Council of the Paper Industry for Air and Stream Improvement, *Volatile Organic Compound Emissions from Non-Chemical Pulp and Paper Mill Sources Test Methods, Quality Assurance/Quality Control Procedures, and Data Analysis Protocol*, The National Council of the Paper Industry for Air and Stream Improvement, Research Triangle Park, NC, 1997, OCLC 38982753.

Orth, R.J. et al., *Trends in the Distribution, Abundance, and Habitat Quality of Submerged Aquatic Vegetation in Chesapeake Bay and its Tidal Tributaries: 1971 to 1991*, Chesapeake Bay Program, U.S. Environmental Protection Agency, Annapolis, MD, 1994, EPA/903/R-95/009.

Puls, R.W., *Groundwater Sampling for Metals (Chapter 14)*, Robert S. Kerr Environmental Research Laboratory, Ada, OK, 1994, EPA/600/A-94/172.

Stribling, J.B. and Gerardi, C., *Generic Quality Assurance Project Plan Guidance for Programs Using Community Level Biological Assessment in Wadable Streams and Rivers*, Tetra Tech, Inc., Owings Mills, MD, Office of Wetlands, Oceans and Watersheds, U.S. Environmental Protection Agency, Washington, 1995, EPA-68-C3-0303; EPA/841/B-95/004

Suter, G.W. and Barnthouse, L.W., *Ecological Risk Assessment*, Lewis Publishers, Chelsea, MI, 1993.

Templeman, B.D., *Quality Assurance for Pams Upper Air Monitoring Sites*, U.S. Environmental Protection Agency, Research Triangle Park, NC, 1995, EPA/600/A-94/256.

Templeman, B.D. et al., *Ground-Based Remote Sensor QA/QC at the Boulder Atmospheric Observatory, National Oceanic and Atmospheric Administration*, Research Triangle Park, NC, 1995, EPA/600/A-95/031.

U.S. Department of Energy, *Streamlined Site Characterization Approach for Early Actions: Impact on Risk Assessment Data Requirements*, Washington, 1994, DOE/EH-231-025/1294.

U.S. Environmental Protection Agency, *Better Assessment Science Integrating Point and Nonpoint Sources (Basins) Modeling System, EPA Region 8 (CO, MT, ND, SD, UT, WY) (Version 2.0)* (on CD-Rom), Office of Science and Technology, Washington, 1999.

U.S. Environmental Protection Agency, *Data Quality Objectives Decision Error Feasibility Trials (DQO/DEFT): User's Guide*, Washington, 1994, EPA/540/C-94/002.

U.S. Environmental Protection Agency, *Data Quality Objectives (DQO) Decision Error Feasibility Trials (DEFT) Version 4.0 (for Microcomputers)*, Office of Emergency and Remedial Response, Washington, 1994, EPA/SW/DK-94/088.

U.S. Environmental Protection Agency, *Data Quality Objectives for Remedial Response Activities, Development Process, Office of Remedial Response and Office of Wastes Programs Enforcement*, Washington, 1987, EPA/540/G-87/003.

U.S. Environmental Protection Agency, *Data Quality Objectives Process for Superfund: Interim Final Guidance*, Office of Emergency and Remedial Response, Washington, 1993, EPA 540-R-93-071.

U.S. Environmental Protection Agency, *Data Quality Objectives Process for Superfund: Workbook*, Office of Solid Waste and Emergency Response, Washington, 1993, EPA 540-R-93-078.

U.S. Environmental Protection Agency, *Designs for Air Impact Assessments at Hazardous Waste Sites (165.16), Student Manual*, Office of Emergency and Remedial Response, Washington, 1998, EPA 540-R-98-036; OSWER-9285.9-33.

U.S. Environmental Protection Agency, *Guidance for Data Quality Assessment: Practical Methods for Data Analysis, EPA QA/G-9, QA96 version*, Washington, 1996, EPA 600/R-96-084.

U.S. Environmental Protection Agency, *Guidance for Data Usability in Risk Assessment (Part A)*, Office of Emergency and Remedial Response, Washington, 1992, OSWER 9285.7-09a.

U.S. Environmental Protection Agency, *Guidance for Data Usability in Risk Assessment (Part B)*, Office of Emergency and Remedial Response, Washington, 1992, OSWER 9285.7-09b.

U.S. Environmental Protection Agency, *Guidance for the Data Quality Objectives Process*, Quality Assurance Mangement Staff, Washington, 1994.

U.S. Environmental Protection Agency, *Guidance for the Data Quality Objectives Process: Final*, Quality Assurance Management Staff, Washington, 1994, EPA QA/G-4.

U.S. Environmental Protection Agency, *Guideline on Data Handling Conventions for the PM NAAQS*, Office of Air Quality Planning and Standards, Research Triangle Park, NC, 1999, EPA-454/R-99-008.

U.S. Environmental Protection Agency, *Innovations in Site Characterization, Case Study: Hanscom Air Force Base, Operable Unit 1 (Sites 1, 2, and 3)*, Office of Solid Waste and Emergency Response, Washington, 1998, EPA/542/R-98/006.

U.S. Environmental Protection Agency, *Quality Assurance Guidance for Conducting Brownfields Site Assessments*, Office of Emergency and Remedial Response, Washington, 1998, EPA 540-R-98-038; OSWER-9230.0-83P.

U.S. Environmental Protection Agency, *Quality Management Plan for the Office of Emergency and Remedial Response*, Office of Emergency and Remedial Response, Washington, 1994, EPA/540/R-94/060; OSWER-9240.1-01.

U.S. Environmental Protection Agency, *Recommended Guidelines for Sampling and Analyses in the Chesapeake Bay Monitoring Program*, Chesapeake Bay Program, Annapolis, 1996, EPA/903/R-96/006; CBP/TRS-148/96.

U.S. Environmental Protection Agency, *SAB Report: Environmental Radiation Ambient Monitoring System (Erams) II*, Radiation Advisory Committee, Washington, 1998, EPA/SAB/RAC/ADV-98/001.

U.S. Environmental Protection Agency, *Visibility Monitoring Guidance*, Emissions, Monitoring, and Analysis Division, Research Triangle Park, NC, 1999, EPA/454/R-99/003.

DRINKING WATER

U.S. Environmental Protection Agency, *Advisory by the Science Advisory Board (SAB) Drinking Water Committee (DWC) Concerning the Health Significance of HPC Bacteria Eluted from POU/POE (Point of Use/Point of Entry) Drinking Water Treatment Devices*, Science Advisory Board, Washington, 1996, EPA-SAB-DWC-ADV-96-002.

U.S. Environmental Protection Agency, *Drinking Water Advisory: Consumer Acceptability Advice and Health Effects Analysis on Methyl Tertiary-Butyl Ether (MTBE)*, Office of Water, Washington, 1997, EPA 822-F-97-008.

U.S. Environmental Protection Agency, *Drinking Water and Health: What You Need to Know*, Office of Water, Washington, 1999, EPA 816-K-99-001.

U.S. Environmental Protection Agency, *Drinking Water Standards and Health Effects*, Advisory Committee on the Revision and Application of the Drinking Water Standards, Washington, National Center for Environmental Health, Atlanta, 1999, EPA 810-F-99-017.

U.S. Environmental Protection Agency, *Health Effects from Exposure to High Levels of Sulfate in Drinking Water Study*, Office of Water, Washington, 1998, EPA 815-R-99-001.

U.S. Environmental Protection Agency, *Health Effects from Exposure to Sulfate in Drinking Water Workshop*, Office of Water, Washington, Centers for Disease Control and Prevention, Atlanta, GA, 1999, EPA/815/R-99/002.

U.S. Environmental Protection Agency, *Health Risk Assessment/Characterization of the Drinking Water Disinfection Byproduct Bromate*, Office of Water, Washington, 1998, EPA/815-B-98-007.

U.S. Environmental Protection Agency, *Health Risk Assessment/Characterization of the Drinking Water Disinfection Byproduct Chloroform*, Office of Water, Washington, 1998, EPA/815/B-98/006.

U.S. Environmental Protection Agency, *Health Risk Assessment/Characterization of the Drinking Water Disinfection Byproducts Chlorine Dioxide & Chlorite*, Office of Water, Washington, 1998, EPA/815/B-98/008.

U.S. Environmental Protection Agency, *Regulatory Impact Analysis and Revised Health Risk Reduction and Cost: Analysis for Radon in Drinking Water, Public Comment Draft*, Office of Water, Washington, 1999, EPA/815/D-99/002.

U.S. Environmental Protection Agency, *Risk Assessment, Management and Communication of Drinking Water Contamination*, Office of Water, Washington, EPA/625/4-89/024.

U.S. Environmental Protection Agency, *Safe Drinking Water Act: One Year Later, Success in Advancing Public Health Protection*, Office of Water, Washington, 1997, EPA 810-F-97-002.

U.S. Environmental Protection Agency, *25 Years of the Safe Drinking Water Act: Protecting Our Health from Source to Tap*, Office of Water, Washington, 1999, EPA 810-K-99-004.

ECOLOGICAL RISK ASSESSMENT

Barnthouse, L.W. and Suter, G.W., *Guide for Developing Data Quality Objectives for Ecological Risk Assessment at Doe Oak Ridge Operations Facilities*, Oak Ridge National Laboratory, Oak Ridge, TN, 1996, ES/ER/TM-185/R1.

Bartell, S.M. et al., *Ecological Risk Estimation*, Lewis Publishers, Chelsea, MI, 1992.

Battelle, Inc., *Risk Analysis to Support Standards for Lead in Paint, Dust, and Soil, Vol. 1, Chapters 1 to 7, Appendix A*, Columbus, OH, Office of Pollution, Prevention, and Toxics, U.S. Environmental Protection Agency, Washington, 1998.

Battelle, Inc., *Risk Analysis to Support Standards for Lead in Paint, Dust, and Soil, Vol. 2, Appendices B to G*, Columbus, OH, Office of Pollution, Prevention, and Toxics, U.S. Environmental Protection Agency, Washington, 1998, EPA/747/R-97/006.

Belluck, D.A. et al., *Defining Scientific Procedural Standards for Ecological Risk Assessment*, in *Environmental Toxicology and Risk Assessment: 2nd Vol., ASTM STP 1173*, Gorsuch, J.W. et al., Eds., American Society for Testing and Materials, Philadelphia, 1993.

Belluck, D.A. et al., Defining scientific procedural standards for ecological risk assessment, in *Second Symposium on Environmental Toxicology and Risk Assessment: Aquatic, Plant, and Terrestrial*, Pittsburgh, 83, 1992.

Bridges, T.S., *Summary of a Workshop on Ecological Risk Assessment and Military-Related Compounds: Current Research Needs*, U.S. Army Engineer Waterways Experiment Station, Vicksburg, MS, 1997.

Brooke, L.T. et al., Eds., *Acute Toxicities of Organic Chemicals to Fathead Minnows (Pimephales Promelas), Vols. 1-5*, Center for Lake Superior Environmental Studies, University of Wisconsin-Superior, Superior, WI, 1984–1990.

California Environmental Protection Agency, *Guidance for Ecological Risk Assessment at Hazardous Waste Sites and Permitted Facilities, Part A: Overview*, Sacramento, 1996.

California Environmental Protection Agency, *Guidance for Ecological Risk Assessment at Hazardous Waste Sites and Permitted Facilities, Part B: Scoping Assessment*, Sacramento, 1996.

Call, D.J. and Geiger, D.L., Eds., *Sub-Chronic Toxicities of Industrial and Agricultural Chemicals to Fathead Minnows (Pimephales Promelas), Vol. 1*, Center for Lake Superior Environmental Studies, University of Wisconsin-Superior, Superior, WI, 1992.

Calow, P., *Handbook of Ecotoxicology*, Blackwell Science, Malden, MA, 1998.

Carey, J.H., *Ecotoxicological Risk Assessment of the Chlorinated Organic Chemicals*, SETAC Press, Pensacola, FL, 1998.

Clark, J.R. et al., *Using Reproductive and Developmental Effects Data in Ecological Risk Assessments for Oviparous Vertebrates Exposed to Contaminants*, National Health and Environmental Effects Research Laboratory, Corvallis, OR, University of North Texas, Denton, TX, Michigan State University, East Lansing, MI, Du Pont De Nemours (E.I.) and Co., Wilmington, DE, Rhone-Poulenc, Shelton, CT, 1998, EPA/600/A-98/133.

Clements, R.G. and Nabholz, J.V., *ECOSAR: a Computer Program for Estimating the Ecotoxicity of Industrial Chemicals Based on Structure Activity Relationships; User's Guide*, Office of Pollution Prevention and Toxics, U.S. Environmental Protection Agency, Washington, 1994, EPA-784-R-93-002.

De Peyster, A. and Day, K., Ecological risk assessment: a meeting of policy and science, in *Proceedings of the SETAC Workshop on Ecological Risk Assessment: a Meeting of Policy and Science*, Society of Environmental Toxicology and Chemistry, Pensacola, FL, 1998.

Desesso, J.M. and Price, F.T., *General Guidance for Ecological Risk Assessment at Air Force Installations*, Brooks Air Force Base, TX, 1990.

Efroymson, R.A. et al., *Preliminary Remediation Goals for Ecological Endpoints*, Oak Ridge National Laboratory, Oak Ridge, TN, 1997, ES/ER/TM-162/R2.

Fairbrother, A. et al., *Report on the Shrimp Virus Peer Review and Risk Assessment Workshop: Developing a Qualitative Ecological Risk Assessment*, Eastern Research Group, Inc., Lexington, MA, Ecological Planning and Toxicology, Inc., Corvallis, OR, Rosenstiel School of Marine and Atmospheric Science, Miami, FL, Menzie-Cura and Associates, Inc., Chelmsford, MA, National Center for Environmental Assessment, U.S. Environmental Protection Agency, Washington, 1999, EPA-68-C6-0041.

Hall, L.W. et al., *Probabilistic Ecological Risk Assessment of Tributyltin in the Chesapeake Bay Watershed Final Report*, Chesapeake Bay Program, U.S. Environmental Protection Agency Annapolis, 1998.

Hansen, D.J. et al., *Sediment Quality Criteria for the Protection of Benthic Organisms: Acenaphthene*, Office of Water, U.S. Environmental Protection Agency, Washington, Applications International Corp., Narragansett, RI, Hydroqual, Inc., Mahwah, NJ, Manhattan College, Bronx, 1993, EPA/822/R-93/013.

Hansen, D.J. et al., *Sediment Quality Criteria for the Protection of Benthic Organisms: Dieldrin*, Office of Water, U.S. Environmental Protection Agency, Washington, Applications International Corp., Narragansett, RI, Hydroqual, Inc., Mahwah, NJ, Manhattan College, Bronx, 1993, EPA/822/R-93/015.

Hansen, D.J. et al., *Sediment Quality Criteria for the Protection of Benthic Organisms: Endrin*, Office of Water, U.S. Environmental Protection Agency, Washington, Applications International Corp., Narragansett, RI, Hydroqual, Inc., Mahwah, NJ, Manhattan College, Bronx, 1993, EPA-822-R-93-016.

Hansen, D.J. et al., *Sediment Quality Criteria for the Protection of Benthic Organisms: Fluoranthene*, Office of Water, U.S. Environmental Protection Agency, Washington, Applications International Corp., Narragansett, RI, Hydroqual, Inc., Mahwah, NJ, Manhattan College, Bronx, 1993, EPA-822-R-93-012

Hansen, D.J. et al., *Sediment Quality Criteria for the Protection of Benthic Organisms: Phenanthrene*, Office of Water, U.S. Environmental Protection Agency, Washington, Applications International Corp., Narragansett, RI, Hydroqual, Inc., Mahwah, NJ, Manhattan College, Bronx, 1993, EPA-822-R-93-014.

Hull, R.N. and Belluck, D.A., Shifting from baseline to life cycle risk assessments to estimate ecological risks and impacts at hazardous waste sites, in *13th Annual Meeting of the Society of Environmental Toxicology and Chemistry*, 1992, Cincinnati, 1992, 233.

Ingersoll, C.G., Dillon, T., and Biddinger, G.R., Ecological risk assessment of contaminated sediments, in *Proceedings of the Pellston Workshop on Sediment Ecological Risk Assessment*, April 23-28, 1995, Pacific Grove, CA, Setac Press, Pensacola, FL, 1997.

Jrgensen, S.E., *Handbook of Estimation Methods in Ecotoxicology and Environmental Chemistry*, Lewis Publishers, Boca Raton, FL, 1998.

Kearney (A.T.), Inc., *Risk Assessment for the Waste Technologies Industries (WTI) Hazardous Waste Incineration Facility (East Liverpool, Ohio), Vol. 5, Human Health Risk Assessment; Evaluation of Potential Risks from Multipathway Exposure to Emissions*, Chicago, ENVIRON International Corp., Arlington, VA, Midwest Research Inst., Kansas City, MO, Earth Tech., Concord, MA, Waste, Pesticides and Toxics Division, U.S. Environmental Protection Agency, Chicago, 1997, EPA-68-W4-0006; EPA/905/R-97/002e.

Kearney (A.T.), Inc., *Risk Assessment for the Waste Technologies Industries (WTI) Hazardous Waste Incineration Facility (East Liverpool, Ohio), Vol. 6, Screening Ecological Risk Assessment*, Chicago, ENVIRON International Corp., Arlington, VA, Midwest Research Inst., Kansas City, MO, Earth Tech., Concord, MA, Waste, Pesticides and Toxics Division, U.S. Environmental Protection Agency, Chicago, 1997, EPA-68-W4-0006; EPA/905/R-97/002f.

Kearney (A.T.), Inc., *Risk Assessment for the Waste Technologies Industries (WTI) Hazardous Waste Incineration Facility (East Liverpool, Ohio), Vol. 8, Additional Analysis in Response to Peer Review Recommendations*, Chicago, ENVIRON International Corp., Arlington, VA, Midwest Research Inst., Kansas City, MO, Earth Tech., Concord, MA, Waste, Pesticides and Toxics Division, U.S. Environmental Protection Agency, Chicago, 1997, EPA-68-W4-0006; EPA/905/R-97/002h.

Kelly, E. et al., *Integrated, Comprehensive Ecological Impact Assessment in Support of Department of Energy Decision Making*, Office of Technical and Environmental Support, U.S. Department of Energy, Washington, No Date.

Lackey, R.T., *Ecological Risk Assessment, Corvallis Environmental Research Laboratory*, Corvallis, OR, U.S. Environmental Protection Agency, Washington, 1996, EPA/600/A-95/151.

Lawrence Berkeley National Laboratory, *Total Risk Integrated Methodology, Implementation of the Trim Conceptual Design through the Trim.Fate Module, a Status Report*, Office of Air Quality Planning and Standards, U.S. Environmental Protection Agency, Research Triangle Park, NC, 1998, EPA-68-D-30094; EPA/452/R-98/001.

Lewis, M.A., Ecotoxicology and risk assessment for wetlands, in *Proceedings from the SETAC Pellston Workshop on Ecotoxicology and Risk Assessment for Wetlands*, SETAC Press, Pensacola, FL, 1999.

Long, E.R. et al., Incidence of adverse biological effects within ranges of chemical concentrations in marine and estuarine sediments, *Environ. Manage.*, 19(1), 81, 1997.

Macdonald, D.D. et al., *The Development of Canadian Marine Environmental Quality Guidelines, Marine Environmental Quality Series, No. 1*, Eco-Health Branch, Environment Canada, Ecosystem Sciences and Evaluation Directorate, Ottawa, Ontario, 1992.

Matanoski, G.M. et al., *Advisory on the Problem Formulation Phase of EPA's Watershed Ecological Risk Assessment Case Studies*, Science Advisory Board, U.S. Environmental Protection Agency, Washington, 1997, EPA-SAB-EPEC-ADV-97-001.

Matanoski, G.E., and Maki, A.W., *SAB Report: Review of the Agency's Draft Ecological Risk Assessment Guidelines*, Science Advisory Board, U.S. Environmental Protection Agency, Washington, 1997, EPA SAB-EPEC-97-002.

McVey, M., *Wildlife Exposure Factors Handbook: Project Summary*, Center for Environmental Research, U.S. Environmental Protection Agency, Cincinnati, 1994, EPA/600/SR-93/187.

McVey, M., *Wildlife Exposure Factors Handbook*, Office of Health and Environmental Assessment, Office of Research and Development, U.S. Environmental Protection Agency, Washington, 1993, EPA/600/R-93/187.

McVey, M. et al., *Wildlife Exposure Factors Handbook, Appendix: Literature Review Database, Vol. 1 of 2*, ICF Kaiser International, Inc., Fairfax, VA, Office of Health and Environmental Assessment, U.S. Environmental Protection Agency, Washington, 1993, EPA-68-C8-0003; EPA-68-W8-0098; EPA/600/R-93/187a.

McVey, M. et al., *Wildlife Exposure Factors Handbook, Appendix: Literature Review Database, Vol. 2 of 2*, ICF Kaiser International, Inc., Fairfax, VA, Office of Health and Environmental Assessment, U.S. Environmental Protection Agency, Washington, 1993, EPA-68-C8-0003; EPA-68-D0-0101; EPA/600/R-93/187b.

Menzie-Cura & Associates Inc., *An Assessment of the Risk Assessment Paradigm for Ecological Risk Assessment*, Commission on Risk Assessment and Risk Management, Washington, 1996.

Molak, V., *Fundamentals of Risk Analysis and Risk Management*, Lewis Publishers, Boca Raton, FL, 1997.

Nabholz, J.V., Environmental hazard and risk assessment under the united states toxic substances control act, *The Science of the Total Environment*, 109/110, 649, 1991.

National Research Council, *Ecological Indicators for the Nation*, Committee to Evaluate Indicators for Monitoring Aquatic and Terrestrial Environments, National Academy Press, Washington, 2000.

Paustenbach, D.J., Ed., *The Risk Assessment of Environmental and Human Health Hazards: a Textbook of Case Studies*, John Wiley & Sons, New York, 1989.

Raccette, P. and Bailey, M., *Developing Risk-Based Cleanup Levels at Resource Conservation and Recovery Act Sites in Region 10, Interim Final Guidance*, Tetra Tech, Inc., U.S. Environmental Protection Agency, Seattle, WA, 1998.

Risk Assessment Forum, *Framework for Ecological Risk Assessment*, U.S. Environmental Protection Agency, Washington, 1992, EPA/630/R-92/001.

Risk Assessment Forum, *Guide for Ecological Risk Assessment*, U.S. Environmental Protection Agency, Washington, 1998, PB97-141303; PB96-193198.

Russo, R.C., *AQUIRE: Aquatic Information Retrieval Toxicity Data Base: Project Description, Guidelines, and Procedures*, Environmental Research Laboratory, Office of Research and Development, U.S. Environmental Protection Agency, Duluth, MN, 1985, EPA/600/8-84/021.

Sample, B.E., et. al., *a Guide to the Ornl Ecotoxicological Screening Benchmarks: Background, Development, and Application*, Oak Ridge National Laboratory, Oak Ridge, TN, 1998, ORNL/TM-13615.

Schuytema, G.S. and Nebeker, A.V., *Amphibian Toxicity Data for Water Quality Criteria Chemicals*, National Health and Environmental Effects Research Laboratory, Corvallis, OR, 1996, EPA/600/R-96/124.

Science Advisory Board, *An SAB Report: Review of the Agency's Draft Ecological Risk Assessment Guidelines*, U.S. Environmental Protection Agency, Washington, 1997, EPA-SAB-EPEC-97-002.

Suter, G.W., *A Framework for Assessing Ecological Risks of Petroleum-Derived Materials in Soil*, Oak Ridge National Laboratory, Oak Ridge, TN, 1997, ORNL/TM-13408.

Suter, G.W., *Ecological Risk Assessment for Contaminated Sites*, Lewis Publishers, Boca Raton, FL, 2000

Suter, G.W., Ecological Endpoints, in *Ecological Assessments of Hazardous Waste Sites: a Field and Laboratory Reference Document*, Warren-Hicks, W. et al., Eds., U.S. Environmental Protection Agency, Washington, 1989, EPA/600/3-89/013.

Suter, G.W., *Guide for Developing Conceptual Models for Ecological Risk Assessments*, Oak Ridge National Laboratory, Oak Ridge, TN, 1996, ES/ER/TM-186.

Suter, G.W., *Guide for Performing Screening Ecological Risk Assessments at Doe Facilities*, Oak Ridge National Laboratory, Oak Ridge, TN, 1995, ES/ER/TM-153.

Suter, G.W., *Risk Characterization for Ecological Risk Assessment of Contaminated Sites*, Oak Ridge National Laboratory, Oak Ridge, TN, 1996, ES/ER/TM-200.

Suter, G.W. et al., *Approach and Strategy for Performing Ecological Risk Assessments for the U.S. Department of Energy's Oak Ridge Reservation: 1995 Revision*, Oak Ridge National Laboratory, Oak Ridge TN, 1995, ES/ER/TM-33/R2.

Tetra Tech, Inc., *Phase I Inventory of Current EPA Efforts to Protect Ecosystems*, Fairfax, VA; Office of Water, U.S. Environmental Protection Agency, Washington, 1995, EPA 841-S-95-001.

U.S. Army, *Tri-Service Procedural Guidelines for Ecological Risk Assessments at U.S. Army Sites. Vol. 1*, Edgewood Research, Development and Engineering Center, Aberdeen Proving Ground, MD, 1996.

U.S. Army Corps of Engineers, *Environmental Quality — Risk Assessment Handbook — Vol. 1, Human Health Evaluation*, Washington, 1999, EM 200-1-4.

U.S. Army Corps of Engineers, *Risk Assessment Handbook, Vol. II: Environmental Evaluation*, U.S. Army Corps of Engineers, Washington, 1996, EM 200-1-4.

U.S. Army Corps of Engineers, *Technical Project Planning (TPP) Process*, Washington, 1998, EM 200-1-2.

U.S. Environmental Protection Agency, *A Compendium of Technologies Used in the Treatment of Hazardous Wastes*, Center for Environmental Research Information, Office of Research and Development, Cincinnati, 1987, EPA/625/8-87/014.

U.S. Environmental Protection Agency, *A Review of Ecological Assessment Case Studies from a Risk Assessment Perspective*, Risk Assessment Forum, Washington, 1993.

U.S. Environmental Protection Agency, *Aquatic Toxicity Information Retrieval Data Base (Aquire for Non-Vms) (On Magnetic Tape)*, Washington, 1995, EPA/DF/MT-95/047.

U.S. Environmental Protection Agency, *Catalogue of Standard Toxicity Tests in Ecological Risk Assessment, Intermittent Bulletin*, Office of Solid Waste and Emergency Response, Washington, 1994, EPA/540- F-94-013.

U.S. Environmental Protection Agency, *Draft Final Guidelines for Ecological Risk Assessment*, Risk Assessment Forum, Washington, 1997.

U.S. Environmental Protection Agency, *Ecological Impact of Land Cleanup and Restoration, Air and Radiation*, 1993, EPA 402-R-93-077.

U.S. Environmental Protection Agency, *Ecological Risk Assessment Guidance for Superfund: Process for Designing and Conducting Ecological Risk Assessments*, Office of Emergency and Remedial Response, Washington, 1997, EPA/540/R-97/006.

U.S. Environmental Protection Agency, *Ecological Risk Assessment for Superfund: Process for Designing and Conducting Ecological Risk Assessments, Interim Final*, Environmental Response Team, Edison, NJ, 1997.

U.S. Environmental Protection Agency, *Ecological Risk Assessment Guidance for RCRA Corrective Action, Region 5, Interim Draft*, Waste Management Division, Chicago, 1994.

U.S. Environmental Protection Agency, *Ecological Risk Assessment in the Federal Government (Computer File)*, Committee on Environment and Natural Resources Research, Washington, 1999, SCERN/5-99/001.

U.S. Environmental Protection Agency, *Ecological Risk Assessment Methods: a Review and Evaluation of Past Practices in the Superfund and RCRA Programs*, Office of Policy Analysis, Washington, 1989, EPA-230-03-89-044.

U.S. Environmental Protection Agency, *Ecological Risk Management in the Superfund and RCRA Programs, Office of Policy Analysis*, Office of Policy, Planning, and Evaluation, Washington, 1989, EPA-230-03-89-045.

U.S. Environmental Protection Agency, *Estimating Releases and Waste Treatment Efficiencies for the Toxic Chemical Release Inventory Form*, Office of Pesticides and Toxic Substances, Washington, 1987, EPA-560/4-88-002.

U.S. Environmental Protection Agency, *Framework for Ecological Risk Assessment*, Risk Assessment Forum, Office of Research and Development, Washington, 1992, EPA/630/R-92/001.

U.S. Environmental Protection Agency, *Guidelines for Deriving Site-Specific Sediment Quality Criteria for the Protection of Benthic Organisms*, Washington, 1993, EPA 822-R-93-017.

U.S. Environmental Protection Agency, *Guidelines for Ecological Risk Assessment*, Risk Assessment Forum, Washington, 1998, EPA/630/R-95/002 F; EPA/630/R-95/002F.

U.S. Environmental Protection Agency, *Inventory of EPA Headquarters Ecosystem Tools Compiled by Program Evaluation Division (2134)*, Office of Policy, Planning and Evaluation, Washington, 1995, EPA 230-S-95-001.

U.S. Environmental Protection Agency, *Issuance of Final Guidance: Ecological Risk Assessment and Risk Mangement Principles for Superfund Sites*, Office of Emergency and Remedial Response, Washington, 1999, EPA/540/F-98/067.

U.S. Environmental Protection Agency, *Priorities for Ecological Protection: an Initial List and Discussion Document for EPA*, Office of Research and Development, Washington, 1997, EPA/600/S-97/002.

U.S. Environmental Protection Agency, *Proposed Guidelines for Ecological Risk Assessment*, Risk Assessment Forum, Washington, 1996, EPA/630/R-95/002b.

U.S. Environmental Protection Agency, *Quality Criteria for Water*, Washington, 1986, EPA 440/5-86-001.

U.S. Environmental Protection Agency, *Regional Guidance for Conducting Ecological Risk Assessment*, Chicago, 1992.

U.S. Environmental Protection Agency, *Risk Assessment Guidelines for Superfund, Vol. II Environmental Evaluation Manual*, Washington, 1989, EPA/540/1-89/001.

U.S. Environmental Protection Agency, *Screening Level Ecological Risk Assessment Protocol for Hazardous Waste Combustion Facilities: Peer Review Draft (Kit)*, Office of Solid Waste and Emergency Response, Washington, 1999.

U.S. Environmental Protection Agency, *Screening Level Ecological Risk Assessment Protocol for Hazardous Waste Combustion Facilities: Peer Review Draft*, 1999, EPA 530-C-99-004.

U.S. Environmental Protection Agency, *Screening Level Ecological Risk Assessment Protocol for Hazardous Waste Combustion Facilities: Peer Review Draft, Vol. 1*, Office of Solid Waste and Emergency Response, Washington, 1999, EPA/530/D-99/001a.

U.S. Environmental Protection Agency, *Screening Level Ecological Risk Assessment Protocol for Hazardous Waste Combustion Facilities, Peer Review Draft, Vol. 2, Appendix A*, Office of Solid Waste and Emergency Response, Washington, 1999, EPA/530/D-99/001b.

U.S. Environmental Protection Agency, *Screening Level Ecological Risk Assessment Protocol for Hazardous Waste Combustion Facilities, Peer Review Draft, Vol. 3, Appendices B-H*, Office of Solid Waste and Emergency Response, Washington, 1999, EPA/530/D-99/001c.

U.S. Environmental Protection Agency, *Summary Report on Issues in Ecological Risk Assessment*, Risk Assessment Forum, Washington, 1991, EPA/625/3-91/018.

U.S. Environmental Protection Agency, *Supplemental Ecological Risk Assessment Guidance for Superfund (EPA Region 10)*, Seattle, 1997, EPA/910/R-97/005.

U.S. Environmental Protection Agency, *Technical Basis for Deriving Sediment Quality Criteria for Nonionic Organic Contaminants for the Protection of Benthic Organisms by Using Equilibrium Partitioning*, Office of Water, Washington, 1993, EPA-822-R-93-011.

U.S. Environmental Protection Agency, *Technical Basis for Deriving Sediment Quality Criteria for Nonionic Organic Contaminants for the Protection of Benthic Organisms by Using Equilibrium Partitioning: Fact Sheet*, Office of Water, Washington, 1993, EPA-822-F-93-012

U.S. Environmental Protection Agency, *Terrestrial Soil Fauna in Environmental Assessment and Environmental Management*, Exposure Assessment Group, Washington, No Date.

U.S. Environmental Protection Agency, *The Nature and Extent of Ecological Risks at Superfund Sites and RCRA Facilities*, Office of Policy Analysis, Washington, 1989, EPA-230-03-89-043.

U.S. Environmental Protection Agency, *Total Risk Integrated Methodology: Technical Support Document for the Trim.Fate Module*, Office of Air Quality Planning and Standards, Research Triangle Park, NC, 1998, EPA/452/D-98/001.

U.S. Environmental Protection Agency, *TRIM: Total Risk Integrated Methodology, Status Report*, Office of Air Quality Planning and Standards, Research Triangle Park, NC, 1999, EPA/453/R-99/010.

U.S. Environmental Protection Agency, *TRIM: Total Risk Integrated Methodology, Trim.Expo Technical Support Document*, Office of Air Quality Planning and Standards, Research Triangle Park, NC, 1999, EPA/453/D-99/001.

U.S. Environmental Protection Agency, *TRIM: Total Risk Integrated Methodology, Trim.Fate: Technical Support Document, Vol. 1, Description of Module, External Review Draft*, Office of Air Quality Planning and Standards, Research Triangle Park, NC, 1999, EPA/453/D-99/002a.

U.S. Environmental Protection Agency, *TRIM: Total Risk Integrated Methodology, Trim.Fate: Technical Support Document, Vol. 2, Description of Chemical Transport and Transformation Algorithms, External Review Draft*, Office of Air Quality Planning and Standards, Research Triangle Park, NC, 1999.

U.S. Environmental Protection Agency, *Watershed Ecological Risk Assessment*, Office of Research and Development, Office of Water, Washington, 1997, EPA/822/F-97-002.

U.S. Fish and Wildlife Service, *Contaminant Hazard Reviews*, Patuxent Wildlife Research Center, Laural, MD, Occasional Series.

U.S. National Science and Technology Council, *Ecological Risk Assessment in the Federal Government*, The Committee on Environment and Natural Resources Research, Washington, 1999, CERN/5-99/001.

van Straalen, N.M. and Lkke, H., *Ecological Risk Assessment of Contaminants in Soil*, Chapman & Hall, New York, 1997.

Warren-Hicks, W. et al., Eds., *Ecological Assessments of Hazardous Waste Sites: a Field and Laboratory Reference Document*, Washington, Environmental Research Laboratory, Corvallis, OR, 1989, EPA/600/3-89/013.

Warren-Hicks, W. and Moore, D.R.J., Uncertainty analysis in ecological risk assessment, in *Proceedings of the Pellston Workshop on Uncertainty Analysis in Ecological Risk Assessment: 23-28*, SETAC Press, Pensacola, FL, 1998.

Wentsel et al., *Tri-Service Guidelines for Ecological Risk Assessment, Vol. 1*, U.S. Army Edgewood Research, Development and Engineering Center, Aberdeen Proving Ground, MD, 1996.

World Health Organization, *Chemical Risk Assessment: Human Risk Assessment, Environmental Risk Assessment and Ecological Risk Assessment: Training Module No. 3*, Inter-Organization Programme for the Sound Management of Chemicals, World Health Organization, Geneva, Switzerland, 1999.

Zamuda, C., *Ecological Risk Assessment Methods: a Review and Evaluation of Past Practices in the Superfund and RCRA (Resource Conservation and Recovery Act) Programs (Executive Summary Included)*, Office of Policy, Planning and Evaluation, U.S. Environmental Protection Agency, Washington, 1989, EPA/230/03-89/044.

Zamuda, C., *Summary of Ecological Risks, Assessment Methods, and Risk Management Decisions in Superfund and RCRA*, Office of Policy Analysis, Office of Policy, Planning, and Evaluation, U.S. Environmental Protection Agency, Washington, 1989, EPA-230-03-89-046.

Zeeman, M.G., *EPA's Framework for Ecological Effects Assessment, in Screening and Testing Chemicals in Commerce*, Office of Technology Assessment, U.S. Congress, Washington, 1995, OTA-BP-ENV-166.

ECONOMIC ANALYSIS

Anderson, R., Dower, R., and Yang, E., *Economic Analysis and Risk Management: an Application to Hazardous Wastes*, Environmental Law Institute, Washington, Municipal Environmental Research Laboratory, U.S. Environmental Protection Agency, Cincinnati, 1984, EPA-R-805920; EPA-600/2-84-001.

Arnold, F.S., *Economic Analysis of Environmental Policy and Regulations*, John Wiley & Sons, New York, 1995.

Berry, M., *Reducing Lead Exposure in Australia: Risk Assessment and Analysis of Economic, Social and Environmental Impacts: Final Report*, Australian Government Publishing Service, Canberra, Australia, 1994.

Bishop, R.C., and Romano, D., *Environmental Resource Valuation: Applications of the Contingent Valuation Method in Italy*, Kluwer Academic Publishers, Boston, 1998.

Brimson, J.A., *Activity Accounting — an Activity-Based Costing Approach*, John Wiley & Sons, New York, 1991.

Clark, L.H., *EPA's Use of Benefit-Cost Analysis 1981–1986*, Office of Policy Analysis, U.S. Environmental Protection Agency, Washington, 1987, EPA-230-05-87-028.

Collins, F., Ed., *Implementing Activity Based Costing*, Executive Enterprises, New York, 1991.

Cothern, C.R., *Comparative Environmental Risk Assessment*, Lewis Publishers, Boca Raton, FL, 1993.

Estes, R.W., *Corporate Social Accounting*, John Wiley & Sons, New York, 1976.

Finn, S. et al., *Cost-Risk Analysis of Protective Actions for a Low-Level Deposition of Radionuclides*, Office of Radiation Programs, U.S. Environmental Protection Agency, Washington, 1980, EPA-68-01-3549.

Freeman, A.M., *Air and Water Pollution Control: a Benefit-Cost Assessment*, John Wiley & Sons, New York, 1982.

Freeman, A.M., *The Benefits of Environmental Improvement: Theory and Practice*, Resources for the Future, Washington, John Hopkins University Press, Baltimore, 1979.

Guerrero, P.F., *Federal Hazardous Waste Sites: Opportunities for More Cost-Effective Cleanups:* Statement for the Record by Peter F. Guerrero, Director, Environmental Protection Issues, Resources, Community, and Economic Development Division, before the Subcommittee on Superfund, Waste Control and Risk Assessment, Committee on Environment and Public Works, General Accounting Office, U.S. Senate, Washington, 1995.

Harrison, L.L., *McGraw-Hill Environmental Auditing Handbook: a Guide to Corporate and Environmental Risk Management*, McGraw-Hill, New York, 1984.

Kahn, E., *Commercialization of Solar Energy by Regulated Utilities Economic and Financial Risk Analysis*, Lawrence Berkeley Laboratory, University of California, Energy & Environment Division, Berkeley, 1980, LBL-11398.

Kneese, A.V., *Measuring the Benefits of Clean Air and Water*, Resources for the Future, Washington, 1984.

Levin, H.M. and Mcewan, P.J., *Cost Effectiveness Analysis*, Sage Publications, Thousand Oaks, CA, 2000.

Logan, S.E., *Development and Application of a Risk Assessment Method for Radioactive Waste Management Final Contract Report*, Office of Radiation Programs, U.S. Environmental Protection Agency, Las Vegas, 1978, EPA 520/6-78-005; 68-01-3256.

Logan, S.E., and Berbano, M.C., *Development and Application of a Risk Assessment Method for Radioactive Waste Management. Vol. I: Generic Description of Amraw-a Model*, New Mexico University, Albuquerque, Office of Radiation Programs, U.S. Environmental Protection Agency, Washington, Technology Assessment Division, 1978, NE-44(77)EPA-394-1-V1; EPA-68-01-3256; EPA/520/6-78/005a.

Logan, S.E. et al., *Development and Application of a Risk Assessment Method for Radioactive Waste Management. Vol. III: Economic Analysis; Description and Implementation of Amraw-B Model*, New Mexico University, Albuquerque, Office of Radiation Programs, U.S. Environmental Protection Agency, Washington, 1978, NE-44(77)EPA-394-1-V3; EPA-68-01-3256; EPA/520/6-78/005c.

Mishan, E.J., *Cost-Benefit Analysis*, Praeger Publishers, New York, 1976.

Moll, K. et al., *Hazardous Wastes: a Risk-Benefit Framework Applied to Cadmium and Asbestos*, Stanford Research Institute, Menlo Park, CA, Office of Research and Development, U.S. Environmental Protection Agency, Washington, 1975, SRI-EGU-3561; EPA-68-01-2915.

Seneca, J. and Taussig, M.K., *Environmental Economics*, 3rd ed., Prentice-Hall, Englewood Cliffs, NJ, 1984.

Smith, C.B., *Energy Management Principles, Applications, Benefits, and Savings*, Pergamon Press, Elmsford, NY, 1981.

Stanford Research Institute, *Chemical Economics Handbook*, SRI International, Menlo Park, CA, 1965.

Templet, P. and Farber, S., The complementarity between environmental and economic risk an empirical analysis, *Ecol. Econ.*, 9(2), 153–165, 1994.

Tietenberg, T., *Environmental Economics and Policy*, 2nd ed., Addison-Wesley, Reading, MA, 1998.

U.S. Environmental Protection Agency, *BEN Model: Calculates Violators' Economic Benefits from Noncompliance (Version 4.4)* (on Diskette), Office of Enforcement, Washington, 1998, PB98-500382.

U.S. Environmental Protection Agency, *Economic Analysis and Cost-Effectiveness Analysis of Proposed Effluent Limitations Guidelines and Standards for Industrial Waste Combustors*, Washington, Engineering and Analysis Division, 1998, EPA/821/B-97/010.

U.S. Environmental Protection Agency, *Economic and Cost-Effectiveness Analysis for Proposed Effluent Limitations Guidelines and Standards for the Landfills Point Source Category*, Office of Water, Washington, 1998, EPA-821-B-97-005.

U.S. Environmental Protection Agency, *Liner Location Risk and Cost Analysis Model, Appendices*, U.S. Pope-Reid Associates, St. Paul, MN, Office of Solid Waste, Washington, 1985, EPA-68-01-6621.

U.S. Environmental Protection Agency, *Regulatory Impact Analysis and Revised Health Risk Reduction and Cost Analysis for Radon in Drinking Water*, Public Comment Draft, Office of Water, Washington, 1999, EPA/815/D-99/002.

U.S. Environmental Protection Agency, *Regulatory Impact Analysis of the Phase 3 Land Disposal Restrictions Final Rule and Addendum: Revised Risk Assessment for Spent Aluminum Potliners*, Office of Solid Waste, Washington, 1996, EPA/530/R-97/021.

U.S. Environmental Protection Agency, *Risk Ranking Project Region 2: Economic/Welfare Ranking and Problem Analysis*, Region II, New York, 1991, PB96-110671.

U.S. Environmental Protection Agency, *Risk Ranking Project, Region 2, Ecological Ranking and Problem Analysis: Health, Ecology and Welfare/Economics*, New York, 1991, PB94-100351.

U.S. Environmental Protection Agency, *Scoper's Notes, an RI/Fs Costing Guide*, Solid Waste and Emergency Response, Washington, 1990, EPA/540/G-90/002.

U.S. Environmental Protection Agency, *Valuing Potential Environmental Liabilities for Managerial Decision-Making: a Review of Available Techniques*, Office of Pollution Prevention and Toxics, Washington, 1996, EPA/742-R-96-003.

Yoe, C.E., *Introduction to Risk and Uncertainty in the Evaluation of Environmental Investments*, Water Resources Support Center, U.S. Army Corps of Engineers, Alexandria, VA, 1996.

ENVIRONMENTAL LAW AND ENFORCEMENT

American Bar Association, *Law of Environmental Justice: Theories and Procedures to Address Disproportionate Risk*, Section of Environment, Energy, and Resources, American Bar Association, Chicago, 1999.

Anon., *Case Law Pertaining to the Personal Liability of Corporate Directors, Officers and Managers for Environmental Violations*, Bell, Boyd & Lloyd, Chicago, 1992.

Baram, M.S. and Partan, D.G., *Corporate Disclosure of Environmental Risks U.S. and European Law*, Butterworth Legal Publishers, Stoneham, MA, 1990.

Benjamin., S.L. and Belluck, D.A., *Evaluating a Pesticide Exposure*, YLD Newsletter, Wisconsin Bar Association, Madison, WI, 1999.

Benjamin, S.L. et al., *Can Desktop Risk Assessments Replace Science in Law?*, Nato Advanced Study Institute, Antalya, Turkey, 1999.

Belluck, D.A. and Benjamin, S.L., Discussion of toxicological risk assessment distortions editorial, *Ground Water*, 28(4), 616, 1990.

Belluck, D.A. and Benjamin, S.L., *Risk Assessment, Policy and Legal Implications of Groundwater Contamination by Atrazine and Its Metabolites*, North American Association for Environmental Education Conference Proceedings, St. Paul, MN, 1991.

Columbia University School of Law, Symposium: Risk Assessment in Environmental Law, *Columbia J. Environ. Law*, Columbia University, New York, 14, 289, 1989.

Daggett, C.J., Hazen, R.E., and Shaw, J.A., Advancing environmental protection through risk assessment, *Columbia J. Environ. Law*, 14, 315, 1989.

Fagin, D., *Toxic Deception: How the Chemical Industry Manipulates Science, Bends the Law, and Endangers Your Health*, Carol Publishing Group, Secaucus, NJ, 1996.

Foster, K.R., *Phantom Risk: Scientific Inference and the Law*, MIT Press, Cambridge, MA, 1993.

Gage & Tucker and Environmental Audit, Inc., *Environmental Due Diligence Handbook*, Government Institutes, Rockville, MD, 1991.

Hancock, W.A., Ed., *Corporate Counsel's Guide to Environmental Compliance and Audits, Business Laws, Inc.*, Chesterland, OH, 1993.

Latin, H., Good Science, Bad Regulation, and Toxic Risk Assessment, *Yale J. on Regulation*, 5, 89, 1988.

Mays, R.H., *Environmental Laws, Impact on Business Transactions*, Bureau of National Affairs, Washington, 1992

Nabholz, J.V., *Environmental Hazard and Risk Assessment under the United States Toxic Substances Control Act, the Science of the Total Environment*, 109/110, 649, 1991.

Nabholz, J.V. et al., Environmental risk assessment of new chemicals under the toxic substances control act (TSCA) section five, in *Environmental Toxicology and Risk Assessment*, Landis, W.G. et al., Eds., American Society for Testing and Materials, Philadelphia, 1993, ASTM STP 1179.

Schnapf, L.P., *Environmental Liability, Law and Strategy for Business and Corporations, Vols. I and II*, Butterworth Legal Publishers, Salem, NH, 1994.

U.S. Congress, *Trade and Environment: Conflict and Opportunities*, Office of Technology Assessment, U.S. Government Printing Office, Washington, 1992, OTA-BP-ITE-94.

U.S. Environmental Protection Agency, *Enforcement Project Management Handbook*, Office of Solid Waste and Emergency Response, Washington, 1993, Directive 9837.2b.

U.S. Environmental Protection Agency, *Environmental Law: a Selective, Annotated Bibliography and Guide to Legal Research*, Library Management Series, Administrative and Resources Management, New York, 1993, EPA 220-B-93-009.

U.S. Environmental Protection Agency, *Guide to Federal Environmental Requirements for Small Governments*, Office of the Administrator, Washington, 1993, EPA 270-K-93-001.

Wargo, J., *Our Children's Toxic Legacy: How Science and Law Fail to Protect Us from Pesticides*, Yale University Press, New Haven, CT, 1998.

Zeeman, M.G., Ecotoxicity testing and estimation methods developed under section five of the toxic substances control act (TSCA), in *Fundamentals of Aquatic Toxicology: Effects, Environmental Fate, and Risk Assessment*, 2nd ed., Rand, G., Ed., Taylor & Francis, Washington, 1995.

Zeeman, M.G. et al., *The Development of SAR/QSAR for Use under EPA's Toxic Substances Control Act (TSCA): an Introduction, in Environmental Toxicology and Risk Assessment, Vol. 2*, Gorsuch, J.W. et al., Eds., American Society for Testing and Materials, Philadelphia, 1993, ASTM STP 523-539.

Zeeman, M.G. and Gilford, J., Biological hazard evaluation and risk assessment under EPA's toxic substances control Act *(TSCA)*: an introduction, in *Environmental Toxicology and Risk Assessment*, Landis, W.G. et al., Eds., American Society for Testing and Materials, Philadelphia, 1993, ASTM STP 1179.

ENVIRONMENTAL STATISTICS

Armitgage, P. and Berry, G., *Statistical Methods in Medical Research*, Blackwell Scientific Publications, London, 1994.

Breslow, N.E. and Day, N.E., *Statistical Methods in Cancer Research, Vol. I: The Analysis of Case-Control Studies*, Scientific Publication Series, No. 32, International Agency for Research on Cancer, Lyon, France, 1980.

Breslow, N.E. and Day, N.E., *Statistical Methods in Cancer Research, Vol. II: The Analysis of Cohort Studies*, Scientific Publication Series, No. 82, International Agency for Research on Cancer, Lyon, France, 1987.

Chiang, C.L., *An Introduction to Stochastic Processes and their Applications*, Robert E. Krieger Publishing Company, Huntington, NY, 1980.

Clayton, D. and Hills, M., *Statistical Models in Epidemiology*, Oxford University Press, New York, 1993.

Cook, J.R. et al., *A Computer Code for Cohort Analysis of Increased Risks of Death (CAIRD)*, U.S. Environmental Protection Agency, Washington, 1978, EPA 520/4-78-012.

Cox, L.H. and Ross, N.P., *Statistical Issues in Environmental Monitoring and Assessment of Anthropogenic Pollution*, Environmental Statistics and Information Division, U.S. Environmental Protection Agency, Research Triangle Park, NC, 1996, EPA/600/A-96/018.

De Boor, C., A practical guide to splines, in *Applied Mathematical Sciences*, Vol. 27, Springer-Verlag, New York, 1978.

Eastern Research Group, Inc., *Report of the Workshop on Selecting Input Distributions for Probabilistic Assessments*, Office of Research and Development, U.S. Environmental Protection Agency, Washington, 1999, EPA/630-R-98-004

Firestone, M., *Guiding Principles for Monte Carlo Analysis*, Risk Assessment Forum, U.S. Environmental Protection Agency, Washington, 1997, EPA/630-R-97-001.

Fritsch, F.N. and Butland, J., *A Method for Constructing Local Monotone Piecewise Cubic Interpolants*, Lawrence Livermore National Laboratory, 1982, UCRL-87559.

Gad, S.C., *Acute Toxicology Testing*, Academic Press, San Diego, 1998

Gad, S.C., *Statistics and Experimental Design for Toxicologists*, CRC Press, Boca Raton, FL, 1999.

Gad, S.C., *Statistics and Experimental Design for Toxicologists*, CRC Press, Boca Raton, 1991.

Gad, S.C. and Chengelis, C.P., *Animal Models in Toxicology*, M. Dekker, New York, 1992

Gad, S.D. and Weil, C.S., Eds., *Statistics and Experimental Design for Toxicologists*, Telford Press, Caldwell, NJ, 1986.

Gart, J.J. et al., *Statistical Methods in Cancer Research, Vol. III: the Analysis of Long-Term Animal Experiments*, International Agency for Research on Cancer, Lyon, France, 1986, Scientific Publication Series No. 79.

Gilbert, R.O., *Statistical Methods for Environmental Pollution Monitoring*, Van Nostrand Reinhold Co., New York, 1987.

Harass, M.C., Regulatory use of ecotoxicity statistics: a U.S. perspective, *Ecotoxicology*, 5, 145, 1996.

Huntsberger, D.V. and Leaverton, P.E., *Statistical Inference in the Biomedical Sciences*, Allyn and Bacon, Boston, 1970.

Kaputska, L.A. et al., *Quantifying Effects in Ecological Site Assessment: Biological and Statistical Considerations*, Environmental Research Laboratory, Corvallis, OR, 1990, EPA/600/D-90/152.

Kocher, D.C. et al., *Uncertainties in Geologic Disposal of High-Level Wastes: Groundwater Transport of Radionuclides and Radiological Consequences*, Office of Nuclear Regulatory Research, U.S. Nuclear Regulatory Commission, Washington, 1983, NUREG/CR-2506; ORNL-5338.

Kume, H., *Statistical Methods for Quality Improvement*, Association for Overseas Technical Scholarship (AOTS), Tokyo, Japan, 1987.

National Center for Health Statistics, *U.S. Decennial Life Tables for 1979-1981, Vol. 1, No. 1, United States Life Tables*, Public Health Service, National Center for Health Statistics, U.S. Nuclear Regulatory Commission, Hyattsville, MD, 1985, PHS 85-1150-1.

National Institutes of Health, *Report of the National Institutes of Health Ad Hoc Working Group to Develop Radioepidemiological Tables*, Superintendent of Documents, U.S. Government Printing Office, Washington, 1985, NIH Publication 85-2748.

National Research Council, *Health Effects of Exposure to Low Levels of Ionizing Radiation (BEIR V)*, Committee on the Biological Effects of Ionizing Radiations, Board of Radiation Effects Research, Commission on Life Sciences, National Research Council, National Academy Press, Washington, 1990.

Public Health Service, *The International Classification of Diseases, 9th Revision*, Clinical Modification (ICD-9-CM), Vol. 1: Diseases: Tabular List, Public Health Service, U.S. Department of Health and Human Services, Health Care Financing Administration, Superintendent of Documents, U.S. Government Printing Office, Washington, 1980, DHHS Publication No. PHS 80-1260.

Sjoreen, A.L. et al., *PREPAR: a User-Friendly Preprocessor to Create Airdos-EPA Input Data Sets*, Oak Ridge National Laboratory, TN, U.S. Environmental Protection Agency, Washington, Department of Energy, Washington, 1984, Conf-840202-17; W-7405-ENG-26.

Snedecor, G.W. and Cochran, W.G., *Statistical Methods*, Iowa State University Press, Ames, IA, 1980.

United Nations, *Glossary of Environment Statistics*, United Nations, Statistical Division, 1997.

U.S. Environmental Protection Agency, *1997 Toxics Release Inventory: Public Data Release: State Fact Sheets*, Office of Pollution Prevention and Toxics, Washington, 1999, EPA 745-F-99-001.

U.S. Environmental Protection Agency, *A Guide to Selected National Environmental Statistics in the U.S. Government*, Office of Policy, Planning and Evaluation, Environmental Statistics and Information Division, Washington, 1993, EPA 230-R-93-003.

U.S. Environmental Protection Agency, *EMAP Statistical Methods Manual*, National Health and Environmental Effect Research Laboratory, Corvallis, OR, 1996, EPA/620/R-96/002.

U.S. Environmental Protection Agency, *Environmental Statistics*, Office of Information Resources Management, Washington, 1990, EPA/IMSD-90/002.

Vaeth, M. and Pierce, D.A., *Calculating Excess Lifetime Risk in Relative Risk Models*, Editorial Office, Radiation Effects Research Foundation, Hiroshima, Japan, 1989, RERF CR 3-89.

EPIDEMIOLOGY

Aldrich, T.E., *Environmental Epidemiology and Risk Assessment*, Van Nostrand Reinhold, New York, 1993.

Allen, B.C. et al., *Investigation of Cancer Risk Assessment Methods, Vol. 1, Introduction and Epidemiology*, Office of Health and Environmental Assessment, U.S. Environmental Protection Agency, Washington, 1987, EPA-68-01-6807; EPA/600/6-87/007b.

Andelman, J.B. and Underhill, D.W., *Health Effects from Hazardous Waste Sites*, University of Pittsburgh, Center for Environmental Epidemiology, U.S. Environmental Protection Agency, Lewis Publishers, Chelsea, MI, 1987.

Bayliss, D., *Health Assessment for 2,3,7,8-Tetrachlorodibenzo-P-Dioxin (TCDD) and Related Compounds, in Epidemiology / Human Data*, Syracuse Research Corporation, NY, Office of Health and Environmental Assessment, U.S. Environmental Protection Agency, Washington, 1992, EPA-68-C0-0043; EPA/600/AP-92/001G.

Blancato, J.N., Hopkins, J., and Rhomberg, L., *Update to the Health Assessment Document and Addendum for Dichloromethane (Methylene Chloride): Pharmacokinetics, Mechanism of Action, and Epidemiology*, External Review Draft, Office of Health and Environmental Assessment, U.S. Environmental Protection Agency, Washington, 1987, EPA/600/8-87/030A.

Castellani, A., *Epidemiology and Quantitation of Environmental Risk in Humans from Radiation and Other Agents*, Plenum Press, New York, 1985.

Chapman, R.S. et al., *Epidemiology of Lung Cancer in Xuan Wei, China: Current Progress, Issues, and Research Strategies*, Health Effects Research Lab, Research Triangle Park, NC, Institute of Environmental Health and Engineering, Beijing, China, Yunnan Province Anti-epidemic Station, Kunming, China, 1988, EPA/600/J-88/261.

Clayton, D. and Hills, M., *Statistical Models in Epidemiology*, Oxford University Press, New York, 1993.

Conference of State and Territorial Epidemiologists, *Proc. Environ. Epidem. and Toxicol.: Prac. Applic. at the State Level*, St. Charles, MO, Astho Foundation, Kensington, MD, 1983.

Draper, W.M., *Environmental Epidemiology Effects of Environmental Chemicals on Human Health*, American Chemical Society, Washington, 1994.

Dufour, A. P., *Health Effects Criteria for Fresh Recreational Waters*, Health Effects Research Laboratory, U.S. Environmental Protection Agency, Research Triangle Park, NC, 1984, EPA-600/1-84-004.

Ferris, B.G. et al., *Exposure Measurement for Air Pollution Epidemiology*, Harvard School of Public Health, Boston, Health Effects Research Laboratory, U.S. Environmental Protection Agency, Research Triangle Park, NC, 1988, EPA-R-811650; EPA/600/D-88/167.

Gilbert, E.S., *Approaches to Quantitative Expression of Dose Response*, Pacific Northwest Laboratory, Richland, WA, 1985, DE85-013498;PNL-SA—13193.

Gordis, L. and Libauer, C.H., *Epidemiology and Health Risk Assessment*, Oxford University Press, New York, 1988.

Graham, J.D., Role of epidemiology in regulatory risk assessment, *Proceedings of the Conference on the Proper Role of Epidemiology in Risk Analysis*, Boston, Elsevier, New York, 1995.

Health Effects Institute, *Particulate Air Pollution and Daily Mortality: Replication and Validation of Selected Studies, the Phase I Report of the Particle Epidemiology Evaluation Project*, Health Effects Institute, Cambridge, MA, 1995.

Keil, J.E. and Propert, D.M., *Epidemiology Feasibility Study: Effects of Noise on the Cardiovascular System, Appendix C, Review of Non-Noise Related Research of Cardiovascular Disease*, South Carolina University, Columbia, SC, Office of Noise Abatement, U.S. Environmental Protection Agency, Washington, 1981, EPA-68-01-6274; EPA-550/9-81-103C.

Kopfler, F.C. and Craun, G.F., *Environmental Epidemiology*, Division of Environmental Chemistry, American Chemical Society, Lewis Publishers, Chelsea, MI, 1986.

Landau, E., *Epidemiology Studies, Task III: Vinyl Chloride*, Office of Toxic Substances, U.S. Environmental Protection Agency, Washington, 1975.

Lilienfeld, A.M. and Lilienfeld, D.E., *Foundations of Epidemiology, 2nd ed.*, Oxford University Press, New York, 1988.

Mcdonnell, W.F. et al., *U.S. Environmental Protection Agency's Ozone Epidemiology Research Program: a Strategy for Assessing the Effects of Ambient Ozone Exposure Upon Morbidity in Exposed Populations*, Health Effects Research Laboratory, U.S. Environmental Protection Agency, Research Triangle Park, NC, 1993, EPA/600/J-94/044.

Norell, S.E., *A Short Course in Epidemiology*, Raven Press, New York, 1992.

Prentice, R.L. and Whittemore, A.S., Environmental epidemiology: risk assessment, *Proceedings of a Conference Sponsored by Siam Institute of Mathematics and Society and Supported by the Department of Energy*, Siam Institute for Mathematics and Society, Siam, Philadelphia, 1982.

Steenland, K. and Savitz, D.A., *Topics in Environmental Epidemiology*, Oxford University Press, New York, 1997.

U.S. Environmental Protection Agency, *Epidemiology and Air Pollution*, National Research Council, Washington, Committee on the Epidemiology of Air Pollutants, Washington, 1985, EPA-68-02-4073.

Underhill, D.W. and Radford, E.P., New and sensitive indicators of health impacts of environmental agents, *Proceedings of the Third Annual Symposium on Environmental Epidemiology*, 1982, Pittsburgh, PA, Center for Environmental Epidemiology, Graduate School of Public Health, University of Pittsburgh, Pittsburgh, PA, 1986.

Ward, J.R., *Use of Emergency Room Patient Populations in Air Pollution Epidemiology*, Health Effects Research Laboratory, U.S. Environmental Protection Agency, Research Triangle Park, 1978.

Whittemore, A.S., *Epidemiology in Risk Assessment for Regulatory Policy*, Department of Family, Community and Preventive Medicine, Stanford University, CA, Health Effects Research Laboratory, U.S. Environmental Protection Agency, Research Triangle Park, NC, 1986, EPA-R-813495; EPA/600/J-86/531.

EXPOSURE ASSESSMENT

Adams, J.D., *Pesticide Assessment Guidelines: Subdivision K, Exposure, Reentry Protection*, Office of Pesticide and Toxic Substances, U.S. Environmental Protection Agency, Washington, 1984, EPA 540/9-84-001.

Agency for Toxic Substances and Disease Registry, *Case Studies in Environmental Medicine*, Public Health Service, Atlanta, Occasional series.

Agency for Toxic Substances and Disease Registry, *Managing Hazardous Materials Incidents, Vol. 3*, Medical Management Guidelines for Acute Chemical Exposures, Atlanta, No Date.

American Industrial Hygiene Association, *Odor Thresholds for Chemicals with Established Occupational Health Standards*, Akron, OH, 1989.

Burns, L.A. et al., *Exposure Analysis Modeling System (EXAMS) User Manual and System Documentation*, Environmental Research Laboratory, Office of Research and Development, U.S. Environmental Protection Agency, Athens, GA 1982, EPA-600/3-82-023.

Burns, L.A. et al., *Exposure Analysis Modeling System: User's Guide for Exams-II*, Office of Research and Development, U.S. Environmental Protection Agency, Athens, GA, 1985, EPA/600/3/85/038.

California Environmental Protection Agency, *Caltox, a Multimedia Total Exposure Model for Hazardous-Waste Sites*, Technical Reports, Office of Scientific Affairs, Sacramento, CA, 1993.

Daisey, J.M. et al., *SAB Report: Development of the Acute Reference Exposure*, U.S. Environmental Protection Agency, 1998, EPA-SAB-EHC-99-005.

Eastern Research Group, Inc., *Inventory of Exposure-Related Data Systems Sponsored by Federal Agencies*, U.S. Environmental Protection Agency, Washington, Centers for Disease Control, Atlanta, U.S. Department of Health and Human Services, Agency for Toxic Substances and Disease Registry, Chamblee, GA, 1992, EPA-600-R-92-078.

Eastern Research Group, Inc., *Report on the Peer Review Workshop on Revisions to the Exposure Factors Handbook*, held in Washington, 1995, Eastern Research Group, Inc., Lexington, MA, Office of Research and Development, U.S. Environmental Protection Agency, Washington, 1996, EPA-68-C1-0030; EPA-68-D5-0028; EPA/630/R-96/003.

General Sciences Corporation, *Draft Triair User's Guide*, Office of Pesticides and Toxic Substances, Exposure Evaluation Division, U.S. Environmental Protection Agency, Washington, 1990.

Harrigan, P., *Guidelines for Completing the Initial Review Exposure Report*, Office of Pollution Prevention and Toxic Substances, U.S. Environmental Protection Agency, Washington, 1994.

Konz, J.J. et al., *Exposure Factors Handbook*, Versar, Springfield, VA, Office of Health and Environmental Assessment, U.S. Environmental Protection Agency, Washington, 1989, EPA-68-02-4254; OHEA-E-286; EPA/600/8-89/043.

McVey, M. et al., *Wildlife Exposure Factors Handbook*, Office of Health and Environmental Assessment, U.S. Environmental Protection Agency, Washington, 1993, EPA/600/R-93/187.

McVey, M. et al., *Wildlife Exposure Factors Handbook, Vol. 1 of 2*, ICF Kaiser International, Fairfax, VA, Office of Health and Environmental Assessment, U.S. Environmental Protection Agency, Washington, 1993, EPA-68-C8-0003; EPA-68-W8-0098; EPA/600/R-93/187A.

McVey, M. et al., *Wildlife Exposure Factors Handbook, Appendix: Literature Review Database, Vol. 2 of 2*, ICF Kaiser International, Fairfax, VA, Office of Health and Environmental Assessment, U.S. Environmental Protection Agency, Washington, 1993, EPA-68-C8-0003; EPA-68-D0-0101; EPA/600/R-93/187B.

O'Bryan, T.R. and Ross, R.H., Chemical scoring system for hazard and exposure identification, *J. Toxicol. Environ. Health*, 1, 119, 1988.

Reinert, J.C. et al., *Pesticide Assessment Guidelines, Subdivision U: Applicator Exposure Monitoring*, Office of Pesticide Programs, U.S. Environmental Protection Agency, Washington, 1986, EPA/540/9-87/127.

Robinson, J.P. et al., *Microenvironmental Factors Related to Californians' Potential Exposure to Environmental Tobacco Smoke (ETS)*, Atmospheric Research and Exposure Assessment Lab, U.S. Environmental Protection Agency, Research Triangle Park, NC, 1994, EPA/600/R-94/116.

Ryan, J.A. and Chaney, R.L., *Regulation of Municipal Sewage Sludge under the Clean Water Act, Section 503: a Model for Exposure and Risk Assessment for MSW-Compost*, U.S. Environmental Protection Agency, Washington, 1992, EPA/600/A-94/023.

Science Advisory Board, *An SAB Report: Development of the Acute Reference Exposure*, U.S. Environmental Protection Agency, Washington, 1998, EPA-SAB-EHC-99-005.

Sharratt, M., Assessing risks from data on other exposure routes, *Reg. Toxicol. Pharmacol.*, 8, 399, 1988.

U.S. Environmental Protection Agency, *Dermal Exposure Assessment: Principles and Applications, Interim Report*, Office of Research and Development, Washington, 1992, EPA/600/8-91/011b.

U.S. Environmental Protection Agency, *Environmental Protection Agency Guidelines for Estimating Exposures*, Office of the Federal Register, National Archives and Records Administration, Washington, 1986.

U.S. Environmental Protection Agency, *EPA Risk Assessment Guidelines for Carcinogenicity, Mutagenicity, Chemical Mixtures, Developmental Effects, and Exposure Assessment*, Science Advisory Board, Washington, 1985.

U.S. Environmental Protection Agency, *Exposure Factors Handbook CDRom*, National Center for Environmental Assessment, Office of Research and Development, Washington, 1999, EPA/600/C-99-001.

U.S. Environmental Protection Agency, *Exposure Factors Handbook*, Office of Research and Development, National Center for Environmental Assessment, Washington, 1997, EPA 600-P-95-002FA; EPA 600-P-95-002FB; EPA 600-P-95-002FC.

U.S. Environmental Protection Agency, *Exposure Factors Handbook, Update to Exposure Factors Handbook*, Exposure Assessment Group, Office of Health and Environmental Assessment, Washingon, 1995, EPA/600/P-95/002A.

U.S. Environmental Protection Agency, *Exposure Factors Handbook: Vol. I — General Factors*, Office of Health and Environmental Assessment, Office of Research and Development, Washington, 1995, EPA/600/P-95/002Ba.

U.S. Environmental Protection Agency, *Exposure Factors Handbook, Vol. I of III — General Factors, Update to Exposure Factors Handbook*, Office of Research and Development, National Center for Environmental Assessment, Washington, 1996, EPA/600/P-95/002Ba.

U.S. Environmental Protection Agency, *Exposure Factors Handbook: Vol. II — Food Ingestion Factors*, Office of Health and Environmental Assessment, Office of Research and Development, Washington, 1995, EPA/600/P-95/002Bb.

U.S. Environmental Protection Agency, *Exposure Factors Handbook: Vol. III — Activity Factors*, Office of Health and Environmental Assessment, Office of Research and Development, Washington, 1995, EPA/600/P-95/002Bc.

U.S. Environmental Protection Agency, *Exposure Factors Handbook, Exposure Assessment Group*, Office of Health and Environmental Assessment, Washington, 1990, EPA/600/8-89/043.

U.S. Environmental Protection Agency, *Exposure Factors Handbook, Update to Exposure Factors Handbook, General Factors*, May 1989, Exposure Assessment Group, Office of Health and Environmental Assessment, Washington, 1995, EPA/600/P-95/002FA.

U.S. Environmental Protection Agency, *Exposure Factors Handbook, Vol. II of III — Food Ingestion Factors, Update to Exposure Factors Handbook*, May 1989, Office of Research and Development, National Center for Environmental Assessment, Washington, 1996, EPA/600/P-95/002Bb.

U.S. Environmental Protection Agency, *Exposure Factors Handbook, Vol. II, Food Ingestion Factors (Final Report)*, National Center for Environmental Assessment, Washington, 1997, EPA-68-D0-0101; EPA-68-D3-0013; NCEAW-0005-V2; EPA/600/P-95/002FB.

U.S. Environmental Protection Agency, *Exposure Factors Handbook, Vol. III of III — Activity Factors, Update to Exposure Factors Handbook*, May 1989, Office of Research and Development, National Center for Environmental Assessment, Washington 1996, EPA/600/P-95/002Bc.

U.S. Environmental Protection Agency, *Guidelines for Exposure Assessment*, Risk Assessment Forum, Office of Research and Development, Washington, 1992, EPA/600/Z-92/001.

U.S. Environmental Protection Agency, *Guidelines for the Use of Anticipated Residues in Dietary Exposure Assessment*, Washington, 1970.

U.S. Environmental Protection Agency, *Human Health Evaluation Manual, Supplemental Guidance: Standard Default Exposure Factors*, Washington, 1991, OSWER Directive 9285.6-03.

U.S. Environmental Protection Agency, *Interim Guidance for Dermal Exposure Assessment*, Office of Research and Development, Washington, 1991, EPA/600/8-91/011A.

U.S. Environmental Protection Agency, *Methods for Exposure-Response Analysis for Acute Inhalation Exposure to Chemicals*, Office of Research and Development, Washington, 1998, EPA/600/R-98/051.

U.S. Environmental Protection Agency, *Report on the Peer Review Workshop on Revisions to the Exposure Factors Handbook*, Office of Research and Development, Washington,, 1996, EPA-68-C1-0030; EPA-68-D5-0028; EPA/630/R-96/003.

U.S. Environmental Protection Agency, *Risk Assessment Guidance for Superfund, Vol. I: Human Health Evaluation Manual Supplemental Guidance, Standard Default Exposure Factors, Interim Final*, Washington, 1991, 9285.6-03; PB91-921314

U.S. Environmental Protection Agency, *Risk Assessment Guidance for Superfund, Vol. I, Human Health Evaluation Manual, Supplemental Guidance: Standard Default Exposure Factors*, Office of Emergency and Remedial Response, Washington, 1991, OSWER-9285.6-03.

U.S. Environmental Protection Agency, *SAB Advisory: the National Human Exposure Assessment Survey (NHEXAS) Pilot Studies*, Science Advisory Board, 1999, EPA-SAB-IHEC-ADV-99-004.

U.S. Environmental Protection Agency, *SAB Report: Review of Draft Final Exposure Assessment Guidelines*, Review of the Office of Health and Environmental Assessment and the Risk Assessment Forum's Draft Final Guidelines for Exposure Assessment by the Indoor Air Quality and Total Human Exposure Committee, Science Advisory Board, Washington, 1992, SAB-IAQC-92-015.

U.S. Environmental Protection Agency, *Selection Criteria for Mathematical Models Used in Exposure Assessments: Surface Water Models*, Office of Health and Environmental Assessment, Washington, 1987, EPA/600/8-87/042.

U.S. Environmental Protection Agency, *Sociodemographic Data Used for Identifying Potentially Highly Exposed Populations*, Office of Research and Development, Washington, 1999, EPA/600/R-99/060.

U.S. Environmental Protection Agency, *Superfund Exposure Assessment Manual, OSWER Directive 9285.5-1*, Office of Emergency and Remedial Response, Washington, 1988, EPA/540/1-88/001.

U.S. Environmental Protection Agency, *Superfund Fact Sheet: Exposure Pathways*, Office of Solid Waste and Emergency Response, Washington, 1992, Publication 9230.0-05FSB.

U.S. Environmental Protection Agency, *Superfund's Standard Default Exposure Factors for the Central Tendency and Reasonable Maximum Exposure*, Risk Assessment Council, Washington, 1993, EPA 600/D-93-901.

Vallero, D.A., *Transport, Transformation and Fate of Endocrine Disruptors: Potential Areas of Exposure Research*, National Exposure Research Laboratory, Research Triangle Park, NC, EPA/600/A-96/124.

Versar, Inc., *Exposure Factors Handbook, Preliminary SAB Review Draft, Vols. 1-3*, Versar, Springfield, VA, Exposure Assessment Div., National Center for Environmental Assessment, U.S. Environmental Protection Agency, Washington, 1996, EPA/600/P-95/002Ba; EPA/600/P-95/002Bb; EPA/600/P-95/002Bc; 68-02-4254; 68-D0-0101; 68-D3-0013; 68-D5-0051.

Wood, P. et al., *Exposure Factors Handbook, Vol. 1 of 3, General Factors*, Versar, Springfield, VA, Exposure Assessment Div., National Center for Environmental Assessment. U.S. Environmental Protection Agency, Washington, 1996, EPA-68-DO-0101; EPA-68-D3-0013; EPA-600/P-95/002Ba.

Wood, P. et al., *Exposure Factors Handbook, Vol. 2, Food Ingestion Factors (Final Report)*, Versar, Springfield, VA, Exposure Assessment Div., National Center for Environmental Assessment, U.S. Environmental Protection Agency, Washington, 1997, EPA-68-DO-0101; EPA-68-D3-0013; NCEA-W-0005-V2; EPA/600/P-95/002FB.

Wood, P. et al., *Exposure Factors Handbook, Vol. 3, Activity Factors (Final Report)*, Versar, Springfield, VA, Exposure Assessment Div., National Center for Environmental Assessment, U.S. Environmental Protection Agency, Washington, 1997, EPA-68-DO-0101; EPA-68-D3-0013; NCEA-W-0005-V3; EPA/600/P-95/002FC.

Yeh, K.W., *Radiation Exposure and Risks Assessment Manual (RERAM)*, Risk Assessment Using Radionuclide Slope Factors, Cohen, S. and Associates, Mclean, VA, Office of Air and Radiation, U.S. Environmental Protection Agency, Washington, 1996, EPA-68d20155; EPA/402/R-96/016.

FATE AND TRANSPORT

American Society for Testing and Materials, *RBCA Fate and Transport Models: Compendium and Selection Guidance*, ASTM, West Conshohocken, PA, 1999.

Barnthouse, L.W., *Assessing the Transport and Fate of Bioengineered Microorganisms in the Environment*, Oak Ridge National Laboratory, Oak Ridge, TN, 1990, CONF-8503163-1; DE85 016964.

Belluck, D.A., *Atrazine Biospheric Cycling, Minnesota Pollution Control Agency 7050 Rule Technical Exhibit*, Minnesota Pollution Control Agency, St. Paul, MN, 1993.

Callahan, M.A. et al., *Water-Related Environmental Fate of 129 Priority Pollutants*, Office of Water Planning and Standards, U.S. Environmental Protection Agency, Washington, 1979, EPA-440/4-79-029a and -029b.

Cho, J.S. and Wilson, J.T., *Exposure and Risk Assessment at Petroleum Contamination Site with Multimedia Contaminant Fate, Transport, and Exposure Model (MMSOILS)*, National Risk Management Research Laboratory, Ada, OK, 1998.

Clark, B. et al., Fugacity analysis and model of organic chemical fate in a sewage treatment plant, *Environ. Sci. Technol.*, 29, 1995.

General Sciences Corporation, *A User's Guide to Environmental Partitioning Model*, 1985.

Hedden, K.F., *Multimedia Fate and Transport Models: an Overview*, Environmental Research Laboratory, U.S. Environmental Protection Agency, Athens, GA, 1984, EPA-600/J-84-060.

Johnson, R.L., *Seminar Publication: Transport and Fate of Contaminants in the Subsurface*, Robert S. Kerr Environmental Research Laboratory, Ada, OK, 1989, EPA/625/4-89/019.

Joint United States-Mexico Conference, *Fate, Transport, and Interactions of Metals, a Joint United States-Mexico Conference*, Public Health Service, National Institutes of Health, National Institute of Environmental Health Sciences, U.S. Department of Health and Human Services, Research Triangle Park, NC, 1995.

Kollig, H.P., Ed., *Environmental Fate Constants for Organic Chemicals under Consideration for EPA's Hazardous Waste Identification Project*, Office of Research and Development, U.S. Environmental Protection Agency, Athens, GA, 1993, EPA/600/R-93/132.

Mackay, D. et al., *Illustrated Handbook of Physical-Chemical Properties and Environmental Fate for Organic Chemicals*, Lewis Publishing, Boca Raton, FL, 1992.

Mackay, D., *Multimedia Environmental Models, the Fugacity Approach*, Lewis Publishers, Chelsea, MI, 1993.

Mulkey, L.A. et al., *Aquatic Fate and Transport Modeling Techniques for Predicting Environmental Exposure to Organic Pesticides and Other Toxicants — a Comparative Study*, Environmental Research Laboratory, Athens, GA, 1986, EPA/600/D-86/122.

Ney, R.E., *Fate and Transport of Organic Chemicals in the Environment: a Practical Guide*, Government Institutes, Rockville, MD, 1998.

Nowell, C.J. et al., *Light Nonaqueous Phase Liquids*, Office of Research and Development, Robert S. Kerr Environmental Research Laboratory, Ada, OK, 1995, EPA/540/S-95/500.

Patterson, M.R. et al., *A User's Manual for UTM-TOX, the Unified Transport Model*, Oak Ridge National Laboratory, Oak Ridge, TN, 1984, ORNL-6064.

Reinbold, K.A. et al., *Adsorption of Energy-Related Organic Pollutants: a Literature Review*, National Research Laboratory, U.S. Environmental Protection Agency, Athens, GA, 1979, EPA 600/3-79-086.

Tinsely, I.J., *Chemical Concepts in Pollutant Behavior*, John Wiley & Sons, New York, 1979.

Trapp, S., *Modeling the Uptake of Organic Compounds into Plants, Fate and Prediction of Environmental Chemicals in Soils, Plants, and Aquatic Systems*, Lewis Publishing, Boca Raton, FL, 1993.

U.S. Environmental Protection Agency, *EMSOFT: Exposure Model for Soil-Organic Fate and Transport* (on Diskette), National Center for Environmental Assessment, Washington, 1997.

U.S. Environmental Protection Agency, *EPA's (Environmental Protection Agency's) Risk Assessment Methodology for Municipal Incinerator Emissions: Key Findings and Conclusions*, Report of the Municipal Waste Combustion Subcommittee Environmental Effects, Transport and Fate Committee, Science Advisory Board, Washington, 1987, SAB/EETFC-87/027.

U.S. Environmental Protection Agency, *EPA's Risk Assessment Methodology for Municipal Incinerator Emissions: Key Findings and Conclusions*, 1987, SAB/EETFC-87/027.

U.S. Environmental Protection Agency, *Fate, Transport and Transformation Test Guidelines: Oppts 835 Series*, Office of Prevention, Pesticides and Toxic Substances, Washington, 1998, PB98-118102.

U.S. Environmental Protection Agency, *Future Risk: Research Strategies for the 1990s*, the Science Advisory Board, Washington, 1988, SAB-EC-88-040.

U.S. Environmental Protection Agency, *Project Summary, Fate, the Environmental Fate Constants Information System Database*, Environmental Research Laboratory, Athens, GA, 1991, EPA/600/S3-91/045.

U.S. Environmental Protection Agency, *The Environmental Fate Constants Information System Database (FATE)*, Environmental Research Laboratory, Athens, GA, 1991.

U.S. Environmental Protection Agency, *Trim: Total Risk Integrated Methodology, Trim, Fate: Technical Support Document, Vol. 2, Description of Chemical Transport and Transformation Algorithms, External Review Draft*, Office of Air Quality Planning and Standards, Research Triangle Park, NC, 1999, EPA/453/D-99/002B.

Vallero, D.A., *Transport, Transformation, and Fate Endocrine Disruptors: Potential Areas of Exposure Research*, National Exposure Research Laboratory, U.S. Environmental Protection Agency, Research Triangle Park, NC, 1996, EPA/600/A-96/124.

FISH

Bigler, J., *Guidance for Assessing Chemical Contaminant Data for Use in Fish Advisories Risk Assessment and Fish Consumption Limits*, Office of Water, Fish Contaminant Section, U.S. Environmental Protection Agency, 1997, EPA 823-B-97-009.

Bigler, J., *Guidance for Assessing Chemical Contaminant Data for Use in Fish Advisories, Vol. II: Risk Assessment and Fish Consumption Limits*, Office of Water, Office of Science and Technology, U.S. Environmental Protection Agency, Washington, 1994, EPA 823-B-94-004.

Bigler, J. and Green, A., *Guidance for Assessing Chemical Contaminant Data for Use in Fish Advisories*, Office of Water, Office of Science and Technology, U.S. Environmental Protection Agency, Washington, 1993, EPA 823-R-93-002.

Dourson, M.L. and Clark, J.M., *Fish Consumption Advisories: Toward a Unified, Scientifically Credible Approach*, Environmental Criteria and Assessment Office, U.S. Environmental Protection Agency, Cincinnati, 1990, EPA/600/J-90/484.

Gardner, G.R., *Field and Laboratory Studies of Chemical Contamination and Environmentally Related Diseases in Fish and Molluscs of New England*, Environmental Research Laboratory — Narragansett, Newport, OR, 1993, EPA/600/A-93/120.

Hnath, J.G. et al., *Great Lakes Fish Disease Control Policy and Model Program*, Great Lakes Fishery Commission, Ann Arbor, MI, 1993.

Hoeting, J.A. and Olsen, A.R., *Are the Fish Safe to Eat, Assessing Mercury Levels in Fish in Maine Lakes*, National Health and Environmental Effects Research Laboratory, Corvallis, OR, Western Ecology Division, Department of Statistics, Colorado State University, Fort Collins, CO, 1997, EPA/600/A-97/062.

Knuth, B.A., *Guidance for Assessing Chemical Contaminant Data for Use in Fish Advisories, Risk Communication*, Office of Water, Office of Science and Technology, U.S. Environmental Protection Agency, Washington, 1995, EPA 823-R-95-001.

Kollig, H.P., *Environmental Fate Constants for Organic Chemicals under Consideration for EPA's Hazardous Waste Identification Projects*, Environmental Research Laboratory, U.S. Environmental Protection Agency, Athens, GA, 1993, EPA/600/R-93/132.

Mahaffey, K. et al., *Mercury Study Report to Congress, Vol. 7, Characterization of Human Health and Wildlife Risks from Mercury Exposure in the United States*, Office of Air Quality Planning and Standards, U.S. Environmental Protection Agency, Research Triangle Park, NC, 1997, EPA/452/R-97/009.

Research Triangle Institute, *National Listing of State Fish and Shellfish Consumption Advisories and Bans*, Research Triangle Institute, Research Triangle Park, NC, Office of Water, U.S. Environmental Protection Agency, Washington, 1993, EPA/823/B-93/005.

Roberts, D.W. et al., *Missouri Chlordane Exposure Study: a Report on Persons Who Consumed Chlordane-Contaminated Fish*, Missouri Department of Health, Jefferson City, MO, Agency for Toxic Substances and Disease Registry, Atlanta, 1992, ATSDR/HS-93/21.

Ryans, R.C., Fish physiology, fish toxicology, and fisheries management, *Proc. Intern. Symp.* Guangzhou, PRC, 1988, Office of Research and Development, Environmental Research Laboratory, Athens, GA, 1990, EPA/600/9-90/011.

U.S. Department of Health and Human Services, *Health Study to Assess the Human Health Effects of Mercury Exposure to Fish Consumed from the Everglades*, Public Health Service, Agency for Toxic Substances and Disease Registry, 1995, H75/ATH496918.

U.S. Environmental Protection Agency, *Consumption Surveys for Fish and Shellfish, a Review and Analysis of Survey Methods*, Office of Water, Washington, 1992, EPA 822/R-92-001.

U.S. Environmental Protection Agency, *Great Lakes Fish Monitoring Program: a Technical and Scientific Model for Interstate Environmental Monitoring*, Office of Water, Washington, 1990, EPA/503/4-90/004.

U.S. Environmental Protection Agency, *Guidance for Assessing Chemical Contaminant Data for Use in Fish Advisories, Overview of Risk Management*, Office of Water, Washington, 1996, EPA 823-B-96-006.

U.S. Environmental Protection Agency, *National Study of Chemical Residues in Fish, Vol. I and II*, Office for Science and Technology, Standards and Applied Science Division, Washington, 1992, EPA 823-R-92-008 A & B.

U.S. Environmental Protection Agency, *Guidance for Assessing Chemical Contaminant Data for Use in Fish Advisories, Vol. 2, Risk Assessment and Fish Consumption Limits*, Office of Water, Washington, 1994, EPA/823/B-94/004.

U.S. Environmental Protection Agency, *Guidance for Assessing Chemical Contaminant Data for Use in Fish Advisories, Vol. 4, Risk Communication*, Office of Water, Washington, 1995, EPA/823/R-95/001.

U.S. Environmental Protection Agency, *Pesticide Fact Sheet: Fish Oil, Biopesticides and Pollution Prevention Division*, Washington, 1998, RPA/730/F-98/004.

U.S. Environmental Protection Agency, *Potential Human Health Effects of Ingesting Fish Which are Taken from Locations Near the Savannah River Site (SRS)*, 1996, EPA 904/R-96/006.

U.S. Environmental Protection Agency, *Proceedings: National Forum on Mercury in Fish, Held in New Orleans, Louisiana, 1994*, Office of Water, Washington, 1995, EPA/823/R-95/002.

U.S. Environmental Protection Agency, *Proceedings of the U.S. Environmental Protection Agency's National Technical Workshop "PCBS in Fish Tissue,"* Office of Water, Washington, 1993, EPA/823/R-93/003

U.S. Environmental Protection Agency, *Special Interest Group (SIG) Forum for Fish Consumption Risk Management, User's Manual, Vol. 10: a Division of the Nonpoint Source Information Exchange Computer Bulletin Board System (NPS BBS)*, Office of the Assistant Administrator for Water, Washington, 1992, EPA/822/B-91/001.

U.S. Environmental Protection Agency, *Update: Listing of Fish and Wildlife and Advisories*, Office of Water, Washington, 1998, EPA/823/F-98/009.

GLOSSARIES

Air Risk Information Support Center, *Glossary of Terms Related to Health, Exposure, and Risk Assessment*, Air RISC, Research Triangle Park, NC, 1989, EPA/450/3-88/016.

Hyatt, D.E., *Environmental Monitoring and Assessment Program: Master Glossary*, Office of Research and Development, U.S. Environmental Protection Agency, Washington, 1993, EPA/620/R-93/013.

U.S. Environmental Protection Agency, *Terms of Environment: Glossary, Abbreviations, and Acronyms, Communications, Education and Public Affairs*, Washington, 1993, EPA 175-B-93-001.

GROUNDWATER

Anderson, H.A. et al., *Public Health Related Groundwater Standards-Cycle 3, Summary of Scientific Support Documentation for NR 140.10*, Wisconsin Department of Health and Social Services, Madison, WI, 1988.

Anderson, H.A. et al., *Public Health Related Groundwater Standards-Cycle 3, Summary of Scientific Support Documentation for NR 140.10*, Wisconsin Department of Health and Social Services, Madison, WI, 1986.

Anderson, H.A. et al., *Public Health Related Groundwater Standards-Cycle 3, Summary of Scientific Support Documentation for NR 140.10*, Wisconsin Department of Health and Social Services, Madison, WI, 1985.

ASTM, *RBCA Fate and Transport Models: Compendium and Selection Guidance*, ASTM, West Conshohocken, PA, 1999.

Barcelona, M.J. et. al., *Practical Guide for Ground-Water Sampling*, Illinois State Water Survey, Department of Energy and Natural Resources, Champaign, IL, Robert S. Kerr Environmental Research Laboratory, U.S. Environmental Protection Agency, Ada, OK, 1985, EPA/600/2-85/104.

Belluck, D.A. and Anderson, H.A., Wisconsin's Risk assessment based numerical groundwater standards program, in *Agrichemicals and Groundwater Protection: Resources and Strategies for State and Local Management*, Proceedings of a Conference, 1988, Freshwater Foundation, Navarre, MN, 1988.

Belluck, D.A. and Benjamin, S.L., Challenges for groundwater risk assessment in the 1990s, *U.S. Water News*, 7, 1990.

Belluck, D.A. and Benjamin, S.L., Groundwater contamination, *Wisconsin Lawyer*, 63(12), 17, 1990.

Belluck, D.A. and Benjamin, S.L., Groundwater initiatives and challenges in the Great Lakes Region, *U.S. Water News*, 1, 1991.

Belluck, D.A. and Benjamin, S.L., *Scientifically Credible Site-Specific Soil Cleanup Levels to Protect Ground Water*, RCRA Policy Forum, 3, 1-9, 1995.

Benjamin, S.L. and Belluck, D.A., *State Groundwater Regulation: Guide to Laws, Standards, and Risk Assessment*, BNA, Washington, 1994.

Brown, J., *Handbook of RCRA Ground-Water Monitoring Constituents, Chemical and Physical Properties (40 CFR Part 264, Appendix 9)*, Office of Solid Waste, U.S. Environmental Protection Agency, Washington, 1992.

Hantush, M.M. et al., *Analytical Tools for Groundwater Pollution Assessment*, National Risk Management Research Laboratory, Ada, OK, Subsurface Protection and Remediation Division, Mantech Environmental Technology, Ada, OK, California University, Davis, CA, 1998, EPA/600/A-98/042.

Montgomery, J.H., *Groundwater Chemicals Desk Reference*, Lewis Publishers, Boca Raton, FL, 1996.

Pettyjohn, W.A., *Regional Assessment of Aquifer Vulnerability and Sensitivity in the Coterminous United States*, Office of Research and Development, U.S. Environmental Protection Agency, Washington, 1991, EPA/600/2-91/043.

U.S. Environmental Protection Agency, *Assessment Framework for Ground-Water Model Applications*, Washington, 1994, EPA/500-B-94-003.

U.S. Environmental Protection Agency, *Groundwater, Handbook*, Robert S. Kerr Environmental Laboratory, Ada, OK, March 1987.

U.S. Environmental Protection Agency, *Handbook of RCRA Ground-Water Monitoring Constituents, Chemical and Physical Properties (40 CFR Part 264, Appendix 9)*, Office of Solid Waste, Washington, 1992, EPA 530-R-92-022.

U.S. Environmental Protection Agency, *Industrial Waste Management Evaluation Model (IWEM): Ground-Water Model: Draft*, Solid Waste and Emergency Response, 1999. SEPA 530-R-99-002.

U.S. Environmental Protection Agency, *Ground Water Resource Assessment*, Office of Ground Water and Drinking Water, Washington, 1993, EPA813-R-93-003.

U.S. Environmental Protection Agency, *Ground Water Sampling — a Workshop Summary*, Office of Research and Development, Washington, 1995, EPA/600/R-94/205.

HAZARDOUS AIR POLLUTANTS

Palazzolo, M.A., *Parametric Evaluation of VOC/HAP (Volatile Organic Compounds-Hazardous/Toxic Air Pollutants) Destruction Via Catalytic Incineration*, Radian Corp., Research Triangle Park, NC, Air and Energy Engineering Research, U.S. Environmental Protection Agency, Washington, 1985, EPA-68-02-3171; EPA-68-02-3515; EPA/600/2-85/041.

Stelling, J.H.E., *Emission Factors for Equipment Leaks of VOC (Volatile Organic Compound) and HAP (Hazardous Air Pollutants)*, Radian Corp., Research Triangle Park, NC, Office of Air Quality Planning and Standards, Research Triangle Park, U.S. Environmental Protection Agency, NC, 1986, EPA-68-02-3889; EPA/450/3-86/002.

U.S. Environmental Protection Agency, *National Emission Standards for Hazardous Air Pollutants (NESHAP) for Pesticide Active Ingredient Production: Summary of Public Comments and Responses*, Office of Air Quality Planning and Standards, U.S. Environmental Protection Agency, 1999, EPA-453/R-98-001.

U.S. Environmental Protection Agency, *Hazardous Air Pollutant Program (HAP-PRO), Version 1 (for Microcomputers)*, Research Triangle Park, NC, 1991, EPA/SW/DK-92/025.

HAZARD ASSESSMENT

American Petroleum Institute, *Management of Process Hazards*, Production and Refining Departments, American Petroleum Institute, Washington, 1990.

Benson, R., Toxicological review of Tributyltin Oxide (Cas No. 56-35-9), in *Support of Summary Information on the Integrated Risk Information System (IRIS)*, National Center for Environmental Assessment, U.S. Environmental Protection Agency, Washington, 1997, EPA/635/R-98/002.

Clansky, K.B., *Suspect Chemicals Sourcebook, a Guide to Industrial Chemicals Covered under Major Federal Regulatory and Advisory Programs*, Roytech, Burlingame, CA, 1988.

Council on Environmental Quality, *Cumulative Effects Handbook*, Washington, 1997.

Federal Emergency Management Agency et al., *Handbook of Chemical Hazard Analysis Procedures*, Technological Hazards Division, Washington, No Date.

Freeman, H.M., *Standard Handbook of Hazardous Waste Treatment and Disposal*, McGraw Hill, New York, 1998.

Foureman, G.L., Toxicological review of Cumene (Cas No. 98-82-8) in *Support of Summary Information on the Integrated Risk Information System*, National Center for Environmental Assessment, U.S. Environmental Protection Agency, Washington, 1997, EPA/635/R-98/001.

Government of Canada, Health and Welfare Canada, and Environment Canada, *Canadian Environmental Protection Act*, Priority Substances List Assessment Report, Canada, Occasional Series.

Gift, J.S., Toxicological review of Methyl Methacrylate (Cas No. 80-62-6) in *Support of Summary Information on the Integrated Risk Information System (IRIS)*, National Center for Environmental Assessment, U.S. Environmental Protection Agency, Washington, 1998, EPA/635/R-98/004.

Greenway, R.A., *Risk Management Planning Handbook: a Comprehensive Guide to Hazard Assessment*, Accidental Release Prevention, and Consequence Analysis, Government Institutes, Rockville, MD, 1998, OCLC 38504361.

International Agency for Research on Cancer (IARC), *IARC Monographs on the Evaluation of Carcinogenic Risk of Chemicals to Humans*, Lyon, France, 1988.

Mayer, F.L. et al., *Predicting Chronic Lethality of Chemicals to Fishes from Acute Toxicity Test Data: Concepts and Linear Regression Analysis, (Hazard Assessment)*, Environmental Research Laboratory, Gulf Breeze, FL, Missouri Agricultural Experiment Station, Columbia, MO, National Fisheries Contaminant Research Center, Columbia, MO, 1994, EPA/600/J-94/281.

Mcgaughy, R.E., Toxicological review of Chlordane (Technical) (Cas No. 12789-03-6) in *Support of Summary Information on Integrated Risk Information System (IRIS)*, National Center for Environmental Assessment, U.S. Environmental Protection Agency, Washington, 1997, EPA/635/R-98/003.

Miller, D.B., *Caveats in Hazard Assessment: Stress and Neurotoxicity*, Neurotoxicology Division, Health Effects Research Laboratory, U.S. Environmental Protection Agency, Research Triangle Park, NC, 1991, EPA/600/D-91/216.

National Fire Protection Association, *Fire Protection Guide on Hazardous Materials*, National Fire Protection Association, Quincy, MA, 1997.

National Institute for Occupational Safety and Health (NIOSH), *NIOSH Recommendations for Occupational Safety and Health*, Compendium of Policy Documents and Statements, Cincinnati, 1992.

National Research Council, *Setting Priorities for Drinking Water Contaminants*, Committee on Drinking Water Contaminants, National Research Council, Washington, 1998.

Owens/Urie Enterprises, *Chemical Substance Hazard Assessment and Protection Guide*, Owens/Urie Enterprises, Wheat Ridge, CO, 1995, OCLC 33431449.

Sax, I.N. and Lewis, R.J., *Dangerous Properties of Industrial Materials, 7th ed.*, Van Nostrand Reinhold, New York, 1989.

Sax, I.N. and Lewis, R.J., *Hazardous Chemicals Desk Reference*, Van Nostrand Reinhold, New York, 1987.

Smith, J.E., *Seminar Publication: Control of Lead and Copper in Drinking Water*, Office of Research and Development, U.S. Environmental Protection Agency, Washington, 1993, EPA/625/R-93/001.

Stribling, J.B. et al., *Biological Assessment Methods, Biocriteria, and Biological Indicators: Bibliography of Selected Technical, Policy, and Regulatory Literature*, Office of Policy, Planning, and Evaluation, U.S. Environmental Protection Agency, Washington, 1996, EPA/230-B-96-001.

Stricoff, R.S., *NIOSH/OSHA Pocket Guide to Chemical Hazards*, U.S. Government Printing Office, Washington, 1995.

U.S. Environmental Protection Agency, *EPCRA Section 313 Questions and Answers: Section 313 of the Emergency Planning and Community Right-To-Know Act*, Toxic Chemical Release Inventory, Office of Pollution Prevention and Toxics, 1999, EPA 745-B-98-004.

U.S. Environmental Protection Agency, *Health Effects Assessment Documents*, Environmental Criteria and Assessment Office, Washington, 1986, EPA/540/S 1-86/059.

U.S. Environmental Protection Agency, *Health Effects Assessment Documents, Project Summary*, Office of Health and Environmental Assessment, Washington, 1991, EPA/600/S8-91/041.

U.S. Environmental Protection Agency, *Integrated Risk Information System (IRIS)* (on Diskette), Office of Science and Technology, Washington, 1998, PB98-591330.

U.S. Environmental Protection Agency, *Integrated Risk Information System (IRIS)*, (For Microcomputers), Office of Health and Environmental Assessment, Washington, 1997, PB97-591330.

U.S. Environmental Protection Agency, *IRIS Background Paper*, Office of Health and Environmental Assessment, Office of Research and Development, Washington, 1993.

U.S. Environmental Protection Agency, *Lead in Drinking Water: an Annotated List of Publications*, Office of Water, Washington, 1993, EPA/812-R-93-001.

U.S. Environmental Protection Agency, *Proceedings of the Solvent Reactivity Conference*, Research Triangle Park, NC, 1974, EPA/650/3-74-010.

U.S. Environmental Protection Agency, *Review of the U.S. Consumer Product Safety Commission's Health Effects and Exposure Assessment Documents on Nitrogen Dioxide*, Report of the Clean Air Scientific Advisory Committee, Science Advisory Board, Washington, 1988, SAB/CASAC-86/021.

U.S. Environmental Protection Agency, *Updated Health Effects Assessment Documents: Project Summary*, Office of Health and Environmental Assessment, Cincinnati, 1991, EPA/600/S8-91/042.

Weiss, G., *Hazardous Chemicals Data Book*, Noyes Data Corporation, Park Ridge, NJ, 1986.

HUMAN HEALTH RISK ASSESSMENT

Anderson, H.A., *Workshop on Cancer Risk Assessment Guidelines Issues, Premeeting Comments*, Office of Research and Development, U.S. Environmental Protection Agency, Washington, 1994, EPA/630/R-94/005B.

Anonymous, The role of quantified risk assessment, *Health and Safety Information Bulletin*, 165, 4, 1989.

Agency for Toxic Substances and Disease Registry, *Public Health Assessment Manual*, Public Health Service, Atlanta, 1992.

Belluck, D.A. et al., *Human and Ecological Health Risks from Heavy Metals and Other Substances Released to the Environment from Metal Shredders*, in Cytotoxic, Mutagenic and Carcinogenic Potential of Metals Related to Human Environment, NATO Advanced Study Institute Series, Hadjiliadis, N.D., Ed., Kluwer Academic Publishers, The Netherlands, 363, 1997.

Beer, T. and Ziolkowski, F., *Environmental Risk Assessment: an Australian Perspective*, Commonwealth of Australia, Supervising Scientist Report 102, 1995.

Center for Risk Analysis, *Historical Perspective on Risk Assessment in the Federal Government*, Harvard School of Public Health, Boston, 1994.

Cohrssen, J.J. and Covello, V.T., *Risk Analysis: a Guide to Principles and Methods for Analyzing Health and Environmental Risks*, Executive Office of the President of the United States, Washington, 1989.

Dyckman, L.J., *Superfund EPA's Use of Risk Assessment in Cleanup Decisions*, statement of Lawrence J. Dyckman, Associate Director, Environmental Protection Issues, Resources, Community, and Economic Development Division, before the Subcommittee on Water Resources and Environment, Committee on Transportation and Infrastructure, General Accounting Office, House of Representatives, Washington, 1995, GAO/T-RCED-95-231.

Eastern Research Group, Inc., *Intergovernmental Public Meeting on Risk Assessment in the Federal Government: Asking the Right Questions*, Meeting Report and Public Comment Summary, Office of Health and Environmental Assessment, U.S. Environmental Protection Agency, Washington, EPA-68-C1-0030; EPA/600/R-94/206.

Eastern Research Group, Inc., *Report on the U.S, EPA Technical Workshop on WTI Incinerator Risk Assessment Issues*, Risk Assessment Forum, U.S. Environmental Protection Agency, Washington, 1996, EPA-68-D5-0028; EPA/630/R-96/001.

Eastern Research Group, Inc., *Report on the Workshop on Cancer Risk Assessment Guidelines Issues*, Center for Environmental Research Information, U.S. Environmental Protection Agency, Washington, 1994, EPA/630/R-94/005a.

Eisler, R., *Handbook of Chemical Risk Assessment: Health Hazards to Humans, Plants, and Animals*, Lewis Publishers, Boca Raton, FL, 2000.

Federal Emergency Management Agency, *Handbook of Chemical Hazard Analysis Procedures*, U.S. Department of Transportation, U.S. Environmental Protection Agency, Washington, 1990.

Greenway, R.A., *Risk Management Planning Handbook: a Comprehensive Guide to Hazard Assessment, Accidental Release Prevention, and Consequence Analysis*, Government Institutes. Rockville, MD, 1998.

Guerrero, P.F., *Superfund: Proposals to Remove Barriers to Brownfield Redevelopment*, Statement by Peter F. Guerrero, Director, Environmental Protection Issues, Resources, Community, and Economic Development Division, before the Subcommittee on Superfund, Waste Control, and Risk Assessment, Committee on Environment and Public Works, General Accounting Office, U.S. Senate, Washington, 1997.

Helfand, J.S. and Mancy, K.H., Carcinogenesis risk assessment model for environmental chemicals, *Wat. Sci. Tech.*, 27, 7-8, 279, 1993.

Kimmel, C.A., Quantitative approaches to human health risk assessment for noncancer health effects, *Neurotoxicology*, 11, 189, 1990.

Meek, M.E. et al., Approach to assessment of risk to human health for priority substances under the Canadian Environmental Protection Act, *Environmental Carcinogenesis and Ecotoxicology Reviews*, C12(2), 105, 1994.

Nabholz, J.V., Environmental hazard and risk assessment under the United States Toxic Substances Control Act, *The Science of the Total Environment*, 109/110, 649, 1991.

National Governors Association, *Risk Assessment in Government Programs*, Washington, 1996.

National Research Council, *Environmental Cleanup at Navy Facilities: Risk-Based Methods*, National Academy Press, Washington, 1999.

National Research Council, *Health Effects of Ingested Fluoride*, Subcommittee on Health Effects of Ingested Fluoride, National Academy Press, Washington, 1993.

National Research Council, *Issues in Risk Assessment*, National Academy Press, Washington, 1993.

National Research Council, *Risk Assessment in the Federal Government: Managing the Process*, National Academy Press, Washington, 1983.

National Research Council, *Science and Judgment in Risk Assessment*, National Academy Press, Washington, 1994.

U.S. Army Corps of Engineers, *Environmental Quality — Risk Assessment Handbook, Vol. I: Human Health Evaluation*, Washington, 1999, EM 200-1-4.

U.S. Army Corps of Engineers, *Risk Assessment Handbook, Vol. II: Environmental Evaluation*, Washington, 1996, EM 200-1-4.

U.S. Congress, *Researching Health Risks*, Office of Technology Assessment (OTA), Washington, 1993.

U.S. Department of the Army, *Health Risk Assessment Guidance for the Installation Restoration Program and Formerly Used Defense Sites*, Washington, 1996, PAM 40-578.

U.S. Environmental Protection Agency, *Assessment of Risks to Human Reproduction and to Development of the Human Conceptus from Exposure to Environmental Substances (ORNL-EIS-197)*, Washington, 1982, EPA/600-98-2001.

U.S. Environmental Protection Agency, *Guidance on Cumulative Risk Assessment. Part I, Planning and Scoping*, Science Policy Council, Washington, 1997.

U.S. Environmental Protection Agency, *Guidance for Data Usability in Risk Assessment, Interim Final*, Office of Emergency and Remedial Response, Washington, 1992, EPA/540/G-90/008.

U.S. Environmental Protection Agency, *Guidance on Risk Characterization for Risk Managers and Risk Assessors*, memorandum from Deputy Administrator for the Office of Solid Waste and Emergency Response, H. Habicht, to Assistant and Regional Administrators, 1992.

U.S. Environmental Protection Agency, *Hazard Ranking System Guidance Manual*, Office of Solid Waste and Emergency Response, Washington, 1992, EPA 540-R-92-026.

U.S. Environmental Protection Agency, *Human Health Evaluation Manual, Supplemental Guidance: Standard Default Exposure Factors*, Washington, 1991, OSWER Directive 9285.6-03.

U.S. Environmental Protection Agency, *Human Health Risk Assessment Protocol for Hazardous Waste Combustion Facilities, Vol. 2, Appendix A*, Office of Solid Waste and Emergency Response, Washington, 1998, EPA/530-D-98-001B.

U.S. Environmental Protection Agency, *Methodology for Assessing Health Risks Associated with Indirect Exposure to Combustor Emissions, Interim Final*, Environmental Criteria and Assessment Office, Office of Health and Environmental Assessment, Office of Research and Development, Cincinnati, 1990, EPA/600/6-90/003.

U.S. Environmental Protection Agency, *Proposed Guidelines for Carcinogen Risk Assessment*, National Center for Environmental Assessment, Washington, 1996, EPA//600/P-92/003C.

U.S. Environmental Protection Agency, *Qualitative and Quantitative Carcinogenic Risk Assessment*, Office of Air Quality Planning and Standards, Research Triangle Park, NC, 1987, EPA 450/5-87-003.

U.S. Environmental Protection Agency, *Risk Assessment Guidance for Superfund, Vol. I: Human Health Evaluation Manual (Part A), Interim Final*, Office of Emergency and Remedial Response, Washington, 1989, EPA/540/1-89/002.

U.S. Environmental Protection Agency, *Risk Assessment Guidance for Superfund, Vol. I: Human Health Evaluation Manual (Part B, Development of Risk-Based Preliminary Remediation Goals), Interim Final*, Office of Emergency and Remedial Response, Washington, 1991, EPA/540/R-92/003.

U.S. Environmental Protection Agency, *Risk Assessment Guidance for Superfund,Vol. I: Human Health Evaluation Manual (Part C, Risk Evaluation of Remedial Alternatives), Interim Final*, Office of Emergency and Remedial Response, Washington, 1992, EPA/540/R-92/004.

U.S. Environmental Protection Agency, *Risk Assessment Guidance for Superfund: Vol. 1. Human Health Evaluation Manual (Part D, Standardized Planning, Reporting and Review of Superfund Risk Assessments) (Interim)*, Office of Emergency and Remedial Response, Washington, 1998, EPA/540/R-97/033.

U.S. Environmental Protection Agency, *Role of the Baseline Risk Assessment in Superfund Remedy Selection Decisions*, Office of Solid Waste and Emergency Response, Washington, 1991, OSWER Directive 9355.0-30.

U.S. Environmental Protection Agency, *Selecting Exposure Routes and Contaminants of Concern by Risk-Based Screening, Region III*, Hazardous Waste Management Branch, Philadelphia, 1993.

U.S. Environmental Protection Agency, *Superfund Today: Focus on Revisions to Superfund's Risk Assessment Guidance*, Office of Solid Waste and Emergency Response, Washington, 1999, EPA/540/F-98/055.

U.S. Environmental Protection Agency, *Supplemental Guidance to RAGS: Calculating the Concentration Term*, Office of Solid Waste and Emergency Response, Washington.

U.S. Environmental Protection Agency, *The Risk Assessment Guidelines of 1986*, Office of Health and Environmental Assessment, Washington, 1987, EPA/600/8-87/045.

U.S. Environmental Protection Agency, *Understanding Superfund Risk Assessment*, Office of Emergency and Remedial Response, Washington, 1992, OSWER-9285.7-06FS.

Walker, B., Perspectives on quantitative risk assessment, *J. Environ. Health*, 55, 15, 1992.

World Health Organization, *Assessing Human Health Risks of Chemicals: Derivation of Guidance Values for Health-Based Exposure Limits*, IPCS (International Programme on Chemical Safety), World Health Organization, Geneva, 1994, Environmental Health Criteria No. 170.

World Health Organization, *Risk Assessment Guidance for Superfund: Vol. I, Human Health Evaluation Manual, Part A*, Office of Emergency and Remedial Response, U.S. Environmental Protection Agency, Washington, 1990, EPA/9285.7-01/FS.

MIXTURES

Brusseau, M.L., *Environmental Research Brief: Complex Mixtures and Ground Water Quality*, Robert S. Kerr Environmental Research Laboratory, Ada, OK, 1993, EPA/600/S-93/004.

Claxton, L.D. et al., *Results of the IPCS Collaborative Study on Complex Mixtures*, Health Effects Research Laboratory, U.S. Environmental Protection Agency, Research Triangle Park, NC, 1992, EPA/600/J-92/058.

Claxton, L.D. et al., *Overview, Conclusions, and Recommendations of the IPCS Collaborative Study on Complex Mixtures*, Health Effects Research Laboratory, U.S. Environmental Protection Agency, Research Triangle Park, NC, 1992, EPA/600/J-92/054.

Demarini, D.M. et al., *Molecular Analysis of Mutations Induced at the "HISD3052" Allele of Salmonella by Single Chemicals and Complex Mixtures*, National Research Council, Washington, 1994, EPA/600/J-94/218.

Houk, V.S. et al., *Use of the Spiral Salmonella Assay to Detect the Mutagenicity of Complex Environmental Mixtures*, Health Effects Research Laboratory, U.S. Environmental Protection Agency, Research Triangle Park, NC, 1991, EPA/600/J-91/059.

International Agency for Research on Cancer, *Complex Mixtures and Cancer Risk*, International Agency for Research on Cancer, Lyon, France, 1990.

Johnson, T.B., *Detection of Genotoxins in Contaminated Sediments: an Evaluation of a New Test for Complex Environmental Mixtures*, Great Lakes National Program Office, U.S. Environmental Protection Agency, Chicago, 1995, EPA 905-R95-002.

Krewski, D. et al., *Mutagenic Potency of Complex Chemical Mixtures Based on the Salmonella/Microsome Assay*, Health Effects Research Laboratory, U.S. Environmental Protection Agency, Research Triangle Park, NC, 1992, EPA/600/J-92/057.

Lewtas, J. et al., *Comparison of DNA Adducts from Exposure to Complex Mixtures in Various Human Tissues and Experimental Systems*, Health Effects Research Laboratory, U.S. Environmental Protection Agency, Research Triangle Park, NC, 1993, EPA/600/J-93/357.

Lewtas, J., *Strategies for Using Bioassay Methods for the Identification of Hazardous Components and Comparative Risk Assessment of Complex Mixtures*, Health Effects Research Laboratory, Office of Research and Development, U.S. Environmental Protection Agency, Research Triangle Park, NC, 1987.

Ma, T-H et al., Synergistic and antagonistic effects on genotoxicity of chemicals commonly found in hazardous waste sites, *Mutation Research*, 270, 71, 1992.

Mcdonnell, W.F., *Utility of Controlled Human Exposure Studies for Assessing the Health Effects of Complex Mixtures and Indoor Air Pollutants*, Health Effects Research Laboratory, U.S. Environmental Protection Agency, Research Triangle Park, NC, 1993, EPA/600/J-94/338.

National Research Council, *Complex Mixtures: Methods for in vivo Toxicity Testing*, National Academy Press, Washington, 1988.

Oregon State University, *Corvallis*, Corvallis Environmental Research Laboratory, OR, 1998, EPA/600/R-98/074.

Schubauer-Berigan, M.K. et al., Behavior and identification of toxic metals in complex mixtures: examples from effluent and sediment pore water toxicity identification evaluations, *Archives of Environmental Contamination and Toxicology*, 24, 298, 1993.

U.S. Environmental Protection Agency, *Detection of Genotoxins in Contaminated Sediments: an Evaluation of a New Test for Complex Environmental Mixtures*, Great Lakes National Program Office, Chicago, 1995, EPA 905-R95-002.

U.S. Environmental Protection Agency, *Guidelines for the Health Risk Assessment of Chemical Mixtures*, Washington, 1986, EPA/630-R-98-002.

U.S. Environmental Protection Agency, *SAB Report: the Cumulative Exposure Project*, Review of the Office of Planning, Policy, and Evaluation's Cumulative Exposure Project (Phase 1) by the Integrated Human Exposure Committee, Science Advisory Board, Washington, 1996, EPA-SAB-IHEC-ADV-96-004.

MODELING

California Environmental Protection Agency, *CALTOX (Trade Name): a Multimedia Total Exposure Model for Hazardous Waste Sites (Version 2.3)*, Office of Scientific Affairs, California Environmental Protection Agency, Sacramento, CA, 1997.

Cho, J.S. and Wilson, J.T., *Exposure and Risk Assessment at Petroleum Contamination Site with Multimedia Contaminant Fate, Transport, and Exposure Model (MMSOILS)*, National Risk Management Research Laboratory, U.S. Environmental Protection Agency, Ada, OK, 1998.

De Broissia, M., *Selected Mathematical Models in Environmental Impact Assessment in Canada*, Canadian Environmental Assessment Research Council, Hull, Quebec, 1986.

Dean, J.D., *Risk of Unsaturated/Saturated Transport and Transformation of Chemical Concentrations (RUSTIC), Vol. 1, Theory and Code Verification*, Environmental Research Laboratory, U.S. Environmental Protection Agency, Athens, GA, 1989, EPA-68-03-6304; EPA/600/3-89/048A.

European Centre for Ecotoxicology and Toxicology of Chemicals, *Monitoring and Modelling of Industrial Organic Chemicals with Particular Reference to Aquatic Risk Assessment*, Ecetoc, Brussels, Belgium, 1999.

Puglionesi, P.S., *Compliance Guidance and Model Risk Management Program for Water Treatment Plants*, American Water Works Association, Denver, CO, 1998.

Richardson, M., *Risk Assessment of Chemicals in the Environment*, Royal Society of Chemistry, London, 1988.

Sheng, Y.P. et al., *Three Dimensional Hydrodynamic Model for Stratified Flows in Lakes and Estuaries (HYDRO3D): Theory, User Guidance, and Applications for Superfund and Ecological Risk Assessments*, National Exposure Research Laboratory, U.S. Environmental Protection Agency, Athens, GA, 1999, EPA-R-814345-01-0; EPA/600/R-99/049.

Sparks, L.E., *RISK: an IAQ Model for Windows*, Air Pollution Prevention and Control Division, U.S. Environmental Protection Agency, Research Triangle Park, NC, 1998, EPA/600/A-98/004.

U.S. Environmental Protection Agency, *EMSOFT: Exposure Model for Soil-Organic Fate and Transport* (on Diskette), National Center for Environmental Assessment, Washington, 1997.

U.S. Environmental Protection Agency, *Pathogen Risk Assessment Model for Land Application of Municipal Sewage Sludge (Landapp Version 4.02)* (for Microcomputers), Office of Research and Development, Washington, 1996.

Weil, J.C. and Mitchell, W.J., *Progress in Developing an Open Burn/Open Detonation Dispersion Model*, National Oceanic and Atmospheric Administration, Boulder, CO; Atmospheric Research and Exposure Assessment Laboratory, Research Triangle Park, NC, 1996, EPA/600/A-96/056.

NEUROTOXICITY

Featherstone, D. et al., *Noninvasive Neurotoxicity Assay Using Larval Medaka*, Environmental Research Laboratory, Duluth, MN, Iowa State University, Ames, IA, Iowa Agricultural and Home Economics Experimentation, Ames, IA, 1993, EPA/600/J-94/019.

Goldey, E.S. et al., *Developmental Neurotoxicity: Evaluation of Testing Procedures with Methylazoxymethanol and Methylmercury*, Health Effects Research Laboratory, U.S. Environmental Protection Agency, Research Triangle Park, NC, 1994, EPA-68-D2-0056; EPA/600/J-94/527.

National Research Council, *Environmental Neurotoxicology*, National Academy Press, Washington, 1992.

Office of Technology Assessment, *Neurotoxicity*, U.S. Congress, Washington, 1990, OTA-BA-436.

Reiter, L.W., *Neurotoxicology in Regulation and Risk Assessment*, Health Effects Research Laboratory, U.S. Environmental Protection Agency, Research Triangle Park, NC, 1987, EPA/600/J-87/198.

U.S. Environmental Protection Agency, *Assessing the Neurotoxic Potential of Chemicals: a Multidisciplinary Approach*, Health Effects Research Laboratory, Research Triangle Park, NC, 1993, EPA/600/J-93/368.

U.S. Environmental Protection Agency, *Guidelines for Neurotoxicity Risk Assessment*, Risk Assessment Forum, Washington, 1998, EPA/630-R-95/001F.

U.S. Environmental Protection Agency, *Workshop Report on Developmental Neurotoxic Effects Associated with Exposure to PCBs*, Research Triangle Park, NC, 1992, Risk Assessment Forum, Washington, May 1993, EPA/630/R-92/004.

PESTICIDES

Adams, J.D., *Pesticide Assessment Guidelines: Subdivision K, Exposure, Reentry Protection*, Office of Pesticide and Toxic Substances, U.S. Environmental Protection Agency, Washington, 1984, EPA 540/9-84-001.

Anon., *Farm Chemicals Handbook 2000*, Meister Publishing Co., Willoughby, OH, 2000.

Belluck, D.A. et al., Aberrant oviposition by the Caddisfly Triaenodes Tardus Milne (Trichoptera: Leptoceridae), *Ent. News*, 91(5), 173, 1980.

Belluck, D.A., *Atrazine Biospheric Cycling*, Minnesota Pollution Control Agency 7050 Rule Technical Exhibit, Minnesota Pollution Control Agency, St. Paul, MN, 1993.

Belluck, D. and Felsot, A., Bioconcentration of pesticides by egg masses of the Caddisfly, Triaenodes Tardus Milne, *Bull. Environ. Contam. Toxicol.*, 26, 299, 1981.

Belluck, D.A., *Caddisfly Egg Toxicity Bioassay*, Natural History Survey Reports, 204, 1, 1981.

Belluck, D.A. et al., Groundwater monitoring for pesticide metabolites and formulation materials: lessons from Wisconsin, *J. Pesticide Reform*, 9(4), 28, 1990.

Belluck, D.A. et al., Groundwater contamination by atrazine and its metabolites: risk assessment, policy, and legal implications, in *Fate and Significance of Pesticide Degradation Products*, Somasundaram, L. and Coats, J.R., Eds., ACS Press, Washington, 459, 254, 1991.

Belluck, D.A. and Benjamin, S.L., Pesticides and human health, defining acceptable and unacceptable risk levels, *J. Environ. Health*, 53(1), 11, 1990.

Belluck, D.A., *Pesticides in the Aquatic Environment*, Ph.D. thesis, University of Illinois, Urbana-Champaign, IL, 1981.

Belluck, D.A. et al., *Preliminary Analysis of the Toxic Effects of Selected Pesticides on Egg Development of the Caddisfly Triaenodes Flavescens (Trichoptera: Leptoceridae)*, 28th Annual Meeting of the North American Benthological Society, Savannah, GA, 1980.

Fiore, M.C. et al., Chronic exposure to aldicarb-contaminated groundwater and human immune function, *Environ. Res.*, 41, 633, 1986.

Furnish, J. et al., Phoretic relationships between Corydalus Cornutus (Megaloptera:corydalidae) and chironomidae in Eastern Tennessee, *Ann. Entomol. Soc. Amer.*, 74(1), 29, 1981.

Kaloyanova, F.P. and El Batawi, M.A., *Human Toxicology of Pesticides*, CRC Press, Boca Raton, FL, 1991.

Kello, D., WHO drinking water quality guidelines for selected herbicides, *Food Additives and Contaminants*, 6, Supplement 1, S79, 1989.

Kinch, R. and Jones, L., *Best Demonstrated Available Technology (BDAT) Background Document for Organic Toxicity Characteristic Wastes D018-D043 and Addendum to Nonwastewater Forms of Pesticide Toxicity Characteristic Wastes D012-D017, Final*, Waste Treatment Branch, U.S. Environmental Protection Agency, Washington, 1994, EPA-530-R-95-031.

Kinch, R. and Jones, L., *Best Demonstrated Available Technology (BDAT) Background Document for Universal Standards, Vol. A: Universal Standards for Nonwastewater Forms of Listed Hazardous Wastes*, Waste Treatment Branch, U.S. Environmental Protection Agency, Arlington, VA, 1994, EPA-530-R-95-032.

Lawless, E.W., *Guidelines for the Disposal of Small Quantitied of Unused Pesticides*, Office of Research and Development, U.S. Environmental Protection Agency, Washington, 1975, EPA/670/2-75/057.

Madrone, B., *Farmworker Protection Resource Guide: For Farmworker Pesticide Outreach and Education*, Healthcare Provider Pesticides Exposure Training and Grantfunding Sources to Support These Activities, U.S. Environmental Protection Agency, Seattle, 1999, EPA 910-B-99-001.

Maroni, M. and Fait, A., *Health Effects in Man from Long-Term Exposure to Pesticides*, International Center for Pesticide Safety, Elsevier Scientific Publishers, Ireland, 1993.

Meisner, L. et al., Cytogenetic effects of alachlor and/or atrazine *in vivo* and *in vitro*, *Environmental and Molecular Mutagenesis*, 19, 77, 1992.

Mirkin, I.R. et al., Changes in T-lymphocyte distribution associated with ingestion of aldicarb-contaminated drinking water: a follow-up study, *Environ. Res.*, 51, 35, 1990.

Morgan, D.P., *Recognition and Management of Pesticide Poisonings, 4th ed.*, Office of Pesticide Programs, U.S. Environmental Protection Agency, Washington, 1990, EPA-540/9-88-001.

Nishioka, M.G. and Andrews, K.D., *Method Validation and Application for Semivolatile Organic Compounds in Dust and Soil: Pesticides and PCBs*, National Exposure Research Laboratory, U.S. Environmental Protection Agency, Research Triangle Park, NC, 1998, EPA-68-D4-0023; EPA/600/R-97/141.

National Research Council, *Pesticides in the Diets of Infants and Children*, National Academy Press, Washington, 1993.

Purdue Pesticide Programs, *Pesticides and Ecological Risk Assessment, History, Science, and Process*, Purdue University Cooperative Extension Service, West Lafayette, IN, 1997, PPP-41.

Purdue Pesticide Programs, *Pesticides Toxicology, Evaluating Safety and Risk*, Purdue University Cooperative Extension Service, West Lafayette, IN, 1997, PPP-40

Reinert, J.C. et al., *Pesticide Assessment Guidelines, Subdivision U: Applicator Exposure Monitoring*, Office of Pesticide Programs, U.S. Environmental Protection Agency, Washington, 1986, EPA/540/9-87/127.

Roloff, B.D. et al., Cytogenetic effects of cyanazine and metolachlor on human lymphocytes exposed *in vitro*, *Mutation Res.*, 281, 295, 1992.

Roloff, B.D. et al., Cytogenetic studies of herbicide interactions *in vitro* and *in vivo* using atrazine and linuron, *Arch. Environ. Contam. Toxicol.*, 22, 267, 1992.

Sherma, J., *Manual of Analytical Quality Control for Pesticides and Related Compounds in Human and Environmental Samples, Second Revision*, Office of Research and Development, U.S. Environmental Protection Agency, Research Triangle Park, NC, 1981, EPA/600/2-81/059.

Smith, S.J., Sharpley, A.N., and Ahuja, L.R. Agricultural chemical discharge in surface water runoff, *J. Environ. Qual.*, 22, 474, 1993.

Thompson, H.M., Interactions between pesticides; a review of reported effects and their implications for wildlife risk assessment, mini review, *Ecotoxicology*, 5, 59, 1996.

U.S. Environmental Protection Agency, *Citizen's Guide to Pest Control and Pesticide Safety, Prevention, Pesticides, and Toxic Substances*, Washington, 1995, EPA 730-K-95-001.

U.S. Environmental Protection Agency, *Development Document for Best Demonstrated Available Technology, Pretreatment Technology, and New Source Performance Technology for the Pesticide Formulating (SIC), Packaging and Repackaging Industry, Proposed*, Office of Water, Washington, 1994, EPA-821-R-94-002.

U.S. Environmental Protection Agency, *Fate, Transport and Transformation Test Guidelines: Oppts 835 Series*, Office of Prevention, Pesticides and Toxic Substances, Washington, 1998, PB98-118102.

U.S. Environmental Protection Agency, *Pesticide Occupational and Residential Cancer Risk Policy Statement*, Office of Prevention, Pesticides, and Toxic Substances, Washington, 1994.

U.S. Environmental Protection Agency, *Pesticide Occupational and Residential Cancer Risk Policy Statement*, Office of Prevention, Pesticides, and Toxic Substances, Washington, 1994.

U.S. Environmental Protection Agency, *Protect Yourself from Pesticides: Guide for Pesticide Handlers*, Office of Prevention, Pesticides and Toxic Substances, Washington, 1993, EPA 735-B-93-003.

Vettorazzi, G., *Carbamate and Organophosphorus Insecticides used in Agriculture and Public Health*, World Health Organization, Geneva, Switzerland, 1975.

Worthing, C.R. and Barrie Walker, S., *Pesticide Manual: a World Compendium*, British Crop Protection Council, Farnham, Surrey, UK, 1997.

PEER REVIEW

Dearfield, K.L. and Flaak, A.R., *Science Policy Council Handbook: Peer Review*, U.S. Environmental Protection Agency, Washington, 1998, EPA/100-B-98-001.

Eastern Research Group, Inc., *Report on the Peer Review Workshop on Revisions to the Exposure Factors Handbook*, 1995, Office of Research and Development, U.S. Environmental Protection Agency, Washington, 1996, EPA-68-C1-0030; EPA-68-D5-0028; EPA/630/R-96/003.

Fairbrother, A. et al., *Report on the Shrimp Virus Peer Review and Risk Assessment Workshop: Developing a Qualitative Ecological Risk Assessment*, 1998, National Center for Environmental Assessment, U.S. Environmental Protection Agency, Washington, 1999.

U.S. Environmental Protection Agency, *Report on the Benchmark Dose Peer Consultation Workshop*, Risk Assessment Forum, Washington, 1996, EPA/630-R-96-011.

U.S. Environmental Protection Agency, *Risk Characterization Materials for Peer Review*, Washington, 1999, EPA/600-R-99-025.

U.S. Environmental Protection Agency, *Science Policy Council Handbook: Peer Review*, Office of Science Policy, Washington, 1998, EPA 100-B-98-001.

RADIATION

Abrahamson, S. et al., *Health Effects Model for Nuclear Power Plant Accident Consequence Analysis, Low Let Radiation, Part II: Scientific Bases for Health Effects Models*, U.S. Nuclear Regulatory Commission, Washington, 1989.

Abrahamson, S. et al., *Health Effects Models for Nuclear Power Plant Accident Consequence Analysis, Modifications of Models Resulting from Recent Reports on Health Effects of Ionizing Radiation, Low Let Radiation, Part II: Scientific Bases for Health Effects Models*, U.S. Nuclear Regulatory Commission, Washington, 1991.

Barnthouse, L.W., *Effects of Ionizing Radiation on Terrestrial Plants and Animals: a Workshop Report*, Oak Ridge National Laboratory, Oak Ridge, TN, 1995, ORNL/TM-1314.

Blaylock, B.G. et al., *Methodology for Estimating Dose Rates to Freshwater Biota Exposed to Radionuclides in the Environment*, Oak Ridge National Laboratory, Oak Ridge, TN, 1993, ES/ER/TM-78.

Goodhead, D.T., An assessment of the role of microdosimetry in radiobiology, *Radiat. Res.*, 91, 45, 1982.

Grosovsky, A.J. and Little, J.B., Evidence for linear response for the induction of mutations in human cells by X-ray exposures below 10 rads, *Proc. Natl. Acad. Sci. U.S.A.*, 82, 2092, 1985.

Howe, G.R., Effects of dose, dose-rate and fractionation on radiation-induced breast and lung cancers: the Canadian Fluoroscopy Study, in *International Conference on Radiation Effects and Protection*, Japan Atomic Energy Research Institute, Tokyo, 1992.

ICRP, The biological basis for dose limitation in *The Skin, Annals of the ICRP*, 22, 2, 1992, ICRP Publication 59.

ICRP, *Radionuclide Release into the Environment: Assessment of Doses to Man*, A Report of Committee 4 of the International Commission on Radiological Protection, Pergamon Press, New York, 1979.

Kellerer, A.M. and Rossi, H.M., The theory of dual radiation action, *Curr. Topics Radiat. Res. Quart.*, 8, 85, 1972.

Land, C.E. and Sinclair, W.K., The relative contributions of different organ sites to the total cancer mortality associated with low-dose radiation exposure, in *Risks Associated with Ionising Radiations, Annals of the ICRP*, 22, 1, 1991.

Lea, D.E., *Actions of Radiation on Living Cells*, Cambridge University Press, Cambridge, UK, 1962.

Little, M.P. and Charles, M.W., Time variations in radiation-induced relative risk and implications for population cancer risks, *J. Radiol. Prot.*, 11, 91, 1991.

Little, M.P. et al., Time variations in the risk of cancer following irradiation in childhood, *Radiat. Res.*, 126, 304, 1991.

National Academy of Sciences, *Health Effects of Exposure to Low Levels of Ionizing Radiation (BEIR V)*, National Academy Press, Washington, 1990.

National Academy of Sciences, *The Effects on Populations of Exposure to Low Levels of Ionizing Radiation (BEIR III)*, National Academy Press, Washington, 1980.

National Academy of Sciences, *Health Risks of Radon and Other Internally Deposited Alpha-Emitters (BEIR IV)*, National Academy Press, Washington, 1988.

National Institutes of Health, *Report of the National Institutes of Health Ad Hoc Working Group to Develop Radioepidemiological Tables*, U.S. Government Printing Office, Washington, 1985, NIH Publication 85-2748.

National Research Council, *Risk Assessment of Radon in Drinking Water*, National Academy Press, Washington, 1999.

Till, J.E. and Meyer, H.R., *Radiological Assessment: a Textbook on Environmental Dose Analysis*, U.S. Nuclear Regulatory Commission, Washington, 1983, NUREG/CR-3332.

U.S. Environmental Protection Agency, *Diffuse NORM Wastes Waste Characterization and Preliminary Risk Assessment*, Office of Radiation and Indoor Air, Washington, 1993, 68-D20-155.

U.S. Environmental Protection Agency, *Radionuclides: Background Information Document for Final Rules, Vol. I*, Office of Radiation Programs, Washington, 1984, EPA 520/1-84-022-1.

U.S. Environmental Protection Agency, *Risk Assessment Methodology, Environmental Impact Statement, NESHAPS for Radionuclides, Background Information Document — Vol. 1*, Office of Radiation Programs, Washington, 1989, EPA/520/1-89-005.

U.S. Environmental Protection Agency, *Technical Support Document for the 1992 Citizen's Guide to Radon*, Office of Radiation Programs, Washington, 1992, EPA 400-R-92-011.

Yeh, K.W., *Radiation Exposure and Risk Assessment Manual (RERAM)*, Office of Air and Radiation, Washington, 1996, EPA 402-R-96-016.

RCRA

Air Pollution Control Association, *Negotiating the Cleanup: CERCLA, RCRA, TSCA*, Air Pollution Control Association, Pittsburgh, PA, 1988.

Raccette, P. and Bailey, M., *Developing Risk-Based Cleanup Levels at Resource Conservation and Recovery Act Sites in Region 10, Interim Final Guidance*, U.S. Environmental Protection Agency, Seattle, 1998, EPA-68-W4-0004; EPA/910/R-98/001.

Turnblom, S.M. et al., *Guidelines for Developing Risk-Based Cleanup Levels at RCRA Sites in Region 10*, U.S. Environmental Protection Agency, Seattle, 1992, EPA-69-W9-0009; EPA/910/9-92/019.

U.S. Environmental Protection Agency, *Commentary on the Ecological Risk Assessment for the Proposed RIA for RCRA Corrective Action Rule*, Science Advisory Board, Washington, 1993, EPA-SAB-EPEC-COM-001; EPA-SAB-EPEC-COM-94-001.

RISK ASSESSMENT RESOURCES GUIDE

U.S. Environmental Protection Agency, *Consideration of RCRA Requirements in Performing Cercla Responses at Mining Waste Sites*, Office of Emergency and Remedial Response, Washington, 1986, EPA/9234.0-04.

U.S. Environmental Protection Agency, *Ecological Risk Assessment Methods: a Review and Evaluation of Past Practices in the Superfund and RCRA Programs*, Office of Policy Analysis, Washington, 1989, EPA-230-03-89-044.

U.S. Environmental Protection Agency, *Ecological Risk Management in the Superfund and RCRA Programs*, Office of Policy Analysis, Office of Policy, Planning, and Evaluation, Washington, 1989, EPA-230-03-89-045.

U.S. Environmental Protection Agency, *Exposure Assessment Guidance for RCRA Hazardous Waste Combustion Facilities: Draft*, Office of Solid Waste and Emergency Response, Washington, 1994, EPA-530-R-94-021.

U.S. Environmental Protection Agency, *Hazardous Waste TSDF: Background Information for Proposed RCRA Air Emission Standards, Vol. 2, Appendices D-F*, Office of Air Quality Planning and Standards, Research Triangle Park, NC, 1991, EPA/450/3-89/023B.

U.S. Environmental Protection Agency, *Hazardous Waste TSDF: Background Information for Proposed RCRA Air Emission Standards, Vol. 3, Appendices G-L*, Office of Air Quality Planning and Standards, Research Triangle Park, NC, 1991, EPA/450/3-89/023C.

U.S. Environmental Protection Agency, *Nature and Extent of Ecological Risks at Superfund Sites and RCRA Facilities*, Office of Policy Analysis, Office of Policy, Planning, and Evaluation, Washington, 1989, EPA-230-03-89-043.

U.S. Environmental Protection Agency, *Preliminary Biennial RCRA Hazardous Waste Report (Based on 1993 Data)*, Solid Waste and Emergency Response, Washington, 1995, EPA-530-R-95-005A;EPA530-R-95-005B;EPA-530-R-95-005C;EPA-530-R-95-005D.

U.S. Environmental Protection Agency, *RCRA Public Participation Manual*, Office of Solid Waste, Washington, 1996, EPA/530-R-96-007.

U.S. Environmental Protection Agency, *RCRA Special Study on Waste Definitions: Sites that Require Additional Consideration Prior to NPL Proposal under the Superfund Amendments and Reauthorization Act*, Office of Emergency and Remedial Response, Washington, 1987, OSWER-9320.1-06.

U.S. Environmental Protection Agency, *RCRA/UST, Superfund, and EPCRA Hotline Training Module, Introduction to Superfund Community Involvement*, Office of Emergency and Remedial Response, Washington, 1996, EPA/540/R-96/010.

U.S. Environmental Protection Agency, *Revised Interim Soil Lead Guidance for CERCLA Sites and RCRA Corrective Action Facilities*, Office of Emergency and Remedial Response, Washington, 1994, EPA/540/F-94/043.

U.S. Environmental Protection Agency, *Summary of Ecological Risks, Assessment Methods, and Risk Management Decisions in Superfund and RCRA*, Office of Policy Analysis, Washington, 1989, EPA-230-03-89-046.

U.S. Environmental Protection Agency, *The Nature and Extent of Ecological Risks at Superfund Sites and RCRA Facilities*, Office of Policy Analysis, Washington, 1989, EPA-230-03-89-043.

Zamuda, C., *Ecological Risk Assessment Methods: a Review and Evaluation of Past Practices in the Superfund and RCRA Programs (Executive Summary Included)*, Office of Policy, Planning and Evaluation, U.S. Environmental Protection Agency, Washington, 1989, EPA/230/03-89/044.

Zamuda, C., *Summary of Ecological Risks, Assessment Methods, and Risk Management Decisions in Superfund and RCRA*, Office of Policy Analysis, Office of Policy, Planning, and Evaluation, U.S. Environmental Protection Agency, Washington, 1989, EPA-230-03-89-046.

RFDS

Barnes, D.G. and Dourson, M., *Reference Dose (RFD): Description and Use in Health Risk Assessments*, Environmental Criteria and Assessment Office, U.S. Environmental Protection Agency, Cincinnati, 1988, EPA/600/J-88/310.

Barnes, D.G. and Dourson, M., Reference dose (RFD): descriptions and uses in health risk assessments, *Reg. Toxicol. Pharmacol.*, 8, 471, 1988.

Jarabek, A.M. et al., *U.S. Environmental Protection Agency's Inhalation RFD Methodology: Risk Assessment for Air Toxics*, Health Effects Research Laboratory, U.S. Environmental Protection Agency, Research Triangle Park, NC, 1989, EPA/600/J-90/357.

RISK CHARACTERIZATION

Burtner, P.A., *Risk Characterization of Groundwater Contamination Sources: a Method for Prioritizing Public Water Supplies in Need of Wellhead Protection*, Water Management Division, U.S. Environmental Protection Agency, Philadelphia, 1992.

Eastern Research Group, Inc., *Technical Approaches to Characterizing and Cleaning up Iron and Steel Mill Sites under the Brownfields Initiative*, Office of Research and Development, U.S. Environmental Protection Agency, Cincinnati, 1998, EPA-68-D7-0001; EPA/625/R-98/007.

Habicht, H.F., *Guidance on Risk Characterization for Risk Managers and Risk Assessors: Memorandum*, Office of the Administrator, U.S. Environmental Protection Agency, Washington, 1992.

Kearney (A.T.), Inc., *Risk Assessment for the Waste Technologies Industries (WTI) Hazardous Waste Incineration Facility (East Liverpool, Ohio), Vol. 3, Characterization of the Nature and Magnitude of Emissions*, Kearney (A.T.), Chicago, ENVIRON International Corp., Arlington, VA, Midwest Research Inst., Kansas City, MO, Earth Tech., Concord, MA, Waste, Pesticides, and Toxics Division, U.S. Environmental Protection Agency, Chicago, 1997, EPA-68-W4-0006; EPA/905/R-97/002c.

U.S. Environmental Protection Agency, *Diffuse Norm Wastes Waste Characterization and Preliminary Risk Assessment*, Office of Radiation and Indoor Air, Washington, 1993, 68-D20-155.

U.S. Environmental Protection Agency, *EPA Risk Characterization Program*, Office of the Administrator, Washington, 1995, EPAX 9503-0006.

U.S. Environmental Protection Agency, *EPA Superfund Site Assessment and Remediation*, Office of Emergency and Remedial Response, Washington, 1998, PB98-963300.

U.S. Environmental Protection Agency, *Guidance on Risk Characterization for Risk Managers and Risk Assessors*, Memorandum from Deputy Administrator for the Office of Solid Waste and Emergency Response, H. Habicht to Assistant and Regional Administrators, 1992.

U.S. Environmental Protection Agency, *Health Risk Assessment/Characterization of the Drinking Water Disinfection Byproduct Bromate*, Office of Water, Washington, 1998, EPA/815-B-98-007.

U.S. Environmental Protection Agency, *Health Risk Assessment/Characterization of the Drinking Water Disinfection Byproducts Chlorine Dioxide & Chlorite*, Toxicology Excellence for Risk Assessment, 1998, EPA/815-B-98-008.

U.S. Environmental Protection Agency, *Health Risk Assessment/Characterization of the Drinking Water Disinfection Byproduct Chloroform*, Office of Water, Washington, 1998, EPA/815/B-98/006.

U.S. Environmental Protection Agency, *Indoor Air Assessment: a Review of Indoor Air Quality Risk Characterization Studies*, Environmental Criteria and Assessment Office, Research Triangle Park, NC, 1991, EPA/600/8-90/044; ECAO-R-0314.
U.S. Environmental Protection Agency, *Quality Assurance Guidance for Conducting Brownfields Site Assessments*, Office of Emergency and Remedial Response, Washington, 1998, EPA 540-R-98-038; OSWER-9230.0-83P.
U.S. Environmental Protection Agency, *Risk Characterization: a Practical Guidance for NCEA-Washington Risk Assessors*, Office of Research and Development, Washington, 1997, NCEA-W-0105.
U.S. Environmental Protection Agency, *Risk Characterization Materials for Peer Review*, 1999, Review Draft, Washington, 1999, EPA/600-R-99-025.

RISK COMMUNICATION AND PUBLIC PARTICIPATION

American Chemical Society, *Understanding Risk Analysis*, Washington, 1998.
Anon., *Public Involvement in Comparative Risk Projects: Principles and Best Practices, a Sourcebook for Project Managers*, Western Center for Environmental Decision-Making, Boulder, CO, 1996.
Benjamin, S.L. and Belluck, D.A., Risk feedback: an important step in risk communication, *American Water Works Association*, 82(11), 50, 1990.
Belluck, D.A. and Benjamin, S.L., Communicating carcinogenic risks from "low-level" exposures: the myth of the vanishing zero, *J. Environmental Health*, 56(3), 33, 1993.
Belluck, D.A. et al., Improving risk assessment credibility, *U.S. Water News*, 23, 1992.
CH2M Hill, Inc., *Public Involvement Strategies: a Manager's Handbook*, AWWA Research Foundation, Denver, CO, 1995, OCLC 34355218.
Chess, C., *Improving Dialogue with Communities a Short Guide for Government Risk Communication*, New Jersey Department of Environmental Protection and Energy, Trenton, NJ, 1991.
Hadden, S.G. and Bales, B.V., *Risk Communication about Chemicals in Your Community, a Manual for Local Officials*, 1989, EPA-230-09-89-066.
Hance, B.J., *Improving Dialogue with Communities, a Risk Communication Manual for Government*, New Jersey Department of Environmental Protection and Energy, Trenton, NJ, 1991.
Hance, B.J., Chess, C., and Sandman, P., *Improving Dialogue with Communities*, New Jersey Department of Environmental Protection, New Brunswick, NJ, 1988.
Johnson, B.B. et al., *Explaining Uncertainty in Health Risk Assessment Effects on Risk Perception and Trust*, Risk Communication Project, U.S. Environmental Protection Agency, Washington, 1994, CR-820522; EPA 230/R-94-903.
Johnson, B.B. and Slovic, P., *Lay Views on Uncertainty in Health Risk Assessment: a Report on Phase II Research*, Office of Policy, Planning, and Evaluation, Risk Communication Project, U.S. Environmental Protection Agency.
National Research Council, *Improving Risk Communication*, National Academy Press, Washington, 1989.
Sandman, P., *Explaining Environmental Risk*, TSCA Assistance Office, U.S. Environmental Protection Agency, Washington, 1986.
Sandman, P.M. and Weinstein, N.D., *Communicating Effectively about Risk Magnitudes: Bottom Line Conclusions and Recommendations for Practitioners*, U.S. Environmental Protection Agency, 1994.

Smith, V.K. et al., *Communicating Radon Risk Effectively: a Mid-Course Evaluation*, Office of Policy Analysis, Washington, 1987.

Stern, P.C. and Fineberg, H.V., Eds., *Understanding Risk: Informing Decisions in a Democratic Society*, National Academy Press, Washington, 1996.

U.S. Department of Labor, *Chemical Hazard Communication*, Occupational Safety and Health Administration, Washington, 1989, OSHA 3084.

U.S. Environmental Protection Agency, *Communicating with the Public about Hazardous Materials: an Examination of Local Practice, Policy, Planning, and Evaluation*, Washington, 1990, EPA 230-04-90-077.

U.S. Environmental Protection Agency, *Community Relations in Superfund: a Handbook*, Office of Emergency and Remedial Response, Washington, 1992, EPA /540/R-92/009.

U.S. Environmental Protection Agency, *Guidance for Community Advisory Groups at Superfund Sites*, Office of Emergency and Remedial Response, Washington, 1995, EPA/540/K-96/001.

U.S. Environmental Protection Agency, *Hazardous Substances in Our Environment: a Citizen Guide to Understanding Health Risks and Reducing Exposure, Policy, Planning, and Evaluation*, Washington, 1990, EPA 230/09/90/081.

U.S. Environmental Protection Agency, *How Safe Am I? Helping Communities Evaluate Chemical Risks*, Environmental Health Center, Washington, 1999, EPA/550-B-99-013.

U.S. Environmental Protection Agency, *Innovative Methods to Increase Public Involvement in Superfund Community Relations*, Office of Emergency and Remedial Response, Washington, 1990, OSWER Directive 9230.0-20.

U.S. Environmental Protection Agency, *RCRA Public Participation Manual*, Office of Solid Waste, Washington, 1996, EPA/530-R-96-007.

U.S. Environmental Protection Agency, *RCRA Public Involvement Manual*, Office of Solid Waste, Washington, 1993, EPA530-R-93-006.

U.S. Environmental Protection Agency, *RCRA/UST, Superfund, and Epcra Hotline Training Module, Introduction to Superfund Community Involvement*, Office of Emergency and Remedial Response, Washington, March 1996, EPA/540/R-96/010.

U.S. Environmental Protection Agency, *Seven Cardinal Rules of Risk Communication*, Washington, 1992, EPA 230-K-92-001.

RISK MANAGEMENT

American Chemical Society, *Understanding Risk Analysis*, Washington, 1998.

Anderson, H. and Belluck, D., *Exercising Discretion, in a Matter of Chance, a Matter of Choice*, Wisconsin Department of Natural Resources, Madison, WI, 1989, 30.

Anon., Setting environmental priorities: the debate about risk, *EPA Journal*, 17, 2, 1991.

Belluck, D.A. et al., Breaking the reactive paradigm: a proactive approach to risk assessment management, *Total Quality Environmental Management*, 1(3), 253, 1992.

Belluck, D.A. et al., *Management of Waste Electronic Appliances*, Minnesota Office of Environmental Assistance, St. Paul, MN, 1995.

County of San Diego, *Site Assessment & Mitigation Manual*, 1994 SA/M Manual, San Diego Department of Health Services, 1994.

Den, A.R. et al., *Risk & Decision Making*, Office of the Senior Science Advisor, U.S. Environmental Protection Agency, San Francisco, CA, 1992.

Fava, J.A., *Generalized Methodology for Conducting Industrial Toxicity Reduction Evaluations (TRES), the Chemicals and Chemical Product Branch*, U.S. Environmental Protection Agency, Cincinnati, 1989, EPA/600/2-88/070.

Federal Focus, Inc., *Toward Common Measures, Recommendations for a Presidential Executive Order on Environmental Risk Assessment and Risk Management Policy*, Washington, 1991.

Gallant, A.L., *Regionalization as a Tool for Managing Environmental Resources*, Environmental Research Laboratory, Corvallis, OR, U.S. Environmental Protection Agency, Washington, 1989, EPA/600/3-89/060.

Kelly, E. et al., *Integrated, Comprehensive Ecological Impact Assessment in Support of Department of Energy Decision Making*, Office of Technical and Environmental Support, U.S. Department of Energy, No Date.

Lyman, W.J. et al., *Survey Study to Select a Limited Number of Hazardous Materials to Define Amelioration Requirements*, U.S. Coast Guard Report, Washington, 1974, CG-D-46-75.

McNelly, G., *Chemical Dehalogenation Treatability Studies under CERCLA: An Overview*, Office of Solid Waste and Emergency Response, U.S. Environmental Protection Agency, Cincinnati, 1992, EPA/540/R-92/013b.

Singhvi, R., *Contaminants and Remedial Options at Pesticide Sites*, U.S. Environmental Protection Agency, Washington, 1995, EPA/600/R-94/202.

U.S. Department of Transportation et al., *2000 Emergency Response Guide Book, a Guidebook for First Responders During the Initial Phase of a Dangerous Goods/Hazardous Materials Incident*, Washington, 2000.

U.S. Environmental Protection Agency, *Compendium of Federal Facilities Cleanup Management Information*, Federal Facilities Restoration and Reuse Office, Washington, 1998, EPA-540-R-98-004.

U.S. Environmental Protection Agency, *A Compendium of Technologies Used in the Treatment of Hazardous Wastes*, Center for Environmental Research Information, Office of Research and Development, Cincinnati, 1987, EPA/625/8-87/014.

U.S. Environmental Protection Agency, *Conference on the Risk Assessment Paradigm After Ten Years: Policy and Practice, Then, Now, and in the Future*, Office of Research and Development, Washington, 1993, EPA/600-R-93-039.

U.S. Environmental Protection Agency, *ECO Update*, Office of Solid Waste and Emergency Response, Washington, Occasional Series.

U.S. Environmental Protection Agency, *Environmental Equity: Reducing Risk for All Communities*, Washington, 1992, EPA/230-R-92-008.

U.S. Environmental Protection Agency, *Environmental Equity: Reducing Risk for All Communities, Vol. 2: Supporting Document*, Washington, 1992, EPA/230-R-92-008A.

U.S. Environmental Protection Agency, *EPA Children's Environmental Health Yearbook*, Office of Children's Health Protection, Washington, 1998, EPA/100-R-98-100.

U.S. Environmental Protection Agency, *General Guidance for Risk Management Programs (40 CFR Part 68)*, Washington, 1998, EPA 550-B-98-003.

U.S. Environmental Protection Agency, *Guidance on Risk Characterization for Risk Managers and Risk Assessors*, Memorandum from Deputy Administrator for the Office of Solid Waste and Emergency Response, H. Habicht to Assistant and Regional Administrators, 1992.

U.S. Environmental Protection Agency, *Guide to Environmental Planning for Small Communities: a Guide for Local Decision-Makers*, Office of Research and Development, Office of Regional Operations and State/Local Relations, Washington, 1994, EPA/625/R-94/009.

U.S. Environmental Protection Agency, *Hazard Ranking System Guidance Manual*, Office of Solid Waste and Emergency Response, Washington, 1992, EPA 540-R-92-026.

U.S. Environmental Protection Agency, *Principles of Environmental Impact Assessment*, Office of Federal Activities, Washington, 1998, EPA/315-B-98-001.

U.S. Environmental Protection Agency, *Risk Management Program Guidance for Offsite Consequences*, Washington, 1999, EPA 550-B-99-009.

U.S. Environmental Protection Agency, *Risk Management Programs, 40 CFR Part 68*, Office of Emergency and Remedial Response, Washington, 1998.

U.S. Environmental Protection Agency, *Risk Ranking Project: Ecological Ranking and Problem Analysis*, Risk Ranking Work Group, New York, 1991.

U.S. Environmental Protection Agency, *Risk Updates*, Region I-New England, Occasional Series.

U.S. Environmental Protection Agency, *Role of the Baseline Risk Assessment in Superfund Remedy Selection Decisions*, Office of Solid Waste and Emergency Response, Washington, 1991.

SOILS AND SEDIMENTS

American Society for Testing and Materials, *Standard Guide for Selection of Methods of Particle Size Analysis of Fluvial Sediments (Manual Methods)*, Method D4822-88, American Society for Testing and Materials, Philadelphia, 1992.

American Society for Testing and Materials, *Standard Guide for Conducting 10-Day Static Sediment Toxicity Tests with Marine and Estuarine Amphipods*, American Society for Testing and Materials, Philadelphia, 1992.

American Society for Testing and Materials, *Standard Guide for Conducting Sediment Toxicity Tests with Freshwater Invertebrates*, American Society for Testing and Materials, Philadelphia, 1992.

Bechtel Jacobs Company LLC, *Biota Sediment Accumulation Factors for Invertebrates: Review and Recommendations for the Oak Ridge Reservation*, Oak Ridge, TN, 1998, BJC/OR-112.

Bechtel Jacobs Company LLC, *Empirical Models for the Uptake of Inorganic Chemicals from Soil by Plants*, Oak Ridge, TN, 1998, BJC/OR-133.

Belluck, D.A. and Benjamin, S.L., *Scientifically Credible Site-Specific Soil Cleanup Levels to Protect Ground Water*, RCRA Policy Forum, 3, 1-9, 1995.

Breckenridge, R.P., Williams, J.R., and Keck, J.F., *Characterizing Soils for Hazardous Waste Site Assessments*, Office of Research and Development, Office of Solid Waste and Emergency Response, U.S. Environmental Protection Agency, Washington, 1991, EPA/540/4-91/003.

Carsel, R.F. et al., *Users Manual for the Pesticide Root Zone Model (PRZM) Release 1*, Environmental Research Laboratory, U.S. Environmental Protection Agency, Athens, GA, 1984, EPA-600-3-84-109.

De Rosa, C.T. et al., Dioxin and dioxin-like compounds in soil, part 1: ATSDR interim policy guideline, *Toxicology and Industrial Health*, 13(6), 759, 1997.

Efroymson, R.A. et al., *Toxicological Benchmarks for Contaminants of Potential Concern for Effects on Soil and Litter Invertebrates and Heterotrophic Processes: 1997 Revision*, Oak Ridge National Laboratory, Oak Ridge, TN, 1997, ES/ER/TM-126/R2.

Jones, D.S. et al., *Toxicological Benchmarks for Screening Contaminants of Potential Concern for Effects on Sediment-Associated Biota: 1997 Revision*, Oak Ridge National Laboratory, Oak Ridge TN, 1997, ES/ER/TM-95/R4.

Kahn, L., *Determination of Total Organic Carbon in Sediment, Environmental Services Division*, Monitoring Management Branch, U.S. Environmental Protection Agency, Edison, NJ, 1988.

Meylan, W. et al., Molecular topology/fragment contribution method for predicting soil sorption coefficients, *Environ. Sci. Technol.*, 26, 1992.

Persaud, D. et al., *Guidelines for the Protection and Management of Aquatic Sediment Quality in Ontario*, Ontario Ministry of the Environment, Ottawa, Ontario, 1993.

U.S. Environmental Protection Agency, *Clarification to the 1994 Revised Interim Soil Lead Guidance for CERCLA Sites and RCRA Corrective Action Facilities*, Office of Emergency and Remedial Response, Washington, 1998, EPA/540/F-98/030; OSWER-9200.4-27P.

U.S. Environmental Protection Agency, *Comparison of EPA's First 30 Draft Generic Soil Screening Levels with State's Soil Levels*, Office of Emergency and Remedial Response, Washington, 1994.

U.S. Environmental Protection Agency, *Description and Sampling of Contaminated Soils, a Field Pocket Guide*, Technology Transfer, Cincinnati, 1991, EPA/625/12-91/002.

U.S. Environmental Protection Agency, *Draft Soil Screening Guidance: Issues Document*, Office of Emergency and Remedial Response, Washington, 1994, EPA-540/R-94/105.

U.S. Environmental Protection Agency, *Ecotox Thresholds*, Office of Solid Waste and Emergency Response, Washington, EPA 540/F-95/038.

U.S. Environmental Protection Agency, *EPA's Contaminated Sediment Management Strategy*, Office of Water, Washington, 1998, EPA/823/R-98/001.

U.S. Environmental Protection Agency, *Innovations in Site Characterization, Case Study: Hanscom Air Force Base, Operable Unit 1 (Sites 1, 2, and 3)*, Office of Solid Waste and Emergency Response, Washington, 1998, EPA/542/R-98/006.

U.S. Environmental Protection Agency, *Interim Sediment Criteria Values for Nonpolar Hydrophobic Organic Contaminants*, Office of Water Regulations and Standards, Criteria and Standards Division, Washington, 1988, SCD #17.

U.S. Environmental Protection Agency, *Methods for Assessing the Toxicity of and Bioaccumulation of Sediment-Associated Contaminants with Freshwater Invertebrates*, Office of Research and Development, Narragansett, RI, Environmental Research Laboratory, Duluth, MN, 1994, EPA/600/R-94/024.

U.S. Environmental Protection Agency, *National Sediment Bioaccumulation Conference*, 11-1996, Bethesda, MD, Proceedings, 1998, Office of Water, Washington, 1998, EPA 823-R-98-002.

U.S. Environmental Protection Agency, *Sediment Classification Methods Compendium*, Sediment Oversight Technical Committee, Washington, 1992, EPA 823-R-92-006.

U.S. Environmental Protection Agency, *Sediment Sampling Quality Assurance User's Guide*, Environmental Monitoring Systems Laboratory, Las Vegas, Washington, 1985, EPA/600/4-85/048.

U.S. Environmental Protection Agency, *Soil Screening Guidance*, Office of Solid Waste and Emergency Response, Washington, 1994, EPA/540/R-94/101.

U.S. Environmental Protection Agency, *Soil Screening Guidance: Fact Sheet*, Office of Solid Waste and Emergency Response, Washington, 1996, EPA/50/F-95/041.

U.S. Environmental Protection Agency, *Soil Screening Guidance: User's Guide*, Office of Solid Waste and Emergency Response, Washington, 1996.

U.S. Environmental Protection Agency, *Soil Screening Guidance: User's Guide, 2nd ed.*, Office of Solid Wastes and Emergency Response, Washington, 1996, Publication 9355.4-23.

U.S. Environmental Protection Agency, *Soil Screening Guidance: User's Guide, Attachment A: Conceptual Site Model Summary*, Office of Solid Waste and Emergency Response, Washington, 1996, EPA/540/R-96/018.

U.S. Environmental Protection Agency, *Soil Screening Guidance: User's Guide, Attachment B: Soil Screening DQOs for Surface Soils and Subsurface Soils*, Office of Solid Waste and Emergency Response, Washington, 1996, EPA/540/R-96/018.

U.S. Environmental Protection Agency, *Soil Screening Guidance: User's Guide, Attachment C: Chemical Properties for SSL Development*, Office of Solid Waste and Emergency Response, Washington, 1996, EPA/540/R-96/018.

U.S. Environmental Protection Agency, *Soil Screening Guidance: User's Guide, Attachment D: Regulatory and Human Health Benchmarks for SSL Development*, Office of Solid Waste and Emergency Response, Washington, 1996, EPA/540/R-96/018.

U.S. Environmental Protection Agency, *Technical Background Document for Soil Screening*, Office of Solid Waste and Emergency Response, Washington, 1995, EPA/540/R-95/128.

Van Ee, J.J., Blume, L.J., and Starks, T.H., *A Rationale for the Assessment of Errors in the Sampling of Soils*, Environmental Monitoring Systems Laboratory, Office of Research and Development, Las Vegas, 1990, EPA/600/4-90/013.

Van Ee, J.J., Blume, L.J., and Starks, T.H., *A Rationale for the Assessment of Errors in the Sampling of Soils*, Environmental Monitoring Systems Laboratory, Office of Research and Development, Las Vegas, 1990, EPA/600/4-90/013.

SUPERFUND

Hoddinott, K.B., *Superfund Risk Assessment in Soil Contamination Studies, Third Vol.*, ASTM, West Conshohocken, PA, 1998, OCLC 39930664.

U.S. Environmental Protection Agency, *Air/Superfund National Technical Guidance Study Series*, Office of Air Quality Planning and Standards, Research Triangle Park, NC, Multi-volume Series.

U.S. Environmental Protection Agency, *Common Contaminants Found at Superfund Sites*, Office of Emergency and Remedial Response, Washington, 1998, EPA/540/R-98/008; OSWER-9203.1-17a.

U.S. Environmental Protection Agency, *Contingency Analysis Modeling for Superfund Sites and Other Sources*, Office of Air Quality Planning and Standards, Research Triangle Park, NC, 1993, EPA-454/R-93-001.

U.S. Environmental Protection Agency, *Ecological Risk Assessment Guidance for Superfund: Process for Designing and Conducting Ecological Risk Assessments*, Office of Emergency and Remedial Response, Edison, NJ, 1997, EPA/540/R-97/006; OSWER-9285.7-25.

U.S. Environmental Protection Agency, *Ecological Risk Assessment Methods: a Review and Evaluation of Past Practices in the Superfund and RCRA Programs*, Office of Policy Analysis, Washington, 1989, EPA-230-03-89-044.

U.S. Environmental Protection Agency, *Ecological Risk Management in the Superfund and RCRA Programs*, Office of Policy Analysis, Washington, 1989, EPA-230-03-89-045.

U.S. Environmental Protection Agency, *EPA Superfund Site Assessment and Remediation*, Office of Emergency and Remedial Response, Washington, 1999, PB98-963300.

U.S. Environmental Protection Agency, *Getting Ready, Scoping the RI/Fs, Solid Waste and Emergency Response (Os-220)*, 1989, No. 9355.3-01FS1.

U.S. Environmental Protection Agency, *Guidance for Performing Site Assessment under Cercla, Interim Final*, Office of Emergency and Remedial Response, Washington, 1992, EPA/540-R-92-021.

U.S. Environmental Protection Agency, *Risk Assessment Guidance for Superfund, Vol. I: Human Health Evaluation Manual (Part A), Interim Final*, Office of Emergency and Remedial Response, Washington, 1989, EPA/540/1-89/002.

U.S. Environmental Protection Agency, *Risk Assessment Guidance for Superfund, Vol. I: Human Health Evaluation Manual (Part B, Development of Risk-based Preliminary Remediation Goals), Interim Final*, Office of Emergency and Remedial Response, Washington, 1991, EPA/540/R-92/003.
U.S. Environmental Protection Agency, *Risk Assessment Guidance for Superfund, Vol. I: Human Health Evaluation Manual (Part C, Risk Evaluation of Remedial Alternatives), Interim Final*, Office of Emergency and Remedial Response, Washington, 1992, EPA/540/R-92/004.
U.S. Environmental Protection Agency, *Risk Assessment Guidance for Superfund: Vol. 1. Human Health Evaluation Manual (Part D, Standardized Planning, Reporting and Review of Superfund Risk Assessments) (Interim)*, Office of Emergency and Remedial Response, Washington, 1998, EPA/540/R-97/033.
U.S. Environmental Protection Agency, *Risk Assessment Guidance for Superfund: Vol. 1. Human Health Evaluation Manual, Supplement to Part A: Community Involvement in Superfund Risk Assessments*, Office of Emergency and Remedial Response, Washington, 1999, EPA-540-R-98-042.
U.S. Environmental Protection Agency, *Risk Assessment Guidance for Superfund, Vol. II, Environmental Evaluation Manual, Interim Final*, Office of Emergency and Remedial Response, Washington, 1989, EPA/540/1-89/001.
U.S. Environmental Protection Agency, *Role of the Baseline Risk Assessment in Superfund Remedy Selection Decisions*, Office of Solid Waste and Emergency Response, Washington, 1991, OSWER Directive 9355.0-30.
U.S. Environmental Protection Agency, *Rules of Thumb for Superfund Remedy Selection*, Office of Emergency and Remedial Response, Washington, 1997, EPA/540/R-97/013; OSWER-9355.0-69.
U.S. Environmental Protection Agency, *Summary of Ecological Risks, Assessment Methods, and Risk Management Decisions in Superfund and RCRA*, Office of Policy Analysis, Washington, 1989, EPA-230-03-89-046.
U.S. Environmental Protection Agency, *Superfund Exposure Assessment Manual*, Office of Remedial Response, Washington, 1988, EPA/540/1-88/001.
U.S. Environmental Protection Agency, *Superfund Fact Sheet: Exposure Pathways*, Office of Solid Waste and Emergency Response, Washington, 1992, Publication 9230.0-05FSb.
U.S. Environmental Protection Agency, *Superfund Today: Focus on Revisions to Superfund's Risk Assessment Guidance*, Office of Solid Waste and Emergency Response, Washington, 1999, EPA/540/F-98/055.
U.S. Environmental Protection Agency, *The Nature and Extent of Ecological Risks at Superfund Sites and RCRA Facilities*, Office of Policy Analysis, Washington, 1989, EPA-230-03-89-043.
U.S. Environmental Protection Agency, *Understanding Superfund Risk Assessment*, Office of Emergency and Remedial Response, Washington, 1992, OSWER-9285.7-06Fs.

TERRESTRIAL PLANTS

Baud-Grasset, F. et al., *Evaluation of the Bioremediation of a Contaminated Soil with Phytotoxicity Tests*, Risk Reduction Engineering Laboratory, U.S. Environmental Protection Agency, Cincinnati, 1993, EPA/600/J-93/166.
Efroymson, R.A. et al., *Toxicological Benchmarks for Screening Contaminants of Potential Concern for Effects on Terrestrial Plants: 1997 Revision*, Oak Ridge National Laboratory, Oak Ridge, TN, 1997, ES/ER/TM-85/R3.

Eisler, R., *Handbook of Chemical Risk Assessment: Health Hazards to Humans, Plants, and Animals*, Lewis Publishers, Boca Raton, FL, 2000.

Fletcher, J.S. et al., *Database Assessment of Phytotoxicity Data Published on Terrestrial Vascular Plants*, Corvallis Environmental Research Laboratory, U.S. Environmental Protection Agency, OR, 1988, EPA/600/J-88/556.

Guderian, R., *Air Pollution: Phytotoxicity of Acidic Gases and its Significance in Air Pollution Control*, Springer-Verlag, New York, 1977.

Ikuma, H., Rapid *Biochemical Technique for Phytotoxicity Modes-of-Action of Herbicides, Part I*, Office of Pesticide Programs, U.S. Environmental Protection Agency, Washington, 1978, EPA-68-01-2482; EPA/540/9-78/001.

Ikuma, H. et al., *Rapid Biochemical Technique for Phytotoxicity Modes-of-Action of Herbicides, Part II*, Office of Pesticide Programs, U.S. Environmental Protection Agency, Washington, 1978, EPA-68-01-1907; EPA/540/9-78/002.

Lower, W.R., *Single Laboratory Evaluation of Phytotoxicity Test*, Environmental Monitoring Systems Laboratory, U.S. Environmental Protection Agency, Las Vegas, 1987, EPA/600/S 4-87/012.

McFarlane, C. et al., *Closed Chambers for Plant Studies of Chemical Uptake, Accumulation and Phytotoxicity (Proceedings)*, Corvallis Environmental Research Laboratory, U.S. Environmental Protection Agency, Corvallis, OR, 1989, EPA/600/D-90/009.

Olszyk, D.M. and Tingey, D.T., *Phytotoxicity of Air Pollutants, Evidence for the Photodetoxification of SO_2 but not O_3*, Corvallis Environmental Research Lab, Corvallis, OR, 1984, EPA-600/J-84-044.

U.S. Environmental Protection Agency, *ECO Update: Field Studies for Ecological Risk Assessment*, Office of Emergency and Remedial Response, Washington, 1994, EPA/540/F-94/014.

U.S. Environmental Protection Agency, *Guide to the Biosolids Risk Assessments for the EPA Part 503 Rule*, Office of Wastewater Management, Washington, 1995, EPA/832/B-93/005.

THRESHOLD LIMIT VALUES (TLVS)

American Conference of Governmental Industrial Hygienists, *1999 TLVS and BEIS: Threshold Limit Values for Chemical Substances and Physical Agents*, Biological Exposure Indices, American Conference of Governmental Industrial Hygienists, Cincinnati, 1999.

American Conference of Governmental Industrial Hygienists, *1997 TLVS and BEIS: Threshold Limit Values for Chemical Substances and Physical Agents*, Biological Exposure Indices, American Conference of Governmental Industrial Hygienists, Cincinnati, 1997.

American Conference of Governmental Industrial Hygienists, *1994–1995 Threshold Limit Values for Chemical Substances and Physical Agents and Biological Exposure Indices*, American Conference of Governmental Industrial Hygienists, Cincinnati, 1994.

American Conference of Governmental Industrial Hygienists, *1993–1994 Threshold Limit Values for Chemical Substances and Physical Agents and Biological Exposure Indices*, American Conference of Governmental Industrial Hygienists, Cincinnati, 1993.

American Conference of Governmental Industrial Hygienists, *1992–1993 Threshold Limit Values for Chemical Substances and Physical Agents and Biological Exposure Indices*, American Conference of Governmental Industrial Hygienists, Cincinnati, 1992.

American Conference of Governmental Industrial Hygienists, *Documentation of the Threshold Limit Values and Biological Exposure Indices*, American Conference of Governmental Industrial Hygienists, Cincinnati, 1991.

American Conference of Governmental Industrial Hygienists, *Documentation of the Threshold Limit Values and Biological Exposure Indices*, American Conference of Governmental Industrial Hygienists, American Conference of Governmental Industrial Hygienists, Cincinnati, 1986.
American Conference of Governmental Industrial Hygienists, *Threshold Limit Values for Chemical Substances and Physical Agents and Biological Exposure Indices for 1990–1991*, American Conference of Governmental Industrial Hygienists, Cincinnati, 1990.
American Conference of Governmental Industrial Hygienists, *Threshold Limit Values and Biological Exposure Indices for 1989–1990*, American Conference of Governmental Industrial Hygienists, Cincinnati, 1989.
American Conference of Governmental Industrial Hygienists, *Threshold Limit Values and Biological Exposure Indices for 1988–1989*, American Conference of Governmental Industrial Hygienists, Cincinnati, 1988.
American Conference of Governmental Industrial Hygienists, *Threshold Limit Values and Biological Exposure Indices for 1987–1988*, American Conference of Governmental Industrial Hygienists, American Conference of Governmental Industrial Hygienists, Cincinnati, 1987.
American Conference of Governmental Industrial Hygienists, *TLVS and BEIS: Threshold Limit Values for Chemical Substances and Physical Agents*, Biological Exposure Indices, ACGIH Worldwide, Cincinnati, 1996.
American Conference of Governmental Industrial Hygienists, *TLVS, Threshold Limit Values for Chemical Substances in the Work Environment*, American Conference of Governmental Industrial Hygienists, Cincinnati, 1986.
American Conference of Governmental Industrial Hygienists, *TLVS, Threshold Limit Values for Chemical Substances in the Work Environment Adopted by ACGIH with Intended Changes for 1985-86*, American Conference of Governmental Industrial Hygienists, American Conference of Governmental Industrial Hygienists, Cincinnati, 1985.
American Conference of Governmental Industrial Hygienists, *TLVS, Threshold Limit Values for Chemical Substances in the Work Environment Adopted by ACGIH for 1984-85*, American Conference of Governmental Industrial Hygienists, Cincinnati, 1984.
American Conference of Governmental Industrial Hygienists, *TLVS, Threshold Limit Values for Chemical Substances and Physical Agents in the Work Environment with Intended Changes for 1983-84*, American Conference of Governmental Industrial Hygienists, Cincinnati, 1983.
American Conference of Governmental Industrial Hygienists, *TLVS, Threshold Limit Values for Chemical Substances in Work Air Adopted by ACGIH for 1982*, American Conference of Governmental Industrial Hygienists, Cincinnati, 1982.
American Conference of Governmental Industrial Hygienists, *TLVS, Threshold Limit Values for Chemical Substances and Physical Agents in the Workroom Environment with Intended Changes for 1982*, American Conference of Governmental Industrial Hygienists, Cincinnati, 1982.
American Conference of Governmental Industrial Hygienists, *TLVS, Threshold Limit Values for Chemical Substances in Workroom Air*, American Conference of Governmental Industrial Hygienists, Cincinnati, 1981.
American Conference of Governmental Industrial Hygienists, *TLVS, Threshold Limit Values for Chemical Substances and Physical Agents in the Workroom Environment with Intended Changes for 1981*, American Conference of Governmental Industrial Hygienists, Cincinnati, 1981.

Grose, E.C. and Graham, J.A., *Cadmium as a Respiratory Toxicant*, Health Effects Research Laboratory, U.S. Environmental Protection Agency, Research Triangle Park, NC, 1987, EPA/600/D-87/020.

Lanier, M.E., *Threshold Limit Values: Discussion and Thirty-Five Year Index with Recommendations*, American Conference of Governmental Industrial Hygienists, Cincinnati, 1984.

TOXICITY ASSESSMENT

Agency for Toxic Substances and Disease Registry, *1997 CERCLA Priority List of Hazardous Substances that Will be the Subject of Toxicological Profiles & Support Document*, U.S. Department of Health and Human Services, Atlanta, 1997.

Agency for Toxic Substances and Disease Registry, *ATSDR's Toxicological Profiles on CD-Rom CRCnetbase 1999 (Computer File)*, CRC Press, Boca Raton, FL, 1999.

Agency for Toxic Substances and Disease Registry, *Criteria for Selecting Toxicological Profiles for Development*, U.S. Department of Health and Human Services, Atlanta, 1993.

Agency for Toxic Substances and Disease Registry, *Toxicological Profiles*, Atlanta, Occasional Series.

Agency for Toxic Substances and Disease Registry, *Toxicological Profile for 1, 2-Dichloroethane: Draft*, U.S. Department of Health and Human Services, Atlanta, 1999.

Agency for Toxic Substances and Disease Registry, *Toxicological Profile for Mercury*, U.S. Department of Health and Human Services, Atlanta, 1999.

Agency for Toxic Substances and Disease Registry, *Toxicological Profile for Methyl Parathion: Draft*, U.S. Department of Health and Human Services, Atlanta, 1999.

Agency for Toxic Substances and Disease Registry, *Toxicological Profile for Pentachlorophenol: Draft*, U.S. Department of Health and Human Services, Atlanta, 1999.

Alberta Environmental Protection and Canadian Network of Toxicology Centers, *Toxicology, an Environmental Education Unit for Secondary Schools and Communities*, Educators Resource Guide, 1995, Revised Printing 1997.

American Industrial Health Council, *Presentation of Risk Assessment of Carcinogens*, Washington, No Date.

Amdur, M.O. et al., Eds., *Cassarett and Doull's Toxicology, the Basic Science of Poisons, 4th ed.*, McGraw Hill, New York, 1991.

Becker, C.D. and Thatcher, T.O., *Toxicity of Power Plant Chemicals to Aquatic Life*, U.S. Atomic Energy Commission, 1973, WASH-1249.

Belluck, D.A. et al., *Single Chemical and Complex Mixture Biomarkers: Making the Case for Regulatory Decision Making and Litigation*, Presentation at NATO Advanced Study Institute, Antalya, Turkey, 1999.

Botts, J.A., *Toxicity Reduction Evaluation Protocol for Municipal Wastewater Treatment Plants*, Risk Reduction Engineering Laboratory, U.S. Environmental Protection Agency, Cincinnati, 1989, EPA/600/2-88/062.

Bruce, R.M., *Toxicological Review of Beryllium and Compounds (Cas. 7440-41-7)*, National Center for Environmental Assessment, U.S. Environmental Protection Agency, Washington, 1998, EPA/635/R-98/008.

Cabelli, V.J., *Health Effects Criteria for Marine Recreational Waters*, Health Effects Research Laboratory, U.S. Environmental Protection Agency, Research Triangle Park, NC, 1983, EPA-600/1-80-031.

California Environmental Protection Agency, *The Toxics Directory, Section 1: Resources*, Updated 1999.

Clayton, G.D. and Clayton, F.E., *Patty's Industrial Hygiene and Toxicology*, 4th ed., John Wiley & Sons, New York, 1994.

Clements, R.G., *Estimating Toxicity of Industrial Chemicals to Aquatic Organisms Using Structure-Activity Relationships*, Office of Pollution Prevention and Toxics, Health and Environmental Review Division, U.S. Environmental Protection Agency, Washington, 1988, EPA-560/6-88-001.

Cogliano, V.J., *PCBS: Cancer Dose-Response Assessment and Application to Environmental Mixtures*, U.S. Environmental Protection Agency, Washington, 1996, EPA/600-P-96-001A.

Crump, K. et al., *The Use of the Benchmark Dose Approach in Health Risk Assessment*, Risk Assessment Forum, U.S. Environmental Protection Agency, Washington, 1995, EPA/630/R-94/007.

Cullen, M.R., *Workers with Multiple Chemical Sensitivities*, Hanley & Belfus, Philadelphia, 1987.

Dorland, W.A.N., *Dorland's Illustrated Medical Dictionary, 28th ed.*, Saunders, Philadelphia, 1994.

Ecobichon, D.J., *The Basis of Toxicity Testing*, CRC Press, Boca Raton, FL, 1992.

Government Institutes, Inc., *Toxicology Handbook*, Government Institutes, Rockville, MD, 1994.

Hayes, A.W., *Principles and Methods of Toxicology, Student Edition*, Raven Press, New York, 1982.

Hodgson, E. et al., *Dictionary of Toxicology*, Van Nostrand Reinhold, New York, 1988.

Hodgson, E. and Guthrie, F.E., *Introduction to Biochemical Toxicology*, Elsevier North Holland, New York, 1980.

Klaassen, C.D. et al., Eds., *Toxicology: the Basic Science of Poisons*, 2nd ed., Macmillan Publishing, New York, 1980.

Lawson, L., *Staying Well in a Toxic World: Understanding Environmental Illness, Multiple Chemical Sensitivities, Chemical Injuries, and Sick Building Syndrome*, Noble Press, Chicago, 1993.

Matanoski, G.M. et al., *Evaluation of Superfund Ecotox Threshold Benchmark Values for Water and Sediment*, Science Advisory Board, U.S. Environmental Protection Agency, 1997, EPA-SAB-EPEC-LTR-97-009.

Meisner, L.F. et al., In vitro effects of n-nitrosoatrazine on chromosome breakage, *Arch. Environ. Contam. Toxicol.*, 24, 108, 1993.

Mitchell, F.L., *Multiple Chemical Sensitivity a Scientific Overview*, Princeton Scientific Publishing, Princeton, NJ, 1995.

Mount, D.I., *Methods for Aquatic Toxicity Identification Evaluation: Phase II Toxicity Identification Procedures*, Office of Research and Development, U.S. Environmental Protection Agency, Washington, 1989, EPA/600/3-88/035.

Mount, D.I., *Methods for Aquatic Toxicity Identification Evaluation: Phase III Toxicity Confirmation Procedures*, Office of Research and Development, U.S. Environmental Protection Agency, Washington, 1989, EPA/600/3-88/036.

Nabholz, J.V. et al., Validation of Structure-Activity Relationships Used by the U.S. EPA's Office of Pollution Prevention and Toxics for the Environmental Hazard Assessment of Industrial Chemicals, *Environmental Toxicology and Risk Assessment*, Vol. 2, Gorsuch, J.W. et al., Eds., American Society for Testing and Materials, Philadelphia, 1993, ASTM STP 1216.

National Institute for Occupational Safety and Health, *NIOSH Current Intelligence Bulletins*, Occasional Series.

National Research Council, *Arsenic in Drinking Water*, National Research Council, Washington, 1999.

National Research Council, *Biologic Markers in Pulmonary Toxicology*, National Research Council, Washington, 1989, EPA-R-812547.

National Research Council, *Biologic Markers in Urinary Toxicology*, National Academy Press, Washington, 1995.

National Research Council, *Monitoring Human Tissues for Toxic Substances*, National Academy Press, Washington, 1991.

National Research Council, *Multiple Chemical Sensitivities: Addendum to Biologic Markers in Immunotoxicology*, National Academy Press, Washington, 1992.

National Research Council, *Toxicological and Performance Aspects of Oxygenated Motor Vehicle Fuels*, National Academy Press, Washington, 1996.

National Research Council, *Toxicity of Military Smokes and Obscurants, Vol. 1*, National Academy Press, Washington, 1997.

National Research Council, *Toxicity of Military Smokes and Obscurants, Vol. 2*, National Academy Press, Washington, 1999.

National Research Council, *Toxicity of Military Smokes and Obscurants, Vol. 3*, National Academy Press, Washington, 1999.

National Toxicology Program, *Annual Report on Carcinogens*, National Institute of Environmental Health Sciences, Rockville, MD, Occasional Publication.

National Toxicology Program, *NTP Technical Report on the Effect of Dietary Restriction on Toxicology and Carcinogenesis Studies in F344/N Rats and B6C3F1 Mice*, U.S. Department of Health and Human Services, Washington, 1997.

National Toxicology Program, *NTP Technical Report on the Toxicology and Carcinogenesis Studies of Acetonitrile (Cas No. 75-05-8) in F344/N Rats and B6C3F1 Mice (Inhalation Studies)*, U.S. Department of Health and Human Services, Washington, 1996.

National Toxicology Program, *NTP Technical Report on the Toxicology and Carcinogenesis Studies of 1-Amino-2,4-Dibromoanthraquinone (Cas No. 81-49-2) in F344/N Rats and B6C3F1 Mice (Feed Studies)*, U.S. Department of Health and Human Services, Washington, 1996.

National Toxicology Program, *NTP Technical Report on the Toxicology and Carcinogenesis Studies of Anthraquinone (Cas No.84-65-1) in F344/N Rats and B6C3F1 Mice (Feed Studies)*, U.S. Department of Health and Human Services, Washington, 1999.

National Toxicology Program, *NTP Technical Report on the Toxicology and Carcinogenesis Studies of T-Buthylhydroquinone [Sic] (Cas No. 1948-33-0) in F344/N Rats and B63C3F1 Mice (Feed Studies)*, U.S. Department of Health and Human Services, Washington, 1997, NTP TR 459.

National Toxicology Program, *NTP Technical Report on the Toxicology and Carcinogenesis Studies of Butyl Benzyl Phthalate (Cas No. 85-68-7) in F344/N Rats (Feed Studies)*, U.S. Department of Health and Human Services, Washington, 1997.

National Toxicology Program, *NTP Technical Report on the Toxicology and Carcinogenesis Studies of Chloroprene (Cas No. 126-99-8) in F344/N Rats and B6C3F1 Mice (Inhalation Studies)*, U.S. Department of Health and Human Services, Washington, 1996, NTP TR/467.

National Toxicology Program, *NTP Technical Report on the Toxicology and Carcinogenesis Studies of Codeine (Cas No. 76-57-3) in F344/N Rats and B6C3F1 Mice (Feed Studies)*, U.S. Department of Health and Human Services, Washington, 1996.

National Toxicology Program, *NTP Technical Report on the Toxicology and Carcinogenesis Studies of D&C Yellow No. 11 (Cas No. 8003-22-3) in F344/N Rats (Feed Studies)*, U.S. Department of Health and Human Services, Public Health Service, Washington, 1997.

National Toxicology Program, *NTP Technical Report on the Toxicology and Carcinogenesis Studies of 1,2-Dihydro-2,2,4-Trimethylquinoline (Cas No. 147-47-7) in F344/N Rats and B6C3F1 Mice (Dermal Studies) and the Initiation/Promotion Study (Dermal Study) in Female Sencar Mice*, U.S. Department of Health and Human Services, Washington, 1997.

National Toxicology Program, *NTP Technical Report on the Toxicology and Carcinogenesis Studies of Emodin (Cas No. 518-82-1) in F344/N Rats and B6C3F1 Mice (Inhalation Studies)*, U.S. Department of Health and Human Services, Washington, 1999.

National Toxicology Program, *NTP Technical Report on the Toxicology and Carcinogenesis Studies of Fumonisin B1 (Cas No.116355-83-0) in F344/N Rats and B6C3F1 Mice (Feed Studies)*, U.S. Department of Health and Human Services, Washington, 1999.

National Toxicology Program, *NTP Technical Report on the Toxicology and Carcinogenesis Studies of Gallium Arsenide (Cas No.1303-00-0) in F344/N Rats and B6C3F1 Mice (Inhalation Studies)*, U.S. Department of Health and Human Services, Washington, 1999.

National Toxicology Program, *NTP Technical Report on the Toxicology and Carcinogenesis Studies of Isobutyraldehyde (Cas No. 78-84-2) in F344/N Rats and B6C3F1 Mice (Inhalation Studies)*, U.S. Department of Health and Human Services,Washington, 1996, NTP TR/472.

National Toxicology Program, *NTP Technical Report on the Toxicology and Carcinogenesis Studies of Isobutyl Nitrite (Cas No. 542-56-3) in F344/N Rats and B6C3F1 Mice (Inhalation Studies)*, U.S. Department of Health and Human Services, 1996.

National Toxicology Program, *NTP Technical Report on the Toxicology and Carcinogenesis Studies of Lauric Acid Diethanolamine Condensate (Cas No.120-40-1) in F344/N Rats and B6C3F1 Mice (Dermal Studies)*, U.S. Department of Health and Human Services, Washington, 1997.

National Toxicology Program, *NTP Technical Report on the Toxicology and Carcinogenesis Studies of Molybdenum Trioxide (Cas No. 1313-27-5) in F344/N Rats and B6C3F1 Mice (Inhalation Studies)*, U.S. Department of Health and Human Services, Washington, 1997, NTP TR 462.

National Toxicology Program, *NTP Technical Report on the Toxicology and Carcinogenesis Studies of Nitromethane (Cas No.75-52-5) in F344/N Rats and B6C3F1 Mice (Inhalation Studies)*, U.S. Department of Health and Human Services,Washington, 1997.

National Toxicology Program, *NTP Technical Report on the Toxicology and Carcinogenesis Studies of Oleic Acid Diethanolamine (Cas No. 93-83-4) Condensate in F344/N Rats and B6C3F1 Mice (Dermal Studies)*, U.S. Department of Health and Human Services, Washington, 1997.

National Toxicology Program, *NTP Technical Report on the Toxicology and Carcinogenesis Studies of Phenolphthalein (Cas No. 77-09-8) in F344/N Rats and B6C3F1 Mice (Feed Studies)*, U.S. Department of Health and Human Services,Washington, 1996.

National Toxicology Program, *NTP Technical Report on the Toxicology and Carcinogenesis Studies of Polyvinyl Alcohol (Molecular Weight = 24,000) (Cas No. 9002-89-5) in Female B6C3F1 Mice (Intravaginal Studies)*, U.S. Department of Health and Human Services, Washington, 1996.

National Toxicology Program, *NTP Technical Report on the Toxicology and Carcinogenesis Studies of Salicylazosulfapyridine (Cas No. 599-79-1) in F344/N Rats and B6C3F1 Mice (Gavage Studies)*, U.S. Department of Health and Human Services, Washington, 1997.

National Toxicology Program, *NTP Technical Report on the Toxicology and Carcinogenesis Studies of Scopolamine Hydrobromide Trihydrate (Cas No. 6533-68-2) in F344/N Rats and B6C3F1 Mice (Gavage Studies)*, U.S. Department of Health and Human Services, Washington, 1997.

National Toxicology Program, *NTP Technical Report on the Toxicology and Carcinogenesis Studies of Tetrafluoroethylene: (Cas No. 116-14-3) in F344/N Rats and B6C3F1 Mice (Inhalation Studies)*, U.S. Department of Health and Human Services, Washington, 1997.

Norberg-King, T., *Methods for Aquatic Toxicity Identification Evaluation: Phase I Toxicity Characterization Procedures*, Environmental Research Laboratory, U.S. Environmental Protection Agency, Duluth, MN, 1991, EPA/600/6-91/003.

Rand, G.M., Ed., *Fundamentals of Aquatic Toxicology, Methods and Application, 2nd ed.*, Taylor & Francis, Washington, 1995.

Rand, G.M. and Petrocelli, S.R., Eds., *Fundamentals of Aquatic Toxicology, Methods and Application*, Hemisphere Publishing, Washington, 1985.

Russell, H. et al., *The Toxics Directory, Section 1, Resources, Pesticide and Environmental Toxicology Section*, California Department of Health Services, Berkeley, CA, 1999.

Stephan, C.E. et al., *Interim Guidance on the Determination and Use of Water-Effect Ratios for Metals*, Office of Water, U.S. Environmental Protection Agency, Washington, 1994, EPA 823-B-94-001.

Sweet, D.V., *Registry of Toxic Effects of Chemical Substances (RTECs), Comprehensive Guide to the RTECs*, U.S. Department of Health and Human Services, Cincinnati, 1993, DHHS (NIOSH) Publication No. 93-130.

U.S. Air Force, *The Installation Restoration Toxicology Guide, Vols. 1-5*, Wright-Patterson Air Force Base, OH, 1989.

U.S. Environmental Protection Agency, *Assessment of Risks to Human Reproduction and to Development of the Human Conceptus from Exposure to Environmental Substances (ORNL-EIS-197)*, Washington, 1982, EPA/600-98-2001.

U.S. Environmental Protection Agency, Chemical carcinogens; a review of the science and its associated principles, 1985, *Federal Register*, 50, 10372, 1985.

U.S. Environmental Protection Agency, *EPA Toxicology Handbook*, Government Institutes, Rockville, MD, 1986.

U.S. Environmental Protection Agency, *Guidelines for Reproductive Toxicity Risk Assessment*, Office of Research and Development, Washington, 1996, EPA/630/R-96/009.

U.S. Environmental Protection Agency, *Integrated Risk Information System (IRIS)* (on Diskette), Office of Science and Technology, Washington, 1998.

U.S. Environmental Protection Agency, *Office of Water Policy and Technical Guidance on Interpretation and Implementation of Aquatic Life Metals Criteria*, Office of Water, Washington, 1993.

U.S. Environmental Protection Agency, *Screening Exposure Assessment Software (SEAS)*, Exposure Assessment Branch, Washington, 1995.

U.S. Environmental Protection Agency, *Special Report on Environmental Endocrine Disruption: an Effects Assessment and Analysis*, Risk Assessment Forum, Washington, 1996, EPA/630-R-96-012.

Wexler, P. and Gad, S.C., *Encyclopedia of Toxicology*, Academic Press, San Diego, 1998.

UNCERTAINTY

Abernathy, C.O. and Poirier, K.A., Uncertainties in the risk assessment of essential trace elements, *Human and Ecological Risk Assessment*, 3, 627, 1997.

Cullen, A.C., *Probabilistic Techniques in Exposure Assessment: a Handbook for Dealing with Variability and Uncertainty in Models and Inputs*, Plenum Press, New York, 1999.

Dawoud, E.A. and Purucker, S.T., *Quantitiative Uncertainty Analysis of Superfund Residential Risk Pathway Models for Soil and Groundwater: White Paper*, Office of Environmental Management, U.S. Department of Energy, Washington, 1996.

Firestone, M., *Guiding Principles for Monte Carlo Analysis*, Office of Research and Development, U.S. Environmental Protection Agency, Washington, 1997, EPA/630/R-97/001.

Hoffman, O.F., *An Introductory Guide to Uncertainty Analysis in Environmental and Health Risk Assessment (Technical Memorandum)*, U.S. Department of Energy, Washington, 1992, ES/ER/TM35/R1.

Johnson, B.B. et al., *Explaining Uncertainty in Health Risk Assessment Effects on Risk Perception and Trust*, Risk Communication Project, U.S. Environmental Protection Agency, Washington, 1994, CR-820522; EPA 230/R-94-903.

Klee, A.J., *Mouse Uncertainty Analysis System, Risk Reduction Engineering Laboratory*, U.S. Environmental Protection Agency, Cincinnati, 1994, EPA/600/A-94/142.

Massachusetts Department of Environmental Protection, *Conservatism & Uncertainty in Risk Assessment in the Context of the Massachusetts Contingency Plan*, 1992.

Rhomberg, L., *Developments at EPA in Addressing Uncertainty in Risk Assessment*, Office of Health and Environmental Assessment, U.S. Environmental Protection Agency, Washington, 1993, EPA/600/J-93/507; OHEA-C-536.

U.S. Environmental Protection Agency, *Modeling Phosphorus Loading and Lake Response under Uncertainty: a Manual and Compilation of Export Coefficients*, Office of Water Regulations and Standards, Washington, 1980, EPA 440-5-80-011.

U.S. Environmental Protection Agency, *Workgroup Summary Report on Methodological Uncertainty in Conducting Sediment Ecological Risk Assessments with Contaminated Sediments*, Corvallis Environmental Research Laboratory, U.S. Environmental Protection Agency, Corvallis, OR, 1996, EPA/600/A-96/098.

Viscusi, W.K., Mortality costs of regulatory expenditures, *J. Risk and Uncertainty*, Kluwer Academic, Boston, 8, 1, 1994.

Warren-Hicks, W. and Moore, D.R.J., *Uncertainty Analysis in Ecological Risk Assessment: Proceedings of the Pellston Workshop on Uncertainty Analysis in Ecological Risk Assessment*, Pellston, MI, Setac Press, Pensacola, FL, 1998.

Yoe, C.E., *Introduction to Risk and Uncertainty in the Evaluation of Environmental Investments*, Water Resources Support Center, U.S. Army Corps of Engineers, Alexandria, VA, 1996.

WILDLIFE

U.S. Environmental Protection Agency, *Wildlife Exposure Factors Handbook*, Office of Research and Development, Washington, 1993, EPA/600/R-93/187.

U.S. Environmental Protection Agency, *Workshop on the Use of Available Data and Methods for Assessing the Ecological Risks of 2,3,7,8-Tetrachlorodibenzo-P-Dioxin to Aquatic Life and Associated Wildlife*, 1993, Risk Assessment Forum, U.S. Environmental Protection Agency, Washington, 1994, EPA/630/R-94/002.

Index

A

Abiotic processes, chemicals in environment altered by, 44
Abrupt catastrophe, 433
Absorption, 59
Acceptable daily intakes (ADIs), 529
Acute-effect level, benchmark for, 88
Acute studies, 55
ADIs, *see* Acceptable daily intakes
ADR, *see* Alternative dispute resolution
Advection, 365
Aerial photographs, sources of, 461
Aerosolization, 37
Agency for Toxic Substances and Disease Control, 342
Air
  emissions, 135
    data, sources of, 381
    permit applications, 137
  model
    predictions, reliability of, 373
    selection, 375
  pathway
    analysis, 183
    fate and transport issues, 383
  pollutants, facility permit to emit, 8
  sampling and evaluation, 464
  stacks, 81
Airborne chemicals, risk assessment of, 465–477
  conceptual site models, 466–472
    developing data quality objectives, 470–472
    DQO process, 472
    indirect exposure pathways, 469
    project manager role in conceptual site model development, 469–470
    estimating chemical concentrations at exposure points, 473
  occupational exposure and risk assessment, 473–476
  risk characterization and informing risk manager, 476
Air toxics dispersion and deposition modeling, 369–388
  collection of emissions data appropriate for site-specific, multi-pathways risk assessments, 386–387
  consultant selection, 372–373
  cutting edge air modeling issues for risk assessment, 383–386
    air pathway fate and transport issues, for contentious multiphase contaminants, 383–385
    atmospheric fate and deposition modeling, 385
    limitations of deposition modeling, 385–386
    micrometeorological effects, 386
  overview of air modeling process for risk assessment, 373–374
  practical air modeling considerations, 374–380
    basic air modeling concepts, 374–376
    deposition modeling, 378–380
    dispersion modeling, 376–378
  regulatory drivers affecting risk assessment modeling studies, 371–372
  sources of air quality models, 380
  sources of data, 380–382
    air quality and meteorological data, 380–381
    evaluating and interpreting air emissions data for risk assessment modeling, 382
    sources of air emissions data, 381–382
Airway irritation, 60
ALAD, *see* delta-Aminolevulinic acid dehydratase

Alcohol consumption, 539
Aliphatic hydrocarbons, 273
Allergens, 57
Alpha radiation, 481
Alta Vista, 449
Alternative dispute resolution (ADR), 165, 253
Ambient concentrations, 33
Ambient media, chemical concentrations in, 93
American Society for Testing and Materials (ASTM), 93, 328
delta-Aminolevulinic acid dehydratase (ALAD), 94
Analytical studies, 54
Anesthetics, 57
Animal
  cell bioassays, 55
  sampling, 321
Applicable or relevant and appropriate requirements (ARARs), 34, 498, 499
Aquaculture, 181
Aquatic species, 92
ARARs, *see* Applicable or relevant and appropriate requirements
Armitage-Doll model, 65
Arsenic, 272
Arsine gas, 475
Artificial respiration, 476
Asphyxiants, 57, 273
Assessment audit, 512
ASTM, *see* American Society for Testing and Materials
Atmospheric deposition, 37
Atmospheric fate and deposition modeling, 385
Atomic absorption spectrophotometry, 284
Atoms, 266
Axonopathy, 61

B

Back-calculation air modeling analyses, 375
Background
  definitions of, 403
  samples, 291
Bad stuff, 407
Baseline risk assessment (BRA), 518, 520
Battery of tests approach, 331
Behavioral toxicants, 57
Benchmark(s)
  acute-effect level, 88
  choosing correct, 262
  chronic-effect level, 88
  low-effect level, 88
  no-effect level, 88
  population-effect level, 88
  values, 157
  wildlife, 90
Bench sheets, 329
Benthic community, 86, 89
Benzene, 545
Beta radiation, 481
Bid solicitation package, 127
BIFs, *see* Boilers and industrial furnaces
Bioaccumulation, 37
Bioassays, 160, 161
Bioconcentration factors, 197
Biokinetic models, 489
Biological degradation reactions, 44
Biologically effective dose, 51
Biological uptake, 37
Biomarkers, 94, 261
Biota sampling and evaluation, 464
Biotic processes, chemicals in environment altered by, 44
Bird sampling, 322
Blank(s)
  Field, 287, 288
  Instrument, 288, 294
  Method, 278, 28
  types of, 288–289
Blind field duplicate sample, 290, 294
Blood-brain barrier, 61
Blood-nerve barrier, 61
Body burden, 59, 194
Boilers and industrial furnaces (BIFs), 371
Box plots, 397
  of contamination levels, 399
  sample, 398
BRA, *see* Baseline risk assessment
Briefing document, 114
Bright line(s)
  approach, 227
  noncarcinogen, 226
Brownian motion, 378
Bulletin boards, 456–458
Burial grounds, 81

C

California Environmental Protection Agency, 459
Cancer
  potency factor, 177
  risk(s), 30, 74
    assessment of chemical, 392
    assessment guidelines, 530
    quantitating, 527
    yardstick, 24

INDEX 639

Carbon monoxide, 272
Carcinogen(s), 30, 57, 421
   quantifiable, 194
   risks, 188, 210
Carcinogenicity group, 177
Cardiovascular system, 63
CDI, *see* Chronic Daily Intake
CD-ROMs, 447, 448, 449
Central Nervous System (CNS), 61
Central Tendency (CT), 72
CERCLA, *see* Comprehensive Environmental Response, Compensation, and Liability Act
CERCLA and RCRA risk assessments, 515–526
   CERCLA, 518–520
   comparison of RCRA and CERCLA risk assessment, 525
   components of baseline risk assessment, 520–523
   overview of risk assessment, 516
   RCRA, 523–524
   risk assessment process, 517–518
      approach, 517–518
      purpose, 517
CERCLIS, *see* Comprehensive Environmental Response, Compensation, and Liability Information System
C.F.R., *see* Code of Federal Register
CGL, *see* Comprehensive General Liability
Chemical(s)
   background concentrations, defining, 33
   cancer risk, assessment of, 392
   concentrations, uncertainties associated with measured, 199
   of concern (COC), 31
      biological transformation of, 46
      body burdens, 194
      cancer-causing, 63
      identification of potential, 343
      list, 39, 75
      screening process, 47
      selection, 193
   disciplines, major, 271
   environmental fate, physical properties affecting, 45–46
   environmental releases of, 277
   identity, 37
   intake, 50
   manufacturing, 140
   measurements, 269
   movement, 43
   physical form of released, 198
   of potential concern (COPC), 31, 314
      comprehensive list of, 197
      list, problems associated with developing, 39
      measurement of, 196
      selection of, 72
   reactions, types of, 267
   release(s)
      minor components in, 193
      by wind erosion, 471
   selection
      process, for assessment, 176
      report, 201, 202
   special case, 73
   -specific risks, 24
   transformations, 44
   trans-media movement of, 373
Chemodynamics, 44
Chemtox.Dialog, 462
Chloracne, 60
Chronic Daily Intake (CDI), 68
Chronic-effect level, benchmark for, 88
Circulatory system, 62
Cirrhosis, 62
Clean Air Act, 83
Clean Air Act Amendments of 1990, 240, 385
Clean areas, 33
Cleanup decisions, 497
Clean Water Act, 83, 499
Clearinghouses, 456–458
Clinical studies, 239
CLP, *see* U.S. EPA Contract Laboratory Program
CMI, *see* Corrective Measures Implementation
CMS, *see* Corrective Measures Study
CNS, *see* Central Nervous System
Coal tar, 544, 545
COC, *see* Chemical of concern
Code of Federal Register (C.F.R.), 234
Code of Federal Regulations, 459
Combination reactions, 267
Communication(s)
   clear, 439
   inadequate, 163
   protocols, 163, 247
   requirements, 162
Compensation terms, misapplication of, 254
Compensative incentives, 253
Compliance monitoring samples, 391
Composite samples, 34
Comprehensive Environmental Response, Compensation, and Liability Information System (CERCLIS), 112
Comprehensive Environmental Response, Compensation, and Liability Act (CERCLA), 371, 502, 515
   baseline risk assessment, 229

liability, 509, 511
municipal landfill sites, 116
Comprehensive General Liability (CGL), 541
Computer
  databases, peer-reviewed, 187
  mathematical models, 373
  models, 156, 261
  programs, calculation of endpoints appropriate to test method using, 328
  simulations, 44
Conceptual models, 178, 359
Conflicts of interest, 109
Consultant(s)
  role of, 17
  selection, 333, 372
  working with, 342
Contaminant(s)
  air modeling, 49
  air pathway fate and transport issues for contentious multiphase, 383
  environmental, 23
  movement, 354
  sediment, 48
  soil, 23
  transport, 365
  water, 23
Contaminants of Potential Ecological Concern (COPEC), 83
Contingency costs, 162
Contract(s)
  closing of, 22
  management strategy, proactive, 18
  negotiation, 21, 152, 153
  Required Detection Limit (CRDL), 295, 296
  Required Quantitation Limit (CRQL), 296
  schedules, 253
  types of, 248, 249
Contract, conclusion of risk assessment, 221–230
  conclusion of report, 221–228
    accepting of final draft, 221–223
    addressing of risk management and risk communication, 223–228
    closing of risk assessment contract, 223
  follow-up studies and activities, 228–229
Contract formation, risk assessment, 245–255
  common contracting pitfalls, 254–255
    contract amendments, 255
    lack of clearly defined scope of work, 254
    misapplication of compensation terms, 254–255
  contract components, 250–253
    compensation, 251–253
    schedule, 251
    scope of services, 250–251
    standard commercial terms and conditions, 253
  contracting philosophy, 246–250
    affected participants, 247
    communication protocols, 247–248
    interim work products, 248–250
    objectives and assumptions, 247
    types of contracts, 248
Contracting pitfalls, 246
Contractor
  costs, 158
  fiscal services staff, 136
  qualifications, soliciting of, 127
  RFP response, information to consider for inclusion on, 133–151
  risk assessor, 100
  selection, 102, 132, 170, 507, 510
Control
  charts, 330
  response, 330
COPC, see Chemical of potential concern
COPEC, see Contaminants of Potential Ecological Concern
Corrective Measures Implementation (CMI), 523
Corrective Measures Study (CMS), 523
Corrosivity, 270
Court cases, 538
CRDL, see Contract Required Detection Limit
Criteria tables, 75
Crop ingestion, 145
CRQL, see Contract Required Quantitation Limit
CT, see Central Tendency
Cumulative probability distribution, 304
Cyanide, 272
Cytochemical markers, 49

D

Daphnid reproduction, 326
Data
  accuracy, 117
  air emissions, 381
  analysis, 309
  collection, 309, 343
    road map for project, 307
    of variable quality, 196
  evaluation, 155, 190
  gaps, methods to address, 175, 183, 186
  HEAST, 346
  indicator, 36
  judged deficient, 313
  management services, 142
  mass balance facility, 342

meteorological, 380
needs, defining, 33
qualifiers, 36
quality
    assessment process, 395
    defining acceptable, 32
    effect of on data usability in risk assessment, 297
    indicators, PARCC, 286
    quality objectives (DQOs), 32, 155, 278, 394
        achievement of, 287
        developing, 470
        documents, U.S. EPA, 410
        process, 472
    reduction, 284
    reporting, 285
    requirements, for predictive models, 364
    sets, surrogate, 160, 176
    sources, 380
        defining, 37
        hierarchy of, 156, 209
    sufficiency, 197
    summary tables, 175
    verification, 336
    water quality, 302
Databases
    government, 462
    private, 462
DBCP, see Dibromochloropropane
DCFs, see Dose conversion factors
DDE, 275
DDT, 274, 275
Decision
    criteria tables, 178
    logic, 4, 75
    -making, key component to environmental, 12
Decomposition reactions, 267
Degradation
    products, 274
    rates, 46
Demonstrable competence, 130
Dendrites, 61
Dense-packed page, 443
Deposition modeling, limitations of, 385
Dermal toxicants, 60
Descriptive studies, 54
Developmental toxicants, 57
Dibromochloropropane (DBCP), 559
Diesel fuel, 273
Dilution
    media, quality of, 330, 334
    water, 327
Dioxin, in soils, 44
Dirty-spots sample, 406
Discretization, 362
Dispersion models, 49, 377
Dissolved organic carbon, 354
Distribution, 59
    cumulative probability, 304
    fitting, 400
    Laplace, 401
    lognormal, 65
    Monte Carlo empirical, 393
    Weibull, 65
DNA, 481
Dockets, 456–468
Dose conversion factors (DCFs), 490
Dose-response
    assessment, 372
    curves, 210
    data, 64
Dot-plot, example, 399
Double-replacement reactions, 267
DQOs, see Data quality objectives
Drinking water, 210, 458
Dry deposition modeling, 378
Duplicate analyses, 294
Dynamic models, 359

E

Ecological effects assessment, 88
Ecological exposure assessment, 84
Ecological risk assessment (ERA), 12, 79–97, 482
    comparisons with other studies, 96
    concluding of ERA, 96
    ecological effects assessment, 88–94
        benthic community, 89
        fish community, 89
        sampling, 90–93
        soil invertebrate and plant communities, 89
        sources of other effects information, 94
        wildlife, 90
    ecological exposure assessment, 84–88
        benthic macroinvertebrate community, 86
        fish community, 86
        soil invertebrate species, 86
        terrestrial plants, 86
        terrestrial wildlife, 86–88
    ecological risk characterization, 94–96
    technical aspects of ecological problem formulation, 80–84
Ecological risk assessment review, 257–263
    effects analysis, 261
    exposure analysis, 260–261
    problem formulation, 258–260
    risk characterization and uncertainty analysis, 261–262

Ecological risk characterization, 94
Ecosystem characteristics, 80, 83
Ecotoxicity testing, in risk assessment, 325–337
    consultant selection, 333–335
        accreditation and certification, 335
        qualifications of consultant, 333
        quality system, 333–335
    technical review, 326–333
        basic concepts, 326–331
        current issues and uncertainties, 331–333
        important tools for implementation, 331
Eddy motion, 378
EEC, *see* Expected Environmental Concentration
Effects analysis, 261
Effluent pipes, 81
EIS, *see* Environmental Impact Statement
Electronic bulletin boards, 134
Electronic media, 448
Electronic transmission facilities, 434
Electrons, valence, 266
Electrostatic attraction, 378
Elements, 266
E-mail, 380
Emission(s)
    data, collection of, 386
    expert, 4
    factor development, 138
    fugitive, 374
    inventories, 137
    monitoring, 139
    rates, point-source, 374
    upsets in, 382
Endangered Species Act, 83, 92
End-users, 108
Engineering judgement, 391
Environmental chemistry, 265–275
    major chemical disciplines, 271–275
        inorganic chemistry, 271–272
        organic chemistry, 272–275
    practical environmental chemistry, 266–271
        chemical measurements, 269
        chemists' shorthand, 267
        physical states, 269–271
        types of chemical reactions, 267–269
Environmental cleanup laws, 502
Environmental contaminants, 23
Environmental Impact Statement (EIS), 229, 237, 443, 444
Environmental law, 235
Environmental regulations, 233
Environmental risk assessment
    introduction to, 6–12
        common terms, 6–7
        risk assessment controversy, 7–12
    multipathway analysis, 82

participants, goals of, 14
Enzyme concentrations, 94
EPC, *see* Exposure point concentration
Epidemiological intervention, determination of need for in risk assessment, 349
Epidemiological studies, 53
Epidemiologic investigations, 239
Epidemiology, *see* Health risk assessment, epidemiology and
EQL, *see* Estimated quantitation limit
ERA, *see* Ecological risk assessment
Estimated quantitation limit (EQL), 295
Excite, 449
Exclusionary risk assessment tools, 41
Excretion, 59
Expected Environmental Concentration (EEC), 314
Expert(s)
    internal, 16
    tasks, technical credentials needed to perform, 12
Exposure
    analysis, 260
    assessment, 31, 41, 179, 220, 344
        conservatism, 51
        ecological, 84
        precautions in, 207
        scoping, sample RAPPD for, 118
    Case, 51
    duration, 50, 59, 327
    equations, 49
    evaluation of, 394
    frequency, 50, 205
    indirect, 42
    inhalation, 41
    measures of, 83
    medium, 202, 467
    occupational, 473
    pathway(s), 192
        analysis, example of, 203
        human health risk ranking for, 315
    point, 467
        concentration (EPC), 72, 207, 523
        monitoring data, 47
    potential, 427
    relevant, 404
    routes, 42, 56, 85, 184, 202, 522
    scenarios, 204
    worst-case, 405
Extraction
    methods, 282
    solid phase, 283
    solvent, 283
    supercritical fluid, 283

INDEX 643

F

Facility risk assessment, 505–514
   preemptive assessment, 511–513
      accepting of proposal, 512
      assessment audit, 512
      contractor selection, 512
      control of results, 512–513
      evaluation of proposals, 512
      responding to results, 513
      writing request for proposal, 511
   RCRA facility assessments, 506–511
      evaluation of results, 510–511
      RCRA compliance assessment, 506–509
      RCRA facility assessment, 509–510
Fatal flaw analyses, 141
Fate and deposition modeling, atmospheric, 385
Fate and transport
   analysis, 42, 44
   diagram, example of, 203
   models, 42
      limitations of, 47
      recommendations, 205, 206
Feasibility study (FS), 518
Federal Emergency Management Agency (FEMA), 111
Federal Insecticide, Fungicide, and Rodenticide Act (FIFRA), 236, 237
Federal Register, 459
Federal Reporting Data System, 462
FEMA, see Federal Emergency Management Agency
Fibrosis producers, 57
Field
   Blank, 287, 288
   Duplicate sample, 290, 294
   equipment, 136
   monitoring emission measurements, 381
   sampling, 91
   surveys, 90, 91
FIFRA, see Federal Insecticide, Fungicide, and Rodenticide Act
Finite difference method, 362
Fish
   community, 86, 89
   ingestion guidelines, 383
   sampling, 92, 322
Fish and Wildlife Conservation Act, 83
Fish and Wildlife Service, 80
FLAA, see Flame atomic absorption
Flame atomic absorption (FLAA), 284
Flood Insurance Rate Maps, 461
Floodplain maps, 111
Flue-gas stack exit velocity, 374

Food
   chain
      exposures, terrestrial, 500
      modeling, 145
      parameter values, 489
      pathways analysis, 484, 486
   exposure modeling, 49
   web, 197
Food and Drug Administration action levels, 35
Formal quantitative uncertainty analysis, 419
FS, see Feasibility study
Fuels, 273
Fugacity Model, 44
Fugitive emissions, 374

G

Garbage in, garbage out (GIGO) principle, 301, 554
Gas chromatography, 284
Gasoline, 273
Gaussian air dispersion models, 376
Gaussian models, 377
Gaussian plume (GP) models, 553
Genotoxicants, 57
Geologist, 280
Geostatistics, 146
GFAA, see Graphite furnace atomic absorption
Ghost page, 444
GIGO, see Garbage in, garbage out principle
GLP, see Good Laboratory Practice
Glycol ethers, 385
Good Laboratory Practice (GLP), 333
Goodness-of-fit test, Kolmogrov-Smirnov, 401
Governmental databases, 462
Government files, 111
Government Printing Office, 459
GP models, see Gaussian plume models
Grab samples, 34
Graphite furnace atomic absorption (GFAA), 284
Gravitational settling, 378
Groundshine, 489
Groundwater
   contamination, 436
   model, 48, 357
   movement, 37
   pathway analysis, 180
   quality monitoring, 367
   sampling and evaluation, 463
Groundwater modeling, in health risk assessment, 357–368
   groundwater modeling reports, 358

technical aspects of contaminant transport, 365–367
  groundwater quality monitoring, 367
  model misuse, limitations, and sources of error, 366
  physical and chemical forces influencing movement, 365–366
technical aspects of groundwater modeling, 358–365
  conceptual model, 359
  dynamic models, 359–361
  modeling process, 363–365
  model selection, 361–363
Ground Water Site Inventory (GWSI), 111
Guidance documents, 25
GWSI, *see* Ground Water Site Inventory

H

Habitat surveys, 322
Half-life, 59
Halogenated hydrocarbon, 274
HAPs, *see* Hazardous air pollutants
Harp trap, 321
Hartley-Seilkin model, 65
Hazard(s)
  assessment, 31, 40, 164, 220
    conducting, 196
    conservatism, 39
  Evaluation, 31
  identification, 31, 390
  index (HI), 69, 391
  noncancer, 73
  quotient (HQ), 69
Hazardous air pollutants (HAPs), 371
Hazardous waste
  land disposal of, 240
  landfill site, 370
  sites, 36, 397
Health Effects Summary Table (HEAST), 39, 176, 450
Health risk assessment, epidemiology and, 339–348
  data collection and evaluation, 343
  exposure assessment, 344–345
  risk characterization, 346–348
  toxicity assessment, 345–346
  using epidemiology in health risk assessments, 340–341
  working with consultants, 342–343
Health Risk Values (HRVs), 227
HEAST, *see* Health Effects Summary Table
Heme production, 272
Henry's law constant, 45
Hepatocytes, 61
HHRA, *see* Human health risk assessment
HI, *see* Hazard index
High pressure liquid chromatography (HPLC), 284
Historical toxicology and risk assessment, 537–549
  alcohol consumption, 539–540
  application of historical toxicology in insurance law, 541–542
  historical toxicology, 539
  historical toxicology in litigation, 542–547
    manufactured gas plant industry, 543–546
    using historical toxicology review in litigation, 546–547
  selection of proper analysis, 538–539
History matching, 364
Hotlines, 456–458
HPLC, *see* High pressure liquid chromatography
HQ, *see* Hazard quotient
HRVs, *see* Health Risk Values
Human health risk assessment (HHRA), 12, 29–77, 482
  concluding HHRA, 70
  contractor work plan outline, example of, 155–157
  data, presenting, 70
  exposure assessment, 41–51
    chemical intake and uptake, 50–51
    concerns for review of, 204–205
    conservatism, 51–52
    exposure equations, 49–50
    fate and transport analysis, 42–49
  hazard assessment, 32–38
    conservatism, 39–41
    defining acceptable data quality, 32–33
    defining acceptable sampling and analytical plan, 34–36
    defining chemical background concentrations, 33–34
    defining data needs, 33
    defining data sources, 37–38
    defining methods for pooling sampling data, 36
    defining quality assurance/quality control methods, 36
  hazard evaluation, concerns for, 197–199
  multipathway analysis, 43
  paradigm, 189
  presenting HHRA data, 70–75
    decision logic or criteria tables, 75
    U.S. EPA's standard tables for Superfund risk assessments, 71–74
    variable selection tables, 75

regulatory toxicity and science of toxicology, 53–67
    absorption, distribution, metabolism, and excretion, 59
    exposure duration, 59
    exposure routes, 56–58
    pharmacokinetic properties, 54–55
    physical and chemical properties, 54
    target organs, 59–63
    types of tests, 53–54
    use of regulatory toxicology in toxicity assessment, 55–56
    using toxicological understandings in toxicity assessment, 63–67
  review, 211, 220
  risk characterization, 67–70, 210–211
  steps of, 31
  toxicity assessment, 53
Hydrogeologist, 280
Hydrolysis, 206
Hypothesis(es)
  testing, 400
  verification/rejection of, 309

I

IARC, see International Agency for Research on Cancer
ICP, see Inductively coupled argon plasma spectrophotometry
IDL, see Instrument detection limit
Ignitability, 270
Illustration clarity, criteria for, 441
Immune system, 63
Immunotoxicants, 58
IMPACT model, 488
In-cloud rainout scavenging, 379
Indirect exposure, 42, 469
Individual Chemical Scores, 38
Individual Lifetime Risk, 67
Individual risk, 347
Inductively coupled argon plasma spectrophotometry (ICP), 284
Industrial facility operational lifetime, 204
Industrial scenario, 84
Information
  general non-EPA sources of, 459–460
  sources of, 455, 528
Ingestion reference dose, 177
Inhalation
  exposure, 41, 380
  reference dose, 177
Inorganic chemistry, 271

Input/output analysis, 213, 217
In-scope task, 154
Insect(s)
  disappearance of, 319
  sampling, 322
Instrument
  Blank, 288, 294
  detection limit (IDL), 295
Insurance law, application of historical toxicology in, 541
Integrated Risk Information System (IRIS), 39, 450
Interim deliverables, approval of, 189
International Agency for Research on Cancer (IARC), 528
Interquartile range (IR), 396
Intuition, reviewer, 212
In vitro toxicological studies, 54
In vivo toxicological studies, 54
Ionizing radiation, 480
IR, see Interquartile range
IRIS, see Integrated Risk Information System
Irritants, 58
Issue statement, 304
Iterative review, 18, 122, 174, 213

J

Jaundice, 61
Jet fuel, 273
Junk science, risk assessment as, 9

K

Kerosene, 273
Kick-off meeting, 21, 107, 250
  hosting of, 170
  for potential customers, 129
Known human carcinogens, 63–64
Kolmogrov-Smirnov (KS) test, 401
KS test, see Kolmogrov-Smirnov test

L

Laboratory reporting
  limit, 295
  unit, 289
Lacrimation, 60
Lagoons, 81, 180, 370
Landfills, 81, 140

Land use, 135
Laplace distribution, 401
Latin hypercube sampling, 421
Laugh test, 409
Law, environmental, 235
Lead, 272
Leaking underground storage tank (LUST), 238
Legal context, of environmental risk assessment, 233–243
  expanse of environmental regulations, 234–235
  how regulations address risk, 236–238
    preventive regulations, 236–237
    reactive regulations, 237–238
  regulatory framework, 233–234
  regulatory methods for addressing risk, 238–243
    numerical standards, 239
    risk assessment, 240–243
    technology-based standards, 240
  risk in environmental regulatory framework, 235–236
Liability assessment, 147
Library resources, 448
Limit of detection (LOD), 295
Limit of quantitation (LOQ), 295
Litigation, historical toxicology in, 542, 546
LOAEL, *see* Lowest-Observed-Adverse-Effect-Level
Local effect, 56
LOD, *see* Limit of detection
LOEC, *see* Lowest-observed-effect-concentration
LOEL, *see* Lowest observed effects levels
Lognormal Distribution model, 65
LOQ, *see* Limit of quantitation
Low-effect level, benchmark for, 88
Lowest-Observed-Adverse-Effect-Level (LOAEL), 66
Lowest-observed-effect-concentration (LOEC), 328
Lowest observed effects levels (LOEL), 345
Lump-sum costs, 252
LUST, *see* Leaking underground storage tank
Lycos, 449

M

MACT, *see* Maximum achievable control technology
Magellan, 449
Mammalian toxicology, 146
Mandated science, risk assessment as, 13
Manufactured Gas Plant (MGP) industry, 542, 543
Maps, sources of, 461
Margin of exposure (MOE), 65
Mass balance facility data, 342
Mathematical models, 360, 361
Matrix spikes, 290, 294
Maximum achievable control technology (MACT), 240
Maximum Contaminant Level Goals (MCLGs), 239
Maximum contaminant levels (MCLs), 234
Maximum tolerated dose (MTD), 532
MCLGs, *see* Maximum Contaminant Level Goals
MCLs, *see* Maximum contaminant levels
MCPA, *see* Minnesota Pollution Control Agency
MDL, *see* Method detection limit
Media needs, 430
Mercury, oxidized, 384
Metabolites, 274
Meteorological data, 380
Method Blank, 287, 288
Method detection limit (MDL), 295
MGP industry, *see* Manufactured Gas Plant industry
Migratory Bird Treaty Act, 83
Minnesota Pollution Control Agency (MPCA), 226
MLTs, *see* Mouse liver tumors
Model(s)
  adjustments, 364
  air, 375
  Armitage-Doll, 65
  based on advection-dispersion equations, 365
  biokinetic, 489
  calculation, best method of checking, 555
  choice of, 206
  classification, 352
  coefficients, single-value estimates of, 414
  computer, 156, 261, 373
  conceptual, 178, 359
  contaminant air, 49
  definition of, 358
  dispersion, 49, 376, 377
  dry deposition, 378
  dynamic, 359
  equality of risk assessment, 10
  fate and transport, 42, 47
  food chain, 145
  Fugacity, 44
  fully justified, 204
  Gaussian, 377
    air dispersion, 376
    plume, 553
  groundwater, 48, 357

INDEX 647

Hartley-Seilkin, 65
IMPACT, 488
intercomparisons, 146
Lognormal Distribution, 65
mathematical, 360, 361
misuse, 366
Monte Carlo, 408
multistage linear, 372
objective, 352
options, 355
predictive, 364
results reconciliation, 141
sediment contaminant, 48
selection, 355, 361
site conceptual, 81, 468, 469
soil contaminant, 48
source term, 485
stochastic, 360
surface water exposure, 48
transport, 473
use of in risk assessment, 552
Weibull Distribution, 65
MOE, see Margin of exposure
Monte Carlo empirical distribution, 393
Monte Carlo model, 408
Monte Carlo sampling, 420
Monte Carlo simulation, 70, 421, 558
Morbidity Ratio, 68
Mountain-valley wind flows, 386
Mouse liver tumors (MLTs), 532, 533
MTD, see Maximum tolerated dose
Municipal landfill sites, CERCLA, 116
Mutagens, 58
Myelinopathy, 61

N

NAAQS, see National Ambient Air Quality Standards
Narcotics, 57
National Academy of Sciences, 527
National Ambient Air Quality Standards (NAAQS), 35
National Ambient Water Quality Criteria (NAWQC), 88–89
National Climatic Data Center (NCDC), 381
National Environmental Protection Act of 1969 (NEPA), 236
National Institute for Occupational Safety and Health recommended exposure limits, 35

National Library of Medicine (NLM), 454
National Priorities List (NPL), 518
National Technical Information Service, 460
National Wetlands Inventory, 460
Natural resources damages, 133
NAWQC, see National Ambient Water Quality Criteria
NCDC, see National Climatic Data Center
Necrosis producers, 58
NEPA, see National Environmental Protection Act of 1969
Network contacts, 430
Neuropathy, 61
Neurotoxicants, 58
Neurotransmitter, release of into synapse, 61
Neutralization reactions, 267, 268
NLM, see National Library of Medicine
NOAEL, see No-Observed-Adverse-Effect-Level
NOEC, see No-observed-effect-concentration
No-effect level, benchmark for, 88
NOEL, see No observed effect levels
Noncancer toxicity data, 73
Nondetects, 402, 403
Nonionizing radiation, 480
No-Observed-Adverse-Effect-Level (NOAEL), 65
No-observed-effect-concentration (NOEC), 328
No observed effect levels (NOEL), 345, 392, 529
Normal distribution statistical curve, 376
NPL, see National Priorities List
Numerical standards, 239

O

OAQPS, see U.S. EPA Office of Air Quality Planning and Standards
Occupational exposure, 473
OECD, see Organisation for Economic Cooperation and Development
Open dumps, 180
Optic nerve damage, 60
Organic chemistry, 272
Organisation for Economic Cooperation and Development (OECD), 528
Organism health criteria, 334
Organochlorine risk assessment, 144
Out-of-scope task, 154
Overland flow, 37
Oxidation, 206
Oxidation-reduction reactions, 267, 268

P

PARCC, *see* Precision, accuracy, representativeness, comparability and completeness
Particle
 density, 378
 radiations, 481
 washout, 379
Particulates, wet deposition of, 379
Partition coefficient, 46
PATHSCAN, 462
PCBs, *see* Polychlorinated Biphenyls
PE, *see* Performance Evaluation
Peer review, 214, 555
Performance
 Evaluation (PE), 290, 294
 goals, 471
 standards, 17, 18
  enforcing rigor through, 125
  teams applying, 20
Periodic Table, 266
Peripheral Nervous System (PNS), 61
Personality traits, reviewer, 212
Pesticide(s)
 assessments, 531
 Information Network (PIN), 452
 inhalation of, 319
 Monitoring Inventory (PMI), 452
 parent, 274
 residue measurements, 397
 risk assessment, 144
Pesticides, international health risk assessment approaches for, 527–535
 differences in policy and objectives, 528–530
 issues affecting risk assessment decisions, 532–533
 performing assessments, 531–532
 sources of information, 528
 status of international harmonization efforts, 533–534
Petroleum
 derivatives, 273
 manufacturing, 140
Pharmacokinetic properties, 54
Pharmacokinetics/toxicokinetics, 59
Phase change, 43
Photolysis, 44, 206
Photooxidation, 44
Physical properties, affecting chemical environmental fate, 45–46
Physical states, 269
PICs, *see* Products of incomplete combustion
Pie chart, 442
PIN, *see* Pesticide Information Network
Pit traps, 322
Plain language, achievement of in writing, 442
Planning-level rigor, 124, 125
PMI, *see* Pesticide Monitoring Inventory
PNS, *see* Peripheral Nervous System
Point-source configurations, 374
Pollution
 prevention strategies, 351
 response costs necessary to cleanup, 541
Polychlorinated Biphenyls (PCBs), 236
Population
 -effect level, benchmark for, 88
 inferences about larger area of, 303
 risk, 67, 347, 348
Possible human carcinogens, 64
Potential to emit estimations, 24
Potentiation, 56
PQL, *see* Practical quantitation limit
Practical quantitation limit (PQL), 289, 295
Precision, accuracy, representativeness, comparability and completeness (PARCC), 278, 286
Predictive models, data requirements for, 364
Preliminary remediation goals (PRGs), 517
PRGs, *see* Preliminary remediation goals
Private databases, 462
Probable human carcinogens, 64
Problem formulation, 258, 259
Process standards, 18, 108
Products of incomplete combustion (PICs), 387
Product standards, 19, 108, 169
Professional trade organizations, 113
Profit margin, 158
Project
 accounting services, 134
 data collection, road map for, 307
 deadlines, 251
 description, 281
 expectations, documenting of, 108, 168
 float, 123
 funding, 126, 170
 limitations, identification of, 109
 management
  decision, 339
  training, 150
 manager(s), 16
  contracting organization, 257
  responsibilities of, 6
  selection, 103, 167
 planning, 101
  questions of, 117
  tables, 115
 proposers, 16
 responsibility, 148
 scheduling, 121

INDEX 649

scoping of, 110
team
    building and managing of, 104
    kick-off meeting, 107
time lines, 130
Project planning, risk assessment, 99–171
    building foundation for contracting of risk assessment, 100
    documents generated prior to beginning of risk assessment report, 100–101
    project planning, 101–166
        building of risk assessment project team, 105–106
        determination of need for risk assessment, 101–103
        documenting of project expectations, 108–109
        funding of project, 126
        hosting of kick-off meeting for potential contractors, 129–132
        identifying of project limitations, 109–110
        negotiating of contract, 152–153
        negotiating of work plan, 153–166
        organizing of project management team, 107–108
        scoping of project, 110–126
        selection of contractor, 132–152
        selection of project manager, 103–104
        soliciting of contractor qualifications or proposals, 127–129
Public relations, 162
Pulmonary edema, 60
Purge and trap, 283

Q

QA/QC, *see* Quality assurance and quality control
QF, *see* Quality factors
Quality assurance manager, 280
Quality Assurance Project Plans (QAPPs), 278, 282, 297
Quality assurance and quality control (QA/QC), 13, 123
    coordinating committee, 149
    Data Validation Report, 201
    individuals in risk assessment project, 280
    measures, 285, 335
    methods, 36, 190
    performance standards, 200
    practices, 157
    report, on analytical data, 178
    requirements, compliance with, 333
    sampling and analysis, 164
    standards, 209
    tools, 279
    work product, discipline checklist for, 336
Quality assurance/quality control, for environmental samples used in risk assessment, 277–299
    effect of data quality on data usability in risk assessment, 297–298
    effective use of analytical QA/QC for risk assessment, 279
    role of analytical QA/QC in risk assessment preparation, review, and management, 279–297
        blanks, 287–296
        choosing laboratory analytical methods, 296
        from sampling to data analysis, 282–287
        project description, 281–282
        quality assurance project plans, 297
        where analytical QA/QC is used in risk assessment reports, 296
Quality control samples, types of, 290
Quality factors (QF), 481
Quantitative uncertainty analysis, 413

R

Radiation(s)
    alpha, 481
    beta, 481
    dose, 489
    exposure analysis, 483
    ionizing, 480
    nonionizing, 480
    particle, 481
    response analysis, 490
    types of, 480
    units, 481
Radiation risk assessment, 479–496
    radiation types and sources, 480–482
        radiation sources, 482
        radiation units, 481
        types of radiation, 480–481
    risk assessment for radioactive substances, 482–493
        problem formulation, 483
        radiation exposure analysis, 483–493
        risk assessment process, 482–483
Radioactive substances, 480, 494
Radioadaptation, 489
Radionuclide(s)
    beta emissions of, 489
    release scenario, 492

risk assessment of, 393
transport analysis, 486
Radio telemetry, 321
Radish seedling germination, 326
RAPPD, *see* Risk Assessment Project Planning Document
RARAS, *see* Risk Assessment Review Accounting System
RCRA, *see* Resource Conservation and Recovery Act
Reactive management, 174
Reactive regulations, 237
Reactivity, 270
REACT loop
    method, 430, 431
    risk team, 434
Reasonable Maximum Exposure (RME), 51
Receptor, 202
Records of Decision (ROD), 521
REDs, *see* Reregistration Eligibility Decision
Reducing agent, 269
Reduction, 206
Reference Dose (RfD), 66
    development, 144
    ingestion, 177
    inhalation, 177
Reference site identification, 147
Reference toxicant(s), 309
    data, 336
    testing, 330
Regulatory science, level of rigor, 124
Regulatory standards and guidelines, common, 35
Regulatory toxicity, use of in toxicity assessment, 55
Relative risk, 68
Relative toxicity rating, 177
Release
    mechanism, 466
    model quality, 198
Relevant exposure, estimation of, 404
Remedial investigation (RI), 241, 518
Remediation
    goals, chemical-specific, 517
    risk assessment, 497–504
        evaluation of remedial alternatives, 501
        residual risk, 501–502
Replacement reactions, 268
Report
    documentation, 137
    scope, 190
Report development, managing of risk assessment, 173–220
    concluding of HHRA review, 220

human health risk assessment review, 211–219
    input/output analysis, 217–219
    peer review of human health risk assessment, 214–216
    risk assessment report checklists, 216–217
managing of project, 174–211
    conducting exposure assessment, 202–208
    conducting final review of draft risk assessment report, 209–211
    conducting hazard assessment, 196–202
    conducting risk characterization, 208–209
    conducting toxicity assessment, 208
    implementing iterative review, comment, and approval of interim deliverables, 189–195
Reproductive system, 62
Request for Proposals (RFP), 20, 21, 248, 506
    assembly of, 132
    process, systematic, 255
    response, information to consider for inclusion on contractor, 133–151
    solicitation, 128, 132, 168
Request for Qualifications (RFQ), 20, 21
Reregistration Eligibility Decision (REDs), 451
Residential scenario, 84
Residual risk, 501
Resource Conservation and Recovery Act (RCRA), 371, 515
    baseline risk assessment, 229
    clean closure, 525
    compliance assessment, 508
    Facility Assessment (RFA), 506, 509, 523
    risk assessments, *see* CERCLA and RCRA risk assessments
Respiratory system damage, 60
Respiratory tract, 59
Response costs, 541
Retinal damage, 60
RFA, *see* RCRA Facility Assessment
RfD, *see* Reference Dose
RFP, *see* Request for Proposals
RFQ, *see* Request for Qualifications
RI, *see* Remedial investigation
Rigor
    enforcement of through performance standards, 125
    regulatory science, 124
    technical, 126
Risk(s)
    advisor, 17, 105
    assessor(s), 6, 51, 280
        contractor, 100
        reliance of on data qualifiers, 36
    cancer, 24, 30, 74, 527

INDEX

carcinogen, 188, 210
characterization, 22, 67, 188, 261, 346
   conducting of, 208
   ecological, 94
   studies, 370
chemical-specific, 24
communicator, 434, 435
conservatism, 52
drivers, 74
estimates, 6
expression of, 67
findings, 6, 30
Individual Lifetime, 67
Management
   CRA baseline, 229
   decisions, 10, 223
   separating risk assessment from, 225
   systematic approach to, 226
   zonal approach to, 228
managers, responsibilities of, 6
neutral language, 192
population, 67, 347, 348
Reduction Tables, 229
regulatory approaches to, 237
relative, 68
residual, 501
societal, 67
underestimation of, 341
Risk assessment(s)
   accepting of, 24
   advanced, 161
   air modeling process for, 373
   assumption, 195
   baseline, 242
   building foundation for contracting, 100
   complexity rating and costing scheme, sample, 122
   contractor, 173
   controversy, 7
   data used in, 10
   determination of need for, 101
   findings, communicating, 426
   formal steps of, 31
   funding of, 21
   historical, 538
   inherent flaws in, 9
   as junk science, 9
   limitations inherent in, 26
   as mandated science, 13
   minimum standards for, 215
   modeling studies, regulatory drivers affecting, 371
   Monte Carlo-based, 409
   need for, 167
   organochlorine, 144

paradigm, 537
pesticide, 144
Planning Form, 25
process, literature on, 5
Project
   critical elements for planning, 167
   early information on, 427
   Planning Document (RAPPD), 115, 118
   team, building of, 105
radiological, 484
relationship of epidemiology and, 340
reliability of, 301
remediation, 498
report(s), 8, 212
   checklists, 216
   credibility of, 66
   critical elements for managing, 175–188
   development, 221
   input/output table used in, 218
   review, process components for, 192–195
   review, product components for, 190–191
Review Accounting System (RARAS), 217
scientifically defensible, 9
scope of work, 130–132
screening-level, 312
semi-quantitative, 124
services, contract formation for, 245
site-specific, 516
software, 11
Superfund, 71
team(s)
   building, 167
   roles in, 15
technical guidance, 109
technical qualifications to produce, 12
   different risk assessments need different experts, 12
   technical credentials needed to perform expert tasks, 12
tools, exclusionary, 41
uncertainty in, 413, 557
waste reutilization, 144
wildlife, 311
work plan, 154
Risk assessment, as multidisciplinary endeavor, 13–20
   mandated science, 13
   roles in risk assessment teams, 15–17
   team establishment of performance standards, 17–20
   team work in risk assessment, 13–15
Risk assessment process, overview of, 20–25
   accepting of risk assessment, 24–25
   after risk assessment, 25
   managing of risk assessment, 23–24

planning of risk assessment, 20-22
Risk Assessment Planning Form, 25
Risk communication, 7, 22, 223, 227, 425-437
   abrupt catastrophe, 433-436
      on-scene priorities, 434-436
      preliminaries, 433-434
   benefits of, 428, 435
   building trust, 429
   communicating risk assessment findings and results, 426-428
      benefits of early risk communication, 428
      creating meaningful dialogue, 427
      overcoming need to know everything, 426
      planning for useful communication, 428
   experts, 429
   managing risk communication, 430
      media needs, 430
      network contacts, 430
   managing, 430
   practical applications, 432-433
   problem discovered as result of testing, 436
   understanding environmental risk, 428-429
   what to look for when hiring consultant, 431-432
      training, 432
      trust, 431-432
RME, *see* Reasonable Maximum Exposure
ROD, *see* Records of Decision

S

Safe Drinking Water Act of 1974 (SDWA), 234
Safety Factors, 531
Sample(s)
   Blind Field Duplicate, 294
   collection equipment, contaminated, 293
   compliance monitoring, 391
   contamination, 285
   dirty-spots, 406
   Field Duplicate, 294
   handling, 293
   matrix, 291
   quantitation limit (SQL), 289, 296
   support, 402, 408
Sampling
   and Analysis Plans (SAPs), 278
   animal, 321
   bird, 322
   data, methods for pooling, 36
   design, 320
      elements and interrelationships in, 306
      goals of risk-based approach to, 311
      team, 302
   field, 91
   fish, 92, 322
   insect, 322
   issues, 90
   Latin hypercube, 421
   location, 177
   Monte Carlo, 420
   plan, design of, 292
   plans, 34
   QA/QC, 164
   reaches, 90
   requirements, 308
   site, 196
   strategies, 406
   vegetation, 321
Sampling, ecological risk assessment, 319-323
   animal sampling, 321
   bird sampling, 322
   diet determination, 323
   fish sampling, 322
   habitat surveys, 322
   insect sampling, 322
   sampling design, 320
   vegetation sampling, 321
Sampling design, environmental, 301-318
   conventional statistical approaches, 304-308
      defining statistical tests needed, 307
      issue statement, 304
      purpose and goals, 304-305
      sampling design, 307-308
      statistical hypotheses to address key questions, 305-307
   risk-based approach to sampling design, 310-315
   sampling and analysis plan, 308-310
      data analysis and verification/rejection of hypotheses, 309
      quality control and quality assurance, 309-310
   sampling design team, 302-304
SAPs, *see* Sampling and Analysis Plans
Scatter plot, 400
Scientific library research, 447-464
   library resources, 448-454
      electronic media, 448-449
      hard copy, 452-454
      surfing the Net for risk assessment data, 449-452
   selected environmental information sources, 455
Scientific rigor requirements, 159
SCM, *see* Site conceptual model
Scope of Work, 106, 110, 117, 254
SCRAM, *see* Support Center for Regulatory Air Models

Screening-level risk assessment (SLRA), 310, 311, 313
SDWA, see Safe Drinking Water Act of 1974
Seabreeze effects, 386
Secondary source, 466
Sediment(s)
  contaminant modeling, 48
  sampling and evaluation, 464
Sensitivity analysis, 210, 364, 418
SF, see Slope factor
Single-replacement reactions, 267
Single-species testing, 331
Site
  conceptual model (SCM), 81, 468, 469
  remediation, 304
  Sampling and Analysis, 196, 200
Slope factor (SF), 64, 65
SLRA, see Screening-level risk assessment
Smoke stacks, gas condensation from, 40
Societal risk, 67
Software, risk assessment, 11
Soil(s)
  contaminant, 23, 48
  dioxin in, 44
  erosion, runoff by, 37
  excavation, worker exposure to VOCs during, 501
  ingestion, 409
  invertebrate species, 86, 89
  particles, adsorption to, 37
  pathway analysis, 182, 183
  sampling and evaluation, 463
  surveys, 143
Solid phase extraction (SPE), 283
Solid waste management units (SWMUs), 242
Solvent extraction, 283
SOPs, see Standard operating procedures
Source
  air testing, 141
  loading, 353
  term development, 485
SPE, see Solid phase extraction
Special topics in risk assessments, 551–561
  uncertainty in risk assessment, 557–560
  use of models in risk assessment, 552–557
    checking equations and calculations, 555–556
    input and output parameters, 554–555
    reality checks, 556–557
    relevance of model, 553–554
Species diet determination, 323
SQL, see Sample quantitation limit
Stack plume impaction concentration impacts, 376
Standardized Ratio, 68

Standard operating procedures (SOPs), 328, 329
Standard tables, 71
State air standards, 35
State Department of Transportation Maps, 461
State fish flesh contaminant advisories, 35
Statistical design, 320
Statistical hypotheses, 305
Statistics, use of in health and environmental risk assessments, 389–4111
  estimation of relevant exposure, 404–408
  evaluation of utility of environmental sampling, 395–404
    contamination exceeding background, 403–404
    distributional fitting and other hypothesis testing, 400–402
    graphical methods, 396–400
    nondetects, 402
    sample support, 402–403
  finding out what is important, 408–409
  statistical thinking and regulatory guidance, 391–395
    data quality objectives, 394–395
    evaluation of exposure, 394
    risk assessment of radionuclides, 393–394
    risk assessments, 391–393
  tools, 409–410
Sticky traps, 322
Stochastic models, 360
STORET, 462
Stressor
  characteristics, 260
  -response relationship, 261
Subcontractors, 151
Sublethal tests, examples of, 326
Sublimation, 44
Subtasks, organizing information by, 251
Summary statistic, 417
Supercritical fluid extraction (SFE), 283
Superfund, 143, 237, 432
  Docket Information Center, 457
  process, risk evaluation of remedial alternatives in, 519
  remedial investigation, 241
  risk assessments, 71, 394, 474
Superposition, 362
Support Center for Regulatory Air Models (SCRAM), 380
Surface water
  exposure modeling, 48
  pathway analysis, 181, 182
  quality standards, 383
  sampling and evaluation, 464
Surface water modeling, 351–356

model classification, 352–354
  degree of detail, 353
  principal model components, 353–354
model objective, 352
model selection, 355
  budget, 355
  function, 355
  options, 355
  resources, 355
  sources of models, 355–356
SWMUs, *see* Solid waste management units
Synapses, 61
Synergism, 56
Systematic error, 416
Systemic effect, 56

T

Target organs, 59
Technical editing, 149
Technical guidance documents, 463–464
Technical rigor, 126
Technology-based standards, 240
Teratogens, 58, 265
Terrestrial habitats, 92
Terrestrial plants, 86
Terrestrial wildlife, 86
T&E species, *see* Threatened and endangered species
Test
  laugh, 409
  organisms, 55, 325, 330
  reproducibility, 332
  sensitivity, 332
Threatened and endangered (T&E) species, 320
Tolerance, 58
Toxic effects, types of, 57–58
Toxicity, 270
  assessment, 22, 53, 164, 191, 220
    conducting of, 208
    using toxicological understandings in, 63
  describing, 474
  tests, 90, 92, 325
  values, 95, 209
Toxic response, categorization of, 55
Toxic Substances Control Act (TSCA), 236, 237
Toxic Substances Docket, 457
TOXSTAT, 328
Tracer studies, 142
Transparent reports, 216
Transport
  mechanism, 467
  models, 473

Treatment, storage, and disposal (TSD) facility, 237–238
TSCA, *see* Toxic Substances Control Act
TSD facility, *see* Treatment, storage, and disposal facility
Typographic criteria, 442
Typography, 439

U

UCL, *see* Upper confidence limit
UF, *see* Uncertainty factors
Uncertainty
  communication of, 421
  factors (UF), 66
  graphical depictions of, 422
Uncertainty analysis, 134, 259, 261, 413–424
  checklist to evaluate adequacy of, 424
  communication of uncertainty, 421–423
  examples of, 493
  Monte Carlo methods, 420–421
  qualitative, 184
  quantitative, 184
  technical aspects of, 415–418
    describing and summarizing data, 416–417
    sensitivity analysis, 418
    sources of uncertainty, 415–416
    uncertainty and variability, 415
  uncertainty in risk assessment, 414–415
Underground storage tanks, leaking, 112
Upper confidence limit (UCL), 38, 523
U.S. Army Corps of Engineers, 355, 460
U.S.C., *see* U.S. Code
U.S. Code (U.S.C.), 234
U.S. Environmental Protection Agency (U.S. EPA), 13
  Contract Laboratory Program (CLP), 285
  drinking water health advisory concentrations, 35
  DQO documents, 410
  environmental information documents, 450
  maximum contaminant level, 35
  Office of Air Quality Planning and Standards (OAQPS), 380
  priority pollutants, 313
  regional staff, 524
  standard tables, 192
  tolerance levels, 35
  water quality criteria, 35
U.S. EPA, *see* U.S. Environmental Protection Agency

INDEX

U.S. Fish and Wildlife Service (USFWS), 323, 460
USFWS, *see* U.S. Fish and Wildlife Service
U.S. Geological Survey (U.S.G.S.), 363, 460
U.S.G.S., *see* U.S. Geological Survey

V

Valence electrons, 266
Value-based issues, 427
Vapor pressure, 46
Variable selection table, 75, 207
Vegetation sampling, 321
VOCs, *see* Volatile organic compounds
Volatile organic compounds (VOCs), 409, 500, 501
Volatilization, 37

W

Waste
   combustion sources, 379
   reutilization risk assessment, 144
   site investigation, 91
Water
   authority, local, 459
   contaminant, 23
   dilution, 327
   laboratory-grade, 287, 288
   movement, 354
   quality, 149
      criteria, aquatic life, 314
      data, 302
   resources, 133
   rights, 135
   sampling, 206
   solubility, 46
WATSTORE, 462
Weibull Distribution model, 65
Weight-of-evidence approach, 94, 208, 262
Well drillers, local, 459
WellFax, 462
Wet deposition, 379
Wetlands, 91, 260
White carbon, 272
Wildlife
   benchmarks, 90
   contaminant intake equation, 87
   risk assessment, 311
Wild and Scenic Rivers Act, 83
Wind erosion, chemicals released by, 471
Work plan
   acceptability, assessing, 164
   development, 153
   dispute resolution, 165
   negotiation of, 153
Work products
   interim, 248
   review of final, 221
Writing, clear communication in risk assessment, 439–445
   clear writing problems, 443–444
   graphics, 442–443
   readability, 440–442

Y

Yahoo, 449

Z

Zonal Risk Management Approach, 227, 228

For Product Safety Concerns and Information please contact our EU representative GPSR@taylorandfrancis.com
Taylor & Francis Verlag GmbH, Kaufingerstraße 24, 80331 München, Germany

www.ingramcontent.com/pod-product-compliance
Ingram Content Group UK Ltd.
Pitfield, Milton Keynes, MK11 3LW, UK
UKHW021426080625
459435UK00011B/178